Take The Collegiate Investment Challenge!
Win Cash Prizes!

Manage a $500,000 simulated portfolio in real-time

20% – 33% Discount when you register with your Access Code.

EXPERIENCE WALL STREET . . . FIRST HAND

Imagine being given half a million dollars to experience the world of investing while assuming absolutely no risk. Now consider the valuable experience you'll gain from buying and selling equities and performing more sophisticated transactions like trading options, short selling and purchasing on margin — all at current market prices. Experience the excitement of Wall Street in the Collegiate Investment Challenge, the nation's most realistic stock market simulation. Form a team, or do it individually, for as little as $29.95!*

Playing the Collegiate Investment Challenge is easy. We'll give you all the resources you'll need to make your experience a hit. By the end of the Challenge you'll be able to walk, talk and act like you work on Wall Street. Who knows? You may even win your share of thousands of dollars in prizes and get recognized in a national newspaper. This semester don't just read about it - experience it.

FREE trial offer for Professors:

The Collegiate Investment Challenge provides all the resources professors need to easily and effectively implement the program into their classrooms:
- Free Professor Portfolio
- Classroom Reports (including ROI, # of trades, types of trades, etc.)
- Ability to view students' accounts
- Toll-Free Professor Support Line
- Professor Chat Room

For More Information:

Challenges are offered each semester. For information, please call The Investment Challenge at 1–800–487–3862 or visit the website at <www.ichallenge.net/prenticehall>.

prices subject to change

Your Discount Student Access Code is: C2WH-KFH-PBC   This access code is for one time use only!

Register with this Serial Number and save 20%–33%. Register on-line at <www.ichallenge.net/prenticehall> or call 1–800–487–3862

How Does It Work?

The Collegiate Investment Challenge mimics the daily operation of a real brokerage firm. With a $500,000 fictional brokerage account, you get to experience the markets just like the professionals. Implement your own investment strategies by buying and selling equities and options listed on all the major exchanges (NYSE, NASDAQ, AMEX, CBOE, PSE, and PHLX). In addition, you can try your hand at purchasing on margin and short selling. The Challenge is so realistic—it even incorporates commission schedules, dividends, and stock splits. Trades can be placed on-line, via the Investment Challenge Website at <www.ichallenge.net/prenticehall>, or by fax, answering machine, or through live brokers using the toll-free number 1-800-487-3862. For maximum authenticity, every trade is placed using real-time prices (i.e., no delay).

If Investing Intimidates You . . . Don't Worry.

The Investment Challenge's staff of brokers are there to answer all your questions by phone or e-mail. They can guide you through the basics—from buying a stock to even the most sophisticated option trades. Whatever your needs or questions, the Investment Challenge staff is there to help.

The Right Account For You.

WebTrader Account
- Exclusive Internet trading
- On-line Investment Manual
- On-line Newsletters
- On-line stock research information
- Full e-mail support

TradersEdge Account
- All WebTrader features plus:
  - Access to live brokers
  - Biweekly mailed newsletters
  - Biweekly mailed account statements

National Competition

Compete against college students from across the country for a chance to win thousands of dollars in prizes:
- Prizes for top performers
- Weekly prizes
- Recognition in a national newspaper
- Special prizes and rankings for Investment Clubs

Your Resume's Secret Weapon

The Collegiate Investment Challenge provides the kind of hands-on experience that looks great on a resume, and makes an excellent topic of conversation in interviews. Whether you are thinking about a career in the financial services industry, or just want to learn more about the stock market, participating in The Collegiate Investment Challenge will be a valuable asset.

Register Today. Call 1-800-487-3862
or register on-line at www.ichallenge.net/prenticehall
and enter your Prentice Hall Student "access code"
Save 33% off WebTrader Accounts
Save 20% off TradersEdge Accounts

©1999, 0-13-021226-1

INVESTMENT CHALLENGE — www.ichallenge.net/prenticehall

Put A New Twist On Time-Value-of-Money, And $5 In Your Pocket.

You will learn a great deal in this book about the Time-Value-of-Money. TI wants to help you learn the "value of time and money" with a great offer on the BAII PLUS™ financial calculator.

Save time. The easy-to-use features of the BAII PLUS will speed you through calculations such as net present value, internal rate of return, time-value-of-money, and more. And because the BAII PLUS is available at most stores where calculators are sold, you won't spend time searching for it.

Save money. You don't have to spend a lot of money because the BAII PLUS is priced to fit your budget. Plus, for a limited time, TI will put an extra $5 in your pocket.

Take advantage of this offer on the BAII PLUS today, and get the most value out of your time and money.

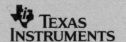
TEXAS INSTRUMENTS

Yes! I would like to receive a $5 rebate on my purchase of the BAII PLUS.

To receive $5, I have included the following:
1. Itemized sales receipt (original or copy) with the calculator purchase circled. Receipt must be dated between January 1, 1999 and April 30, 2001. (Offer valid for purchases made in the U.S. and Canada.)
2. UPC bar code from the back of the BAII PLUS package.
3. This completed mail-in rebate coupon. Original only. (Rebate will be sent to addresses only in the U.S. and Canada.)

I have read, understood, and complied with all the terms and conditions of this offer. Rebate terms on back.

Signature (required) _____ Date _____

Make check payable to: (Please print)

Name _____

Address _____ City, State, Zip Code _____

Daytime Phone _____ Serial Number (located on back of the calculator) _____

College/University _____ Major _____

Mail the completed coupon, sales receipt, and UPC bar code to:
    Texas Instruments Incorporated
    BAII PLUS™ Rebate
    P.O. Box 650311 MS 3962
    Dallas, TX 75265

Your envelope must be postmarked by May 31, 2001. Claims postmarked after this date will be returned as ineligible.

Please allow 8 to 10 weeks for delivery.

BAII PLUS Rebate Terms and Conditions

This offer is valid only for BAII PLUS purchases between January 1, 1999 and April 30, 2001. All claims must be postmarked by May 31, 2001. Allow 8 to 10 weeks for processing. All purchases must be made in the U.S. or Canada. Rebates will be sent only to addresses in the U.S. and Canada and paid in U.S. dollars. Not redeemable at any store. Send this completed form along with the cash register receipt (original or copy) and the UPC bar code to the address indicated. This original mail-in certificate must accompany your request and may not be duplicated or reproduced. Offer valid only as stated on this form. Offer void where prohibited, taxed, licensed, or restricted. Limit one rebate per household or address. Texas Instruments reserves the right to discontinue this program at any time and without notice.

©1999 TI. ™Texas Instruments Incorporated

Yes! I Want $5 Back On My Purchase of the BAII PLUS.

FINANCIAL MANAGEMENT
Principles and Practice
Second Edition

Timothy J. Gallagher
Professor of Finance,
Colorado State University

Joseph D. Andrew, Jr.
Senior Financial Analyst,
BIA Consulting, Inc.

Prentice Hall
Upper Saddle River, New Jersey 07458

The Finance Center - A FREE

The Prentice Hall Finance Center is one of the premier sites for Finance. By visiting the Finance Center, students can obtain a wealth of ON-LINE material that is directly linked to the book, such as math practice, careers exploration, and current events associated with each chapter to name a few.

The Finance Center CD is FREE with each copy of Gallagher/Andrew: Financial Management: Principles and Practice, 2/e and includes . . .

- ## The Web Center

(PHLIP/CW — Prentice Hall Learning on the Internet Partnership/Companion Web site) Accessed directly from the Finance Center CD or remotely at **www.prenhall.com/gallagher**, this is the most advanced, text-specific site for finance available on the Web! Developed *by* professors *for* professors and their students, this easy-to-use design allows both instructors and students to utilize a wealth of ON-LINE learning resources, including a FREE on-line study guide, current events, downloadable spreadsheets, a chat room, on-line help, and much more!

Place to Learn Finance

• Career Center

Explore the working world of Finance! "A Day in the Life" profiles how finance professionals use the concepts that students are currently learning, improve personal interviewing skills, survey the job market, and more!

• The Math Practice Center

Created by Dr. Puneet Handa of the University of Iowa, **FinCoach** has been proven to improve students' grades by allowing them to learn and practice the math of finance on their own time and at their own speed.

FinCoach — the most sophisticated financial math practice software available anywhere!

Acquisitions Editor: Paul Donnelly
Associate Editor: Gladys Soto
Editorial Director: James C. Boyd
Editor-in-Chief: PJ Boardman
Marketing Manager: Lori Braumberger
Associate Managing Editor: Cynthia Regan
Manufacturing Supervisor: Paul Smolenski
Manufacturing Manager: Vincent Scelta
Designer: Kevin Kall
Design Manager: Patricia Smythe
Interior/Cover Design: Karen Quigley
Photo Research Supervisor: Melinda Lee Reo
Image Permission Supervisor: Kay Dellosa
Photo Research: Teri Stratford
Illustrator (Interior): York Production Services
Cover Illustration/Photo: Greg Smith/Index Stock Imagery, Inc.
Compostion: York Production Services

Photo Credits

Page   1: Steven Gottlieb/FPG International LLC
Page   9: Courtesy of Melanie J. Rosen
Page 19: Benjamin Shearn/FPG International LLC
Page 20: John Abbott Photography
Page 41: Steve Smith/FPG International LLC
Page 59: Dennis Hallinan/FPG International LLC
Page 60: Adam Nadel/AP/Wide World Photos
Page 83: Jook Leung/FPG International LLC
Page 99: Courtesy of Lee Anne Schuster
Page 117: David Waldorf/FPG International LLC
Page 143: Dennie Cody/FPG International LLC
Page 175: Micheal Simpson/FPG International LLC
Page 207: Art Montes de Oca/FPG International LLC
Page 208: Courtesy of Jennifer Owen
Page 235: Ron Chapple/FPG International LLC
Page 252: Courtesy of Jim Bruner
Page 273: Micheal Simpson/FPG International LLC
Page 295: David McGlynn/FPG International LLC
Page 307: Courtesy of Fred Higgins
Page 323: M. Antman/The Image Works
Page 351: Telegraph Colour Library/FPG International LLC
Page 377: Frank LaBua/Pearson Education/PH College
Page 381: Courtesy of Christopher E. Heller
Page 401: Gary Buss/FPG International LLC
Page 421: David Noble/FPG International LLC
Page 431: Courtesy of Michael J. Coleman
Page 439: John Gajda/FPG International LLC
Page 445: Courtesy of Karen Noble
Page 463: Telegraph Colour Library/FPG International LLC
Page 482: Courtesy of Joann K. Jones
Page 495: Arthur Tilley/FPG International LLC
Page 517: Chip Simons/FPG International LLC
Page 527: Courtesy of Donald N. Burton

Copyright ©2000, 1997 by Prentice-Hall, Inc.
Upper Saddle River, New Jersey 07458

All rights reserved. No part of this book may be reproduced, in any form or by any means, without written permission from the Publisher.

Library of Congress Cataloging–in–Publication Data
Gallagher, Timothy James
  Financial management : principles and practice / Timothy J. Gallagher, Joseph D. Andrew, Jr. — 2nd ed.
    p. cm.
  Includes bibliographical references and index.
  ISBN 0–13–012696–9 (pbk.)
  1. Corporations—Finance. I. Andrew, Joseph D. II. Title.
HG4026.G348 2000
658.15—dc21   99-12133
                                                                                CIP

Prentice-Hall International (UK) Limited, London
Prentice-Hall of Australia Pty. Limited, Sydney
Prentice-Hall Canada, Inc., Toronto
Prentice-Hall Hispanoamericana, S.A., Mexico
Prentice-Hall of India Private Limited, New Delhi
Prentice-Hall of Japan, Inc., Tokyo
Prentice-Hall (Singapore) Pte. Ltd.
Editora Prentice-Hall do Brasil, Ltda., Rio de Janeiro

Printed in the United States of America

10 9 8 7 6 5 4 3 2 1

To Susan and Emily
—T.J.G., J.D.A.

BRIEF CONTENTS

PART ONE: THE WORLD OF FINANCE

Chapter 1: Finance and the Firm   1
Chapter 2: Financial Markets and Interest Rates   19
Chapter 3: Financial Institutions   41

PART TWO: ESSENTIAL CONCEPTS IN FINANCE

Chapter 4: Review of Accounting   59
Chapter 5: Analysis of Financial Statements   83
Chapter 6: Forecasting for Financial Planning   117
Chapter 7: Risk and Return   143
Chapter 8: The Time Value of Money   175
Chapter 9: Bond and Stock Valuation   207

PART THREE: LONG-TERM FINANCIAL MANAGEMENT DECISIONS

Chapter 10: Capital Budgeting Decision Methods   235
Chapter 11: Estimating Incremental Cash Flows   273
Chapter 12: The Cost of Capital   295
Chapter 13: Capital Structure Basics   323
Chapter 14: Corporate Bonds, Preferred Stock, and Leasing   351
Chapter 15: Common Stock   377
Chapter 16: Dividend Policy   401

PART FOUR: SHORT-TERM FINANCIAL MANAGEMENT DECISIONS

Chapter 17: Working Capital Policy   421
Chapter 18: Managing Cash   439
Chapter 19: Accounts Receivable and Inventory   463
Chapter 20: Short-Term Financing   495

PART FIVE: FINANCE IN THE GLOBAL ECONOMY

Chapter 21: International Finance   517

CONTENTS

Preface   xviii
About the Authors   xxviii

PART ONE: THE WORLD OF FINANCE

CHAPTER 1: FINANCE AND THE FIRM   1

The Field of Finance   2
  Finance Career Paths   3
Financial Management   3
  The Role of the Financial Manager   3
The Basic Financial Goal of the Firm   5
  In Search of Value   5
Legal and Ethical Challenges in Financial Management   8
  Agency Issues   8
   ▶ *Finance at Work:* Melanie Rosen, Electronic Media, The New York Times   9
  The Interests of Other Groups   10
  The Interests of Society as a Whole   11
Forms of Business Organization   11
  The Proprietorship   10
  The Partnership   12
  The Corporation   13
What's Next   14
Summary   15
Self-Test   16
Review Questions   16
Build Your Communication Skills   17
Answers to Self-Test   17

CHAPTER 2: FINANCIAL MARKETS AND INTEREST RATES   19

The Financial System   21
  Securities   21
  Financial Intermediaries   21
Financial Markets   22
  The Primary Market   23
  The Secondary Market   23
  The Money Market   23
  The Capital Market   23
  Security Exchanges   23
  The Over-the-Counter (OTC) Market   24
  Market Efficiency   24
Securities in the Financial Marketplace   25
  Securities in the Money Market   25
   ▶ *Ethical Connections:* The Dog Ate My Investment   26
  Securities in the Capital Market   27
   ▶ *Financial Management and You:* How to Buy Treasury Securities without Using a Broker   30
Interest   31
  Determinants of Interest Rates   31
  The Yield Curve   34
What's Next   35
Summary   35
Self-Test   36
Review Questions   37
Build Your Communication Skills   37
Problems   38
Answers to Self-Test   38

CHAPTER 3: FINANCIAL INSTITUTIONS   41

Financial Intermediation   42
Denomination Matching   42
Absorbing Credit Risk   43
Commercial Banks   44
Bank Regulation   44
Commercial Bank Operations   44
> ▶ *Ethical Connections:* How Ethical Is Your Financial Institution?   46

The Federal Reserve System   47
Organization of the Fed   47
Controlling the Money Supply   47
The Discount Window   49

Savings and Loan Associations   49
Legislation Affecting S&Ls   49
Regulation of S&Ls   50
Mutual Companies versus Stockholder-Owned Companies   50
The Problem of Matching Loan and Deposit Maturities   51
S&Ls' Real Assets   52

Credit Unions   52
The Common Bond Requirement   52
Members as Shareholders   52
Credit Unions Compared to Banks   52
Credit Union Regulation   53

Finance Companies, Insurance Companies, and Pension Funds   53
Types of Finance Companies   53
Insurance Companies   54
> ▶ *Financial Management and You:* The Social Security System   56

What's Next   56
Summary   56
Self-Test   57
Review Questions   58
Build Your Communication Skills   58
Answers to Self-Test   58

PART TWO: ESSENTIAL CONCEPTS IN FINANCE

CHAPTER 4: REVIEW OF ACCOUNTING   59

Review of Accounting Fundamentals   61

Basic Accounting Financial Statements   61
The Income Statement   61
> ▶ *Financial Management and You:* A Special Earnings Category: EBITDA   63
> ▶ *Ethical Connections:* Box Office Hits—Who Profits?   64

The Balance Sheet   66
The Statement of Cash Flows   68

Depreciation   71
Calculating the Amount of Depreciation Expense   72

Income Taxes   73
> ▶ *Financial Management and You:* Your Personal Tax Rates   74
> ▶ *Finance at Work:* Business Changes in the Taxpayer Relief Act of 1997   75

What's Next   76
Summary   76
Self-Test   77
Review Questions   78
Build Your Communication Skills   78
Problems   79
Answers to Self-Test   82

CHAPTER 5: ANALYSIS OF FINANCIAL STATEMENTS   83

Assessing Financial Health   84
Misleading Numbers   84
Financial Ratios   85

The Basic Financial Ratios   86
Calculating the Ratios   87

▶ *Finance at Work:* Lee Anne Schuster, Kitchell Contractors, Marketing 99

Trend Analysis and Industry Comparisons 100
Trend Analysis 100
Industry Comparisons 101

Summary Analysis: Trend and Industry Comparisons Together 101

Locating Information about Financial Ratios 103

What's Next 105
Summary 105
Equations Introduced in This Chapter 106
Self-Test 107
Review Questions 108
Build Your Communication Skills 108
Problems 108
Answers to Self-Test 116

CHAPTER 6: FORECASTING FOR FINANCIAL PLANNING 117

Why Forecasting Is Important 119
Forecasting Approaches 119
Why Forecasts Are Sometimes Wrong 120

Forecasting Sales 120

Forecasting Financial Statements 122
Budgets 122
Producing *Pro Forma* Financial Statements 123
Analyzing Forecasts for Financial Planning 131
▶ *Finance at Work:* Keith Ender, Customer Service Representative, James River Paper Company 132

What's Next 134
Summary 134
Self-Test 135
Review Questions 135
Build Your Communication Skills 135
Problems 136
Answers to Self-Test 141

CHAPTER 7: RISK AND RETURN 143

Risk 144
Risk Aversion 144
The Risk-Return Relationship 144

Measuring Risk 145
Using Standard Deviation to Measure Risk 145
Using the Coefficient of Variation to Measure Risk 149

The Types of Risks Firms Encounter 151
Business Risk 151
Financial Risk 153
Portfolio Risk 155

Dealing with Risk 161
Risk-Reduction Methods 161
Compensating for the Presence of Risk 162
▶ *Financial Management and You:* Mutual Funds and Risk 162

What's Next 165
Summary 165
Equations Introduced in This Chapter 166
Self-Test 167
FinCoach Practice Exercises 168
Review Questions 168
Build Your Communication Skills 168
Problems 169
Answers to Self-Test 172

CHAPTER 8: THE TIME VALUE OF MONEY 175

Why Money Has Time Value 177

Measuring the Time Value of Money 178
The Future Value of a Single Amount 178
The Sensitivity of Future Values to Changes in Interest Rates or the Number of Compounding Periods 181

The Present Value of a Single Amount 181
The Sensitivity of Present Values to Changes in k and n 182a

Working with Annuities 182c
Future Value of an Ordinary Annuity 182c
The Present Value of an Ordinary Annuity 183
▶ *Ethical Connections:* When a Million Isn't a Million. Taking a Chance on the Time Value of Money 185
Future and Present Values of Annuities Due 185
Perpetuities 187
Present Value of an Investment with Uneven Cash Flows 187

Special Time Value of Money Problems 189
Finding the Interest Rate 189
Finding the Number of Periods 192
Solving for the Payment 193

Compounding More Than Once per Year 195

What's Next 198
Summary 199
Equations Introduced in This Chapter 199
Self-Test 201
FinCoach Practice Exercises 202
Review Questions 202
Build Your Communication Skills 202
Problems 202
Answers to Self-Test 205

CHAPTER 9: BOND AND STOCK VALUATION 207

The Importance of Bond and Stock Valuation 208
A General Valuation Model 209

Bond Valuation 210
Semiannual Coupon Interest Payments 212
The Yield to Maturity of a Bond 212
The Relationship between Bond YTM and Price 215

Preferred Stock Valuation 215
Finding the Present Value of Preferred Stock Dividends 216
The Yield on Preferred Stock 217

Common Stock Valuation 218
Common Stock Going Concern Valuation Models 218
Balance Sheet Valuation Approaches 221
Deciding Which Stock Valuation Approach to Use 222
The Yield on Common Stock 222

What's Next 222
Summary 223
Equations Introduced in This Chapter 224
Self-Test 226
FinCoach Practice Exercises 226
Review Questions 226
Build Your Communication Skills 227
Problems 227
Answers to Self-Test 230

Appendix 9A: Common Stock Valuation: Supernormal Growth 232

PART THREE: LONG-TERM FINANCIAL MANAGEMENT DECISIONS

CHAPTER 10: CAPITAL BUDGETING DECISION METHODS 235

The Capital Budgeting Process 236
Decision Practices 236
Types of Projects 237
Capital Budgeting Cash Flows 237
Stages in the Capital Budgeting Process 237

Capital Budgeting Decision Methods 238
The Payback Method 238
The Net Present Value (NPV) Method 239
The Internal Rate of Return (IRR) Method 245
Conflicting Rankings between the NPV and IRR Methods 248

Capital Rationing 249

Risk and Capital Budgeting 251
Measuring Risk in Capital Budgeting 251
▶ *Finance at Work:* Jim Bruner, Former Maricopa County Supervisor, State of Arizona 252
Adjusting for Risk 255

What's Next 255
Summary 256
Equations Introduced in This Chapter 257
Self-Test 257
FinCoach Practice Exercises 258
Review Questions 258
Build Your Communication Skills 258
Problems 259
Answers to Self-Test 265

Appendix 10A: Wrinkles in Capital Budgeting 267
Nonsimple Projects 267
Multiple IRRs 268
Mutually Exclusive Projects with Unequal Project Lives 270
  Comparing Projects with Unequal Lives 271
Equations Introduced in This Appendix 272

CHAPTER 11: ESTIMATING INCREMENTAL CASH FLOWS 273
Incremental Cash Flows 274
Types of Incremental Cash Flows 274
  Initial Investment Cash Flows 275
  Operating Cash Flows 276
  Cash Flows at the End of a Project's Life 279
  Incremental Cash Flows of an Expansion Project 280
  ▶ *Financial Management and You:* The Incremental Costs of Studying Abroad 281
An Asset Replacement Decision 285
Financing Cash Flows 286
What's Next 287
Summary 287
Self-Test 288
FinCoach Practice Exercises 289
Review Questions 289
Build Your Communication Skills 289
Problems 289
Answers to Self-Test 293

CHAPTER 12: THE COST OF CAPITAL 295
The Cost of Capital 296
Sources of Capital 297
  The Cost of Debt 297
  The Cost of Preferred and Common Stock Funds 298
  The Weighted Average Cost of Capital (WACC) 303
The Marginal Cost of Capital (MCC) 306
  ▶ *Finance at Work:* Interview with Fred Higgins: Minit Mart Foods, Inc., CEO 307
  The Firm's MCC Schedule 307
The MCC Schedule and Capital Budgeting Decisions 311
  The Optimal Capital Budget 313
  The Importance of MCC to Capital Budgeting Decisions 314
What's Next 314
Summary 314
Equations Introduced in This Chapter 315
Self-Test 317
FinCoach Practice Exercises 317
Review Questions 317
Build Your Communication Skills 318
Problems 318
Answers to Self-Test 322

CHAPTER 13: CAPITAL STRUCTURE BASICS 323
Breakeven Analysis and Leverage 324
  Constructing a Sales Breakeven Chart 324
  Applying Breakeven Analysis 327
Leverage 330
  Operating Leverage 330
  Financial Leverage 333
  Combined Leverage 335
LBOs 336

▶ *Ethical Connections:* Let's Rip off the Bondholders? 337

Capital Structure Theory 338
Tax Deductibility of Interest 338
Modigliani and Miller 339
Toward an Optimal Capital Structure 339

What's Next 341
Summary 341
Equations Introduced in This Chapter 343
Self-Test 344
Review Questions 345
Build Your Communication Skills 345
Problems 345
Answers to Self-Test 349

CHAPTER 14: CORPORATE BONDS, PREFERRED STOCK, AND LEASING 351

Bond Basics 352
Features of Bond Indentures 353
Security 353
Plans for Paying Off the Bond Issue 354
Call Provisions 354
Restrictive Covenants 355
The Independent Trustee of the Bond Issue 357

Types of Bonds 357
Secured Bonds 357
Unsecured Bonds (Debentures) 358
Convertible Bonds 358
Variable-Rate Bonds 361
Putable Bonds 361
Junk Bonds 362
▶ *Ethical Connections:* SEC Charges 50 Municipalities in Bid to Stop Fraud in Muni–Bond Market 363
International Bonds 363
Super Long-Term Bonds 364

Preferred Stock 364
Preferred Stock Dividends 364
Preferred Stock Investors 365

Leasing 365
Genuine Leases versus Fakes 365
Operating and Financial (Capital) Leases 366

▶ *Finance At Work:* Hybrid Financing—Mezzanine Capital 366
Lease or Buy? 368

What's Next 368
Summary 368
Equations Introduced in This Chapter 369
Self-Test 369
Review Questions 370
Build Your Communication Skills 370
Problems 370
Answers to Self-Test 372

Appendix 14A: Bond Refunding 372
A Sample Bond Refunding Problem 373

CHAPTER 15: COMMON STOCK 377

The Characteristics of Common Stock 378
Stock Issued by Private Corporations 379
Stock Issued by Publicly Traded Corporations 380
Institutional Ownership of Common Stock 380
▶ *Finance at Work:* Chris Heller, Corporate Communications Consultant 381

Voting Rights of Common Stockholders 382
Proxies 382
Board of Directors Elections 382

The Pros and Cons of Equity Financing 384
Disadvantages of Equity Financing 385
Advantages of Equity Financing 385

Issuing Common Stock 386
The Function of Investment Bankers 387
Pricing New Issues of Stock 387

Rights and Warrants 389
Preemptive Rights 390
Warrants 392

What's Next 394
Summary 394

Equations Introduced in This Chapter 395
Self-Test 396
Review Questions 396
Build Your Communication Skills 396
Problems 397
Answers to Self-Test 399

CHAPTER 16: DIVIDEND POLICY 401

Dividends 402
Why a Dividend Policy Is Necessary 403
Factors Affecting Dividend Policy 403
Need for Funds 403
Management Expectations and Dividend Policy 403
Stockholders' Preferences 404
Restrictions on Dividend Payments 404

Cash versus Earnings 404
Leading Dividend Theories 406
The Residual Theory of Dividends 406
The Clientele Dividend Theory 407
The Signaling Dividend Theory 408
The Bird-in-the-Hand Theory 408
Modigliani and Miller's Dividend Theory 409

The Mechanics of Paying Dividends 409
Dividend Reinvestment Plans 410

Alternatives to Cash Dividends 410
Stock Dividends and Stock Splits 410
▶ *Financial Management and You:* Dividend Reinvestment Records Can Avoid Tax Headaches 411

What's Next 415
Summary 415
Equations Introduced in This Chapter 416
Self-Test 416
Review Questions 417
Build Your Communication Skills 417
Problems 417
Answers to Self-Test 419

PART FOUR: SHORT-TERM FINANCIAL MANAGEMENT DECISIONS

CHAPTER 17: WORKING CAPITAL POLICY 421

Managing Working Capital 422
Why Businesses Accumulate Working Capital 423
Fluctuating Current Assets 423
Permanent and Temporary Current Assets 424

Liquidity versus Profitability 425
Establishing the Optimal Level of Current Assets 425
Managing Current Liabilities: Risk and Return 426
Three Working Capital Financing Approaches 426
The Aggressive Approach 427
The Conservative Approach 428
The Moderate Approach 429

Working Capital Financing and Financial Ratios 429
What's Next 430
▶ *Finance at Work:* Interview with Michael Coleman, Vice President of Tek Soft 431

Summary 432
Self-Test 433
Review Questions 433
Build Your Communication Skills 434
Problems 434
Answers to Self-Test 437

CHAPTER 18: MANAGING CASH 439

Cash Management Concepts 440
Determining the Optimal Cash Balance 441

The Minimum Cash Balance 441
The Maximum Cash Balance 442
Determining the Optimal Cash
    Balance 443
  ▶ *Finance at Work:* *Karen Noble, Professional Golfer* 445

Forecasting Cash Needs 446
Developing a Cash Budget 447

Managing the Cash Flowing In and Out of the Firm 451
Increasing Cash Inflows 451
Decreasing Cash Outflows 454
Speeding Up Cash Inflows 454
Slowing Down Cash Outflows 455

What's Next 456
Summary 456
Equations Introduced in This Chapter 457
Self-Test 457
Review Questions 457
Build Your Communication Skills 458
Problems 458
Answers to Self-Test 461

CHAPTER 19: ACCOUNTS RECEIVABLE AND INVENTORY 463

Why Firms Accumulate Accounts Receivable and Inventory 465

How Accounts Receivable and Inventory Affect Profitability and Liquidity 465

Finding Optimal Levels of Accounts Receivable and Inventory 467
The Optimal Level of Accounts
    Receivable 467
The Optimal Level of Inventory 472

Inventory Management Approaches 478
The ABC Inventory Classification
    System 478
Just-in-Time Inventory Control (JIT) 478

Making Credit Decisions 479
Collection Policies to Handle
Bad Debts 480
  ▶ *Finance at Work:* *Joann K. Jones, CEO of Capital Electric Supply* 482

What's Next 483
Summary 483
Equations Introduced in This Chapter 484
Self-Test 485
Review Questions 485
Build Your Communication Skills 486
Problems 486
Answers to Self-Test 490

CHAPTER 20: SHORT-TERM FINANCING 495

The Need for Short-Term Financing 496

Short-Term Financing versus Long-Term Financing 497

Short-Term Financing Alternatives 497
Short-Term Loans from Banks
    and Other Institutions 497
Trade Credit 499
Commercial Paper 500

How Loan Terms Affect the Effective Interest Rate of a Loan 502
The Effective Interest Rate 502
Discount Loans 503
Compensating Balances 503
Loan Maturities Shorter Than One
    Year 505
A Comprehensive Example 506
Computing the Amount to Borrow 507

Collateral for Short-Term Loans 507
Accounts Receivable as Collateral 508
Inventory as Collateral 508
  ▶ *Financial Management and You:* *Easy Come, Easy Go: The Cost of Credit* 509

What's Next 510

Summary 510
Equations Introduced in This Chapter 511
Self-Test 512
Review Questions 512
Build Your Communication Skills 513
Problems 513
Answers to Self-Test 514

PART FIVE: FINANCE IN THE GLOBAL ECONOMY

CHAPTER 21: INTERNATIONAL FINANCE 517

Multinational Corporations 518
Financial Advantages of Foreign Operations 518
Ethical Issues Facing Multinational Corporations 519
Comparative Advantage 520

Exchange Rates and Their Effects 521
Fluctuating Exchange Rates 521
Cross Rates 522
Exchange Rate Effects on MNCS 524
Exchange Rate Effects on Foreign Stock and Bond Investments 524

Managing Risk 525
Hedging 525
Diversification Benefits of Foreign Investments 525

American Depository Receipts 526
▶ **Finance at Work:** Interview with Don Burton, International Import–Export Institute 527

Exchange Rate Theories 528
Purchasing Power Parity Theory 528
International Fisher Effect 528
Interest Rate Parity Theory 529
Other Factors Affecting Exchange Rates 529
Government Intervention in Foreign Exchange Markets 530

Political and Cultural Risks Facing MNCs 530
Political Risk 530
Cultural Risk 531

International Trade Agreements 532
NAFTA 532
GATT 532
European Union 533
Free Trade versus Fair Trade 533

Summary 534
Equations Introduced in This Chapter 535
Self-Test 535
Review Questions 535
Build Your Communication Skills 536
Problems 536
Answers to Self-Test 537

Appendix   A-1
Glossary   G-1
Index   I-1

PREFACE

THE CHALLENGE

The prospect of taking the Introductory Finance course is daunting to some students. Maybe this is because students often have the mistaken belief that the introductory finance course requires an understanding of high level mathematics and is irrelevant to their career plans. Also, some students mistakenly believe finance is an area in which they will not need competency. Finance concepts often seem far removed from daily life. In spite of this, almost every major in a college of business, and many majors in other colleges, require the principles of finance course. As a result, many of the students who find themselves sitting in finance class on the first day of the semester do not want to be there.

We do not believe that this needs to be the case. Finance is important, dynamic, interesting, and fun. The challenge we take head-on in *Financial Management: Principles and Practice* is to convince students of this. In order to learn, students must want to learn. If they can see the usefulness of what is presented to them they will work hard and they will learn.

Our many years of teaching experience have taught us that the introductory financial management course can be one that students enjoy and that they see as having added considerable value to their educational experiences. Finance is, after all, central to any business entity. More CEOs have come up through the finance ranks than any other discipline. Students need to know that the principles and practices of financial management apply to any business unit from the very large multinational corporation to the very smallest proprietorship, even the family. Financial ratios tell a story; they are not numbers to be calculated as an end unto themselves. Risk is important and can be managed. Time value of money has meaning and is understood as the central tool of valuation. Funds have a cost and different sources of funds have different costs. Financial performance and condition can be assessed. Amortized loan payments, rates of return on investment, future value of investment programs, present value of payments to be received from bonds and stocks can be calculated. The opportunities and special challenges of international operations can be understood.

OUR APPROACH

We believe that students should walk out of the room after taking the final exam for a finance course believing that they have learned something useful. They should see a direct benefit to themselves personally, rather than just the belief that some set of necessary job skills has been mastered.

In writing the first edition of *Financial Management: Principles and Practice,* our author mandate was to always write from a perspective with the student in mind. We consider students to be our real customers. They are the ones who must be inspired to learn and gain a working facility with the subject. For this reason our examples have been carefully chosen to focus on organizations, topics, and people with which students are familiar. With the second edition, we have further enhanced our goal of student understanding with the addition of the PRENTICE HALL FINANCE CENTER CD. This unique companion software provides the student with a powerful finance mathematics tutorial, FINCOACH; a broad and informative look at the world of finance through interactive interviews with a vast array of people working in finance; and a direct link to PHLIP (Prentice Hall Learning on the Internet Partnership). PHLIP provides current articles geared to specific chapters in the text as well as a FREE On-line Study Guide where students can test their skills and understanding. Our intention is to present finance in such a way that it inspires students to learn.

Two key characteristics of *Financial Management: Principles and Practice* are theoretical currency and relevance. One of the authors of the text is an academic with over twenty years of teaching experience and the other is a full-time financial practitioner. This combination of backgrounds results in a text that presents the latest in financial theory while retaining a strong "real-world" connection. No other textbook on the market enjoys this balance of academic and practitioner perspectives.

There are a few cartoons in this book. Do not allow their presence to mislead you. This book is serious about learning and the cartoons serve a serious purpose. They tend to "lighten up" the presentation in order to capture the students' interest. Although our style is lighthearted, the writing itself is substantive, concise, clear, and easy to understand.

Distinctive Focus

Although there are many other introductory financial management books on the market, none contain the unique style and content of *Financial Management: Principles and Practice,* second edition. Many texts focus mostly on accounting with little presentation of the economic theory that underlies the financial techniques presented. Others assume that the students remember all that was learned in the accounting course that is usually a prerequisite for this course. Still others claim to take a "valuation approach" but present their topics in a straight accounting framework. In this book we are serious about focusing on what creates value. We are consistent in this approach throughout the book, addressing such issues as what creates value, what destroys it, and how value and risk are related. In so doing we maximize the value of the finance course to the student.

Organization of the Text

The book is organized into five major parts as follows:

Part I. *The World of Finance* contains chapters on the structure and goals of firms, the role of financial managers, and an examination of the financial environment.

Part II. *Essential Concepts in Finance* presents chapters on accounting statements and their interpretation, forecasting, risk and return, the time value of money, and security valuation.

Part III. *Long-Term Financial Management Decisions* are included in chapters on capital budgeting, incremental cashflow estimation, capital structure, bonds, preferred stock, and leasing, common stock, and dividend policy.

Part IV. *Short-Term Financial Management Decisions* includes chapters on working capital policy, cash and marketable securities, accounts receivable and inventory, and short-term financing.

Part V. *Finance in a Global Economy* is where international finance topics are covered, in addition to those international topics that are woven throughout the book.

SPECIAL FEATURES IN THE TEXT
Real-World Examples

Each chapter in *Financial Management: Principles and Practice* begins with a real-world example that illustrates the concept to be addressed in that chapter. This serves to give the student a reason to learn this material and to show its practical application.

Special Boxes

Three different types of special boxes are integrated into the chapters. *Ethical Connections* boxes identify many financial decisions that have ethical dimensions to them. *Financial Management and You* boxes take financial management concepts intended for use within a firm and show how these same concepts can be used by individuals for personal financial decision making. This makes these topics less distant. *Finance At Work* boxes are designed to demonstrate how finance relies on, contributes to, and interacts with other functional areas of the firm. Not every student taking this course is a finance major and the understanding of the cross-functional role of finance within the firm is important.

Take Note

Located in the margins of the text pages, these helpful notes contain learning tips, offer additional examples, or show how material links to business practice.

Flipback

These margin icons provide the student with a convenient way to review perquisite information previously covered and necessary for the comprehension of material later in the text. These icons become especially convenient if certain chapters of the text are not assigned or read prior to the chapter under study.

Calculator Solutions

Financial calculator solutions to all general time value of money and specific security valuation problems are clearly illustrated within the text using keystroke functions to a TI BAII Plus Calculator solution. However, differing preferences as to the use of other calculators is accommodated with the "Calculator Guide for Financial Management" card provided with each text. This card provides keystroke procedures for the Texas Instruments BAII Plus, BA II, and BA-35 as well as the Hewlett-Packard 12C.

Summaries

The summary for each chapter specifically describes how the learning objectives have been achieved and it also provides a bridge to the next chapter.

Key Terms

Each chapter has **bolded** key terms that are defined in the chapter and in the glossary. There are self-test questions and problems at the end of chapters, along with their solutions, so that students can check their grasp of the material presented.

Practice Questions and Problems

Study questions and an abundant number of end-of-chapter problems are included in the appropriate chapters. Self-Test Questions and Problems are included with answers provided in the text. Answers to Review Questions and Problems are provided in the Solutions Manual and on the PHLIP Web site.

Computer Spreadsheet Supported Problems

A number of end-of-chapter problems are marked with the special computer problem logo shown above. This indicates that a downloadable Excel spreadsheet template for the problem is available at the PHLIP-Financial Management: Principles and Practice Web Center. Access the PHLIP on the World Wide Web at **http://www.prenhall.com/gallagher**. Once at the Web site, proceed to the applicable text chapter and locate the spreadsheet template corresponding to the problem number in the text. Follow the instructions to download the file to your computer.

FinCoach Drills

Following each chapter requiring the understanding of quantitative skills, the student is directed to the respective practice drills in the FinCoach software program contained in the Prentice Hall Finance Center CD provided on the inside back cover of this test.

Communication Skills

Suggested assignments to build students' written and oral communication skills are included in each chapter.

Cartoons

Cartoons are added for fun, and occasionally to make a financial point.

CHANGES IN THE SECOND EDITION

- Color has been integrated to assist the pedagogy and to make the book easy to read.
- Regulatory, legal, and industry practice changes have been thoroughly updated in this new edition. These include changes in reserve requirements, capital gains rates, and ex-dividend date policy to name a few.
- Time-sensitive examples have been updated. The number of companies students can relate to such as Yahoo, Lucent Technologies, and GeoCities has been increased.
- Almost all first edition chapter openers were replaced with newer, real-world vignettes or interviews that make the students want to read the content of that chapter.
- EVA, MVA, and EBITDA coverage was added because of their extensive real-world usage in financial analysis.
- Real options are covered in the capital budgeting chapters.
- End-of-chapter problems were expanded, including some of increased difficulty, along with changes that were made in some so as to increase clarity and to correct some errors.
- Fincoach: The Financial Management Math Practice Program which has helped thousands of students learn financial mathematics has been integrated into the pedagogy of the text. Each quantitative chapter contains an end-of-chapter Fin-Coach Drill Section. Students will find the Fincoach software contained in the Prentice Hall Finance Center CD accompanying the text.
- Substantial updates were made to the international coverage. This includes the financial troubles in Asia, the EU and EMU, and the introduction of the euro. End-of-chapter problems and examples were similarly changed to reflect the new material. More international content was woven throughout all of the chapters in addition to the changes made to the dedicated international finance chapter.
- Most of the Ethical Connections boxes and Financial Management and You boxes were rewritten or updated.
- A multilevel quality control program was implemented for the text and supplements. The authors personally oversaw these activities and checked each item. This quality control program included:

  Ferreting out and fixing errors in the first edition.
  Rewriting and verifying the testbank and solutions manual.
  Enhancing and updating PowerPoint slides as well as all spreadsheet templates.

FEATURES RETAINED FROM THE FIRST EDITION

- The book is still written in the student-friendly style that was extremely popular in the first edition. The concise, easy-to-understand presentation loved by student users is maintained.

- The book still maintains the level of rigor professors demand. When professors get past the friendly style, they find all the rigor and all the mainstream topics they expect in a book of this type. For example, if you are not already a Gallagher/Andrew user, does your book:

    Cover real options?
    Cover EVA, MVA, and EBITDA?
    Use a value added (NPV) approach to the inventory and accounts receivable investment coverage rather than the outmoded return on investment ratio approach?
    Use examples of high-tech and Internet companies so as to hold student interest?
    Include the euro in its international coverage and exchange rate problems?
    Have supplements that were not "farmed out" to subcontractors but that instead have the authors' hands-on participation?

- Attempts to expand the book, and to make it longer, have been resisted. The topics that professors actually teach are here. Those that are most likely to be taught in the second course in financial management are left out. Students don't have to buy more than they need.

The Technology Learning Package

Financial Management: Principles and Practice integrates the most advanced technology available to assist the student and the instructor in not only making their financial management course come alive with the most current information, but also to enhance a total understanding of all tools and concepts necessary in mastering the course.

For The Student

The Prentice Hall Finance Center CD Contained in the inside back cover of this text is the Prentice Hall Finance Center CD. This robust learning tool contains the following features all designed to increase student awareness of what finance professionals do, to ensure comprehension and mastery of the financial mathematics contained in the text, and to supply a direct link to PHLIP—the Prentice Hall Learning on the Internet Partnership:

- *Careers Center*—Introduces the student to a vast array of professional opportunities in finance through video interviews with professionals and insights into what they do on the job in an average day. Here the student will meet an options trader, a mutual fund manager, investment analysts, a CFO, and others. Also accessible are features for personal development, résumé writing, interviewing techniques, and career-planning information.
- *FINCOACH*—*The Financial Math Practice Center*

Contains more than 5 million problems and self-tests in virtually all math areas covered in this text and financial management. Save problems, review them, and print them. This is a step-by-step guide to solve any corporate finance mathematics problem and to allow the student to rapidly gain mastery in all mathematical challenges. The program algorithm changes the sample problem numbers each time the program is accessed so that students have an infinite number of practice problems to work on along with the help features of the program.

PHLIP—Financial Management: *Principles and Practice Web Center* PHLIP (Prentice Hall Learning on the Internet Partnership) can be accessed either directly from the CD or remotely at *http://www.prenhall.com/gallagher*. Here's what the student can do on PHLIP:

> See current news items that are from the popular business press and are directly related to chapters.
> Use chapter terminology.
> Link to more related information.
> Find discussion questions and projects for assignments by the instructor.
> Download Excel spreadsheet templates related to specific chapter problems.
> Access additional career information.
> Learn study skills, writing skills, engage in conferences with other students studying financial management.

- **Case Connect**—Also linked through the PHLIP Web site and applicable to specific chapters in the text are cases using interactive and linked features to provide the student and the instructor a means of understanding concepts through real-world applications. Generally, these cases focus on a broad area of corporate finance and are appropriate for the level of this text.
- *FREE On-Line Study Guide Companion Web Site*—Another feature accessible by the student on the PHLIP site and designed specifically for this text. This companion Web site allows students to test their understanding of concepts within each chapter and relay their answers via e-mail to their instructor. Instructors will find this custom site useful for allocating student assignments and on-line quizzes, and with its built-in grading feature can grade exams and provide students with immediate feedback.
- **Student Lecture Notes**—This handy lecture aid for the student contains PowerPoint slides and space for lecture note taking for each chapter in the text.

Other Supplements Available for the Student

- *The Collegiate Investment Challenge*—Following the inside front cover of each copy of *Financial Management: Principles and Practice* students will find a discount coupon for the widely popular INVESTMENT CHALLENGE. This highly creative real-time on-line portfolio management simulation allows the student or team of students to manage a $500,000 portfolio. It mimics the daily operation of a real brokerage firm and allows the teaching professor to monitor trades and progress of all participating students. It is an excellent teaching companion for the investments sections in the text.
- *Student Study Guide and Workbook*—Each chapter begins with an overview of the key points in the respective text chapter and serves as a useful review. Additional problems with detailed solutions and self-tests can be used as an aid in preparing for examinations or for preparation of outside assignments.

For the Instructor

- *Instructor's Manual with Solutions*—This manual contains concise chapter orientations, detailed chapter outlines, complete answers to end-of-chapter questions and worked-out solutions to all problems in the text. An Instructor's Guide

to using PHLIP (Prentice Hall Learning on the Internet Partnership) and a Guide to Using the Prentice Hall Finance Center are also contained in this manual.
- *Investment Challenge*—The Collegiate Investment Challenge provides all the resources professors need to easily and effectively implement this dynamic real-time portfolio simulation program into their classrooms:

  FREE Professor Portfolio
  Classroom Reports (including ROI, 3 of trades, types of trades, etc.)
  Ability to view student accounts
  Toll-Free Professor Support Line
  Professor Chat Room

- *Test Item File*—Completely revised and updated by the authors, the second ediition test bank contains over 2000 true/false, multiple choice, and short-answer questions.
- *PH Custom Tests*—Available for both Windows and Macintosh, PH Custom Test is the computerized version of the test item file. It permits instructors to edit, add or delete questions from the test item file, and generate their own custom exams.
- *FinCoach Test Manager and FinCoach Instructor's Manual*—Test Manager software has been developed to allow instructors to generate tests based on FinCoach—*The Financial Management Math Practice Program,* contained within the Prentice Hall Finance Center CD available to all students using *Financial Management: Principles and Practice*. The program algorithm changes the sample problem numbers each time the program is accessed so that instructors have an infinite number of test problems. In addition, an Instructor's Manual for using FinCoach Test Manager and FinCoach in the course has been developed and is included with the Test Manager software.
- *PowerPoint Presentation*—Over 50 slides per chapter of new lecture notes have been prepared by Professors Vickie Bajtelsmit and Tim Gallagher of Colorado State University. These electronic transparencies allow instructors to make full-color presentations coordinated with *Financial Management: Principles and Practice,* second edition, *Student Lecture Notes*. To encourage more active learning, the slides include sample problems for students to solve in class. The PowerPoint files are available from the Prentice Hall PHLIP Web site (http://www.prenhall.com/financecenter). ID and Password designations are available from your local Prentice Hall representative.
- *PHLIP Faculty Web Site* (Go to: http://www.prenhall.com/financecenter)

You can access your FREE PHLIP/CW On–Line Study Guide, Current Events, and Spreadsheet Problem templates, which may correspond to this chapter either through the Prentice Hall Finance Center CD–ROM or by directly going to: **http://www.prenhall.com/gallagher.**

PHLIP/CW — *The Prentice Hall Learning on the Internet Partnership–Companion Web Site*

Instructors will need to acquire a Password and ID code from their local Prentice Hall representative in order to open the Faculty site and gain access to the following materials:

Downloadable PowerPoint Presentations—average 50 per chapter.
Downloadable Instructor's Manual.
Downloadable Excel spreadsheet solutions to selected end-of-chapter problems including solutions to Integrated Problems.
"Talk To Team"—faculty chat room.
"Teaching Archive"—resources for enhancing lecture materials and doing research on the Web.
"Help With Computers"—provides tips and access to getting answers to tricky computer problems.
Solutions and answers to all cases on the Case Connect Web site.
Financial Management: Principles and Practice, Second Edition, On-line Companion Web site for creating syllabuses and assignments and assessing student progress via e-mail relay of answers given on the On-line Study Guide available to all students using the text.

- ***Color Transparencies***—All figures and tables from the text are available as full-color images on 8½ by 11 acetate transparencies.

In Conclusion

We believe that students will understand the very important finance concepts and master necessary problem-solving skills when they complete the course in which this text is used. "Students first" is our philosophy and this belief is evident throughout the second edition text and package. Professors will find they have more enthusiastic students, and students who grasp the important content, both conceptual and problem solving. This will make classroom experiences more rewarding. Our goal is to help make this happen. That is how we determine if we have succeeded in achieving our visions for *Financial Management: Principles and Practice*.

Acknowledgments

Financial Management: Principles and Practice, Second Edition, is much more than just a book. It is a coordinated marriage of text and the latest in multimedia technology. We are extraordinarily grateful to all of those who provided the expertise and talent necessary to help us pull it all together.

We are particularly indebted to a number of colleagues who contributed to this second edition as key reviewers of the material as it was being developed. They are Andrew M. Atkinson, Francene Feldbrugge, Ron Filante, Zhenu Jin, Clarke Maxam, David Minars, Dianne R. Morrison, and Wendy L. Pirie. Additionally, Vickie Bajtelsmit, Joe Brocato, Sue Hine, John Ellis, Lise Graham, John Olienyk, Rob Schwebach, Ed Prill, Ralph Switzer, Elaine Worzala, Kent Zumwalt, Karen Hallows, and Jeff Loudermilk provided guidance to help us with improving both the text and its accompanying package. We would like to specifically recognize Shyam Bhandari who provided specific suggestions for both the text and the entire package.

We thank all these people sincerely for their effort and advice in helping us make this edition an improvement over the last one. We are also very grateful to

Henry Hecht, who rechecked all the material in the text and solutions manual for accuracy. Thanks for your sharp eye, Henry!

Dan Cooper at Marist College, who is the Director of the Prentice Hall Learning on the Internet Partnership Web site, has earned our highest level of appreciation as well as the respect of thousands of faculty and students who use this site in conjunction with the text every day. Also, David Durst at the University of Akron has created a superb On-line Companion Study Guide which is FREE to student and faculty users of the second edition. We are grateful to him for this remarkable job.

The production team at Prentice Hall is, in our opinion, simply the best in the business. Cynthia Regan, Associate Managing Editor, kept the book on schedule and the authors on their toes. To the design team of Pat Smythe and Kevin Kall, we simply say "wow," and we couldn't think of a more appropriate word to express our feelings for their talents.

Lori Braumberger, Senior Marketing Manager at Prentice Hall, and Brian Kibby, Director of Marketing, not only gave the text the marketing support it needed to help make it a winner, they made the campaign fun! To Lori and Brian we owe an enormous expression of thanks. To the editorial staff—Jim Boyd, Editorial Director, and PJ Boardman, Editor-in-Chief, thanks for believing in us. To Gladys Soto, Associate Editor, and Jodi Hirsh, Editorial Assistant, thanks for always being there when we needed you. To Paul Donnelly, Senior Acquisitions Editor–Finance, thanks for making every piece fit together, having faith in us throughout, and organizing an amazing package to facilitate student learning.

Finally, we would like to give special thanks to our friends and business colleagues who suffered through the often hectic development process, and often worked extra hours to take up the slack when we were occupied with producing the book. These include Jerome Fowlkes, Amando Madan, Camilla Jensen, and Jennifer Owen at BIA Consulting, and Carl Richards and Dara Van Dijk at Webster University. Last, but not least, we are most especially grateful for the support of our family members, Susan Shattuck, Emily Adams, and Emily Andrew.

ABOUT THE AUTHORS

Timothy J. Gallagher is professor of finance and Chair of the Department of Finance & Real Estate at Colorado State University. Tim received his Ph.D. in finance in 1978 from the University of Illinois at Urbana-Champaign.

Professor Gallagher has taught undergraduate and graduate finance courses for 20 years, including courses in financial management, markets and institutions, and investments. He has taught traditional and nontraditional students at all levels, including executive MBAs, and in all types of classroom settings—large lecture, small seminar, and distance learning.

Tim has been published in journals such as *The Journal of Money, Credit, and Banking; The Journal of Portfolio Management; Financial Management;* and *The Financial Review;* among others.

Joseph D. Andrew, Jr. is a senior financial analyst with BIA Consulting, Inc., a financial consulting firm in Chantilly, Virginia, which specializes in fair-market valuations of broadcasting, cable, and telecommunications properties and the preparation of financing support packages for communications clients. As an analyst with BIA, Joe performs asset and stock valuations, business plan analyses, industry studies, and litigation support functions for companies in the radio, television, paging, cellular phone, and related industries. He also participates in specialized strategic research projects and impact studies covering various aspects of the communications industry. Joe has been with BIA since April 1997.

Prior to joining BIA, Joe served as the chief financial officer for X-Change Software, Inc., a startup software development firm in Oakton, Virginia. As CFO he was responsible for corporate investment analysis, cash flow planning, forecasting and analysis, receivables and payables management, and supervision of the bookkeeping and accounting functions.

In addition to managing corporate finances, Joe teaches graduate courses for Webster University of St. Louis, Missouri, dividing his time between the university's Washington DC campus and its international campus in Hamilton, Bermuda. His published works include this book and *Effective Writing: A Handbook for Finance People* with Dr. Claire May of the Art Institute of Atlanta and Dr. Gordon May of the University of Georgia.

Joe's past experience includes teaching assignments for National-Louis University in McLean, Virginia, the University of Southern Colorado in Colorado Springs, Chapman University and Webster University in Colorado Springs, and McMurry College in Abilene, Texas. He also served 23 years as a missile maintenance specialist in the U.S. Air Force, retiring at the rank of chief master sergeant in 1982.

Tim and Joe's partnership in this book's creation represents a unique opportunity for readers to experience the best of both worlds—Tim's development of the theory and logic of financial principles and Joe's real-world financial orientation. Their combined experience with students ensures that readers will learn theory and practice in an innovative, up-to-date, accurate manner.

PART 1:
THE WORLD OF FINANCE

CHAPTER 1

FINANCE AND THE FIRM

The race is not always to the swift, nor the battle to the strong, but that is the way to bet.
—Rudyard Kipling

FINANCE GRABS THE BUSINESS HEADLINES

Headlines from the front page of some recent *Wall Street Journal* editions:

- Daimler-Benz merger with Chrysler approved . . .
- Volvo confirms holding talks with VW . . .
- EU backs Euro peg to African currency . . .
- Pepsi posts strong earnings while Coke's fizzle . . .
- Lockheed terminates its $8.3 billion plan for Northrup merger . . .
- Cablevision Systems has $500 million junk deal . . .
- China's exporters press for Yuan's devaluation . . .

What do all these stories have in common? They deal with finance. Companies buying other companies, companies selling other companies, governments concerned about the value of their currencies, earnings reports—these are just a sampling of business stories involving finance that appear every day in the press.

The amount of money involved in business financial issues is staggering. In our one-day sample, nearly $20 billion changed hands. Imagine the amount of money that flows through businesses and governments around the world in a year!

CHAPTER OBJECTIVES

After reading this chapter, you should be able to:

- Describe the field of finance.
- Discuss the duties of financial managers.
- Identify the basic goal of a business firm.
- List factors that affect the value of a firm.
- Discuss the legal and ethical challenges financial managers face.
- Identify the different forms of business organization.

Finance is at the heart of business management. No business firm—or government, for that matter—can exist for long without following at least the basic principles of financial management.

This book is designed to introduce you to basic financial management principles and skills. Some of these concepts and skills are surprisingly straightforward; others are quite challenging. All, however, will help you in the business world, no matter what career you choose.

Chapter Overview

In this chapter we will introduce financial management basics that provide a foundation for the rest of the course. First, we will describe the field of finance and examine the role of financial management within a business organization. Then we will investigate the financial goal of a business firm, and the legal and ethical challenges financial managers face. We will end with a description of the three forms of business in the U.S. economy: sole proprietorship, partnership, and corporation.

The Field of Finance

In business, financial guidelines determine how money is raised and spent. Although raising and spending money may sound simple, financial decisions affect every aspect of a business—from how many people a manager can hire, to what products a company can produce, to what investments a company can make.

For example, when IBM needed software development expertise, it decided to acquire Lotus, a company well known for its spreadsheet software. IBM executives knew that Lotus was valuable because of its talented software engineers, who created and developed products such as Lotus Notes, a successful groupware product. To recruit these skilled employees, IBM worked out complex financial packages that included retirement and benefits options. IBM succeeded—at a cost of $3.5 billion—in purchasing Lotus and retaining the top Lotus software engineers.[1]

Money continually flows through businesses. It may flow in from banks, from the government, from the sale of stocks, and so on, and it may flow out for a variety of reasons—to invest in bonds, to buy new equipment, or to hire top-notch employees. Businesses must pay constant attention to ensure that the right amount of money is available at the right time for the right use.

In large firms it may take a whole team of financial experts to track the firm's cash flows and to develop financial strategies. For instance, when the Walt Disney Company acquired Capital Cities/ABC in 1996, teams of financial analysts worked on every detail of the $19 billion decision to determine how Disney could raise the money to buy the other company.[2]

---

[1]Judith H. Dobrzynski, "The Art of the Hostile Deal," *The New York Times* (June 22, 1995): D1, D2.
[2]Steve McClellan, "Shareholders Approve Disney/CapCities Merger," *Broadcasting & Cable*, 126, iss. 2 (January 8, 1996): 17.

TABLE 1-1

Careers in the Field of Finance

Financial management	Manage the finances of a business firm. Analyze, forecast, and plan a firm's finances; assess risk; evaluate and select investments; decide where and when to find money sources, and how much money to raise; and determine how much money to return to investors in the business.
Financial markets and institutions	Handle the flow of money in financial markets and institutions, and focus on the impact of interest rates on the flow of that money.
Investments	Locate, select, and manage money-producing assets for individuals and groups.

Finance Career Paths

Finance has three main career paths: financial management, financial markets and institutions, and investments. Financial management, the focus of this text, involves managing the finances of a business. Financial managers—people who manage a business firm's finances—perform a number of tasks. They analyze and forecast a firm's finances, assess risk, evaluate investment opportunities, decide when and where to find money sources and how much money to raise, and decide how much money to return to the firm's investors.

Bankers, stockbrokers, and others who work in financial markets and institutions focus on the flow of money through financial institutions and the markets in which financial assets are exchanged. They track the impact of interest rates on the flow of that money. People who work in the field of investments locate, select, and manage income-producing assets. For instance, security analysts and mutual fund managers both operate in the investment field.

Table 1-1 summarizes the three main finance career paths.

FINANCIAL MANAGEMENT

Financial management is essentially a combination of accounting and economics. First, financial managers use accounting information—balance sheets, income statements, and so on—to analyze, plan, and allocate financial resources for business firms. Second, financial managers use economics principles to guide them in making financial decisions that are in the best interest of the firm. In other words, finance is an applied area of economics that relies on accounting for input.

Because finance looks closely at the question of what adds value to a business, financial managers are central to most businesses. Let's take a look at what financial managers do.

The Role of the Financial Manager

Financial managers measure the firm's performance, determine what the financial consequences will be if the firm maintains its present course or changes it, and recommend how the firm should use its assets. Financial managers also locate external financing sources and recommend the most beneficial mix of financing sources, and they determine the financial expectations of the firm's owners.

All financial managers must be able to communicate, analyze, and make decisions based on information from many sources. To do this, they need to be able to analyze financial statements, forecast and plan, and determine the effect of size, risk, and timing of cash flows. We'll cover all of these skills in this text.

Finance in the Organization of the Firm   Financial managers work closely with other types of managers. For instance, they rely on accountants for raw financial data and on marketing managers for information about products and sales. Financial managers coordinate with technology experts to determine how to communicate financial information to others in the firm. Financial managers also provide advice and recommendations to top management.

Figure 1-1 shows how finance fits into a typical business firm's organization.

The Organization of the Finance Team   In most medium-to-large businesses, a chief financial officer (CFO) supervises a team of employees who manage the financial activities of the firm. One common way to organize a finance team in a medium-to-large business is shown in Figure 1-2.

In Figure 1-2 we see that the **chief financial officer** (CFO) directs and coordinates the financial activities of the firm. The CFO supervises a treasurer and a controller. The **treasurer** generally is responsible for cash management, credit management, and financial planning activities, whereas the **controller** is responsible for cost accounting, financial accounting, and information system activities. The treasurer and the controller of a large corporation are both likely to have a group of junior financial managers reporting to them.

At a small firm, one or two people may perform all the duties of the treasurer and of the controller. In very small firms, one person may perform all functions, including finance.

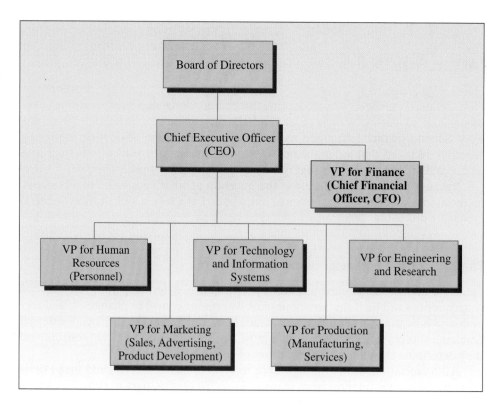

Figure 1-1   The Organization of a Typical Corporation

Figure 1-1 shows how finance fits into a typical business organization. The vice president for finance, or chief financial officer, operates with the vice presidents of the other business teams.

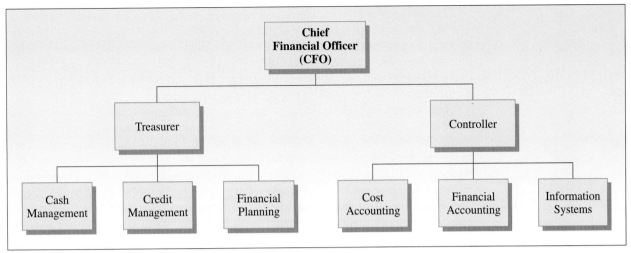

Figure 1-2 **An Example of How to Organize a Finance Team**
This chart shows how to organize a finance team in a medium-to-large business. Most teams include both a finance function (on the left) and an accounting function (on the right). The chief financial officer usually reports to the CEO as shown in Figure 1-1.

The Basic Financial Goal of the Firm

The financial manager's basic job is to make decisions that add value to the firm. When asked what the basic goal of a firm is, many people will answer, "to make a lot of money" or "to maximize profits." Although no one would argue that profits aren't important, the single-minded pursuit of profits is not necessarily good for the firm and its owners. We will explain why this is so in the sections that follow. For now, let's say that a better way to express the primary financial goal of a business firm is to "maximize the wealth of the firm's owners." This is an extremely important, even crucial, point, so we will say it again: *The primary financial goal of the business firm is to maximize the wealth of the firm's owners.*

Everything the financial manager does—indeed, all the actions of everyone in the firm—should be directed toward this goal, subject to legal and ethical considerations that we will discuss in this chapter and throughout the book.

Now, what do we mean by wealth? **Wealth** refers to value. If a group of people owns a business firm, the contribution that firm makes to that group's wealth is determined by the market value of that firm.

This is a very important point: We have defined wealth in terms of *value*. The concept of value, then, is of fundamental importance in finance. Financial managers and researchers spend a lot of time measuring value and figuring out what causes it to increase or decrease.

In Search of Value

We have said that the basic goal of the business firm is to maximize the wealth of the firm's owners—that is, to maximize the value of the firm. The next question, then, is how to measure the value of the firm.

The value of a firm is determined by whatever people are willing to pay for it. The more valuable people think a firm is, the more they will pay to own it. Then the existing owners can sell it to investors for more than their original purchase price, thereby increasing current stockholder wealth. The financial manager's job is to make decisions that will cause people to think more favorably about the firm and, in turn, to be willing to pay more to purchase the business.

For companies that sell stock to the general public, stock price can indicate the value of a business because *stockholders*—people who purchase corporate shares of stock—become part owners of the corporation. (We will discuss stock in greater detail in chapter 2.) People will pay a higher price for stock—that is, part ownership of a business—if they believe the company will perform well in the future. For instance, Yahoo, a developer of Internet access software, originally sold its stock for $3.25 (adjusted for two 2 for 1 stock splits) per share. (A share is one unit of ownership.) Because of the potential of its Internet services it was worth over $152 per share in February 1999. This isn't too bad for a company that had losses for most of its history until it turned a slight quarterly profit in 1998. Investors must be betting on a bright future for Yahoo.

For businesses that sell stock publicly, then, the financial manager's basic role is to help make the firm's stock more valuable. Although some businesses do not sell stock to the general public, we will focus on stock price as a measure of the value of the firm. Keep in mind, however, that investing in one share (one unit) of stock means the investor only owns one small piece of a firm. Many firms sell hundreds of thousands or millions of shares of stock, so the total value of the firm is the equivalent of the sum of all the stock shares.

Next, let's look closely at three factors that affect the value of a firm's stock price: cash flow, timing, and risk.

The Importance of Cash Flow In business, cash is what pays the bills. It is also what the firm receives in exchange for its products and services. Cash is, therefore, of ultimate importance, and the expectation that the firm will generate cash in the future is one of the factors that gives the firm its value.

We use the term *cash flow* to describe cash moving through a business. Financial managers concentrate on increasing cash *in*flows—cash that flows into a business—and decreasing cash *out*flows—cash that flows away from a business. Cash outflows will be approved if they result in cash inflows of sufficient magnitude and if those inflows have acceptable timing and risk.

It is important to realize that sales are not the same as cash inflows. Businesses often sell goods and services on credit, so no cash changes hands at the time of the sale. If the cash from the sale is never collected, the sale cannot add any value to the firm. Owners care about actual cash collections from sales—that is, cash inflows.

Likewise, businesses may buy goods and services to keep the firm running, but may make the purchase on credit—so no cash changes hands at that time. However, bills always come due sooner or later, so owners care about cash expenditures for purchases—cash outflows. For any business firm (assuming other factors remain constant), the higher the expected cash inflows and the lower the expected cash outflows, the higher the firm's stock price will be.

The Effect of Timing on Cash Flows The timing of cash flows also affects a firm's value. To illustrate, consider this: Would you rather receive $100 cash today and $0 one year from now, or would you rather receive $0 cash today and $100 one year from now? The two alternatives follow:

	TODAY	ONE YEAR FROM NOW
Alternative A	+ $100	$0
Alternative B	$0	+ $100

Both alternatives promise the same total amount of cash, but most people would choose Alternative A because they realize they could invest the $100 received today and earn interest on it during the year. By doing so they would end up with more money than $100 at the end of the year. For this reason we say that—all other factors being equal—cash received sooner is better than cash received later.

Owners and potential investors look at when firms can expect to receive cash and when they can expect to pay out cash. All other factors being equal, the sooner companies expect to receive cash and the later they expect to pay out cash, the more valuable the firm and the higher its stock price will be.

The Influence of Risk We have seen that the size of a firm's expected cash inflows and outflows and the timing of those cash flows influence the value of the firm and its stock price. Now let us consider how risk affects the firm's value and its stock price.

Risk affects value because the less certain owners and investors are about a firm's expected future cash flows, the lower they will value the company. The more certain owners and investors are about a firm's expected future cash flows, the higher they will value the company. In short, companies whose expected future cash flows are doubtful will have lower values than companies whose expected future cash flows are virtually certain.

What isn't nearly as clear as the *way* risk affects value is *how much* it affects it. For example, if one company's cash flows are twice as risky as another company's cash flows, is its stock worth half as much? We can't say. In fact, we have a tough time quantifying just how risky the companies are in the first place.

We will examine the issue of risk in some detail in chapter 7. For now, it is sufficient to remember that risk affects stock price—as risk increases, stock price goes down; and conversely, as risk decreases, stock price goes up.

Table 1-2 summarizes the influences of cash flow size, timing, and risk on stock prices.

Profits versus Cash Flow Earlier in the chapter we said that the single-minded pursuit of profits is not necessarily good for the firm's owners. Indeed, the firm's owners view stock value, not profit, as the appropriate measure of wealth. Stock value depends on future cash flows, their timing, and their riskiness. Profit calculations do not consider these three factors. Profit, as defined in accounting, is simply the difference between sales revenue and expenses. If all we were interested in were profits, we could simply start using high-pressure sales techniques, cut all expenses to the bone, and then point proudly to the resulting increase in profits. For the moment, anyway. In all probability, managers practicing such techniques would find

Take Note

The point about cash received sooner being better than cash received later works in reverse, too: It is better to pay out cash later rather than sooner (all other factors being equal, of course).

TABLE 1-2

Accomplishing the Primary Financial Goal of the Firm

The Goal:	Maximize the wealth of the firm's owners
Measure of the Goal:	Value of the firm (measured by the price of the stock on the open market for corporations)

FACTOR	EFFECT ON STOCK PRICE
Size of expected future cash flows	Larger future cash inflows raise the stock price. Larger cash outflows lower the stock price. Smaller future cash inflows lower the stock price. Smaller future cash outflows raise the stock price.
Timing of future cash flows	Cash inflows expected sooner result in a higher stock price. Cash inflows expected later result in a lower stock price. (The opposite effect occurs for future cash outflows.)
Riskiness of future cash flows	When the degree of risk associated with future cash flows goes down, stock price goes up. When the degree of risk associated with future cash flows goes up, stock price goes down.

their firm out of business later, when the quality of the firm's products, services, and work force dropped, eventually leading to declining sales and market share.

It is true that more profits are generally better than less profits. But when the pursuit of short-term profits adversely affects the size of future cash flows, their timing, or their riskiness, then these profit maximization efforts are detrimental to the firm. Concentrating on stock value, not profits, is a better measure of financial success.

Legal and Ethical Challenges in Financial Management

Several legal and ethical challenges influence financial managers as they pursue the goal of wealth maximization for the firm's owners. Examples of legal considerations include environmental statutes mandating pollution control equipment, workplace safety standards that must be met, civil rights laws that must be obeyed, and intellectual property laws that regulate the use of others' ideas.

Ethical concerns include fair treatment of employees, customers, the community, and society as a whole. Indeed, many businesses have written ethics codes that articulate the ethical values of the business organization.

Three legal and ethical influences of special note include the agency problem, the interests of nonowner stakeholders, and the interests of society as a whole. We will turn to these issues next.

Agency Issues

The financial manager, and the other managers of a business firm, are agents for the owners of the firm. An **agent** is a person who has the implied or actual authority to act on behalf of another. The owners whom the agents represent are the **principals**. For example, the board of directors and senior management of IBM are agents for the IBM stockholders, the principals. Agents have a legal and ethical responsibility to make decisions that further the interests of the principals.

The interests of the principals are supposed to be paramount when agents make decisions, but this is often easier said than done. For example, the managing director of a corporation might like the convenience of a private jet on call 24 hours a day,

FINANCE AT WORK

SALES·RETAIL·SPORTS·MEDIA TECHNOLOGY·PUBLIC RELATIONS·PRODUCTION·EXPORTS

Melanie Rosen, Electronic Media, *The New York Times*

Melanie Rosen

Melanie Rosen is the Director of Electronic Media for The New York Times Informational Services Group. She runs a new company within *The New York Times* that will produce educational products for schools. One product is called *Live from the Past*. *Live from the Past* takes historical articles from *The New York Times* and converts them to digital format so that the articles and videotapes from actual events can be used in the classroom, on-line, from a CD, or in any other electronic format. Melanie is a natural for this type of endeavor because she was a communications/graphics major in college, and then earned an M.B.A.

Q. What are some of your daily responsibilities?

Two main aspects of my business are production and finance. Production involves figuring out how to combine many types of media to create the best product possible. To create such a product, though, I have to know the financial impact of what I invest in—will the company benefit from my production decisions?

Q. *In effect, you are starting a new company within* **The New York Times.** *How would you describe what you do?*

I function as a general manager, financial manager, product manager, and sales manager, all under one hat. About 50 percent to 60 percent of my time is devoted to financial management. That is, I constantly evaluate the finances of the product line. I try to assess how much money it will cost to design and produce a product, how much it will cost to market it, how much of the product we will produce, what our risks are, and what our revenues will be.

I budget and monitor my group's performance against its budgeted, or forecasted, performance. I'm also responsible for keeping management informed about how much money our group has invested so far, how much more we expect to invest, how much money we expect to generate, and when.

Q. How supportive is *The New York Times* **of your "new company"?**

When a product like *Live from the Past* is in start-up mode, the company is always supportive. Management tries to provide the opportunity to make as much money as possible. But there are constraints. Most corporations like to see new businesses become profitable in the third year. In fact, many corporations try to push for a two-year time frame. That's because companies are concerned about cash flowing into investments without seeing an accompanying return from the investment.

Q. What basic financial "tools" would you suggest future business people should master?

I'd recommend students should understand some basic financial concepts when they leave college. Business people need to be able to read and understand financial statements, including the statement of cash flows. Only then will they understand how much money the business has before and after taxes, where the money is going, and what the timing is of those cash flows. Too many businesses fail because of poor cash flow.

Source: Interview with Melanie Rosen.

---

but do the common stockholder owners of the corporation receive enough value to justify the cost of a jet? It looks like the interests of the managing director (the agent) and the interests of the common stockholder owners (the principals) of the corporation are in conflict in this case.

The Agency Problem When the interests of the agents and principals conflict, an **agency problem** results. In our jet example, an agency problem occurs if the managing director buys the jet, even though he knows the benefits to the stockholders do not justify the cost.

Another example of an agency problem occurs when managers must decide whether to undertake a project with a high potential payoff but high risk. Even if the project is more likely to be successful than not, managers may not want to take a risk that owners would be willing to take. Why? An unsuccessful project may result in such significant financial loss that the managers who approved the project lose their jobs—and all the income from their paycheck. The stockholder owners, however, may have a much smaller risk because their investment in company stock represents only a small fraction of their financial investment package. Because the risk is so much larger to the manager as compared to the stockholder, a promising but somewhat risky project may be rejected even though it was likely to benefit the firm's owners.

The agency problem can be lessened by tying the managers' compensation to the performance of the company and its stock price. This tie brings the interests of the managers and those of the firm's owners closer together. That is why companies often make shares of stock a part of the compensation package offered to managers, especially top executives. If managers are also stockholders, then the agency problem should be reduced.

Agency Costs Sometimes firms spend time and money to monitor and reduce agency problems. These outlays of time and money are **agency costs.** One common example of an agency cost is an accounting audit of a corporation's financial statements. If a business is owned and operated by the same person, the owner would not need an audit—she could trust herself to report her finances accurately. Most companies of any size, however, have agency costs because managers, not owners, report the finances. Owners audit the company financial statements to see whether the agents have acted in the owners' interests by reporting finances accurately.

The Interests of Other Groups

Stockholders and managers are not the only groups that have a stake in a business firm. There are also nonmanager workers, creditors, suppliers, customers, and members of the community where the business is located. These groups are also **stakeholders**—people who have a "stake" in the business. Although the primary financial goal of the firm is to maximize the wealth of the owners, the interests of these other stakeholders can influence business decisions.

One example of outside stakeholder influence is pressure from political lobbyists or consumer groups. For instance, Magic Johnson, basketball star of the Los Angeles Lakers, partnered with Sony to assist a neighborhood in Los Angeles that had been severely damaged in the events following the original Rodney King beating trial. No company was willing to run the risk of operating a movie theater in the area because the risk of property damage was so high.[3] So—despite the risk of monetary loss—Magic Johnson and Sony built a clean, safe movie theater in the neighborhood, offering job opportunities as well as first-run films.

---

[3]Interview with Taylor Braugh, public relations director of Magic Johnson Enterprises, Inc., January 10, 1996; Kenneth Noble, "Magic Johnson Finding Success in a New Forum," *The New York Times* (January 8, 1996): 8.

It was such a successful venture, both for the L.A. neighborhood and its founders, that Magic and Sony plan to build similar theaters in the inner cities of Atlanta, New York, Chicago, Washington, and Houston.[4] In March 1998 Magic, singer Janet Jackson, and music executive Jheryl Busby bought a majority interest in Founders National Bank of Los Angeles. They paid $2.5 million. This black-owned bank is seeking deposits from African-American leaders and from residents of minority communities.[5]

The Interests of Society as a Whole

Sometimes the interests of a business firm's owners are not the same as the interests of society. For instance, the cost of properly disposing of toxic waste can be so high that companies may be tempted to simply dump their waste in nearby rivers. In so doing, the companies can keep costs low and profits high, and drive their stock prices higher (if they are not caught). However, many people suffer from the polluted environment. This is why we have environmental and other similar laws: so that society's best interests take precedence over the interests of individual company owners.

When businesses take a long-term view, the interests of the owners and society often (but not always) coincide. When companies encourage recycling, sponsor programs for disadvantaged young people, run media campaigns promoting the responsible use of alcohol, and contribute money to worthwhile civic causes, the goodwill generated as a result of these activities causes long-term increases in the firm's sales and cash flows, which translate into additional wealth for the firm's owners.

Although the traditional primary economic goal of the firm is to maximize shareholder wealth, the unbridled pursuit of value is too simplistic a view of this goal. Firms often take into account ethical factors, the interests of other stakeholders, and the long-term interests of society.[6]

Figure 1-3 summarizes the various influences that financial managers may consider in their pursuit of value.

FORMS OF BUSINESS ORGANIZATION

Businesses can be organized in a variety of ways. The three most common types of organization are proprietorships, partnerships, and corporations. The distinguishing characteristics give each form its own advantages and disadvantages.

The Proprietorship

The simplest way to organize a business is to form a **proprietorship,** a business owned by one person. An individual raises some money, finds a location from which

---

[4]Noble, "Magic Johnson": 8.
[5]Ann Brown, "Creating a Little "Magic" at Founders," *Black Enterprise* (August 1998): 17.
[6]Not everyone agrees with this approach. Milton Friedman, Nobel laureate in economics, claims that any action taken by a manager that is not legally mandated, and that reduces the value available to the owners, is theft.

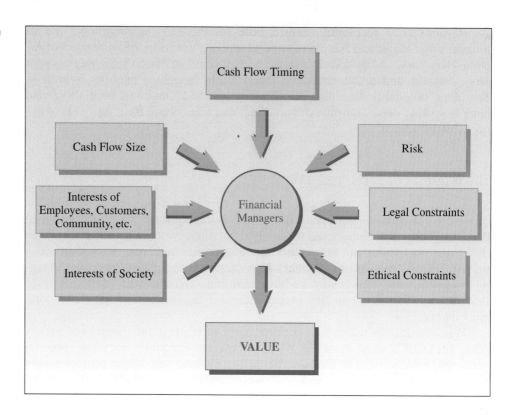

Figure 1-3 Influences on Financial Managers

to operate, and starts selling a product or service. The profits or losses generated are reported on a form called Schedule C of the individual's form 1040 income tax return. The sole proprietor is responsible for any tax liability generated by the business, and the tax rates are those that apply to an individual.

The sole proprietor has *unlimited liability* for matters relating to the business. This means that the sole proprietor is responsible for all the obligations of the business, even if those obligations exceed the amount the proprietor has invested in the business. If a customer is injured on the company premises and sues for $1 million, the sole proprietor must pay that amount if the court awards that amount to the plaintiff customer. This is true even if the total amount invested by the sole proprietor in the business is only $10,000.

Although unlimited liability is a major disadvantage of a proprietorship, liability insurance is often available to reduce the risk of losing business and non-business assets. However, the risk always remains that the business will be unsuccessful and that the losses incurred will exceed the amount of the proprietor's money invested. The other assets owned by the proprietor will then be at risk.

The Partnership

Two or more people may join together to form a business as a **partnership.** This can be done on an informal basis without a written partnership agreement, or a contract can spell out the rights and responsibilities of each partner. This written contract is called the *articles of partnership* and is strongly recommended so as to lessen the likelihood of disputes between partners.

The articles of partnership contract generally spells out how much money each partner will contribute, what the ownership share of each partner will be, how profits and losses will be allocated among partners, who will perform what work for the business, and other matters of concern to the partners. The percent of ownership for each partner does not have to be the same as the percent each partner invests in the partnership.

Each partner in a partnership is usually liable for the activities of the partnership as a whole.[7] This is an important point. Even if there are 100 partners, each one is technically responsible for all the debts of the partnership. If 99 partners declare personal bankruptcy, the hundredth partner still is responsible for all the partnership's debts.

Some partnerships contain two different classes of partners, *general partners* and *limited partners*. General partners are almost always active participants in the management of the business, whereas limited partners are not. As a result, general partners usually contract for a more favorable allocation of ownership, profits, and losses compared to limited partners. General partners have unlimited liability for the partnership's activities. Limited partners are only liable for the amount they invest in the partnership. If you are a limited partner who invests $5,000 in the business, then $5,000 is the most you could lose. For this reason, every partnership must have at least one general partner (a partnership could have all general partners, but it could not have all limited partners).

A partner's profits and losses are reported on Schedule K-1, which is attached to each partner's individual form 1040 income tax return. The partners pay any taxes owed. The partnership itself is not taxed because the income merely passes through the partnership to the partners, where it is taxed.

The Corporation

The third major form of business organization is the **corporation.** Unlike proprietorships and partnerships, corporations are legal entities separate from their owners. To form a corporation, the owners specify the governing rules for the running of the business in a contract known as the *articles of incorporation*. They submit the articles to the government of the state in which the corporation is formed, and the state issues a charter that creates the separate legal entity.

Corporations are taxed as separate legal entities. That is, corporations must pay their own income tax just as if they were individuals.[8] This is where the often-discussed "double taxation" of corporate profits comes into play. First, a corporation pays income tax on the revenue it receives. Then the corporation may distribute to the owners the profits that are left after paying taxes. These distributions, called dividends, count as ordinary income for the owners and are taxed on the individual owners' income tax returns. Thus, the IRS collects taxes twice on the same income.

Double taxation of dividends is bad news for the owners of corporations, but there is good news as well. Stockholders, the corporation's owners, have limited liability for the corporation's activities. They cannot lose more than the amount they paid to buy the stock. This makes the corporate form of organization very

---

[7] But see the discussion that follows on general and limited partners.
[8] Corporations file their income tax returns using form 1041.

TABLE 1-3

Characteristics of Business Ownership Forms

	PROPRIETORSHIP	**PARTNERSHIP**	**CORPORATION**	**LLC**
Ease of formation	Very easy	Relatively easy	More difficult	Relatively easy
Owners' liability	Unlimited	Unlimited for general partners	Limited	Limited
Life of firm	Dies with owner, unless heirs continue operating or sell the business	Surviving partners must deal with the deceased partner's heirs	Can live beyond owners' lifetimes	Dies with owner, heirs may continue operating the business
Separate legal entity?	No	No	Yes	Yes
Degree of control by owners	Complete	May be limited for individual partner	May be very limited for individual stockholder	Depends on the number of owners

attractive for owners who desire to shelter their personal assets from creditors of the business.

Corporations have other benefits too. For example, because they exist separately from their owners, they can "live" beyond the death of their original owners. Another benefit of the corporate form of business is that corporations generally have a professional management team and board of directors, elected by the owners. It is the board's job to look out for the interests of the owners (the stockholders). Stockholders, especially in the case of large corporations, usually do not take an active role in the management of the business, so it is the board of directors' job to represent them.

Subchapter S Corporations and LLCs In a special type of corporation called a **Subchapter S corporation,** the owners are taxed as if they were members of a partnership—thereby avoiding the double-taxation problem faced by the owners of other corporations. Subchapter S corporations are generally very small; in fact, this form of business ownership was created to relieve small businesses, as specified in the federal tax code, from some of the rules that large Subchapter C corporations have to follow. S corporations can have no more than 75 shareholders. The stockholders of Subchapter S corporations also have limited liability.

Another option is to organize as a limited liability company (LLC). LLCs avoid the double taxation C corporations face while shielding the owners from personal liability. LLCs are similar to S corporations but have fewer restrictions. For example, LLCs can have more than 75 owners.

Table 1-3 summarizes the advantages and disadvantages of the various forms of business ownership.

WHAT'S NEXT

In this book we will look at how firms raise and allocate funds, and how firms invest in assets that generate returns at a reasonable risk. In part 1 of the text, we discuss the environment in which financial managers make decisions. In chapter 2 we will examine how funds are raised in the financial marketplace. In chapter 3 we'll explore how financial institutions and interest rates affect financial decisions.

Summary

1. Describe the field of finance.

Finance is important to business people. Financial decisions about how to raise, spend, and allocate money can affect every aspect of a business—from personnel to products. Finance also offers career opportunities in three main areas: financial management, financial markets and institutions, and investments. Financial management focuses on managing the finances of a business.

2. Discuss the duties of financial managers.

Financial managers use accounting information and economics principles to guide their financial decisions. They measure the firm's financial condition, forecast, budget, raise funds, and determine the financial goals of the firm's owners. They also work closely with other managers to further the firm's goals.

At medium and large firms, more than one person usually handles the financial management duties. In some firms a chief financial officer (CFO) supervises the financial activities, including cash and credit management, financial planning, and accounting.

3. Identify the basic goal of a business firm.

The basic goal of the business firm is to maximize the wealth of the firm's owners by adding value; it is not to maximize profits. The value of a firm is measured by the price investors are willing to pay to own the firm. For businesses that sell stock to the general public, stock price indicates the firm's value because shares of stock are units of ownership. So the basic financial goal of such firms is to maximize the price of the firm's stock.

4. List factors that affect the value of a firm.

The value of a firm is affected by the size of future cash flows, their timing, and their riskiness.

- Cash inflows increase a firm's value whereas cash outflows decrease it.
- The sooner cash flows are expected to be received, the greater the value. The later those cash flows are expected, the less the value.
- The less risk associated with future cash flows, the higher the value. The more risk, the lower the value.

5. Discuss the legal and ethical challenges financial managers face.

Legal and ethical considerations include the agency problem, the interests of other stakeholders, and the interests of society as a whole.

- The agency problem exists when the interests of a firm's managers (the agents) are in conflict with those of the firm's owners (the principals).
- Other stakeholders whose interests are often considered in financial decisions include employees, customers, and members of the communities in which the firm's plants are located.
- Concerns of society as a whole—such as environmental or health problems—often influence business financial decisions.

6. Identify the three different forms of business organization.

The three most common forms of business organization are the proprietorship, the partnership, and the corporation.

- Proprietorships are businesses owned by one person. The owner is exposed to unlimited liability for the firm's debts.
- Partnerships are businesses owned by two or more people, each of whom is responsible for the firm's debts. The exception is a limited partner, a partner who contracts for limited liability.
- Corporations are separate legal entities. They are owned by stockholders who are responsible for the firm's debts only to the extent of their investment.

You can access your FREE PHLIP/CW On–Line Study Guide, Current Events, and Spreadsheet Problem templates, which may correspond to this chapter either through the Prentice Hall Finance Center CD–ROM or by directly going to: **http://www.prenhall.com/gallagher.**

PHLIP/CW — *The Prentice Hall Learning on the Internet Partnership–Companion Web Site*

Self-Test

ST-1. What are the three main areas of career opportunities in finance?

ST-2. What are the primary responsibilities of a person holding the title of treasurer at a large corporation?

ST-3. Who is a "principal" in an agent–principal relationship?

ST-4. What legal and ethical factors may influence a firm's financial decisions?

ST-5. What is a Subchapter S corporation?

Review Questions

1. How is finance related to the disciplines of accounting and economics?
2. List and describe the three career opportunities in the field of finance.
3. Describe the duties of the financial manager in a business firm.
4. What is the basic goal of a business?
5. List and explain the three financial factors that influence the value of a business.
6. Explain why accounting profits and cash flows are not the same thing.
7. What is an agent? What are the responsibilities of an agent?
8. Describe how society's interests can influence financial managers.
9. Briefly define the terms *proprietorship, partnership, LLC,* and *corporation.*
10. Compare and contrast the potential liability of owners of proprietorships, partnerships (general partners), and corporations.

Build Your Communication Skills

CS-1. Divide into small groups. Each small group should then divide in half. The first group should defend the idea that managers of a firm should consider only the interests of stockholders, subject to legal constraints. The other group should argue that businesses should consider the interests of other stakeholders of the firm and society at large.

CS-2. Assume you work for WealthMax Corporation in New York City. You've noticed that managers who work late charge the corporation for their dinners and transportation home. You've also noticed that almost all employees from these managers' departments take office supplies, ranging from pens to computer software, for personal use at home. You estimate the costs of this pilfering at a shocking $150,000 a year. Your boss, the chief financial officer for WealthMax, asks you to write a memo to the offending managers describing why their actions and those of their employees violate their duties and conflict with the goal of the firm. Write this memo.

Answers to Self-Test

ST-1. Financial management, financial markets and institutions, and investments.

ST-2. The treasurer of a large corporation is responsible for cash management, credit management, and financial planning.

ST-3. A principal in an agent–principal relationship is the person who hires the agent to act on the principal's behalf. The principal is the person to whom the agent owes a duty.

ST-4. Legal and ethical factors influence businesses. Examples of legal constraints include environmental, safety, and civil rights laws. Examples of ethical considerations include fair treatment of workers, environmental sensitivity, and support for the community.

ST-5. A Subchapter S corporation is a small corporation that is taxed as if it were a partnership. As a result, the owners of a Subchapter S corporation avoid double taxation of corporate income paid to stockholders.

CHAPTER 2

FINANCIAL MARKETS AND INTEREST RATES

It is better to give than to lend, and it costs about the same.
—Philip Gibbs

TRACKING LUCENT'S BOLD ASCENT: TELECOMMUNICATIONS FIRM SOARS BUT ANALYSTS EXPECT PACE TO SLOW

From the day it was spun off by AT&T Corp. in 1996, Lucent Technologies Inc.'s potential as a winning investment was clear.

The New Jersey–based company, the largest manufacturer of telecommunications equipment in North America, was poised to profit from the surging demand for telephone lines, switching devices, and other communications services. And Wall Street was so excited about its long-term prospects that an initial public offering yielded $3 billion, the most ever to that point in U.S. history.

But not even the most enthusiastic analysts expected Lucent's stock to streak as high as it did in early 1998. Its shares skyrocketed more than 81 percent, from a split-adjusted $38.66 at the end of 1997 to $70 in April 1998.

"I set my price target based on rational valuation reasons. Obviously, the market ignored my recommendation," said Salomon Smith Barney analyst, Steven Levy.

"We're in the right industry at the right time," said Bill Price, a spokesman who likened Lucent to Intel Corp. and Microsoft Corp. in its desire to dominate its market. "We intend to be the leader in a new category: communications networking."

CHAPTER OBJECTIVES

After reading this chapter, you should be able to:

- Describe how the U.S. financial system works.

- Define financial securities.

- Explain the function of financial intermediaries.

- Identify the different financial marketplaces.

- Describe the securities traded in the money and capital markets.

- Identify the determinants of the nominal interest rate.

- Construct and analyze a yield curve.

Richard McGinn (left), CEO, and Henry Schacht, chairman (right), both of Lucent Technologies.

Analysts said Lucent's ascent has come about largely because the company has made good on its promises.

Lucent chairman and chief executive Richard A. McGinn noted that Lucent has turned in eight straight quarters of record-setting profits. That's due in part to sales of wireless equipment and software to telephone companies, and despite a dampening of sales overseas caused by the economic troubles in Asia.

Lucent has been aggressive in acquiring companies that can help round out its services. In 1997, for instance, it paid $1.8 billion for Octel Communications Corp., a voice-messaging specialist. It also spent hundreds of millions to buy Prominet Corp., which manufactures network switching gear, and other companies that produce high-speed data networking equipment.

But for all the company's pluses, analysts don't believe Lucent's stock can continue to rise at its first-quarter pace. The company has to improve in several areas, including the development of its overseas business, where revenue has not kept pace with its domestic growth, according to Nikos Theodosopoulos, managing director of UBS Securities Inc. "They have pleasantly surprised us," he said, but added: "I don't think you're still going to see the slope that you've seen these last few months."

Source: Robert O'Harrow, Jr., "Tracking Lucent's Bold Ascent." The Washington Post Company Web page, April 19, 1998 ⟨http://www.washingtonpost.com/wp-s...te⟩. Copyright 1998, *The Washington Post.* Reprinted with permission.

Chapter Overview

The preceding story about Lucent Technologies illustrates the importance of financial markets to business firms. Most businesses do not obtain millions each time they go to the financial markets, but they certainly would face difficulties if markets were not available for raising funds and investing.

One of the central duties of a financial manager is to acquire capital—that is, to raise funds. Few companies are able to fund all their activities solely with funds from internal sources. Most find it necessary at times to seek funding from outside sources. For this reason, all business people need to know about financial markets.

As we see in this chapter, there are a number of financial markets, and each offers a different kind of financial product. In this chapter we discuss the relationship between firms and the financial markets, and briefly explain how the financial system works, including the role of *financial intermediaries*—investment bankers, brokers, and dealers. Next, we explore the markets themselves and describe financial products, ranging from government bonds to corporate stocks. Finally, we examine interest rates.

The Financial System

In the U.S. economy, several types of individuals or entities generate and spend money. We call these *economic units*. The main types of economic units include governments, businesses, and households (households may be one person or more than one person). Some economic units generate more income than they spend and have funds left over. These are called **surplus economic units.** Other economic units generate less income than they spend and need to acquire additional funds in order to sustain their operations. These are called **deficit economic units.**

The purpose of the financial system is to bring the two groups—surplus economic units and deficit economic units—together for their mutual benefit.

The financial system also makes it possible for participants to adjust their holdings of financial assets as their needs change. This is the *liquidity function* of the financial system—that is, the system allows funds to flow with ease.

To enable funds to move through the financial system, funds are exchanged for financial products called *securities*. A clear understanding of securities is essential to understanding the financial system. So before we go further, let's examine what securities are and how they are used.

Take Note

The words surplus *and* deficit *do not imply something good or bad; they simply mean that some economic units need funds, and others have funds available. If Disney needs $2 billion to build a theme park in a year when its income is "only" $1.5 billion, Disney is a deficit economic unit that year. Likewise, if a family earns $40,000 in a year but spends only $36,000, it is a surplus economic unit for that year.*

Securities

Securities are documents that represent the right to receive funds in the future. The person or organization that holds a security is called a **bearer.** A security certifies that the bearer has a *claim* to future funds. For example, if you lend $100 to someone and the person gives you an IOU, you have a security. The IOU is your "claim check" for the $100 you are owed. The IOU may also state *when* you are to be paid, which is referred to as the **maturity date** of the security. When the date of payment occurs, we say the security matures.

Securities have value because the bearer has the right to be paid the amount specified. So a bearer who wanted some money right away could sell the security to someone else for cash. Of course, the new bearer could sell the security to someone else too, and so on down the line. When a security is sold to someone else, the security is being *traded*.

Business firms, as well as local, state, and national governments, sell securities to the public to raise money. After the initial sale, investors may sell the securities to other investors. As you might suspect, this can get to be a complicated business. **Financial intermediaries** facilitate this process. Markets are available for the subsequent traders to execute their transactions.

Financial Intermediaries

Financial intermediaries act as the grease that enables the machinery of the financial system to work smoothly. They specialize in certain services that would be difficult for individual participants to perform, such as matching buyers and sellers of securities. Three types of financial intermediaries are investment bankers, brokers, and dealers.

Investment Bankers Institutions called **investment banking firms** exist to help businesses and state and local governments sell their securities to the public. For example, when Lucent Technologies wanted to sell 112 million shares of stock to the public, the investment banking firm of Morgan Stanley handled the arrangements.

Investment bankers arrange securities sales on either an *underwriting basis* or a *best efforts basis*. The term **underwriting** refers to the process by which an investment banker (usually in cooperation with other investment banking firms) purchases all the new securities from the issuing company and then resells them to the public.

Investment bankers who underwrite securities face some risk because occasionally an issue is overpriced and can't be sold to the public for the price anticipated by the investment banker. The investment banker has already paid the issuing company or municipality its money up front, and so must absorb the difference between what it paid the issuer and what the security actually sold for. To alleviate this risk, investment bankers sometimes sell securities on a **best efforts basis.** This means the investment banker will try its best to sell the securities for the desired price, but there are no guarantees. If the securities must be sold for a lower price, the issuer collects less money.

Brokers Brokers—often account representatives for an investment banking firm—handle orders to buy or sell securities. **Brokers** are agents who work on behalf of an investor. When investors call with an order, brokers work on their behalf to find someone to take the other side of the proposed trade. If investors want to buy, brokers find sellers. If investors want to sell, brokers find buyers. Brokers are compensated for their services when the person whom they represent—the investor—pays them a commission on the sale or purchase of securities.

Dealers **Dealers** make their living buying securities and reselling them to others. They operate just like car dealers who buy cars from manufacturers for resale to others. Dealers make money by buying securities for one price, called the *bid price,* and selling them for a higher price, called the *ask (or offer) price*. The difference, or *spread,* between the bid price and the ask price represents the dealer's fee.

Financial Markets

> **Take Note**
>
> *Do not confuse financial markets with financial institutions. A financial market is a forum in which financial securities are traded (it may or may not have a physical location). A financial institution is an organization that takes in funds from some economic units and makes them available to others.*

As we have pointed out, the financial system allows surplus economic units to trade with deficit economic units. The trades are carried out in the *financial markets*.

Financial markets are categorized according to the characteristics of the participants and the securities involved. In the *primary market,* for instance, deficit economic units sell new securities directly to surplus economic units and to financial institutions. For instance, Lucent Technologies sold its newly issued stock shares to buyers in the primary market. In the *secondary market,* investors trade securities among themselves. Primary and secondary markets can be further categorized as to the maturity of the securities traded. Short-term securities—securities with a maturity of one year or less—are traded in the *money market;* and long-term securities—securities with a maturity of more than one year[1]—are traded in the *capital market*. A number of other financial markets exist, but we are mainly concerned with these four. In the following sections, we examine each of these markets in turn.

---

[1]The dividing line between "short term" and "long term" is arbitrarily set at one year.

The Primary Market

When a security is created and sold for the first time in the financial marketplace, this transaction takes place in the **primary market.** Thus, when Lucent Technologies went to Morgan Stanley and arranged to sell its 112 million shares of stock to the public, the stock sale took place in the primary market. In this market the issuing business or entity sells its securities to investors (the investment banker simply assists with the transaction).

The Secondary Market

Once a security has been issued, it may be traded from one investor to another. Many of the investors who originally bought Lucent's stock, for example, have since traded their shares to other investors. The **secondary market** is where previously issued securities—or "used" securities—are traded among investors. Suppose you called your stockbroker to request that she buy 100 shares of stock for you. The shares would usually be purchased from another investor in the secondary market. Secondary market transactions occur thousands of times daily as investors trade securities among themselves.

The Money Market

Short-term securities (a maturity of one year or less) are traded in the **money market.** Networks of dealers operate in this market. They use phones and computers to make trades rapidly among themselves and with the issuing organizations. The specific securities traded in the money market include Treasury bills, negotiable certificates of deposit, commercial paper, and other short-term debt instruments.

We discuss these securities later in the chapter on pages 25–27.)

The Capital Market

Long-term securities (maturities over one year) trade in the **capital market.** Federal, state, and local governments, as well as large corporations, raise long-term funds in the capital market. Firms usually invest proceeds from capital market securities sales in long-term assets like buildings, production equipment, and so on. Initial offerings of securities in the capital market are usually large deals put together by investment bankers, although after the original issue, the securities may be traded quickly and easily among investors. The two most widely recognized securities in the capital market are bonds and stocks.

Bonds are discussed on pages 27–29. Stocks are discussed on pages 29–31.

Security Exchanges

Security exchanges, such as the New York Stock Exchange (NYSE), are organizations that facilitate trading of stocks and bonds among investors. Corporations arrange for their stock or bonds to be *listed* on an exchange so that investors may trade the company's stocks and bonds at an organized trading location. Corporations list their securities on exchanges because they believe that having their securities traded at such a location will make them easier to trade and, therefore, boost the price. Exchanges accept listings because they earn a fee for their services.

Each exchange-listed stock is traded at a specified location on the trading floor called *the post*. The trading is supervised by specialists who act either as brokers (bringing together buyers and sellers) or as dealers (buying or selling the stock themselves).

Prominent international securities exchanges include the New York Stock Exchange (NYSE), the American Stock Exchange (AMEX),[2] and major exchanges in Tokyo, London, Amsterdam, Frankfurt, Paris, Hong Kong, and Mexico.

The Over-the-Counter (OTC) Market

In contrast to the organized exchanges, which have physical locations, the **over-the-counter market** has no fixed location—or, more correctly, it is everywhere. The over-the-counter market, or OTC, is a network of dealers around the world who maintain inventories of securities for sale. Say you wanted to buy a security that is traded OTC. You would call your broker, who would then shop among competing dealers who have the security in their inventory. After locating the dealer with the best price, your broker would buy the security on your behalf.

Most dealers in the OTC market are connected through a computer network called NASDAQ (National Association of Securities Dealers Automated Quote system). But many others, especially those dealing in securities issued by very small companies, simply buy and sell over the telephone.

Market Efficiency

The term **market efficiency** refers to the ease, speed, and cost of trading securities. In an efficient market, securities can be traded easily, quickly, and at low cost. Markets lacking these qualities are considered inefficient.

The major stock markets are generally efficient because investors can trade thousands of dollars worth of shares in minutes simply by making a phone call and paying a relatively small commission. In contrast, the real estate market is relatively inefficient because it might take you months to sell a house and you would probably have to pay a real estate agent a large commission to handle the deal.

The more efficient the market, the easier it is for excess funds in the hands of surplus economic units to make their way into the hands of deficit economic units. In an inefficient market, surplus economic units may end up with excess funds that are idle. When this happens, economic activity and job creation will be lower than it could be, and deficit economic units may not be able to achieve their goals because they could not obtain needed funds.

Financial markets help firms and individual investors buy and sell securities efficiently. So far, we have discussed the various markets in which securities are traded. Now let's turn to the securities themselves.

---

[2]The NASDAQ and American stock exchanges formally agreed merge in March 1998. NASDAQ is the nation's second-largest stock market and AMEX is the third-largest stock market. NASDAQ and AMEX agreed to operate as separate markets under the management of the NASDAQ-AMEX Market Group.

Securities in the Financial Marketplace

Securities are traded in both the money and capital markets. Money market securities include Treasury bills, negotiable certificates of deposit, commercial paper, Eurodollars, and banker's acceptances. Capital market securities include bonds and stock. We describe each of these securities briefly in the following discussion.

Securities in the Money Market

Governments, corporations, and financial institutions that want to raise money for a short time issue money market securities. Buyers of money market securities include governments, corporations, and financial institutions that want to park surplus cash for a short time, and other investors who want the ability to alter or cash in their investments quickly.

Money market securities are very liquid; that is, they mature quickly and can be sold for cash quickly and easily. Money market securities also have a low degree of risk because purchasers will only buy them from large, reputable issuers (investors don't want to spend a long time checking the issuers' credit for an investment that may only last a few days). These two characteristics, liquidity and low risk, make money market securities the ideal parking place for temporary excess cash.

Let's take a closer look at the main money market securities.

Treasury Bills Every week the United States Treasury issues billions of dollars of **Treasury bills** (T-bills). These money market securities are issued to finance the federal budget deficit (if any) and to refinance the billions of dollars of previously issued government securities that come due each week. After the T-bills are initially sold by the U.S. government, they are traded actively in the secondary market. At maturity, the government pays the face value of the T-bill.

Treasury bills are considered the benchmark of safety because they have essentially no risk. This is because obligations of the U.S. government are payable in U.S. dollars—and, theoretically, the U.S. government could print up all the dollars it needs to pay off its obligations.

Negotiable Certificates of Deposit You may already be familiar with the certificates of deposit (CDs) that you can purchase from your local bank. They are simply pieces of paper that certify that you have deposited a certain amount of money in the bank, to be paid back on a certain date with interest. Small-denomination consumer CDs are very safe investments and they tend to have low interest rates.

Large-denomination CDs (of $100,000 to $1 million or more), with maturities of two weeks to a year, are **negotiable CDs** because they can be traded in the secondary market after they are initially issued by a financial institution. Large corporations and other institutions buy negotiable CDs when they have cash they wish to invest for a short period of time; they sell negotiable CDs when they want to raise cash quickly.

Commercial Paper **Commercial paper** is a type of short-term promissory note—similar to an IOU—issued by large corporations with strong credit ratings. Commercial paper is *unsecured*, meaning that the issuing corporation does not provide any property as collateral that the lender (the one who buys the commercial paper note) can take instead of a payment if the issuing corporation defaults on the note. That is why commercial paper is only issued by financially strong, reliable firms.

ETHICAL CONNECTIONS

The Dog Ate My Investment

When you play around in the financial marketplace, you can expect to be burned every now and then. But if you get burned, whom should you blame? It seems there is a growing trend among America's business leaders to blame everyone but themselves.

For example, Gibson Greetings Inc. of Cincinnati and Orange County, California, recently lost millions of dollars on derivatives, complex financial dealings that some people complain even Einstein couldn't understand. When Gibson lost millions, its financial officers argued that it had been swindled by slick New York sophisticates who never should have let the company engage in such risky deals.

Robert L. Citron, who served for more than two decades as Orange County's treasurer and was in charge of its investments, argued the same thing. "I was an inexperienced investor" who relied on "financial professionals," Citron said.

But Paul Critchlow, a spokesman for Merrill Lynch, said in response that Citron's statement rang "hollow when he claims he didn't know what he was doing," and that he "chose to ignore repeated warnings" that his $20 billion investment fund was "vulnerable to large losses as interest rates grew."

When investors don't get paid, investment funds are also ready with excuses. A missing minus sign is the reason Fidelity Investments gave for a $2 billion-plus mistake in calculating Magellan Fund shareholders' payments, saying an accountant omitted the minus sign from a spreadsheet and no one noticed.

It seems that excuse making has become so prevalent in financial markets that it has even spawned consultants who tell corporations why they shouldn't make excuses. One such consultant is Edward P. Wolfram, Jr., who served ten years in prison for embezzling $47 million from Bell & Beckwith, a brokerage firm in Toledo. His embezzlement put the firm out of business.

When it comes to excuses, Wolfram says he has heard them all. "Look, some things are illegal and inexcusable," he said, "like embezzlement, although I admit that at first I tried to blame anyone but myself for what I did. But we're talking about excuses becoming a way of life for folks who make bad judgments and don't want to take responsibility for their mess."

There may be a fine madness behind the recent outbreak of excuse making. If you admit that your company was at fault, you may be facing a lawsuit from angry stockholders. To the extent that lawyers, the Securities and Exchange Commission, or others say they'll scrutinize a company when things go badly and go after its managers with a class-action or regulatory club, the market for excuses will grow, explains Clifford W. Smith, Jr., a professor at the University of Rochester's Simon School of Business Administration.

Meanwhile, if you invest in financial markets and you lose a bundle, look on the bright side. There must be someone around that you can blame it on.

Questions to Consider

- Why do you think excuse making surges when a financial crisis occurs? What can a business do to avoid excuse making?
- If you invest in a financial market on the advice of a broker, is it your broker's fault when you lose money? Explain.
- In what circumstances would it benefit a firm to accept responsibility for a financial problem that was not the firm's fault?

Source: Margot Slade, "We Forgot to Write a Headline. But It's Not Our Fault," *The New York Times* (February 19, 1995): F5.

Commercial paper is considered to be a safe place to put money for a short period of time. The notes themselves are issued and traded through a network of commercial paper dealers. Most of the buyers are large institutions.

Banker's Acceptances A **banker's acceptance** is a short-term debt instrument that is guaranteed for payment by a commercial bank (the bank "accepts" the responsibility to pay). Banker's acceptances, thus, allow businesses to avoid problems associated with collecting payment from reluctant debtors. They are often used when firms are doing business internationally because they eliminate the worry that the lender will have to travel to a foreign country to collect on a debt.

Securities in the Capital Market

When governments, corporations, and financial institutions want to raise money for a long period of time, they issue capital market securities. In contrast to money market securities, capital market securities may not be very liquid or safe. They are not generally suitable for short-term investments.

The two most prominent capital market securities are bonds and stocks. We'll examine these two securities in some depth now.

Bonds Bonds are essentially IOUs that promise to pay their owner a certain amount of money on some specified date in the future—and in most cases, interest payments at regular intervals until maturity. When companies want to borrow money (usually a fairly large amount for a long period of time), they arrange for their investment bankers to print up the IOUs and sell them to the public at whatever price they can get. In essence, a firm that issues a bond is borrowing the amount that the bond sells for on the open market.

Bond Terminology and Types. Although many types of bonds exist, most bonds have three special features: face value, maturity date, and coupon interest.

- *Face value:* The amount that the bond promises to pay its owner at some date in the future is called the bond's **face value,** or **par value,** or **principal.** Bond face values range in multiples of $1,000 all the way up to more than $1 million. Unless otherwise noted, assume that all bonds we discuss from this point forward have a face value of $1,000.
- *Maturity date:* The date on which the issuer is obligated to pay the bondholder the bond's face value.
- *Coupon interest:* The interest payments made to the bond owner during the life of the bond.[3] Some bonds pay coupon interest once a year; many pay it twice a year. Some bonds don't pay any interest at all. These bonds are called **zero-coupon bonds.**

The percentage of face value that the coupon interest payment represents is called the *coupon interest rate.* For example, assuming the face value of the bond was $1,000, a bond owner who received $80 interest payments each year would own a bond paying an 8 percent coupon interest rate:

$$\$80 \,/\, \$1{,}000 = .08, \text{ or } 8\%$$

The major types of bonds include Treasury bonds and notes, issued by the federal government; municipal bonds, issued by state and local governments; and

---

[3]The name originated decades ago when holders of bearer bonds would actually tear off coupons from their bond certificates and mail them to the bond issuer to get their interest payments, hence, the name *coupon* interest. Today, bonds are sold on a "registered" basis, which means the bonds come with the owner's name printed on them. Interest payments are sent directly to the owner.

corporate bonds, issued by corporations. The significant differences among these types of bonds are described in the following sections.

Treasury Notes and Bonds. When the federal government wants to borrow money for periods of more than a year, it issues Treasury notes or Treasury bonds (T-notes and T-bonds for short). T-notes and T-bonds are identical except for their maturities. T-notes have maturities from one to ten years, and T-bonds have maturities greater than ten years. Both T-notes and T-bonds pay interest semiannually, in addition to the principal, which is paid at maturity.

T-notes and T-bonds are auctioned by the Treasury every three months to pay off old maturing securities and to raise additional funds to finance the federal government's new deficit spending.

Although Treasury securities, such as T-notes and T-bonds, are extremely low risk, they are not risk free. Without the congressional authority to print money, the U.S. Treasury cannot legally pay its obligations, including the interest and principal on Treasury securities. In late 1995 and early 1996, Congress refused to raise the legal ceiling on outstanding Treasury debt unless President Clinton agreed to sign a seven-year balanced budget bill that restructured and eliminated many federal programs. President Clinton vetoed several budget bills because he opposed some of the budget measures.[4] The confrontation between the executive and congressional branches ended before the Treasury defaulted on its debts, but such a confrontation could conceivably result in default.

The U.S. Treasury began selling Treasury bonds that are inflation-indexed in January 1997. Interest rates adjust and are pegged to the inflation rate. This protects the investors from erosion of the purchasing power of the dollars they receive from this investment. In September 1998 the Treasury began selling securities directly from its Web site at www.publicdebt.treas.gov.

Municipal Bonds. The bonds issued by state and local governments are known as **municipal bonds** or "munis." Many investors like municipal bonds because their coupon interest payments are free of federal income tax.

Municipal bonds come in two types: general obligation bonds (GOs) and revenue bonds. They differ in where the money is expected to come from to pay them off. General obligation bonds are supposed to be paid off with money raised by the issuer from a variety of different tax revenue sources. Revenue bonds are supposed to be paid off with money generated by the project the bonds were issued to finance—such as using toll bridge fees to pay off the bonds to finance the toll bridge.

Figure 2-1 shows a general obligation bond issued by the State of Illinois.

Corporate Bonds. Corporate bonds are similar to T-bonds and T-notes except they are issued by corporations. Like T-bonds and T-notes, they pay their owner interest during the life of the bond and repay principal at maturity. Unlike T-bonds and T-notes, however, corporate bonds sometimes carry substantial risk of default. As a last resort, the U.S. government can print money to pay off its Treasury bill, note, and bond obligations; but when private corporations run into trouble, they have no such latitude. Corporations' creditors may get paid late or not at all.

---

[4]Adam Clymer, "Treasury Takes Retirement Funds to Avert Default," *The New York Times* (November 16, 1995): A1.

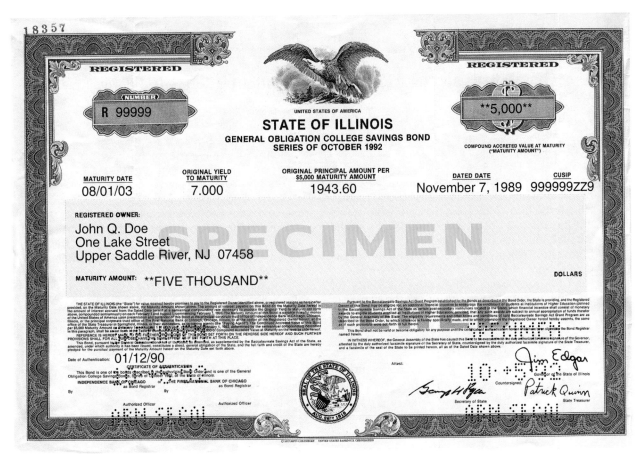

Figure 2-1 **A General Obligation Bond Issued by the State of Illinois**
This is a picture of a general obligation bond from the State of Illinois. The face value of the bond is $5,000, it promises to pay the face value to the bearer, and it matures on August 1, 2003. The bond was purchased on November 7, 1989 for $1,943.60. (Bond provided courtesy of The First National Bank of Chicago.)

Relatively safe bonds are called *investment-grade bonds*. Many financial institutions and money management firms are required to invest only in those corporate bonds that are investment grade. Relatively risky bonds are called *junk bonds*.[5] Junk bonds are generally issued by troubled companies, but may be issued by financially strong companies that later run into trouble.

This completes our introduction to bonds. Now let's turn our attention to the other major security in the capital market, corporate stock.

Corporate Stock   Rather than borrowing money by issuing bonds, a corporation may choose to raise money by selling shares of ownership interest in the company. Those shares of ownership are **stock.** Investors who buy stock are called stockholders.

---

[5]The term *junk bond* is a slang term that is now widely accepted. Firms trying to sell junk bonds hate the term, of course. They would prefer to call such bonds *high yield*.

FINANCIAL MANAGEMENT AND YOU

How to Buy Treasury Securities without Using a Broker

Most people use a broker to buy Treasury securities, even though purchasing them is not difficult. If you are interested in investing in Treasury securities, follow these four simple steps:

1. Contact your regional Federal Reserve Bank or its branch and request a Tender Application Form.
2. Fill out the form, indicating that you wish to submit a noncompetitive tender, and the particular security you wish to buy. Treasury bills, Treasury notes, and Treasury bonds come in minimum denominations of $1,000.
3. Send the form back, along with your check for the amount of the face value of the securities you selected. A cashier's check must be sent for Treasury bills, whereas a personal check is accepted for Treasury notes and bonds. Your investment in Treasury notes, bills, or bonds will be in book-entry form on the computers of the U.S. Treasury. You will receive a confirmation receipt verifying your investment.
4. In the case of Treasury bills, you will receive a check within a few weeks for the amount of the face value minus the price determined at the auction. At maturity all bills, notes, and bonds will pay the face value. Notes and bonds will also pay semiannual interest until the maturity date.

As a source of funds, stock has an advantage over bonds: The money raised from the sale of stock doesn't ever have to be paid back, and the company doesn't have to make interest payments to the stockholders.

A corporation may issue two types of corporate stock, *common stock* and *preferred stock*. Let's look at their characteristics.

Common Stock. Common stock is so called because there is nothing special about it. The holders of a company's common stock are simply the owners of the company. Their ownership entitles them to the firm's earnings that remain after all other groups having a claim on the firm (such as bondholders) have been paid.

Each common stockholder owns a portion of the company represented by the fraction of the whole that the stockholder's shares represent. Thus, if a company issued 1 million shares of common stock, a person who holds one share owns one-millionth of the company.

Common stockholders receive a return on their investment in the form of common stock **dividends,** distributed from the firm's profits, and from **capital gains,** realized when they sell the shares.[6]

Preferred Stock. **Preferred stock** is so called because if dividends are declared by the board of directors of a business, they are paid to preferred stockholders first. If any funds are left over, they may be paid to the common stockholders. So why would anyone buy common stock? First, preferred stockholders normally don't get to vote on how the firm is run. Second, common stock normally provides a slightly higher expected return because it is a more risky investment—that is, dividends are not guaranteed. However, common stockholders—as owners—are entitled to all the

---

[6]Of course, there is no guarantee that a common stockholder's stock will increase in price. If the price goes down, the stockholder will experience a capital loss.

residual income of the firm, and there is no upper limit on how great the residual income of the firm might be.

INTEREST

No one lends money for free. When people lend money to other people, a number of things could happen that might prevent them from getting all their money back. Whenever people agree to take risk, compensation is required before they will voluntarily enter into an agreement. In financial activities, we refer to this compensation as interest. **Interest** represents the return, or compensation, a lender demands before agreeing to lend money. When we refer to interest, we normally express it in percentage terms, called the *interest rate*. Thus, if you lend a person $100 for one year, and the return you require for doing so is $10, we would say that the interest rate you are charging for the loan is $10/$100 = .10, or 10 percent.

Determinants of Interest Rates

The prevailing rate of interest in any situation is called the **nominal interest rate.** In the preceding example the nominal interest rate for the one year $100 loan is 10 percent.

The nominal interest rate is actually the total of a number of separate components, as shown in Figure 2-2. We will explore each of these components in the following sections.

The Real Rate of Interest Lenders of money must postpone spending during the time the money is loaned. Lenders, then, lose the opportunity to invest their money for that period of time. To compensate for the cost of losing investment opportuni-

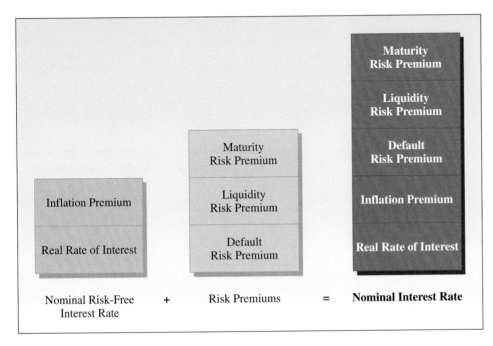

Figure 2-2 **Components of the Nominal Interest Rate**

The nominal interest rate is composed of the real interest rate plus a number of premiums. The nominal risk-free interest rate is the real rate plus an inflation premium. When risk premiums are added, the result is the total nominal interest rate.

ties while they postpone their spending, lenders demand, and borrowers pay, a basic rate of return—the **real rate of interest.** The real rate of interest does not include adjustments for any other factors, such as the risk of not getting paid back. We'll describe this in a moment.

Let's continue with the example on page 31, in which you lent a person $100. The total interest rate that you charged was 10 percent (the nominal interest rate). The portion of the total nominal rate that represents the return you demand for forgoing the opportunity to spend your money now is the real rate of interest. In our example, assume the real rate of interest is 2 percent.

Additions to the real rate of interest are called *premiums*. The major premiums are the inflation premium, the default risk premium, the liquidity risk premium, and the maturity risk premium.

The Inflation Premium   Inflation erodes the purchasing power of money. If inflation is present, the dollars that lenders get when their loans are repaid may not buy as much as the dollars that they lent to start with. Therefore, lenders who anticipate inflation during the term of a loan will demand additional interest to compensate for it. This additional required interest is the **inflation premium.**

If, when you lent $100, you thought that the rate of inflation was going to be 4 percent a year during the life of the loan, you would add 4 percent to the 2 percent real rate of interest you charged for postponing your spending. The total interest rate charge—so far—would be 6 percent.

The Nominal Risk-Free Rate   The interest rate that we have built so far, containing the real rate of interest and a premium to cover expected inflation, is often called the **nominal risk-free rate of interest,** as shown earlier in Figure 2-2. It is called this because it does not include any premiums for the uncertainties associated with borrowing or lending. The yield on short-term U.S. Treasury bills is often used as a proxy for the risk-free rate because the degree of uncertainty associated with these securities is very small.

> **Take Note**
>
> *You can find the yield on U.S. Treasury bills very easily just by looking in* **The Wall Street Journal** *on page C-1. Yields on T-bills and a number of other securities are published there every business day.*

Risk Premiums   The remaining determinants of the nominal interest rate represent extra charges to compensate lenders for taking risk. Risks in lending come in a number of forms. The most common are default risk, liquidity risk, and maturity risk.

The Default Risk Premium.   A *default* occurs when a borrower fails to pay the interest and principal on a loan on time. If a borrower has a questionable reputation or is having financial difficulties, the lender faces the risk that the borrower will default. The **default risk premium** is the extra compensation lenders demand for assuming the risk of default.

In our $100 loan example, if you weren't completely sure that the person to whom you had lent $100 would pay it back, you would demand extra compensation—let's say, two percentage points—to compensate for that risk. The total interest rate demanded so far would be 2 percent real rate of interest + 4 percent inflation premium + 2 percent default risk premium = 8 percent.

The Liquidity Risk Premium. Sometimes lenders sell loans to others after making them. (This happens often in the mortgage business in which investors trade mortgages among themselves.) Some loans are easily sold to other parties and others are not. Those that are easily sold are *liquid,* and those that aren't sold easily are considered *illiquid.* Illiquid loans have a higher interest rate to compensate the lender for the inconvenience of being stuck with the loan until it matures. The extra interest that lenders demand to compensate for the lack of liquidity is the **liquidity risk premium.**

You will probably not be able to sell your $100 loan to anyone else, and will have to hold it until maturity. Therefore, you require another 1 percent to compensate for the lack of liquidity. The total interest rate demanded so far is 2 percent real rate of interest + 4 percent inflation premium + 2 percent default risk premium + 1 percent liquidity risk premium = 9 percent.

The Maturity Risk Premium. If interest rates rise, lenders may find themselves stuck with long-term loans paying the original rate prevailing at the time the loans were made, while other lenders are able to make new loans at higher rates. On the other hand, if interest rates go down, the same lenders will be pleased to find themselves receiving higher interest rates on their existing long-term loans than the rate at which other lenders must make new loans. Lenders respond to this risk that interest rates may change in the future in two ways:

- If lenders think interest rates might rise in the future, they may increase the rate they charge on their long-term loans now and decrease the rate they charge on their short-term loans now to encourage borrowers to borrow short term.
- Conversely, if lenders think interest rates might fall in the future, they may decrease the rate they charge on their long-term loans now and increase the rate they charge on their short-term loans now to encourage borrowers to borrow long term (locking in the current rate).

This up or down adjustment that lenders make to their current interest rates to compensate for the uncertainty about future changes in rates is called the **maturity risk premium.** The maturity risk premium can be either positive or negative.

In our example, if you thought interest rates would probably rise before you were repaid the $100 you lent, you might demand an extra percentage point to compensate for the risk that you would be unable to take advantage of the new higher rates. The total rate demanded is now 10 percent (2 percent real rate of interest + 4 percent inflation premium + 2 percent default risk premium + 1 percent liquidity risk premium + 1 percent maturity risk premium = 10 percent, the nominal interest rate).

The total of the real rate of interest, the inflation premium, and the risk premiums (the default, liquidity, and maturity risk premiums) is the nominal interest rate, the compensation lenders demand from those who want to borrow money.

Next, we will consider the *yield* curve—a graph of a security's interest rates depending on the time to maturity.

The Yield Curve

A yield curve is a graphical depiction of interest rates for securities that differ only in the time remaining until their maturity. Yield curves are drawn by plotting the interest rates of one kind of security but with various maturity dates. The curve depicts the interest rates of these securities at a given point in time.

Yield curves of U.S. Treasury securities are most common because with Treasury securities it is easiest to hold constant the factors other than maturity. All Treasury securities have essentially the same default risk (almost none) and about the same degree of liquidity (excellent). Any differences in interest rates observed in the yield curve, then, can be attributed to the maturity differences among the securities because other factors have essentially been held constant.

Figure 2-3 shows a Treasury securities yield curve for April 27, 1998. The yield curve in Figure 2-3 is upward sloping and would be called a normal yield curve. Although less common, yield curves may also be downward sloping, or inverted.

Making Use of the Yield Curve The shape of the yield curve gives borrowers and lenders useful information for financial decisions. Borrowers, for example, tend to look for the low point of the curve, which indicates the least expensive loan maturity. Lenders tend to look for the highest point on the curve, which indicates the most expensive loan maturity.

Finding the most advantageous maturity is not quite as simple as it sounds because it depends on more factors than cost. For instance, the least expensive maturity is not always the most advantageous for borrowers. If a firm borrows short term, for example, it may obtain the lowest interest rate, but the loan may mature in a short time and have to be renewed at a higher rate if interest rates have risen in the

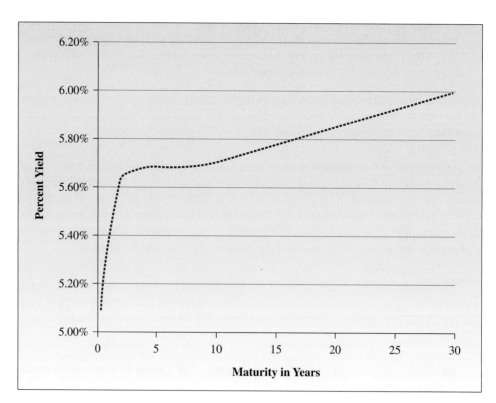

Figure 2-3 **Treasury Yield Curve for April 27, 1998**

This yield curve shows interest rates on April 27, 1998, for treasury securities having maturities from one to thirty years.

Data Source: *Bloomberg Online.* ⟨www.bloomberg.com/markets/c13.html⟩.

interim. Borrowing for a longer term may cost a borrower more at the outset but less in the long run because the interest rate is locked in.

Lenders face the opposite situation. Granting long-term loans at relatively high interest rates may look attractive now; but if short-term rates rise, the lenders may miss profitable opportunities because their funds have already been invested. Both borrowers and lenders must balance their desire for return with their tolerance for risk.

What's Next

In this chapter we investigated financial markets, securities, and interest rates. In the next chapter, we will look at another part of the financial environment, financial institutions.

Summary

1. **Describe how the U.S. financial system works.**

The financial system is made up of surplus economic units, entities and individuals that have excess funds, and deficit economic units, entities and individuals that need to acquire additional funds. The financial system provides the network that brings these two groups together so that funds flow from the surplus economic units to the deficit economic units.

2. **Define financial securities.**

Securities are documents that represent a person's right to receive funds in the future. Firms issue securities in exchange for funds they need now, and investors trade securities among themselves.

3. **Explain the function of financial intermediaries.**

Financial intermediaries act to put those in need of funds in contact with those who have funds available. Investment banking firms help businesses acquire funds from the public by issuing securities in the financial marketplace. Brokers help members of the public trade securities with each other. Dealers buy and sell securities themselves.

4. **Identify the different financial markets.**

Financial markets are forums in the financial system that allow surplus economic units to transact with deficit economic units and for portfolio adjustments to be made. Securities change hands in financial markets. The financial markets include the primary market, in which new securities are issued for the first time; the secondary market, in which previously issued securities are traded among investors; the money market, in which securities with maturities of less than one year are traded; and the capital market, in which securities with maturities longer than one year are traded. Some securities are traded on organized exchanges, like the New York Stock Exchange, and others are traded over the counter (OTC) in a network of securities dealers.

5. **List and define the securities traded in the money and capital markets.**

Securities traded in the money market include:

- *Treasury bills:* short-term debt instruments issued by the U.S. Treasury that are sold at a discount and pay face value at maturity.
- *Negotiable certificates of deposit (CDs):* certificates that can be traded in financial markets and represent amounts deposited at banks that will be repaid at maturity with a specified rate of interest.

- *Commercial paper:* unsecured short-term promissory notes issued by large corporations with strong credit ratings.
- *Banker's acceptances:* documents that signify that a bank has guaranteed payment of a certain amount at a future date if the original promisor doesn't pay.

The two major securities traded in the capital market include:

- *Bonds:* long-term securities that represent a promise to pay a fixed amount at a future date, usually with interest payments made at regular intervals. Treasury bonds are issued by the U.S. government, corporate bonds are issued by firms, and municipal bonds are issued by state and local governments.
- *Stocks:* shares of ownership interest in corporations. Preferred stock comes with promised dividends but usually no voting rights. Common stock may come with dividends, paid at the discretion of the board, but do have voting rights. Common stockholders share in the residual profits of the firm.

6. **Identify the determinants of the nominal interest rate.**

The nominal interest rate has three main determinants:

- *The real rate of interest:* the basic rate lenders require to compensate for forgoing the opportunity to spend money during the term of the loan.
- *An inflation premium:* a premium that compensates for the expected erosion of purchasing power due to inflation over the life of the loan.
- *Risk premiums:* premiums that compensate for the risks of default (the risk that the lender won't be paid back), liquidity (the risk that the lender won't be able to sell the security in a reasonable time at a fair price), and maturity (the risk that interest rates may change adversely during the life of the security).

7. **Construct and analyze a yield curve.**

A yield curve is a graphical depiction of interest rates on securities that differ only in the time remaining until their maturity. Lenders and borrowers may use a yield curve to determine the most advantageous loan maturity.

You can access your FREE PHLIP/CW On–Line Study Guide, Current Events, and Spreadsheet Problem templates, which may correspond to this chapter either through the Prentice Hall Finance Center CD–ROM or by directly going to: **http://www.prenhall.com/gallagher.**

PHLIP/CW — *The Prentice Hall Learning on the Internet Partnership–Companion Web Site*

Self-Test

ST-1. To minimize risk, why don't most firms simply finance their growth from the profits they earn?

ST-2. What market would a firm most probably go to if it needed cash for 90 days? If it needed cash for 10 years?

ST-3. If your company's stock was not listed on the New York Stock Exchange, how could investors purchase the shares?

ST-4. What alternatives does General Motors, a very large and secure firm, have for obtaining $3 million for 60 days?

ST-5. Assume Treasury security yields for today are as follows:

> One-year T-notes 5.75%
> Two-year T-notes 5.5%
> Three-year T-notes 5.25%
> Five-year T-notes 5.0%
> Ten-year T-notes 4.75%
> Twenty-year T-bonds 4%
> Thirty-year T-bonds 3.25%

Draw a yield curve based on these data.

Review Questions

1. What are financial markets? Why do they exist?
2. What is a security?
3. What are the characteristics of an efficient market?
4. How are financial trades made on an organized exchange?
5. How are financial trades made in an over-the-counter market? Discuss the role of a dealer in the OTC market.
6. What is the role of a broker in security transactions? How are brokers compensated?
7. What is a Treasury bill? How risky is it?
8. Would there be positive interest rates on bonds in a world with absolutely no risk (no default risk, maturity risk, and so on)? Why would a lender demand, and a borrower be willing to pay, a positive interest rate in such a no-risk world?

Build Your Communication Skills

CS-1. Imagine the following scenario:

Your firm has decided to build a new plant in South America this year. The plant will cost $10 million and all the money must be paid up front. Your boss has asked you to brief the board of directors on the options the firm has for raising the $10 million.

Prepare a memo for the board members outlining the pros and cons of the various financing options open to the firm. Divide into small groups. Each group member should spend five minutes presenting his or her financing option suggestions to the rest of the group members, who should act as the board members.

CS-2. Prepare an IOU, or "note," that promises to pay $100 one year from today to the holder of the note.

   a. Auction this note off to someone else in the class, having the buyer pay for it with a piece of scratch-paper play money.
   b. Compute the note buyer's percent rate of return if he or she holds the note for a year and cashes it in.
   c. Ask the new owner of the note to auction it off to someone else. Note the new buyer's rate of return based on his or her purchase price.
   d. Discuss the operation of the market the class has created. Note the similarities between it and the bond market in the real world.

Problems

2-1. If the real rate of interest is 2 percent, inflation is expected to be 3 percent during the coming year, and the default risk premium, liquidity risk premium, and maturity risk premium for the Bonds-R-Us corporation are all 1 percent each, what would be the yield on a Bonds-R-Us bond?

2-2. Assume Treasury security yields for today are as follows:

>Three-month T-bills 6.5%
>Six-month T-bills 6.0%
>One-year T-notes 5.75%
>Two-year T-notes 5.5%
>Three-year T-bonds 5.25%
>Five-year T-bonds 5.0%
>Ten-year T-bonds 4.75%
>Thirty-year T-bonds 4.5%

Draw a yield curve based on these data. Discuss the implications if you are:

a. a borrower
b. a lender

Answers to Self-Test

ST-1. In most cases, profits are insufficient to provide the funds needed, especially with large projects. Financial markets provide access to external sources of funds.

ST-2. To obtain cash for 90 days, a business firm would most probably go to the money market in which it would sell a 90-day security. To obtain cash for 10 years, a firm would sell a security in the capital market.

ST-3. Investors would simply purchase the shares on another exchange, or over the counter from a dealer. (Investors simply call their brokers to purchase stock. Brokers decide where to get it.)

ST-4. General Motors could:

- obtain a 60-day loan from a financial institution
- delay payments to its suppliers
- sell commercial paper notes in the money market

ST-5. The yield curve follows:

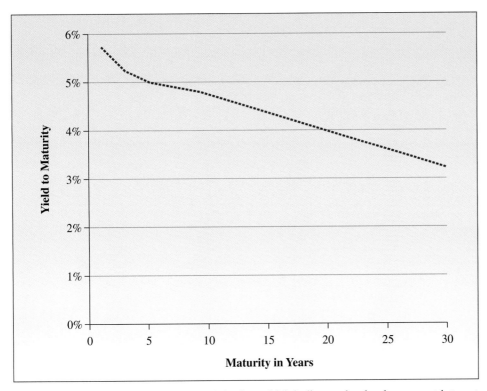

Notice that this yield curve is downward sloping, which indicates that lenders expect interest rates to fall in the future. (See the discussion about the maturity risk premium on page 33.)

CHAPTER 3

FINANCIAL INSTITUTIONS

A banker is a fellow who lends his umbrella when the sun is shining and wants it back the minute it begins to rain.
—Mark Twain

WORLD'S FIRST INTERNET BANK

On October 18, 1995, Security First Network Bank (SFNB) became the world's first fully transactional Internet bank. It is insured by the Federal Deposit Insurance Corporation (FDIC) and received permission to operate on the Web from the Office of Thrift Supervision (OTS).

SFNB offers checking accounts, savings accounts, certificates of deposit, and credit cards. This opened the door for other banks to follow. The Web is rapidly becoming a place for financial service firms. Your home has become a bank branch. Many brokerage firms, such as E*Trade and Ameritrade, are also on-line. Because bricks and mortar are a large part of the assets of a typical bank, doing business with a minimum of these assets has its advantages. A bank is clearly better off if it can provide most of the essential services customers expect while investing a much smaller amount of money to provide those services.

Source: Security First Network Bank web page, September 19, 1998 (http://www.sfnb.com).

CHAPTER OBJECTIVES

After reading this chapter, you should be able to:

- Explain financial intermediation and the role of financial institutions.

- Define commercial banks and explain how reserve requirements influence their operations.

- Describe how the Federal Reserve regulates financial institutions.

- Explain how savings and loan associations differ from commercial banks.

- Describe how credit unions operate.

- Distinguish among finance companies, insurance companies, and pension funds.

Chapter Overview

In the preceding chapter we discussed how the financial system makes it possible for deficit and surplus economic units to come together, exchanging funds for securities to their mutual benefit. In this chapter we will examine how financial institutions help channel available funds to those who need them. We will also see the important role the Federal Reserve plays in regulating the financial system, protecting both deficit and surplus economic units.

Financial Intermediation

The financial system makes it possible for surplus and deficit economic units to come together, exchanging funds for securities, to their mutual benefit. When funds flow from surplus economic units to a financial institution to a deficit economic unit, the process is known as **intermediation.** The financial institution acts as an intermediary between the two economic units.

Surplus economic units can channel their funds into financial institutions by purchasing savings accounts, checking accounts, life insurance policies, casualty insurance policies, or claims on a pension fund. The financial institutions can then pool the funds received and use them to purchase claims issued by deficit economic units, such as Treasury, municipal, or corporate bonds; and common or preferred stock. (The institutions may purchase real assets too, such as real estate or gold.)

At first glance it might seem that intermediation complicates things unnecessarily. Why do the surplus and deficit economic units need a middle person? The answer is that financial institutions can do things for both that they often can't do for themselves. Here are some examples of the services that financial institutions provide.

Denomination Matching

Members of the household sector (net surplus economic units) often have only a small amount of funds available to invest in securities. Although, as a group, they are net suppliers of funds and have a large amount of funds available, this is often not the case for individuals or families. Businesses and government entities (net deficit economic units) usually need large amounts of funds. Thus, it is often difficult for these surplus and deficit economic units to come together on their own to arrange a mutually beneficial exchange of funds for securities. The surplus economic units typically want to supply a small amount of funds, whereas the deficit economic units typically want to obtain a large amount of funds.

A financial institution can step in and save the day. A bank, savings and loan, or insurance company can take in small amounts of funds from many individuals, form a large pool of funds, and then use that large pool to purchase securities from individual businesses and governments. This pooling of funds is depicted in Figure 3-1.

The typical surplus economic unit likes to make funds available to others for a short period of time. Most people, for example, would like to get their money back on short notice if the need were to arise. They would prefer to buy securities that have a short maturity. Most businesses and government entities, on the other hand,

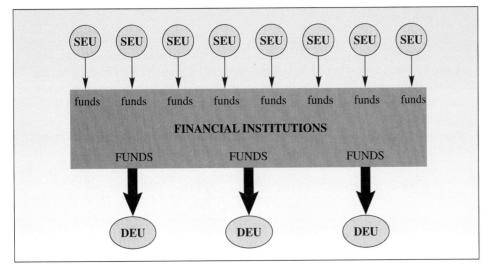

Figure 3-1 Maturity Matching

Figure 3-1 shows how small amounts of funds from many small surplus economic units (SEUs) can be pooled and channelled into the hands of a relatively small number of deficit economic units (DEUs). The financial institution provides each what it needs.

want to make use of funds for a long period of time. The new plants, roads, airports, and the like that businesses and governments buy and build are long-term projects that often require long-term financing. They would prefer to sell securities that have a long maturity.

Here's the problem: How can exchanges agreeable to both sides be arranged when the surplus economic units want the right to get their funds back quickly and the deficit economic units want to keep the funds for a long time? Remember, a financial institution has many different surplus economic units buying its securities (savings accounts, checking accounts, insurance policies, and so on). The number that will want their funds back on any given day is likely to be small, and they will probably withdraw only a very small percentage of the total funds held in the financial institution. So a large percentage of the funds held by the financial institution can be invested in the long-term securities of deficit economic units, with little danger of running out of funds.

Absorbing Credit Risk

Credit risk is the risk that the issuer of a security may fail to make promised payments to the investor at the times specified. When surplus and deficit economic units try to arrange for a direct transfer of funds for securities, this problem is often a large one. Surplus economic units do not usually have the expertise to determine whether deficit economic units can and will make good on their obligations, so it is difficult for them to predict when a would-be deficit economic unit will fail to pay what it owes. Such a failure is likely to be devastating to a surplus economic unit that has lent a proportionately large amount of money. In contrast, a financial institution is in a better position to predict who will pay and who won't. It is also in a better position, having greater financial resources, to occasionally absorb a loss when someone fails to pay.

Now let us turn to the various types of financial institutions. We'll start with commercial banks, which are regulated by various government entities. We'll also discuss the Federal Reserve System, which plays a major role in bank regulation and in overseeing the financial system.

Commercial Banks

Commercial banks are financial institutions that exist primarily to lend money to businesses. Banks also lend to individuals, governments, and other entities, but the bulk of their profits typically comes from business loans. Commercial banks make money by charging a higher interest rate on the money they lend than the rate they pay on money lent to them in the form of deposits. This rate charged to borrowers minus the rate paid to depositors is known as the **interest rate spread.**

Banking is different from many other types of business in that it must have a charter before it can open its doors. A bank charter—much more difficult to obtain than a city license needed to open another type of business—is an authorization from the government granting permission to operate. Commercial bank charters are issued by the federal government or the government of the state where the bank is located. You can't just rent some office space, buy a vault and some office furniture, put up a sign that says "Joe's Bank," and begin taking in deposits and making loans.

Banks can't operate without a charter because banking is a business intimately involved in the payment system and money supply of the economy. To protect individual economic units and the economy as a whole, the government has decided to control entry into this business and to regulate it, too.

Bank Regulation

After a bank has been granted a charter, government entities continue to scrutinize it. To begin with, all banks with federal charters must be members of the Federal Reserve System (commonly known as "the Fed"). State-chartered banks may apply for membership in the Federal Reserve System but are not required to do so. All members of the Federal Reserve System must also belong to the Federal Deposit Insurance Corporation (FDIC), which insures customer deposits at participating institutions for up to $100,000. Nonmember banks, along with other types of financial institutions, may belong to the FDIC also. Almost all banks—whether federally or state chartered, members of the Fed or not—have FDIC insurance for their depositors.

So many agencies regulate banks that it can be difficult to sort them out. To lessen the potentially extensive overlap of authority, bank regulating entities have worked out an agreement. The Office of the Comptroller of the Currency (OCC) has primary responsibility for examining national banks, ensuring that they meet accepted standards. The Fed has primary responsibility for examining state-chartered member banks. The FDIC assumes primary responsibility for examining state nonmember banks having FDIC insurance. State banking authorities have primary examining authority over state nonmember banks with no FDIC coverage for their depositors. Figure 3-2 shows the main examination authority structure for commercial banks.

Commercial Bank Operations

Commercial banks operate with more government oversight than most businesses, but they are managed just like other companies. Commercial banks have stockholders, employees, managers, equipment, and facilities, and the primary financial goal of such banks is to maximize value for their stockholders. The banks do most of their business by receiving funds from depositors and lending the funds to those who

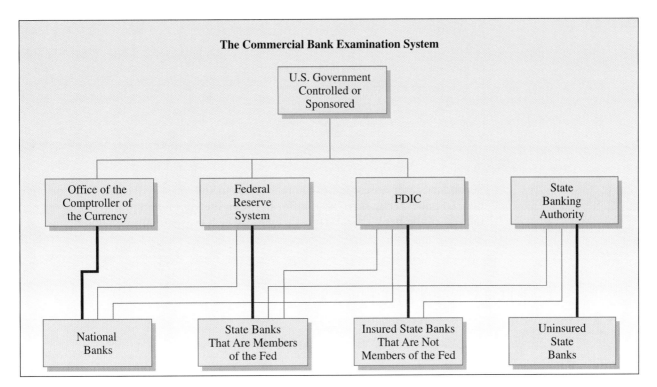

Figure 3-2 **The Commercial Bank Examination System**

Figure 3-2 shows the primary and potential examining authorities for different kinds of banks.

need them. Commercial banks also occasionally issue long-term bonds to raise funds, borrow from the Federal Reserve, or borrow deposits kept by other financial institutions in Federal Reserve banks in what is known as the *federal funds market*.

Commercial Bank Reserves Commercial banks are not allowed to lend all the funds they get from depositors. The Federal Reserve requires all commercial banks to keep a minimum amount of reserves on hand. *Reserves* are cash assets, vault cash, and deposits at the Fed that are available to a bank to meet the withdrawal demands of its depositors and to pay other obligations as they come due. Actually, the reserve requirement is set more with monetary policy in mind than to ensure that banks meet their depositors' withdrawal requests.

The required level of reserves a bank must hold is determined by applying a certain percentage to the average weekly deposits held by the bank. The exact percentage of deposits a bank must hold in reserve, called the **required reserve ratio,** depends on the type of deposit and the size of the bank. It varies from time to time as determined by the Federal Reserve, subject to certain statutory limits (see Table 3-1).

Table 3-1 shows the amount of reserves financial institutions are required to keep, depending on the amount of different kinds of deposits. Vault cash and deposits in the bank's account at the Fed are used to satisfy these reserve requirements; they are called **primary reserves.** These primary reserves are non-interest-earning assets held by financial institutions.

In addition to primary reserves, commercial banks generally hold some **secondary reserves**—assets that can be quickly and easily sold and converted into cash.

TABLE 3-1

Reserve Requirements as of July 1998

3 percent on transaction accounts $0 to $47.8 million
10 percent on transaction accounts above $47.8 million
0 percent on nonpersonal time deposits
0 percent on Eurocurrency liabilities

Source: *Federal Reserve Bulletin* (July 1998): A8.

Secondary reserves consist of short-term securities such as Treasury bills or commercial paper. They serve as a buffer between the very liquid primary reserves and the rest of the bank's assets (mostly loans), which are generally less liquid.

ETHICAL CONNECTIONS

How Ethical Is Your Financial Institution?

The financial world cringed when 28-year-old trader Nicholas Leeson lost $1 billion of his bank's money in bad investments, causing Barings Bank to go belly-up.

Yet there are hints that Leeson was not working in the most ethical environment. In fact, there is considerable evidence that Barings, Britain's oldest merchant bank, skirted rules with some regularity.

For instance, in January and February 1995, the bank routed a staggering $890 million to cover the costs of maintaining the futures and options contracts into which Leeson had entered. That sum was far more than Barings should have used to cover the contracts according to British banking regulations. Those regulations forbade Barings from committing more than 25 percent of its money to one investment.

In Japan, Barings allegedly dodged salary taxes for its employees by paying many of them "offshore"—that is, outside of Japan. For example, large portions of Barings's employees' income, often including bonuses, were earmarked for employees but set aside until the employees left Japan. That way, employees didn't have to pay full income tax in Japan.

Plus, Barings engaged in other questionable practices, such as giving commission "rebates" to valuable customers. Certain valued customers were charged negotiated commissions and then received part of the money back as a rebate. The commission fees, then, depended on the customer.

In fact, the collapse of Barings may have occurred because bank officials violated a banking rule that no bank can hold more than 1,000 contracts that bet on the same market movement. Barings had well over 1,000 contracts betting that the Japanese stock market would move in a certain direction.

The bottom line is that the entire Barings company culture seemed to wink at employees who were clever enough to skirt the rules and regulations set down by England and Japan. It should have come as no surprise to top bank executives, then, that one of their employees bent the rules, bet the bank, and lost.

Questions to Consider

▶ Who is responsible for the ethical culture of an organization? The managers, the owners, the employees, or society?

▶ If Leeson had won his futures and options bet, he would have made millions for the bank. How do you think the bank managers and owners would have viewed Leeson's infractions of the rules in that instance?

▶ What ethical message do you think Barings's managers sent its employees? How can a business do a better job communicating its code of ethics? Why should owners be interested in corporate ethics?

Source: Marcus W. Brauchili, "Many Who Knew Barings PLC Suggest Firm Wasn't Model Corporate Citizen," *The Wall Street Journal* (March 8, 1995): A16.

The Federal Reserve System

The **Federal Reserve System** serves as the central bank of the United States. It regulates the nation's money supply, makes loans to member banks and other financial institutions, and regulates the financial system, as described in the previous section.

Open-market purchases of government securities, making loans to financial institutions, and decreasing reserve requirements all lead to an increase in the money supply. Open-market sales of government securities, receiving payments on loans made to financial institutions, and increasing reserve requirements all lead to a decrease in the money supply.

Organization of the Fed

The Fed is made up of twelve district Federal Reserve banks spread throughout the country, as shown in Figure 3-3, along with a seven-member Board of Governors and a Federal Open Market Committee (FOMC) that has twelve voting members. Both the Board of Governors and the Federal Open Market Committee are located in Washington, D.C. They hold most of the power of the Fed.

The seven members of the Board of Governors are appointed by the president of the United States, subject to confirmation by the United States Senate. The governors serve staggered 14-year terms, partly to insulate them from political influences. It would be naive to believe that these members are not subject to some political influences, but a president would normally have to be well into a second (and final) term before successfully replacing a majority of the Fed members.

The twelve voting members of the Federal Open Market Committee (FOMC) are the seven members of the Board of Governors plus five of the twelve presidents of the district Federal Reserve banks. The district bank presidents take turns serving as voting members of the FOMC, although the president of the Federal Reserve Bank of New York is always one of the five. The nonvoting presidents of the district Federal Reserve Banks usually attend the FOMC meetings and participate in discussions.

Controlling the Money Supply

The main focus of the FOMC is to recommend **open-market operations** that the Fed should implement to increase or decrease the money supply. Open-market operations are purchases and sales of government securities and foreign currencies conducted in the Federal Reserve Bank of New York at a trading desk that exists for this purpose.

When the Fed *buys* government securities or currencies, it increases bank reserves and the money supply. When the Fed *sells* government securities or currencies, it decreases bank reserves and the money supply.

When the Fed wishes to increase the money supply, it instructs its traders at the Federal Reserve Bank of New York to buy government securities (primarily T-bills) on the open market. The traders contact government securities dealers (that are mostly commercial banks) around the country and buy the required amount of securities. These dealers have accounts at the Fed. When the Fed buys government securities from a dealer, it credits that dealer's account at the Fed. This action

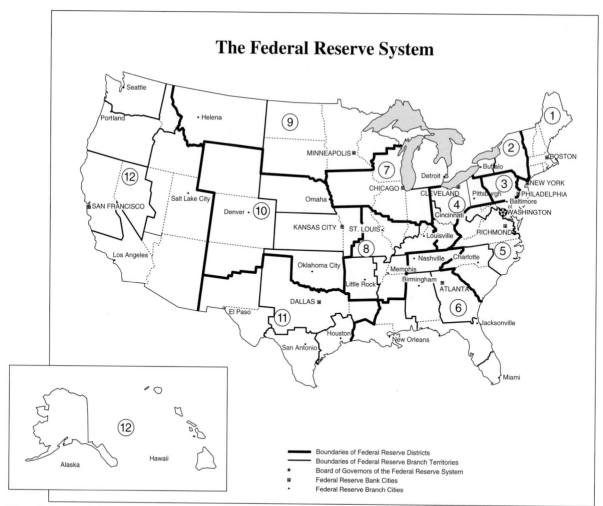

Figure 3-3 **The 12 Fed Districts in the United States**

increases the amount of funds held by the dealer banks and enables them to make additional loans and investments. When the additional loans and investments are made, the supply of money in circulation increases, thus accomplishing the Fed's objective.

The exact opposite occurs when the Fed wishes to decrease the money supply. The Fed calls its traders at the Federal Reserve Bank of New York and instructs them to sell government securities on the open market. The traders contact government securities dealers around the country and sell the required amount of securities to them. When the dealers receive their securities, their accounts are debited and the amount of funds held by these banks decreases. The amount of loans and investments, then, that these banks can support also decreases. Some maturing loans are not renewed and some marketable security investments are not replaced because of the loss of funds. The result is a decrease in the supply of money in circulation.

Why, you might ask, would the Fed want to increase or decrease the money supply? The answer is simple: to influence economic activity. When the members of the

FOMC feel that the economy is growing too slowly, the Fed increases the money supply, thus increasing liquidity in the economy and stimulating growth. When the economy is growing too fast and inflation seems imminent, the Fed decreases the money supply (or slows down its growth). This causes the economy to "cool off" because liquidity has decreased.

Although the government securities and currency markets are very large and efficient, the Fed is like a large elephant: People notice when it enters the market. It buys and sells in huge amounts; so when the Fed buys or sells securities, its actions can't help but affect prices (and interest rates) across the whole market.

The Discount Window

The 12 district Federal Reserve Banks lend money to financial institutions at the *discount window*. Originally, only member banks in a Federal Reserve Bank's district came to the discount window for loans. Since 1980, however, the district Federal Reserve banks have extended loans to nonmember banks and to nonbank financial institutions, too.

The district Federal Reserve banks also provide clearing services—collecting and paying for checks written on and deposited in banks. The Fed charges fees for the services it provides. The fees collected, interest earned on government securities held, and other sources of income provide the funds the Fed needs to operate. The Fed does not require appropriations from Congress. In fact, if excess profits are left over, as is usually the case, they are turned over to the U.S. Treasury. There are not many federal government entities that turn money over *to* the Treasury.

Savings and Loan Associations

Like commercial banks, **savings and loan associations (S&Ls)** are in business to take in deposits and lend money, primarily in the form of mortgage loans. *Mortgage loans are loans that are secured by real property such as real estate.* If a borrower defaults on a mortgage loan, the lender can take legal possession of the property. The property can then be sold and the lender keeps the proceeds from the sale up to the amount owed. S&Ls make a profit by charging a higher interest rate on the money they lend than the rate paid on deposits they take in.

Like banks, S&Ls can borrow from the Federal Reserve and from other financial institutions. S&Ls can also borrow from one of the 12 Federal Home Loan banks to meet some of their funding needs. The Office of Thrift Supervision is the primary regulator of federally chartered S&Ls.

Legislation Affecting S&Ls

The Depository Institutions Deregulation and Monetary Control Act of 1980 (MCA)[1] took the cap off interest rates that S&Ls may pay to depositors. It also brought S&Ls under the control of the Fed with regard to reserve requirements. The

---

[1]This mouthful is also referred to simply as the Monetary Control Act (MCA).

MCA authorized S&Ls to raise funds from new sources such as negotiable orders of withdrawal (NOW accounts).

The 1982 Garn-St. Germain Act authorized S&Ls to offer money market accounts. It also expanded the lending and investing powers of S&Ls. The S&L crisis in the latter years of the 1980s demonstrated that these new powers were not always exercised wisely. Probably the most widely known S&L failure was that of Charles Keating's Lincoln Savings and Loan of California. That one failure cost the federal government about $2 billion.

In 1989 Congress passed the Financial Institutions Reform, Recovery, and Enforcement Act (FIRRE Act) to clean up the mess made when hundreds of S&Ls failed because of bad loans, bad investments, a recession, and fraud. The FIRRE Act created the Resolution Trust Corporation (RTC) to preserve the remaining value of failed S&Ls, liquidate those that were hopelessly insolvent, and sell some failed S&Ls to other institutions where possible. The RTC often purchased poor-quality assets itself to allow a healthy institution to merge with a failing S&L.

It is often said that Congress, having created the RTC, bailed out the failed S&Ls. This is not true. The owners of these failed S&Ls lost most or all their investments. It was the *depositors* who were protected. If Congress had not authorized the $200 billion or so needed to keep S&L depositors from taking losses, thousands of individuals, including many of modest means, would have lost substantial savings.

Regulation of S&Ls

Like commercial banks, savings and loan associations must apply for either a federal or a state charter that authorizes them to operate. All federally chartered S&Ls are regulated by the Office of Thrift Supervision (OTS), and almost all S&Ls have their deposits insured by the Savings Association Insurance Fund (SAIF), which is part of the FDIC.

Savings and loan associations also have to keep reserves based on their size and the amount and type of deposit. The OTS dictates the reserve requirement for S&Ls insured by the FDIC–SAIF.

Mutual Companies versus Stockholder-Owned Companies

Some savings and loan associations are owned by stockholders, just as commercial banks and other corporations are owned by their stockholders. Other S&Ls, called **mutuals,** are owned by their depositors. In other words, when a person deposits money in an account at a mutual S&L, that person becomes a part owner of the firm. The mutual S&L's profits (if any) are put into a special reserve account from which dividends are paid from time to time to the owner/depositors.

On the one hand, mutual S&L owner/depositors do not face as much risk as regular stockholder owners: If the mutual S&L loses money, the loss isn't taken out of the owner/depositors' accounts. (Regular stockholder owners, of course, may well see the value of their holdings decline in bad times.) On the other hand, mutual S&L owner/depositors do not enjoy as much reward potential as regular stockholder owners. For example, unlike regular stockholders, who might be able to sell their stock for a profit, mutual S&L owner/depositors can't sell their deposits to other investors at all.

The Problem of Matching Loan and Deposit Maturities

Most of the mortgage loans made by S&Ls have very long maturities (the 30-year mortgage is most common, although 15-year mortgages are becoming increasingly popular). However, most of the deposits that provide the money for these loans have zero or short maturities (passbook savings accounts have zero maturity because the depositor can withdraw at any time; CDs come in maturities of up to five years). This gap between the 15- to 30-year maturity of the S&Ls' major assets and the zero- to five-year maturity of their deposits creates a problem if market interest rates rise. Consider the following example.

Suppose an S&L wanted to make a 30-year, fixed-rate mortgage loan for $100,000 at 7 percent interest. To raise cash for the loan, the S&L sells a one-year $100,000 CD at 3 percent interest. This creates a favorable spread (7% − 3% = +4%) as long as interest rates stay where they are. Table 3-2 shows the S&L's profit during the first year of the loan.

At the end of the first year, the CD matures and the S&L must pay the CD holder $100,000 plus 3 percent interest ($3,000). So the S&L sells another one-year CD for $100,000, giving the proceeds to the first CD holder. Then it uses $3,000 of its interest income from the loan to pay the interest due on the first CD. At the end of the second year and thereafter, the cycle repeats itself with the S&L selling a new one-year CD each year and using the profits from the loan to pay the interest due on the old CDs. You can see that as long as each new CD is issued for 3 percent interest, the S&L will net a yearly profit of $4,000 ($7,000 income from the loan minus $3,000 paid to the CD holder).

What happens, however, if interest rates rise during the first year, such that at the end of the year the S&L must pay 9 percent to get anyone to buy a new one-year CD? Now the S&L is in trouble. It has to sell a new CD to pay the $100,000 owed to the holder of the first CD, but it can only do so by offering an interest rate two points higher than its mortgage loan is paying. So at the end of the second year, when the S&L must pay the interest to the CD holder, it must pay $9,000 instead of $3,000 and suffers a $2,000 loss for the year. Table 3-3 summarizes the situation.

Of course, market interest rates can go down too, creating extra profits for the S&L, but S&Ls face a risk of loss when market interest rates move against them.

TABLE 3-2 First-Year Profit for an S&L with a 7% Loan Financed by a 3% CD

Interest received from the loan:	$100,000 × .07 = $7,000
Interest paid out to the CD holder:	$100,000 × .03 = $3,000
	Net income: $4,000

Note: For simplicity in this example, we assume the loan's terms allow the borrower to make only interest payments each year, deferring payment of the principal until the end of the loan's term.

TABLE 3-3 Second-Year Loss for an S&L with a 7% Loan Financed by a 9% CD

Interest received from the loan:	$100,000 × .07 = $7,000
Interest paid out to the CD holder:	$100,000 × .09 = $9,000
	Net income: ($2,000)

We discussed the agency problem on pages 8–10.

S&Ls' Real Assets

S&Ls also own buildings and equipment that are needed to conduct business. These assets, which do not earn an explicit rate of return, are supposed to be kept to a low level subject to the needs of the institution. As the fraud of the 1980s showed, however, that has not always been the case. Many S&L executives spent much of their companies' money on private business jets, luxurious offices, and even vacation retreats—a clear example of the agency problem discussed in chapter 1.

CREDIT UNIONS

Credit unions are member-owned financial institutions. They pay interest on shares bought by, and collect interest on loans made to, the members. Members are individuals rather than businesses or government units.

Credit unions are able to make relatively low interest loans to their members because they are cooperative organizations. They don't have to charge extra to make a profit and they don't pay federal income taxes. Also, they make loans *only* to members, who are presumed to be somewhat better credit risks than the general population.

The Common Bond Requirement

To help ensure that credit union members actually are better credit risks than the general population, credit union members must have a *common bond* with one another. This could mean that all members work for the same company, belong to the same labor union, or perhaps just live in the same town as the other members. The theory is that people who all belong to the same group, sharing common values, will be less likely to default on loans supported by money from their fellow group members.

Members as Shareholders

Credit unions are owned by their members. So when credit union members put money in their credit union, they are not technically "depositing" the money. Instead, they are purchasing *shares* of the credit union.

Like owners of other businesses, credit union members are entitled to any income the credit union has after debts and expenses have been paid. This residual income may be distributed in the form of extra dividends paid on the members' shares, or by a rebate on interest paid on loans.

Credit Unions Compared to Banks

Traditionally, credit unions were small institutions that did not compete much with banks. However, they have grown rapidly in recent years and now provide most of the same services as do commercial banks. Because banks now see credit unions as more of a threat, the banking lobby is pressuring Congress to treat credit unions more like banks, including the way they are taxed.

Credit Union Regulation

Credit unions must have charters giving them authority to operate, just like banks and S&Ls. They obtain these charters either from the state where they are located or from the federal government. The federal chartering and regulatory body for credit unions is the National Credit Union Administration (NCUA). The NCUA also oversees the National Credit Union Share Insurance Fund (NCUSIF). This fund insures share accounts up to $100,000. All federally chartered credit unions have NCUSIF insurance. State-chartered credit unions may also apply for NCUSIF insurance for the share accounts of their members.

If credit unions have emergency borrowing needs, they can turn to the Central Liquidity Facility (CLF), which was created by Congress in 1978 and is administered by the National Credit Union Administration. Credit unions can also turn to the Federal Reserve Bank in the district where the credit union is located.

Finance Companies, Insurance Companies, and Pension Funds

Finance companies are nonbank firms that make short-term and medium-term loans to consumers and businesses. They often serve those customers who don't qualify for loans at other financial institutions.

Like banks and S&Ls, finance companies operate by taking in money and lending it out to their customers at a higher interest rate. A major difference between finance companies and other financial institutions, however, lies in the source of finance company funds. Banks and S&Ls receive most of their funds from individuals and businesses that deposit money in accounts at the institutions. Finance companies generally get their funds by borrowing from banks, or by selling commercial paper.

Types of Finance Companies

There are three main types of finance companies: consumer, commercial, and sales. In the following sections, we will explain the characteristics and functions of each type.

Consumer Finance Companies   Consumer finance companies, sometimes known as small-loan companies, make small loans to consumers for car purchases, recreational vehicles, medical expenses, vacations, and the like. Consumer finance companies often make loans to customers with less than perfect credit. Because the customers are a higher risk, the interest rates charged on loans are usually a little higher to compensate for the greater risk.

Commercial Finance Companies   These firms concentrate on providing credit to other business firms. A special type of commercial finance company is called a *factor*. Factoring is the buying of a business firm's accounts receivable, thus supplying needed funds to the selling firm. Commercial finance companies also make loans to businesses, usually with accounts receivable or inventory pledged as collateral. This type of financing will be examined in detail in chapter 20.

Sales Finance Companies   The mission of sales finance companies is to help the sales of some corporation (indeed, many are subsidiaries of the corporation whose sales they are promoting). In the automotive industry, for example, customers are more likely to buy cars from dealers that offer financing on the spot than from dealers who have no financing programs.

A finance company generally gives its retail dealers a supply of loan contract forms, which the dealers fill out at the time of sale. The contract is immediately sold to the finance company (at a slightly reduced price, of course, to allow for the finance company's profit). General Motors Acceptance Corporation (GMAC) and the Ford Motor Credit Company are prominent examples of sales finance companies.

Insurance Companies

Insurance companies are firms that, for a fee, will assume risks for their customers. They collect fees, called premiums, from a large number of customers. Then they draw on the pool of funds collected to pay those customers who suffer damages from the perils they have insured against.

There are two main types of insurance companies: life insurance companies, and property and casualty insurance companies.

Life Insurance Companies *Life insurance companies* sell policies that pay the beneficiaries of the insured when the insured person dies. You might ask how life insurance companies make any money because *everybody* dies sooner or later. If the risk life insurance companies were taking in return for the premium received were the risk of their customers dying, it is true that none of them would make any money. The real risk they are taking, however, is that the insured person will die *sooner than expected.*

To help assess this risk, insurance companies employ **actuaries.** Actuaries help calculate the premium for a life insurance policy for a person of a given age, gender, and state of health, so that the insurance company can pay the insurance benefit, cover expenses, and make a profit. Actuaries cannot tell specifically *who* is going to die when; but they can predict, with a high degree of accuracy, *how many* in a group of 100,000 healthy 40-year-old males will die during the coming year.

Life insurance companies function as financial intermediaries essentially the same way commercial banks or savings and loan associations do. They take money in from surplus economic units in the form of policy premiums and channel it to deficit economic units in the form of investments in common stock, corporate bonds, mortgages, and real estate. Their payout can be predicted with a high degree of accuracy, so they need only a small amount of liquid assets.

Property and Casualty Insurance Companies *Property and casualty insurance companies* insure against a wide range of hazards associated with person and property. These include theft, weather damage from hurricanes, tornadoes and floods, fire, and earthquakes. Two close relatives of property and casualty companies are health insurance companies, which insure people against injuries and illnesses, and disability insurance companies, which insure people against loss of income from being unable to work.

One special hazard that these companies insure against is a policyholder's own negligence. This kind of insurance is called **liability insurance.** Most people are familiar with automobile liability insurance, but liability coverage can also be purchased for such things as medical malpractice, dog bites, and falls by visitors on your property.

The risks protected against by property and casualty companies are much less predictable than are the risks insured by life insurance companies. Hurricanes, fires, floods, and trial judgments are all much more difficult to predict than the number of 60-year-old females who will die this year among a large number in this risk class. This means that property and casualty insurance companies must keep more liquid assets than do life insurance companies.

Pension funds Pension funds are set up by companies, governments, and unions to pay retirement benefits for their employees. They are essentially savings plans. Employees generally contribute money to the funds now in order to draw it out later, on retirement. Employers usually contribute money on behalf of the employees, too. All the money is pooled and invested, and the investment returns are added to the pot. It is always possible, of course, that the sponsor (the company, government, or union) will not be able to pay promised benefits. If this happens the pension fund is said to have failed, and the worker may not collect all the promised benefits.

Pension funds invest so much money that they are the country's greatest source of long-term capital. They have *trillions* of dollars invested in a wide range of securities and other assets, such as real estate. Pension fund officials often hire money management firms just to manage the fund's investments. Bank trust departments and insurance companies also manage pension fund money.

Pension funds generally use one of two types of procedures for determining benefits for retired workers: a defined benefit plan and a defined contribution plan. In a defined benefit plan, retirement benefits are determined by a formula that usually considers the worker's age, salary, and years of service. The employee and/or the firm contribute the amounts necessary to reach the goal. In a defined contribution plan, the contributions to be made by the employee and/or employer are spelled out, but retirement benefits depend on the total accumulation in the individual's account at the retirement date.

Annuities An *annuity* is a series of equal payments that are made at regular time intervals, such as monthly, for a specified period of time. Pension fund benefits are often paid out in the form of annuities. Sometimes the sponsor of a pension fund will use the funds accumulated during the retiring person's working years to purchase an annuity from an insurance company. This provides the retired person's benefits. Insurance companies also sell annuities to investors. In return for the amount paid to the insurance company, the investor receives payments (usually monthly) for the remainder of his or her life. A person who receives annuity payments is called an **annuitant.** The size of the payments depends on how much money is paid to the insurance company at the time of the employee's retirement, along with factors such as the age, gender (if allowed by law), and state of health of the annuitant. If the pension fund investments made on behalf of a given employee earned a high return, a large amount of money will be available to purchase a large annuity. If the defined contribution pension fund investments performed poorly, the retired employee will be able to purchase only a small annuity.

Sometimes the investments made on an employee's behalf will be paid out in a lump sum at retirement. It is then up to the retired employee to invest this money wisely so as to generate the needed income during retirement.

Financial Management and You

The Social Security System

Most of us are participants in the Social Security System. It is the largest pension vehicle in the United States. It is not a pension fund, but a system in which those contributing money today are paying the retirement and disability benefits for those collecting today. Although in recent years the fund has been running a surplus, that will not be the case for much longer.

With a pension fund, you put in money that is set aside to pay future claims. With the Social Security System, in essence you put in babies who will become taxpayers to provide the needed funds for beneficiaries having claims in the future. When the older generation collecting benefits is very large and the younger generation paying the taxes funding those benefits is very small, some difficult choices have to be made. Taxes on the young have to be increased, or benefits to the old have to be decreased. In 1960 there were 5.1 contributors for each beneficiary. In 1998 this ratio was 2.8 to 1, and the value of this ratio will continue to decline as the "baby-boomer" generation retires and stops paying into the Social Security System, and begins collecting from it.

What does this suggest to you about your own financial planning? Polls show that more young adults believe in flying saucers than believe that Social Security will be there when they retire. Other polls show that young people are beginning to save more money when they start working than their parents' generation did at the same age.

What's Next

In this chapter we have seen how financial institutions help to bring together suppliers and users of funds to the benefit of both and to the economy. Commercial banks, savings and loan associations, credit unions, finance companies, insurance companies, and pension funds do this in different ways, and have different constituents, but all assist in this efficient flow of funds.

In the following two chapters we will review financial statements and learn how to analyze them from the perspective of a financial manager.

Summary

1. ***Explain financial intermediation and the role of financial institutions.***

Financial institutions act as intermediaries between surplus and deficit economic units. They coordinate the flow of funds, absorbing differences between the amount of funds offered and needed, between the length of time funds are offered and needed, and between the degree of risk that surplus economic units are willing to bear and the risk that is inherent in the securities offered by the deficit economic units.

2. ***Define commercial banks and explain how reserve requirements influence their operations.***

Commercial banks are financial institutions that are owned by stockholders, that take in deposits, and that make loans (primarily to businesses). Reserve requirements force banks to maintain minimum levels of reserves (vault cash and deposits at the Fed) based on the level and type of deposits they have. Reserves are nonearning assets, so, although they provide liquidity for the bank, they limit its ability to make a profit.

3. Describe how the Federal Reserve regulates financial institutions.

The Federal Reserve is the central bank of the United States. It has seven members on its Board of Governors and a 12 voting-member Federal Open Market Committee. The 12 district Federal Reserve Banks make loans to financial institutions and perform other functions to assist member banks and other financial institutions. The Federal Reserve sets reserve requirements and influences the money supply through its open-market purchases and sales of government securities and foreign currencies. The Fed regulates financial institutions and uses its powers to try to maintain stability in the financial system.

4. Explain how savings and loan associations differ from commercial banks.

Savings and loan associations (S&Ls) are financial institutions that take in deposits and make primarily mortgage loans. The Office of Thrift Supervision is the primary authority for overseeing S&Ls. S&Ls primarily make mortgage loans (to consumers). Banks primarily make commercial loans to businesses.

5. Describe how credit unions operate.

Credit unions are financial institutions that take in funds by selling shares to members and make loans to those members. People are eligible for membership in a credit union if they meet the requirement of having a common bond with the other members. This might be working for a given company, belonging to a certain union, or living in a specified area.

6. Compare and contrast finance companies, insurance companies, and pension funds.

Finance companies take in funds primarily by selling commercial paper and make personal loans. Insurance companies sell policies, collecting premiums and paying beneficiaries if the insured-against event the insurance covers occurs. Pension funds take in funds, usually contributed by both the employer and employee, and invest those funds for future payment to the worker when he or she retires. This retirement benefit may be determined by a formula (a defined benefit plan), or by how much is in the investment fund at the time of retirement (a defined contribution plan).

You can access your FREE PHLIP/CW On–Line Study Guide, Current Events, and Spreadsheet Problem templates, which may correspond to this chapter either through the Prentice Hall Finance Center CD–ROM or by directly going to: **http://www.prenhall.com/gallagher.**

PHLIP/CW — *The Prentice Hall Learning on the Internet Partnership–Companion Web Site*

Self-Test

ST-1. Why is intermediation sometimes needed to bring together surplus and deficit economic units?

ST-2. Is it better to be a surplus economic unit or a deficit economic unit? Explain.

ST-3. Define secondary reserves that are held by a bank.

ST-4. What is the difference, if any, between the way commercial banks and credit unions are taxed?

ST-5. What is the common bond requirement that credit union members must have to be eligible for membership?

ST-6. What is a Federal Reserve discount window loan?

ST-7. What are Federal Reserve open-market operations?

Review Questions

1. Define *intermediation*.
2. What can a financial institution often do for a surplus economic unit (SEU) that the SEU would have difficulty doing for itself if the SEU were to deal directly with a deficit economic unit (DEU)?
3. What can a financial institution often do for a deficit economic unit (DEU) that the DEU would have difficulty doing for itself if the DEU were to deal directly with an SEU?
4. What are a bank's *primary reserves*? When the Fed sets reserve requirements, what is its primary goal?
5. Compare and contrast mutual and stockholder-owned savings and loan associations.
6. Who owns a credit union? Explain.
7. Which type of insurance company generally takes on the greater risks: a life insurance company or a property and casualty insurance company?
8. Compare and contrast a *defined benefit* and a *defined contribution* pension plan.

Build Your Communication Skills

CS-1. Women live longer than men, on average, but some insurance regulators are forcing insurance companies to ignore this fact when setting rates. Do you think it is ethical to charge women and men, who are otherwise similar in age and other risk factors, different amounts for life insurance? Have two groups of students debate this issue.

CS-2. Read three articles about the Federal Reserve's current monetary policy. Use sources such as *The Wall Street Journal, Fortune* magazine, the *Federal Reserve Bulletin,* or sources available on the Internet or CD databases. Write a brief report (two or three pages) summarizing the Fed's current monetary policy. What issues seem to be influencing the Fed's actions the most? What actions are being taken by the Fed to achieve the goals it has defined for itself?

Answers to Self-Test

ST-1. Intermediation is sometimes needed when surplus and deficit economic units cannot agree as to the denomination, maturity, and risk of the security offered and bought. Financial intermediaries can often give each side what it needs by stepping into the middle of the exchange of funds for securities.

ST-2. There is nothing inherently good or bad in the classification of either surplus economic unit or deficit economic unit.

ST-3. Secondary reserves are short-term liquid securities that a bank can sell quickly and easily to obtain cash that can be used to satisfy primary reserve requirements.

ST-4. Commercial banks pay federal income taxes on their profits whereas credit unions do not.

ST-5. Credit union members are required to have some common bond with the other members before the request for membership is approved. Examples of common bonds required by various credit unions include working for a given company, belonging to a certain union, or living in a certain area.

ST-6. A Federal Reserve discount window loan is a loan made by one of the 12 district Federal Reserve banks to a financial institution from that district.

ST-7. Federal Reserve open-market operations are the purchasing and selling of U.S. government securities and foreign currencies by the Federal Reserve. This is done to affect the amount of reserves in the banking system.

PART 2: ESSENTIAL CONCEPTS IN FINANCE

CHAPTER 4

REVIEW OF ACCOUNTING

Anyone who isn't confused here doesn't really understand what is going on.
—Anonymous

DANGEROUS GAMES: DID "CHAINSAW AL" DUNLAP MANUFACTURE SUNBEAM'S EARNINGS IN 1997?

Albert Dunlap likes to tell how confidants warned him in 1996 that taking the top job at the small-appliance maker Sunbeam Corp. would likely be his Vietnam. For a time, the 60-year-old West Point graduate seemingly proved the Cassandras wrong. As the poster boy of nineties-style corporate cost-cutting, he delivered exactly the huge body counts and punishing airstrikes that Wall Street loved. He dumped half of Sunbeam's 12,000 employees by either laying them off or selling the operations where they worked. In all, he shuttered or sold about 80 of Sunbeam's 114 plants, offices, and warehouses.

Sunbeam's sales and earnings responded, and so did its stock price, rising from $12.50 a share the day Dunlap took over in July 1996 to a high of 53 in early March 1998.

But in May 1998 Sunbeam suffered a reversal of fortune that was as sudden and traumatic for Dunlap as the Viet Cong's Tet offensive was to U.S. forces in 1968. After several mild warnings of a possible revenue disappointment, Sunbeam shocked Wall Street by reporting a loss of $44.6 million for the first quarter on a

CHAPTER OBJECTIVES

After reading this chapter, you should be able to:

- Explain how financial managers use the three basic accounting financial statements: the income statement, the balance sheet, and the statement of cash flows.

- Discuss how depreciation affects cash flow and compute depreciation expense.

- Explain how taxes affect a firm's value and calculate marginal and average tax rates.

For Al Dunlap, Sunbeam may be the last hurrah and an unpleasant one at that.

sales decline of 3.6 percent. In a trice, the Sunbeam cost-cutting story was dead, along with "Chainsaw Al" Dunlap's image as the supreme maximizer of shareholder value. By June Sunbeam stock had fallen more than 50 percent from its peak to 22.

We can't say we are surprised by Sunbeam's current woes. In a cover story in 1997 entitled "Careful, Al" (June 16), we cast a skeptical eye at Dunlap's growth objectives in the low-margin, cutthroat small-appliance industry. We also pointed out the yawning gap between Sunbeam's performance claims and reality. We took special note of Sunbeam's accounting gimmickry, which appeared to have transmogrified through accounting wizardry the company's monster 1996 restructuring charge ($337 million before taxes) into 1997's eye-popping sales and earnings rebound. But to no avail. Wall Street remained impressed by Sunbeam's earnings, and the stock continued to rise from a price of 37 at the time of the story.

Sunbeam's financials under Dunlap looked like an exercise in high-energy physics, in which time and space seemed to fuse and bend. They were a veritable cloud chamber. Income and costs moved almost imperceptibly back and forth between the income statement and balance sheet like charged ions, whose vapor trail had long since dissipated by the end of any quarter, when results were reported. There were also some signs of other accounting shenanigans and puffery, including sales and related profits booked in periods before the goods were actually shipped or payment received. Booking sales and earnings in advance can comply with accounting regulations under certain strict circumstances.

Yet, sad to say, the earnings from Sunbeam's supposed breakthrough year appear to be largely manufactured. That, at least, is our conclusion after close perusal of the company's recently released 10-K, with a little help from some people close to the company.

The already-ailing company now has to struggle under $2 billion of additional debt and a negative tangible net worth of $800 million. And Al Dunlap's enemies, including disenchanted shareholders, angry securities analysts, and bitter former employees, are growing in number and circling ever closer to the company's headquarters in Delray Beach. Of course, Dunlap could always escape by using the building's flat roof to chopper out, should it come to that. One can only hope he'll remember to take the American flag with him.

Source: Jonathan R. Lang, Barron's Online, June 8, 1998. Reprinted by permission of *Barron's*, © 1998 Dow Jones & Company, Inc. All rights reserved worldwide.

Chapter Overview

Accounting plays an important role in a firm's financial success. Accountants prepare financial statements that financial managers use to analyze the condition of a firm and to plan for its future. Financial managers must understand, then, how to analyze and interpret financial statements as they make decisions. The financial manager who knows how to use financial statements can help create value for the firm's owners.

In this chapter we will review the three major financial statements: the income statement, the balance sheet, and the statement of cash flows. We will also study how depreciation and taxes affect a firm's cash flows.

Review of Accounting Fundamentals

All public corporations in the United States must follow certain accounting guidelines known as Generally Accepted Accounting Principles (GAAP), which require that public corporations prepare financial statements that comply with GAAP rules. The Financial Accounting Standards Board (abbreviated FASB and pronounced Fahz-bee), a private, professional accounting body, publishes these rules governing how public corporations must account for their business activities.

The Securities and Exchange Commission (SEC) requires all public corporations to file financial statements, and make them available to the public, on 10-K and 10-Q reports. The **10-K reports** are audited financial statements submitted annually to the SEC for distribution to the public. The **10-Q reports** are unaudited financial statements submitted quarterly, also for public distribution.

The following basic accounting equation is central to understanding the financial condition of a firm:

$$\text{Assets} = \text{Liabilities} + \text{Equity}$$

Assets are the items of value a business owns. *Liabilities* are claims on the business by nonowners, and equity is the owners' claim on the business. The sum of the liabilities and equity is the total **capital** contributed to the business. Capital contributions come from three main sources: creditors (including bondholders and banks), preferred stockholders, and common stockholders.

Basic Accounting Financial Statements

You can get a good picture of how a firm is doing by looking at its financial statements. The three basic financial statements are the *income statement*, the *balance sheet*, and the *statement of cash flows*. Each of these statements gives a slightly different view of the firm. Let's look at these financial statements and how they interrelate.

The Income Statement

We can compare the **income statement** to a video: It measures a firm's profitability over a period of time. The firm can choose the length of the reporting time period. It can be a month, a quarter, or a year. (By law, a publicly traded corporation must report its activities at least quarterly but may report more frequently.)

The income statement shows *revenues, expenses,* and *income.* Revenues represent gross income the firm earned during a particular period of time (usually from sales). Expenses represent the cost of providing goods and services during a given period of time. Net income is what is left after expenses are subtracted from revenues.

Figure 4-1 shows an income statement for Acme Corporation, a firm that manufactures birdseed, anvils, rockets, explosives, and giant springs. Acme Corporation is primarily a mail-order company with many customers in the southwestern United States. The income statement is for the year ended December 31, 1999. This income statement describes sales, expenses, and net income for Acme Company from the beginning of the business day on January 1, 1999, until the end of the business day on December 31, 1999.

Revenues As Figure 4-1 shows, Acme's sales totaled $15 million during 1999. Generally, the income statement does not distinguish between cash and credit sales. As a result, we are not sure how much actual *cash* came into the firm from the $15 million in reported sales.

Expenses Expenses include costs incurred while conducting the operations of the firm and financial expenses, such as interest on debt and taxes owed. These items are matched to the revenues that were generated as the expenses were incurred.

Cost of Goods Sold. The first expense subtracted from sales is *cost of goods sold*, which consists of the labor, materials, and overhead expenses allocated to those goods and services sold during the year.

Subtracting cost of goods sold of $5 million from sales of $15 million gives Acme's *gross profit*, which equals $10 million.

Selling and Administrative Expenses. From gross profit, we next subtract Acme's *selling and administrative expenses* ($800,000). Selling expenses include marketing and salespeoples' salaries. Administrative expenses are for expenses that are difficult to associate directly with sales for a specified time period. These would include office support, insurance, and security.

> **Take Note**
> *Notice that* cost of goods sold *is not called* cost of goods produced. *Only cost of goods sold is reported on this year's income statement. Goods produced but not sold are considered inventory.*

Figure 4-1 Acme Corporation Income Statement for the Year Ended December 31, 1999

Net Sales	$15,000,000
Cost of Goods Sold	5,000,000
Gross Profit	10,000,000
Selling and Administrative Expenses	800,000
Depreciation	2,000,000
Operating Income (EBIT)	7,200,000
Interest Expense	1,710,000
Earnings Before Taxes (EBT)	5,490,000
Income Taxes (40%)	2,196,000
Net Income (NI)	$ 3,294,000
Preferred Dividends	$ 110,000
Earnings Available to Common Stockholders	$ 3,184,000
Earnings per Share (EPS) (3 million shares)	$1.06
Common Dividends Paid	$400,000
Addition to Retained Earnings	$ 2,784,000

Depreciation Expense. *Depreciation expense* is subtracted next—$2 million for Acme in 1999. Depreciation expense is the year's allocation of the cost of plant and equipment that have been purchased this year and in previous years. Because assets provide their benefits to the firm over several years, accountants subtract the cost of long-lived assets a little at a time over a number of years. The allocated cost of a firm's assets for the income statement's period of time is the depreciation expense.[1]

Operating Income and Interest Expense. When we subtract selling and administrative expenses and depreciation expense from gross profit, we are left with Acme's *earnings before interest and taxes (EBIT),* also known as operating income. This figure is $7,200,000.

Gross Profit	=10,000,000
Selling and Administrative Expenses	− 800,000
Depreciation Expense	− 2,000,000
Earnings Before Interest and Taxes (EBIT, or Operating Income)	=7,200,000

EBIT is the profit that the firm receives from its business operations before subtracting any financing expenses. From EBIT, we then subtract *interest expense* associated with any debts of the company to arrive at Acme's *earnings before taxes (EBT).* Acme has $1,710,000 in interest expense. When we subtract this figure from the $7,200,000 EBIT figure, we find Acme had earnings before taxes (EBT) of $5,490,000.

Financial Management and You

A Special Earnings Category: EBITDA

Financial analysts often make use of another measure of a company's earnings called earnings before interest, taxes, depreciation, and amortization, or EBITDA (pronounced "ee-bid-dah"). EBITDA is found by adding depreciation expense and amortization expense back to EBIT. Because depreciation and amortization are noncash expenses, which will be discussed later in this chapter, the result of adding depreciation and amortization back into EBIT is a figure that represents revenues minus cash expenses, or approximately the amount of cash earned by the daily operations of a business.

Although EBITDA is of great interest to financial analysts, it is not required to be reported by the Financial Accounting Standards Board and is, thus, not usually shown as a specific line item on most income statements. As a result, it must usually be calculated manually. Acme's EBITDA for 1999 is:

Operating Income (EBIT)	$7,200,000
+ Depreciation and Amortization	2,000,000
= EBITDA	$9,000,000

Having made this calculation, a financial analyst would proceed with the knowledge that Acme's normal business operations threw off approximately $9 million in cash during 1999.

---

[1] We will discuss the role of depreciation, and depreciation rules, later in this chapter.

Ethical Connections

Box Office Hits—Who Profits?

Imagine you wrote a best-selling book that was made into a hit movie that grossed over $657 million. If you were Winston Groom, author of *Forrest Gump,* you would be licking your chops in anticipation of a big check. But it wasn't to be.

In fact, according to Hollywood accounting practices, the movie has yet to show a net profit. Paramount Pictures' accountants claim that the movie has lost $62 million. These types of accounting practices would even make a country boy like Forrest a bit suspicious.

Warner Pictures, a division of Time Warner, Inc., also claims that it does not owe any money to the estate of Jim Garrison, author of the book *On the Trail of the Assassin.* Garrison's book was the basis for Oliver Stone's 1991 movie *J.F.K.* That movie grossed more than $150 million. But again, the studio claims *J.F.K.* did not make a net profit.

Many experts have long cast suspicious glances at Hollywood accounting practices, which usually ensure that even the most successful movies don't report a net profit. Net profits in Hollywood are typically defined as the studio's share of the receipts of the film after subtracting production costs, distribution fees, prints, advertising, overhead, and promotion costs. Because a studio can take anywhere from 30 percent to 60 percent in distribution fees alone, it is a well-known maxim in Hollywood that a share of net profits is often worthless.

In contrast to authors, actors and directors can walk away from a successful film as millionaires. For instance, *Forrest Gump* star Tom Hanks and its director, Robert Zemeckis, earned over $30 million apiece from the movie because they received a percentage of the film company's share of the box office receipts.

When all is said and done, movie accounting is like a box of chocolates. When you open it, you never know what you might find.

Questions to Consider

▶ Is it ethical for actors to get more money than the author of the book on which a movie is based?

▶ Is it ethical for any business to cook the books, so that expenses outweigh the revenue? Or is this just a smart way to avoid paying taxes?

▶ Should a firm have to disclose how its accounting practices may affect the people it hires to provide services? Could disclosure of its practices conflict with the primary goal of a firm? Explain.

Sources: "Jim Garrison's Estate Sues Film Studios on Accounting," *The New York Times* (November 20, 1995): D2; John Lippman, "Author of 'Gump,' Paramount in Talks Over Net Profits," *The Wall Street Journal* (Thursday, May 25, 1995): B6.

All the expenses subtracted so far are tax deductible. In other words, the Internal Revenue Service (IRS) allows firms to subtract these expenses from their gross income before computing the tax they owe. We will discuss income taxes later in the chapter.

Net Income Finally, after we subtract all operating expenses, financing expenses, and taxes from revenues, we arrive at the firm's *net income* (NI). For Acme, net income in 1999 was $3,294,000. This is the firm's accounting net profit for the year.

Preferred Stock Dividends. After the net income entry, the income statement shows dividends the corporation paid to its preferred stockholders in 1999. Acme's preferred stock dividends amount to $110,000. After we subtract this amount from the $3,294,000 in net income, we're left with $3,184,000 for the common stockholders.

Earnings per Share (EPS). Acme's stockholders are very interested in their individual share of the corporation's earnings. Therefore, under the entry *earnings per* *share* (EPS), the income statement shows total earnings available to common stockholders divided by the number of shares of common stock outstanding. The earnings available to common stockholders figure comes straight from the income statement.

Net Income (NI)	=3,294,000
Preferred Stock Dividends Paid	− 110,000
Earnings Available to Common Stockholders	=3,184,000
Earnings per Share (EPS) (3 million shares)	= $1.06

The number of shares outstanding comes from the balance sheet. Acme has 3 million shares of common stock outstanding.

To calculate EPS, divide earnings available to common stockholders by the number of outstanding common stock shares. For Acme, we calculate EPS as follows:

$$\text{EPS} = \frac{\text{Earnings Available to Common Stockholders}}{\text{Number of Shares of Common Stock Outstanding}}$$

$$= \frac{\$3{,}184{,}000}{3{,}000{,}000}$$

$$= \$1.06$$

Acme's EPS, then, is $1.06, as shown on the income statement in Figure 4-1.

Common Stock Dividends and Retained Earnings A company has two options as to what to do with earnings available to common stockholders. It can pay stockholder dividends, or it can retain the earnings. Retaining the earnings will likely lead to greater future growth in sales and net income as new assets are purchased or existing liabilities are paid. In 1999 Acme has chosen to pay $400,000 (12.6 percent of its available earnings) in dividends to its common stockholders. The remaining $2,784,000 is an *addition to retained earnings* of the firm. As shown in Figure 4-1, this amount is to be added to the earnings retained from past years.

In addition to the income statement, many firms prepare a short **statement of retained earnings,** as shown in Figure 4-2, that records dividend and retained earnings information. Assuming that Acme's end of 1998 retained earnings were $7,216,000, Acme's accountants add the 1999 earnings available to common stockholders less the 1999 dividends paid to those stockholders ($3,184,000 − $400,000 = $2,784,000) to the end of 1998 retained earnings. The result is $10 million ($7,216,000 + $2,784,000 = $10,000,000).

If the amount remaining after paying dividends had been **negative,** *then we would have* **subtracted** *our number from retained earnings instead of adding to it. For Acme, there is a positive number to add to retained earnings—$2,784,000.*

Retained Earnings, December 31, 1998	$7,216,000
+ 1999 Earnings Available to Common Stockholders	+ 2,784,000
− 1999 Dividends Paid to Common Stockholders	− 400,000
Retained Earnings, December 31, 1999	$10,000,000

Figure 4-2 Acme Corporation Statement of Retained Earnings for the Year Ended December 31, 1999

The Balance Sheet

If the income statement is like a video, a balance sheet is like a still photograph. The **balance sheet** shows the firm's assets, liabilities, and equity at a given point in time. This snapshot of a company's financial position tells us nothing about the firm's financial position before or after that point in time. Let's examine the end-of-1999 balance sheet for Acme. Figure 4-3 shows the balance sheet for Acme as of the end of the business day, December 31, 1999.

On the balance sheet, the firm's assets are listed in order of their liquidity. As we learned in chapter 1, liquidity is the ease with which you can convert an asset to cash. This means that cash and near-cash assets called **current assets** are listed first. Assets that are difficult to convert to cash are listed later. On the other side of the balance sheet, the liabilities that are due earliest, **current liabilities,** are listed first. Current liabilities are almost always due within one year. The liabilities due later, such as long-term debt, are listed later on the balance sheet.

The equity section lists the claims of the owners (Acme's common stockholders). The owners' claims include both the amount the owners contributed when the common stock was first issued, and the total earnings retained by the firm at the time of the balance sheet.

Figure 4-3 Acme Corporation Balance Sheet for the Year Ended December 31, 1999

Assets:	
Cash	$10,000,000
Marketable Securities	8,000,000
Accounts Receivable	1,000,000
Inventory	10,000,000
Prepaid Expenses	1,000,000
Total Current Assets	$30,000,000
Fixed Assets, Gross	28,000,000
Less Accumulated Depreciation	(8,000,000)
Fixed Assets, Net	20,000,000
Total Assets	$50,000,000
Liabilities and Equity:	
Accounts Payable	$ 4,000,000
Notes Payable	3,000,000
Accrued Expenses	2,000,000
Total Current Liabilities	9,000,000
Long-Term Debt	15,000,000
Total Liabilities	24,000,000
Preferred Stock	1,000,000
Common Stock (3 million shares)	3,000,000
Capital in Excess of Par	12,000,000
Retained Earnings	10,000,000
Total Equity	26,000,000
Total Liabilities and Equity	$50,000,000

The Asset Accounts Acme has both current and fixed assets. Current assets provide short-term benefits, whereas fixed assets provide long-term benefits to the firm.

Current Assets. Acme has $10 million in *cash* at the end of 1999. *Marketable securities*—securities that can quickly and easily be converted to extra cash—are listed next. Acme has $8 million in these securities. Customers owe the company $1.0 million, the amount of *accounts receivable*.

The company has $10 million of *inventory* and $1 million in *prepaid expenses*. The inventory figure reflects the amount of goods produced, but not yet sold to customers. The prepaid expense figure represents future expenses that have been paid in advance. An example of a prepaid expense is the premium paid on an insurance policy. You pay the premium in advance, so the insurance coverage is "owed" to you until the term of coverage expires. Because prepaid expenses, such as insurance premiums, have been paid for but not yet received, they are owed to the company and are considered assets.

The sum of all current assets, including cash, marketable securities, net accounts receivable, inventory, and prepaid expenses, is often referred to as **working capital**. For Acme, this figure is $30 million.

Fixed Assets. Next to be listed are the fixed assets of the firm. **Fixed assets** are assets that are expected to provide a benefit to the firm for more than one year. These assets are generally less liquid than the current assets. Acme has $28 million of gross plant and equipment, which is listed at the original cost of these assets. The *accumulated depreciation* figure is the sum of all the depreciation expenses ever taken on the firm's income statements, for the assets still carried on the books. Acme's accumulated depreciation figure is $8 million. To find the net plant and equipment figure—or net fixed assets—subtract the amount of accumulated depreciation from gross equipment ($28 million minus $8 million). The result is $20 million.

The $30 million in current assets plus the $20 million in net fixed assets are the total assets of the firm. At the end of 1999, Acme's total assets were $50 million.

The Liabilities and Equity Accounts The liability and equity section of the balance sheet shows how the company's assets were financed. The funds come from those who have liability (debt) claims against the firm or from those who have equity (ownership) claims against the firm.

Liabilities. In the liability section of the balance sheet, current liabilities are listed first. Acme has accounts payable at the end of 1999 of $4 million. *Accounts payable* represent money a business owes to suppliers that have sold the firm materials on account.

Notes payable are $3 million for this company. Notes payable are legal IOUs that represent the debt of the borrower (Acme) and the claim the lender has against that borrower. Acme also has accrued expenses of $2 million. Accrued expenses are business expenses that have not been paid yet. For example, universities often make professors work for a full month before they are paid. The universities accrue wages payable for the month before the payroll checks are finally issued. Acme's accounts payable, notes payable, and accrued expenses add up to $9 million in current liabilities.

Net working capital is current assets minus current liabilities. For Acme, this would be $21 million ($30 million current assets − $9 million current liabilities).

Next, long-term liabilities are listed. Long-term liabilities are liabilities that are not due within one year. Acme has $15 million in long-term bonds payable that mature in the year 2015. The $15 million figure listed on the balance sheet refers only to the principal on these bonds.

Preferred Stock, Common Stock, and Retained Earnings. The equity section of the balance sheet shows that Acme has *preferred stock* of $1 million. The common stock equity section of the balance sheet contains three items: common stock, capital in excess of par, and retained earnings. The common stock entry shows that Acme's *common stock* is $3 million, reflecting the 3 million shares issued to investors, each with a $1 par value. **Par value** is the stated value printed on the stock certificate. This figure is almost always very low, sometimes even zero.

The next common stock equity entry is *capital in excess of par*. Capital in excess of par is the original market price per share value of the stock sold minus that stock's par value times the number of shares issued. If Acme originally sold 3 million shares of common stock to the public for $5 each, and the par value of each share was $1, then the price of the stock was $4 in excess of par. Multiplying 3 million shares times $4 gives Acme's $12 million figure for capital in excess of par. You can see that the common stock and the capital in excess of par values together represent the equity capital raised when new common stock was sold.

The last entry in the common stock equity section is retained earnings. The *retained earnings* figure represents the sum of all the earnings available to common stockholders of a business during its entire history, minus the sum of all the common stock dividends that it has ever paid.[2] Those earnings that were not paid out were, by definition, retained. The retained earnings figure for Acme at the end of 1999 is $10 million.

The Statement of Cash Flows

The third major financial statement required of all publicly traded corporations by the Financial Accounting Standards Board (FASB) is the *statement of cash flows*. This statement, like the income statement, can be compared to a video: It shows how cash flows into and out of a company over a given period of time.

We construct the statement of cash flows by adjusting the income statement to distinguish between *income* and *cash flow,* and by comparing balance sheets at the beginning and at the end of the relevant time period. The statement of cash flows shows cash flows in operating, investing, and financing activities, as well as the overall net increase or decrease in cash flow for the firm. You can see Acme's statement of cash flows for the year 1999 in Figure 4-4.

Operating Activities Operating activities on the statement of cash flows shows that Acme had $3,294,000 in net income for the year 1999. This number represents what was left after Acme paid all the firm's expenses for that year. We adjust that number to determine the operating cash flows for 1999.

Adjustment for Depreciation Expense Although depreciation expense is a legitimate reduction of income for accounting purposes, it is not a cash outlay. In other

---

> **Take Note**
>
> *Do not fall into the trap of thinking that the retained earnings account contains cash. Remember, equity accounts, including this one, represent owners'* **claims** *on assets. They are not assets themselves. The earnings not paid out as dividends have already been used to accumulate additional assets or to pay off liabilities.*

> **Take Note**
>
> *Be careful not to confuse the statement of cash flows with the income statement. The income statement lists revenues and expenses over a period of time. The statement of cash flows lists where* **cash** *came from and what it was used for during a period of time.*

---

[2]There are exceptions. If a company pays a dividend in the form of new common stock instead of cash, then there could be a transfer from retained earnings to the other common stock equity accounts. We will skip this exception for now. The use of stock dividends instead of cash dividends, and the resulting accounting treatment, will be examined in chapter 16.

Cash Received from (Used in) Operations:	
Net Income	$3,294,000
Depreciation	2,000,000
Decrease (Increase) Marketable Securities	1,000,000
Decrease (Increase) Accounts Receivable	(300,000)
Decrease (Increase) Inventory	7,300,000
Decrease (Increase) Prepaid Expenses	0
Increase (Decrease) Accounts Payable	(3,000,000)
Increase (Decrease) Accrued Expenses	(1,000,000)
Total Cash from Operations	**$9,294,000**
Cash Received from (Used for) Investments:	
New Fixed Asset Purchases	($14,000,000)
Total Cash from Investments	**($14,000,000)**
Cash Received from (Used for) Financing Activities:	
Proceeds from New Long-Term Debt Issue	$ 4,216,000
Proceeds from New Common Stock Issue	4,000,000
Short-Term Notes Paid Off	(1,000,000)
Preferred Stock Repurchases	(1,000,000)
Preferred Dividends	(110,000)
Common Dividends	(400,000)
Total Cash from Financing	**$5,706,000**
Net Change in Cash Balance	**$1,000,000**
Beginning Cash Balance	**$9,000,000**
Ending Cash Balance	**$10,000,000**

Figure 4-4 Acme Corporation Statement of Cash Flows for the Year Ended December 31, 1999

words, firms record depreciation expense on financial statements but do not write checks to pay it. We must add the $2 million in depreciation expense because net income was reduced by this amount—even though depreciation is a noncash expense.

Changes in Balance Sheet Accounts Changes in asset accounts on the balance sheet indicate changes in the company's cash flow. Because firms must pay cash to accumulate new assets, any increase in an asset account between the time one balance sheet is published and the time the next balance sheet is published indicates a cash outflow. Likewise, because firms sell assets to raise cash, any decrease in an asset account indicates a cash inflow. For Acme, balance sheet changes in marketable securities, accounts payable, accounts receivable (net), and inventory are shown in the operations section of Figure 4-4.

Changes in the liability and equity section of the balance sheet also signal cash flow changes. Because firms must use cash to pay off obligations, any decrease in liability, preferred stock, or common stock equity accounts between the time one balance sheet is published and the time the next balance sheet is published indicates a cash outflow during those time periods. To raise additional cash, firms can incur debt or equity obligations; so any increase in liability, preferred stock, or common stock items indicates a cash inflow.

> **Take Note**
>
> *We do not mention the change in accumulated depreciation in the statement of cash flows. The additional accumulated depreciation of $2 million is already included in the depreciation expense figure on the income statement. We don't want to count this twice.*

Figure 4-5 shows two balance sheets for Acme side by side. We can compare them and note where the cash inflows and outflows that appear on the statement of cash flows came from.

Operating Activities In the asset section of the balance sheet, we see that accounts receivable rose from $700,000 to $1 million, a $300,000 increase. In effect, Acme had a cash outflow of $300,000 in the form of funds recognized as revenue but not collected from its credit customers. In contrast, inventory *decreased* from $17.3 million to $10 million, which represents a $7.3 million *source* of cash in the form of inventory items sold that Acme did not have to make or buy. Similarly, marketable securities decreased by $1 million, signaling that Acme sold some marketable securities to generate a cash inflow of $1 million.

In the liabilities and equity section of the balance sheet, observe that accounts payable decreased by $3 million. Acme must have paid $3 million in cash to its suppliers to decrease the amount owed by that amount; therefore, this represents a cash outflow. Likewise, the accrued expenses account decreased by $1 million indicating that Acme used $1 million in cash to pay them.

Figure 4-5 Acme Corporation Balance Sheet Changes Between December 31, 1998 and December 31, 1999

	12/31/98	12/31/99	CHANGE
Assets:			
Cash	$ 9,000,000	$10,000,000	+1,000,000
Marketable Securities	9,000,000	8,000,000	−1,000,000
Accounts Receivable	700,000	1,000,000	+300,000
Inventory	17,300,000	10,000,000	−7,300,000
Prepaid Expenses	1,000,000	1,000,000	0
Total Current Assets	37,000,000	30,000,000	−7,000,000
Fixed Assets, Gross	14,000,000	28,000,000	+14,000,000
Less Accumulated Depreciation	(6,000,000)	(8,000,000)	−2,000,000
Fixed Assets, Net	8,000,000	20,000,000	+12,000,000
Total Assets	$45,000,000	$50,000,000	+5,000,000
Liabilities and Equity:			
Accounts Payable	$ 7,000,000	$ 4,000,000	−3,000,000
Notes Payable	4,000,000	3,000,000	−1,000,000
Accrued Expenses	3,000,000	2,000,000	−1,000,000
Total Current Liabilities	14,000,000	9,000,000	−5,000,000
Long-Term Debt	10,784,000	15,000,000	+4,216,000
Total Liabilities	24,784,000	24,000,000	−784,000
Preferred Stock	2,000,000	1,000,000	−1,000,000
Common Stock	1,000,000	3,000,000	+2,000,000
Capital in Excess of Par	10,000,000	12,000,000	+2,000,000
Retained Earnings	7,216,000	10,000,000	+2,784,000
Total Equity	20,216,000	26,000,000	+5,784,000
Total Liabilities and Equity	$45,000,000	$50,000,000	+5,000,000

Investment Activities The investments section of the statement of cash flows shows investing activities in long-term securities or fixed assets. Increasing investments require a cash outflow, and decreasing investments signal a cash inflow. For instance, observe in Figure 4-5 that Acme's fixed asssets increased to $28 million in 1999, up from $14 million in 1998. This $14 million increase reflects a cash outlay used to buy additional assets.

Financing Activities The financing section of the statement of cash flows shows financing activities related to the sales and retirement of notes, bonds, preferred and common stock, and other corporate securities. The retirement (i.e., buying back) of previously issued securities signals a cash outflow. The issuing of securities is a cash inflow. On the Acme balance sheet, for example, preferred stock decreased from $2 million to $1 million. The decrease shows that the firm spent $1 million to retire outstanding preferred stock.

Further down in the liabilities and equity section of the balance sheet (Figure 4-5) we see that the common stock and capital in excess of par accounts each increased by $2 million. These increases are the result of $4 million in cash received from a new issue of 2 million shares of common stock at $2 per share. Because the par value of the stock is $1 per share, $2 million is credited to the common stock account, and the remainder is credited to the capital in excess of par account.

In the common stock equity section of the balance sheet, we see that retained earnings increased from $7,216,000 to $10 million. Although this $2,784,000 increase in retained earnings represents a cash inflow to the firm, it is not recorded on the statement of cash flows. Why? Because the cash inflow it represents was recorded on the statement of cash flows as net income ($3,294,000) less preferred stock dividends ($110,000) less common stock dividends ($400,000). To include the increase in retained earnings again would result in double counting.

Net Cash Flow during the Period We have now completed the adjustments necessary to convert Acme's net income for 1999 into actual cash flows. Figure 4-4 shows that the cash inflows exceeded the cash outflows, resulting in a net cash inflow of $1 million for Acme in 1999. (Notice in Figure 4-5 that Acme's cash balance of $10 million on December 31, 1999 is $1 million higher than it was on December 31, 1998.)

DEPRECIATION

Depreciation is important to financial managers because it directly affects a firm's tax liabilities—which, in turn, affect cash flows. Here's how: Taxes paid are negative cash flows. Tax savings realized by deducting expenses generate more cash for the firm—the equivalent of a cash inflow.

Accounting depreciation is the allocation of an asset's initial cost over time. Let's look at why it is important to depreciate fixed assets over time. Suppose Acme bought a piece of equipment in 1999 that was expected to last seven years. If Acme paid $5 million cash for the asset, and the entire cost were charged as an expense in 1998, the transaction would wipe out all but $490,000 of Acme's earnings before taxes for the year ($5,490,000 earnings before taxes − $5,000,000 fixed asset cost). Nothing would show that Acme had acquired an asset worth $5 million. Then, for the next six years, Acme's income statements would show increases in profits generated by the asset—but there would be no corresponding accounting for the cost of that asset. In effect, it would look like Acme spent a lot of money in 1999, got nothing for it, and increased its profits for no reason over the next six years. This would clearly be misleading.

To get around the problem, accountants apply the *matching principle:* Expenses should be matched to the revenues they help to generate. In other words, if you buy an asset, and the asset is expected to last for seven years, then you should recognize the cost of the asset over the entire seven-year period. The cost is *amortized,* or spread out, over the seven-year period. In that way, the value of the asset will be properly shown on the financial statements in proportion to its contribution to profits.

Accounting depreciation is very different from economic depreciation. The latter attempts to measure the change in the value of an asset. Because this involves making value estimates that may turn out to be wrong, accountants use an established set of rules to determine the amount of depreciation to allocate over a certain time period.

Calculating the Amount of Depreciation Expense

Depreciation expense for a given period is determined by calculating the total amount to be depreciated (the *depreciation basis*), and then calculating the percentage of that total to be allocated to a given time period (the *depreciation rules*).

The total amount to be depreciated over the accounting life of the asset is known as the **depreciation basis.** It is equal to the cost of the asset, plus any setup or delivery costs that may be incurred.[3]

Depreciation Methods The cost of an asset can be allocated over time by using any of several sets of depreciation rules. The two most common depreciation rules used in tax reporting are *straight-line depreciation* and *MACRS.*

Straight-Line Depreciation. The simplest method of depreciation is the *straight-line depreciation (SL)* method. To use the **straight-line depreciation** method, you divide the cost of the asset by the number of years of life for the asset according to classification rules, and charge the result off as depreciation expense each year. For instance, if the managers at Acme bought a $5 million piece of equipment that belonged to the seven-year-asset class, then straight-line depreciation for the asset would be computed as follows:

Asset's initial cost: $5,000,000
Divided by length of service: 7 years
Equals depreciation expense each year: $714,286[4]

---

[3] Although in financial statements prepared by public corporations for reporting purposes, salvage value—the value of the asset if sold for salvage—may be subtracted in arriving at the depreciation basis, it is not considered part of the depreciation basis for tax reporting purposes.
[4] To be more precise, we would use what is known as the *half-year convention* in determining the annual depreciation. One-half a year's depreciation would be taken the year the asset was put into service and one-half in the final year. For example, for the preceding asset with a stated seven-year life, depreciation would in fact be spread over eight years. In this case, $357,143 in years 1 and 8, and $714,286 in years 2 through 7.

TABLE 4-1

1999 MACRS
Modified Accelerated Cost Recovery System Personal Property

Asset Class	3-Year Research, Equipment, and Special Tools	5-Year Computers, Copiers, Cars, and Similar Assets	7-Year Furniture, Fixtures, and Most Manufacturing Equipment	10-Year Equipment for Tobacco, Food, and Petroleum Production
Year	\multicolumn{4}{l}{Depreciation Percentages}			
1	33.3%	20%	14.3%	10%
2	44.5%	32%	24.5%	18%
3	14.8%	19.2%	17.5%	14.4%
4	7.4%	11.5%	12.5%	11.5%
5		11.5%	8.9%	9.2%
6		5.8%	8.9%	7.4%
7			8.9%	6.6%
8			4.5%	6.6%
9				6.5%
10				6.5%
11				3.3%

Table 4-1 lists MACRS asset class depreciation percentages for three-year to ten-year assets, and examples of assets in each class.

The Modified Accelerated Cost Recovery System (MACRS). Since the Tax Reform Act of 1986, Congress has allowed firms to use the modified accelerated cost recovery system, or **MACRS** (pronounced "makers") to compute depreciation expense for tax purposes. MACRS specifies that some percentage of the cost of assets will be charged each year as depreciation expense during the asset's life. Table 4-1 shows the 1999 MACRS percentages for various classes of personal property assets.[5]

So, under the MACRS, Acme's $5 million, seven-year asset would be depreciated 14.3 percent during its first year of life, 24.5 percent during the second year, and so on. Note that a seven-year asset is depreciated over eight years because of the half-year convention built into the table. Note also that MACRS is an accelerated depreciation method—greater percentages of the depreciation basis are subtracted from income in the early years, compared to the percentage applied in the later years. The acceleration is important because the more quickly firms can write off the cost of an asset, the sooner they save taxes from the tax-deductible expenses.

INCOME TAXES

Tax rates are set by Congress. The Internal Revenue Service (IRS) determines the amount of federal income tax a firm owes. Federal tax rules dictate that different rates apply to different blocks of income. For instance, the first $50,000 of taxable

---

[5]The half-year convention is built into the MACRS depreciation percentages shown in Table 4-1.

Financial Management and You

Your Personal Tax Rates

If you filed a federal income tax return, you can calculate your own personal average tax rate and marginal tax rate for last year.

1. Review your federal income tax return from last year. Determine the total amount of tax that you owed and divide by the amount of your taxable income (income after deductions). The answer is your average tax rate.
2. Now look up the amount of taxable income in the tax table that applied to you. It will be labeled Single, Married filing a joint return, Married filing a separate return, or Head of Household. (The Head of Household is the classification for a single person with a child claimed as a dependent on the taxpayer's return.) You will note that there are different percentages that apply to different blocks of income. The percentages increase, in steps, as your taxable income increases.
3. Determine the percentage that would have applied to the next dollar of taxable income you would have earned. This is your marginal tax rate for that year.
4. How do your average and marginal tax rates compare?

---

income is taxed at a 15 percent rate, whereas the next $25,000 of taxable income is taxed at a 25 percent rate.

The tax rate that applies to the next dollar of taxable income earned, the **marginal tax rate,** changes as the level of taxable income changes. This pattern—tax rate increases as taxable income increases—reflects the **progressive tax rate structure** imposed by the federal government.[6]

Table 4-2 shows the marginal tax rates for corporations as of June 1998.

TABLE 4-2

Corporate Marginal Tax Rates as of June 1998

Earnings before Taxes (EBT)	Tax Rate
$0–$50,000	15%
$50,001–$75,000	25%
$75,001–$100,000	34%
$100,001–$335,000	39%
$335,001–$10,000,000	34%
$10,000,001–$15,000,000	35%
$15,000,001–$18,333,333	38%
over $18,333,333	35%

Table 4-2 shows marginal tax rates for corporations. The marginal tax rates do not increase continuously for higher brackets because Congress has established rates that take away certain tax benefits for higher-income corporations.

---

[6]The federal corporate tax rate schedule is not strictly progressive for every tax bracket, as shown in Table 4-2. Generally speaking, as a corporation's taxable income increases, its marginal tax rate increases. The exceptions, seen in the tax rate schedule, usually apply to firms in upper-income tax brackets.

FINANCE AT WORK
SALES-RETAIL-SPORTS-MEDIA TECHNOLOGY-PUBLIC RELATIONS-PRODUCTION-EXPORTS

Business Changes in the Taxpayer Relief Act of 1997

On July 31, 1997, Congress passed the Taxpayer Relief Act of 1997 and President Clinton signed the legislation on August 5, 1997. The act included hundreds of changes and marked the first significant federal tax reduction since Ronald Reagan signed the Economic Recovery Tax Act in 1981. Some of the key measures of the act that affect businesses are summarized here.

Employee Education
The rule allowing tax-free employer payments for up to $5,250 of annual education expenses has been retroactively restored and extended through May 31, 2000. The exclusion had expired for courses beginning after May 31, 1997. Unfortunatley, graduate courses remain ineligible for this benefit.

Home Office Deduction Rules Eased
Effective in 1999, home office expenses can be deducted as long as the office is used regularly and exclusively to perform substantial administrative or management functions (such as billing customers) and no fixed space is available at work locations. Deductions are allowed even when the income-earning activities of the business take place elsewhere. This change effects a repeal of the Supreme Court's *Soliman* decision, which disallowed deductions for many self-employeds because their home offices did not qualify as the "principal place of business."

Reductions in Capital Gains Rates
Many consider the act's reduction in the maximum capital gains tax rate (to 20 percent) to be the centerpiece of the legislation. Unfortunately, the act did not include any capital gains tax rate relief for C corporations—they continue to be required to pay the same rate on both capital gains and ordinary income.

Alternative Minimum Tax Changes
C corporations with average annual receipts of less than $5 million for the first tax year beginning after 1996 (based on results for the preceding three tax years) are exempt from the corporate alternative minimum tax (AMT) as long as they also average less than $7.5 million in annual receipts for tax years beginning after 1996. This AMT exemption is effective for tax years beginning after 1997.

Net Operating Loss Provisions
Effective for tax years beginning after August 5, 1997, net operating losses (NOLs) can generally be carried back two years or forward twenty years (versus three and fifteen years under the old rules). However, the three-year carryback rule is retained for certain casualty losses and small business and farming losses attributable to presidentially declared disasters. For tax credits arising in tax years beginning after 1997, the carryback period is one year and the carryforward period is twenty years (versus three and fifteen years under the old law).

Source: Scoggins, McKelvey, Almon, & Malone, L.L.P., 720 East Harrison, Harlingen, TX 78550, (956)423-2111 ⟨http://www.smamcpa.com/taxbus.htm⟩.

Acme's EBT is $5,490,000, so its marginal tax bracket is 34 percent. Its next dollar of taxable income would be taxed at 34 percent. The federal income tax bill in 1999 would be:

$$
\begin{aligned}
&\$50,000 \times .15 &&= \quad \$7,500 \\
+\ &\$25,000 \times .25 &&= \quad \$6,250 \\
+\ &\$25,000 \times .34 &&= \quad \$8,500 \\
+\ &\$235,000 \times .39 &&= \quad \$91,650 \\
+\ &\$5,155,000 \times .34 &&= \underline{\$1,752,700}
\end{aligned}
$$

Taxable Income = $5,490,000   Taxes = $1,866,600

Financial managers use marginal tax rates to estimate the future after-tax cash flow from investments. Often managers want to know how dollars generated by a new investment will affect tax rates. If a new investment results in a huge jump in the company's tax rate, the project will be less desirable.

Average Tax Rates. Financial managers use effective, or **average tax rates,** to determine what percentage of the firm's total before-tax income is owed to the government. Average tax rates are calculated by dividing tax dollars paid by earnings before taxes, or EBT. Acme's effective, or federal average tax rate, for 1999 is:

$$\frac{\$1,866,600 \text{ (taxes paid)}}{\$5,490,000 \text{ (EBT)}} = .34, \text{ or } 34\%$$

Often the marginal tax rate will be greater than the average tax rate. Sometimes, however, the marginal and average tax rates are the same, as is the case for Acme. This occurred because the marginal tax rate changed from 34 percent to 39 percent and dropped to 34 percent again when Acme's tax bill was calculated.

Taxes are paid with cash. Because cash flow affects the value of a business, taxes are an important financial consideration. Financial managers need to understand marginal tax rates to see the marginal impact of taxes on cash flows.

In the income statement shown in Figure 4-1, Acme Corporation's effective tax rate is shown as 40 percent, resulting in taxes of $2,196,000 on before-tax income of $5,490,000. The figure represents **both** *Acme's federal and state tax obligations.*

What's Next

In this chapter we reviewed basic accounting principles and explained how financial managers use accounting information to create value for the firm. In chapter 5 we will discuss how to analyze financial statements.

Summary

1. *Explain how financial managers use the three basic accounting financial statements: the income statement, balance sheet, and statement of cash flows.*

Financial managers need to understand the key elements of financial statements to analyze a firm's finances and plan for its future.

- The income statement shows the amount of revenues, expenses, and income a firm has over a specified period of time.
- The balance sheet describes the assets, liabilities, and equity values for a company at a specific point in time.
- The statement of cash flows describes a firm's cash inflows and outflows over a period of time.

2. *Discuss how depreciation affects cash flow and compute depreciation expense.*

Depreciation is a noncash, tax-deductible expense. Because depreciation is tax deductible, it affects cash flow—the greater a firm's depreciation, the greater its cash flow. Cash flow, in turn, affects the value of the firm. The more cash a firm has, the greater its value.

To allocate the cost of an asset over time, accountants use different depreciation methods, such as straight-line depreciation or the modified accelerated cost recovery system (MACRS). Whatever method is used, accountants must first find the depreciation basis—the total cost of the asset plus setup and delivery costs. Then they calculate the percentage of that total allocated for the time period at issue, as determined by either the straight-line depreciation method or MACRS.

3. **Explain how taxes affect a firm's cash flow and calculate marginal and average tax rates.**

The marginal tax rate—the rate that would apply to the next dollar of taxable income—aids financial decisions because financial managers use it to assess the impact that a new investment will have on cash flow. If the new investment results in such a big jump in taxes that cash flow is affected negatively, the investment would likely be rejected. The amount of taxes owed is computed by multiplying each bracket of income by the corresponding tax rate set by Congress and adding the totals for each income bracket.

The average tax rate is calculated by dividing the amount of taxes paid by earnings before taxes.

You can access your FREE PHLIP/CW On–Line Study Guide, Current Events, and Spreadsheet Problem templates, which may correspond to this chapter either through the Prentice Hall Finance Center CD–ROM or by directly going to: **http://www.prenhall.com/gallagher.**

PHLIP/CW — *The Prentice Hall Learning on the Internet Partnership–Companion Web Site*

Self-Test

ST-1. Brother Mel's Bar-B-Q Restaurant has $80,000 in assets and $20,000 in liabilities. What is the equity of this firm?

ST-2. Cantwell Corporation has sales revenue of $2 million. Cost of goods sold is $1,500,000. What is Cantwell Corporation's gross profit?

ST-3. Adams Computer Store had accumulated depreciation of $75,000 at the end of 1999, and at the end of 1998 this figure was $60,000. Earnings before interest and taxes for the year 1999 were $850,000. Assuming that no assets were sold in 1999, what was the amount of depreciation expense for 1999?

ST-4. Shattuck Corporation had operating income (EBIT) of $2,500,000 in 1999, depreciation expense of $500,000, and dividends paid of $400,000. What is Shattuck's operating cash flow (EBITDA) for 1999?

ST-5. Bubba's Sporting Goods Company had retained earnings of $3 million at the end of 1998. During 1999, the company had earnings available to common stockholders of $500,000 and of this paid out $100,000 in dividends. What is the retained earnings figure for the end of 1999?

ST-6. Ron's In-Line Skating Corporation had retained earnings at the end of 1999 of $120,000. At the end of 1998 this figure was $90,000. If the company paid $5,000 in dividends to common stockholders during 1999, what was the amount of earnings available to common stockholders?

ST-7. Hayes Company recently bought a new computer system. The total cost, including set-up, was $8,000. If this is five-year asset class equipment, what would be the amount of depreciation taken on this system in year 2 using MACRS rules?

ST-8. If Burns Corporation has taxable income of $800,000, how much federal income taxes are owed?

ST-9. If Badeusz Quarry Corporation has taxable income of $4 million, what is the average tax rate for this company?

ST-10. If Parmenter Corporation has taxable income of $20 million, what is the marginal tax rate for this company?

Review Questions

1. Why do total assets equal the sum of total liabilities and equity? Explain.
2. What are the time dimensions of the income statement, the balance sheet, and the statement of cash flows? Hint: Are they videos or still pictures? Explain.
3. Define depreciation expense as it appears on an income statement. How does depreciation affect cash flow?
4. What are retained earnings? Why are they important?
5. Explain how earnings available to common stockholders and common stock dividends paid, as shown on the current income statement, affect the balance sheet item retained earnings.
6. What is accumulated depreciation?
7. What are the three major sections of the statement of cash flows?
8. How do financial managers calculate the average tax rate?
9. Why do financial managers calculate the marginal tax rate?
10. Identify whether the following items belong on the income statement or the balance sheet.
    - **a.** Interest Expense
    - **b.** Preferred Stock Dividends Paid
    - **c.** Plant and Equipment
    - **d.** Sales
    - **e.** Notes Payable
    - **f.** Common Stock
    - **g.** Accounts Receivable
    - **h.** Accrued Expenses
    - **i.** Cost of Goods Sold
    - **j.** Preferred Stock
    - **k.** Long-Term Debt
    - **l.** Cash
    - **m.** Captial in Excess of Par
    - **n.** Operating Income
    - **o.** Depreciation Expense
    - **p.** Marketable Securities
    - **q.** Accounts Payable
    - **r.** Prepaid Expenses
    - **s.** Inventory
    - **t.** Net Income
    - **u.** Retained Earnings
11. Indicate to which section the following balance sheet items belong (current assets, fixed assets, current liabilities, long-term liabilities, or equity).
    - **a.** Cash
    - **b.** Notes Payable
    - **c.** Common Stock
    - **d.** Accounts Receivable
    - **e.** Accrued Expenses
    - **f.** Preferred Stock
    - **g.** Plant and Equipment
    - **h.** Capital in Excess of Par
    - **i.** Marketable Securities
    - **j.** Accounts Payable
    - **k.** Prepaid Expenses
    - **l.** Inventory
    - **m.** Retained Earnings

Build Your Communication Skills

CS-1. Interview a manager or owner of an accounting firm. Ask that person what kinds of oral communication skills he or she needs to communicate financial information. Also ask what kinds of writing skills are required. What kinds of communications skills does this accounting firm executive look for when hiring a new person to do accounting or finance work for the firm? Report your findings to the class.

CS-2. Write a report that describes the best sources of financial information about publicly traded corporations. Discuss where you can find the basic financial statements for a corporation, which sources are the easiest to use, and what information sources—the library, the corporation, a brokerage firm, or the Internet—were most useful.

Problems

4-1. You are interviewing for an entry-level financial analyst position with Zeppelin Associates. Monte Rutledge, the senior partner, wants to be sure all the people he hires are very familiar with basic accounting principles. He gives you the following data and asks you to fill in the missing information. Each column is an independent case.

Financial Statement Connections

	Case A	Case B
Revenues	200,000	————
Expenses	————	70,000
Net Income	————	————
Retained Earnings, Jan 1	300,000	100,000
Dividends Paid	70,000	30,000
Retained Earnings, Dec 31	270,000	————
Current Assets, Dec 31	80,000	————
Noncurrent Assets, Dec 31	————	180,000
Total Assets, Dec 31	————	410,000
Current Liabilities, Dec 31	40,000	60,000
Noncurrent Liabilities, Dec 31	————	————
Total Liabilities, Dec 31	140,000	————
CS and Capital in Excess of Par, Dec 31	520,000	100,000
Total Stockholders' Equity, Dec 31	————	210,000

4-2. Fill in the following missing income statement values. The cases are independent.

Financial Statement Connections

	Case A	Case B
Sales	————	250,000
COGS	200,000	————
Gross Profit	————	150,000
Operating Expenses	60,000	60,000
Operating Income (EBIT)	————	————
Interest Expense	10,000	————
Income Before Taxes (EBT)	————	80,000
Tax Expense (40%)	92,000	————
Net Income	————	————

4-3. Lightning, Inc. has earnings before taxes of $48,000.

Tax Rates

    a. Using the progressive tax rate schedule from Table 4-2, calculate the tax obligation for Lightning, Inc.
    b. What is Lightning's average (effective) tax rate?

4-4. Thunder, Inc. has earnings before taxes of $150,000.

Tax Rates

    a. Using the progressive tax rate schedule from Table 4-2, calculate the tax obligation for Thunder, Inc.
    b. What is Thunder's average (effective) tax rate?

4-5. Following is a portion of Hitchcock Haven, Inc.'s balance sheet.

Equity

Common Stock ($1 par; 400,000 shares authorized; 200,000 shares issued)	$200,000
Capital in Excess of Par	$400,000
Retained Earnings	$100,000

What was the market price per share of the stock when it was originally sold?

Depreciation

4-6. A portion of Hitchcock Haven, Inc.'s comparative balance sheet follows. What is the amount of depreciation expense you would expect to see on the 1999 income statement? No assets that were on the books at the end of 1998 were sold or otherwise disposed of in 1999.

HITCHCOCK HAVEN, INC.
Balance Sheet as of December 31

	1998	1999
Plant and Equipment	$200,000	$250,000
Less: Accumulated Depreciation	$60,000	$70,000
Net Plant and Equipment	$140,000	$180,000

Income Statement

4-7. The following financial data correspond to Callahan Corporation's 1999 operations.

Cost of Goods Sold	$200,000
Selling and Administrative Expenses	40,000
Depreciation Expense	85,000
Sales	440,000
Interest Expense	40,000
Applicable Income Tax Rate	40%

Calculate the following income statement items.

a. Gross Profit
b. Operating Income (EBIT)
c. Earning Before Taxes (EBT)
d. Income Taxes
e. Net Income

Depreciation

4-8. Wet Dog Perfume Company (WDPC), a profit-making company, purchased a process line for $131,000 and spent another $12,000 on its installation. The line was commissioned in January 1998 and it falls into MACRS seven-year class life. Applicable income tax rate for WDPC is 40 percent and there is no investment tax credit. Calculate the following:

a. 1999 depreciation expense for this process line
b. Amount of tax savings due to this investment

Taxes

4-9. In 1999, Goodwill Construction Company purchased $130,000 worth of construction equipment. Goodwill's taxable income for 1999 without considering the new construction equipment would have been $400,000. The new equipment falls into the MACRS five-year class. Assume the applicable income tax rate is 34 percent.

a. What is the company's 1999 taxable income?
b. How much income tax will Goodwill pay?

Income Statement

4-10. Last year Johnson Flow Measurement Systems, Inc. had an operating profit of $600,000, paid $50,000 in interest expenses and $63,000 in preferred stock dividends. The applicable income tax rate for the year was 34 percent. The company had 100,000 shares of common stock outstanding at the end of last year.

a. What was the amount of Johnson's earnings per share last year?
b. If the company paid $1.00 per share to its common stockholders, what was the addition to retained earning last year?

Use the comparative figures of Pinewood Company and Subsidiaries to answer questions 4-11 through 4-15 that follow.

PINEWOOD COMPANY AND SUBSIDIARIES
As of December 31

	1998	1999
Assets:		
Cash	$ 5,534	$ 9,037
Marketable Securities	952	1,801
Accounts Receivable (gross)	14,956	16,110
Less: Allowance for Bad Debts	211	167
Accounts Receivable (net)	14,745	15,943
Inventory	10,733	11,574
Prepaid Expenses	3,234	2,357
Plant and Equipment (gross)	57,340	60,374
Less: Accumulated Depreciation	29,080	32,478
Plant and Equipment (net)	28,260	27,896
Land	1,010	1,007
Long-Term Investments	2,503	4,743
Liabilities:		
Accounts Payable	3,253	2,450
Notes Payable	—	—
Accrued Expenses	6,821	7,330
Bonds Payable	2,389	2,112
Stockholders' Equity:		
Common Stock	8,549	10,879
Retained Earnings	45,959	51,587

(Applies to problems 4-11 through 4-15 below.)

4-11. Compute the following totals for the end of 1998 and 1999.   *Balance Sheet*

  a. Current Assets
  b. Total Assets
  c. Current Liabilities
  d. Total Liabilities
  e. Total Stockholders' Equity

4-12. Show whether or not the basic accounting equation is satisfied in problem 4-11.   *Basic Accounting Equation*

4-13. Calculate the cash flows from the changes in the following from the end of 1998 to the end of 1999. Indicate inflow or outflow.   *Cash Flows*

  a. Accumulated Depreciation
  b. Accounts Receivable
  c. Inventories
  d. Prepaid Expenses
  e. Accounts Payable
  f. Accrued Expenses
  g. Plant and Equipment
  h. Marketable Securities
  i. Land
  j. Long-Term Investments
  k. Common Stock
  l. Bonds Payable

4-14. Prepare a statement of cash flows in proper form using the inflows and outflows from question 4-13. Assume net income (earnings after taxes) from the 1999 income statement was $10,628 and $5,000 in common stock dividends were paid. Ignore the income tax effect on the change in depreciation.   *Statement of Cash Flows*

4-15. Show whether or not your net cash flow matches the change in cash between the end of 1998 and end of 1999 balance sheets.   *Financial Statement Connections*

Answers to Self-Test

ST-1. $80,000 assets − $20,000 liabilities = $60,000 equity

ST-2. Cost of goods sold = $1,500,000
Gross profit = $2,000,000 sales revenue − $1,500,000 cost of goods sold = $500,000

ST-3. $75,000 end of 1999 accumulated depreciation − $60,000 end of 1998 accumulated depreciation = $15,000 1999 depreciation expense

ST-4. $2,500,000 EBIT + $500,000 = $3,000,000 cash flow from operations (Dividend payments are not operating cash flows; they are financial cash flows)

ST-5. $3,000,000 end of 1998 retained earnings + $500,000 earnings available to common stockholders − $100,000 dividends paid = $3,400,000 end of 1999 retained earnings

ST-6. Beginning retained earnings + net income − dividends paid = ending retained earnings
Therefore:
Net income = ending retained earnings − beginning retained earnings + dividends paid
So, for Ron's In-Line Skating Corporation:
Net income = $120,000 − $90,000 + $5,000
Net income = $35,000

ST-7. $8,000 depreciation basis × .32 (second-year MACRS depreciation % for a five-year class asset) = $2,560 year 2 depreciation expense

ST-8. ($50,000 × .15) + ($25,000 × .25) + ($25,000 × .34) + ($235,000 × .39) + ($465,000 × .34) = $272,000 federal income taxes owed

Note: $50,000 + $25,000 + $25,000 + $235,000 + $465,000 = $800,000 taxable income

ST-9. ($50,000 × .15) + ($25,000 × .25) + ($25,000 × .34) + ($235,000 × .39) + ($3,665,000 × .34) = $1,360,000 taxes owed; $1,360,000 ÷ $4 million = .34 = 34% average tax rate

Note: $50,000 + $25,000 + $25,000 + $235,000 + $3,665,000 = $4 million taxable income

ST-10. Taxable income over $18,333,333 is taxed at a 35% rate ∴ the marginal tax rate at $20 million in taxable income is 35%

CHAPTER 5

ANALYSIS OF FINANCIAL STATEMENTS

Money is better than poverty, if only for financial reasons.
—Woody Allen

MAKING SENSE OF THE NUMBERS WITH THE HELP OF THE WEB AND DATABASES

Accounting data do not mean much unless the numbers can be put into the proper context. Financial ratios help the financial analyst to do this. Changes in financial data need to be tracked and comparisons to similar companies made. Fortunately, the data to do this analysis are much more readily available than in the past.

Financial management students used to have to hike to the library to find financial statements on microfiche to do the financial analysis assignments often given by their professors. These were the Dark Ages of financial analysis. Your professor probably remembers these days. Today a little Web surfing will get you the data you need in minutes. Audited annual income statements, balance sheets, and statements of cash flows are readily available from the 10-K reports on EDGAR, the Web site of the Securities and Exchange Commission that receives this information from all publicly traded companies.

Many companies provide financial data about themselves on their own Web sites. Some sites provide data that are free, others provide data for a fee. The availability of such data is constantly changing. Some Web sites to check include

CHAPTER OBJECTIVES

After reading this chapter, you should be able to:

- Explain how financial ratio analysis helps financial managers assess the health of a company.

- Compute profitability, liquidity, debt, asset utilization, and market value ratios.

- Compare financial information over time and among companies.

- Locate ratio value data for specific companies and industries.

Bloomberg, Compustat, Dow Jones Retrieval, Value-Line, and Comstock. Use your search engine to find these sites. The URLs (Web addresses) are constantly changing.

Your library may also have financial databases available on CD-ROM, DVD, or some other medium. Often you can search for certain types of companies by screening the database by SIC code. This is the standard industry classification, and it allows you to identify the primary line of business of a given company. If your professor is old enough, ask him or her to tell you stories about microfiche, computer cards, slide rules, and other extremely interesting tools.

Chapter Overview

In chapter 4 we reviewed the major financial statements, the primary sources of financial information about a business. In this chapter we will learn how to interpret these financial statements in greater detail. All business owners, investors, and creditors use financial statements and ratio analysis to investigate the financial health of a firm. We will see how financial managers calculate ratios that measure profitability, liquidity, debt, asset activity, and market performance of a firm. We will then explore how financial experts use ratios to compare the firm's performance to managers' goals, the firm's past and present performance, and the firm's performance to similar firms in the industry. We will also discuss sources of financial information.

Assessing Financial Health

Medical doctors assess the health of people. Financial managers assess the health of businesses. When you visit a doctor for an examination, the doctor may check your blood pressure, heart rate, cholesterol, and blood sugar levels. The results of each test should fall within a range of numbers considered "normal" for your age, weight, gender, and height. If they don't, the doctor will probably run additional tests to see what, if anything, is wrong.

Like doctors, financial managers check the health of businesses by running basic tests—such as a financial ratio analysis—to see whether the firm's performance is within the normal range for a company of that type. If it is not, the financial manager runs more tests to see what, if anything, is wrong.

Misleading Numbers

Both medical doctors and financial managers must interpret the information they have and decide what additional information they need to complete an analysis. For instance, suppose a doctor examines a six-foot, 230-pound, 22-year-old male named Dirk. The doctor's chart shows that a healthy male of that age and height should normally weigh between 160 and 180 pounds. Because excess weight is a health risk, the numbers don't look positive.

Before the doctor prescribes a diet and exercise program for Dirk, she asks follow-up questions and runs more tests. She learns that Dirk, a starting fullback for his college football team, has only 6 percent body fat, can bench-press 380 pounds, runs a 40-yard dash in 4.5 seconds, has a blood pressure rate of 110/65, and a resting heart rate of 52 beats per minute. This additional information changes the doctor's initial health assessment. Relying on incomplete information would have led to an inaccurate diagnosis.

Like doctors, financial managers need to analyze many factors to determine the health of a company. Indeed, for some firms the financial statements do not provide the entire picture.

In 1998 the attorneys general of many states sued the major tobacco companies alleging liability for smoking-related illness and for advertising allegedly aimed at underage people. Congress also was seeking legislation that would extract hundreds of billions of dollars over several years from these companies. Because the outcome of the pending lawsuits and legislation was uncertain, the (potential) liabilities did not appear on the balance sheet. Financial analysis based only on the financial statements, then, gave a faulty impression of the companies' health.

As the tobacco company example demonstrates, accounting conventions may prevent factors affecting a firm's finances from appearing on financial statements. Just as Dirk's doctor looked beyond the obvious, financial managers using ratio analysis must always seek complete information before completing an analysis. In the sections that follow, we discuss ratios based on financial statements, ratios that use market information, and outside information sources.

Financial Ratios

Financial managers use ratio analysis to interpret the raw numbers on financial statements. A **financial ratio** is a number that expresses the value of one financial variable relative to another. Put more simply, a financial ratio is the result you get when you divide one financial number by another. Calculating an individual ratio is simple, but each ratio must be analyzed carefully to effectively measure a firm's performance.

Ratios are comparative measures. Because the ratios show relative value, they allow financial analysts to compare information that could not be compared in its raw form.[1] Ratios may be used to compare:

- one ratio to a related ratio
- the firm's performance to management's goals
- the firm's past and present performance
- the firm's performance to similar firms

For instance, say a company reaped huge revenues from one investment, but the cost of the investment was high. A financial manager could use a ratio to compare that investment to another that did not generate such high revenues but had low cost. Take James Cameron's *Titanic*. That blockbuster movie grossed more than $1.75 billion—the number-one grossing movie for 1998. Compare *Titanic's* total revenues to the $23.6 million revenues that Ang Lee's *The Wedding Banquet* generated.[2] Looking only at the total revenue figures, *Titanic* looks like a better investment than *The Wedding Banquet*.

---

[1] Financial managers who analyze the financial condition of the firms they work for act as financial analysts. The term *financial analyst*, however, also includes financial experts who analyze a variety of firms.
[2] *Colorado Springs Gazette-Telegraph* (January 15, 1994): A2.

Source: *DILBERT © 1996 United Features Syndicate. Reprinted by permission.*

However, analysts in the movie industry use a return on cost ratio (total revenues divided by total cost) to find a movie's net return per $1 invested. Using that ratio we see that *Titanic,* at a cost of $200 million, had a return-on-cost ratio of 8.75 ($1,750,000,000 ÷ $200,000,000 = 8.75). *The Wedding Banquet,* at a cost of $1 million, had a return-on-cost ratio of 23.6 ($23,600,000 ÷ $1,000,000 = 23.6). Although *Titanic* made more total revenue, *The Wedding Banquet* made more money relative to its cost than did *Titanic.*

Financial managers, other business managers, creditors, and stockholders all use financial ratio analysis. Specifically, creditors may use ratios to see whether a business will have the cash flow to repay its debt and interest. Stockholders may use ratios to see what the long-term value of their stock will be. For example, in 1995 Euro Disney reported a yearly profit, its first ever since the Paris Disney theme park opened in 1992. However, analysts predicted that compared to its total costs—costs that included average interest payments of $30 million per year—the revenues were too low to guarantee Euro Disney's future financial success. Despite the reported profits, stock price dropped by 14 percent.[3]

The Basic Financial Ratios

Financial ratios are generally divided into five categories: profitability, liquidity, debt, asset activity, and market value. The ratios in each group give us insights into different aspects of a firm's financial health.

- *Profitability ratios* measure how much company revenue is eaten up by expenses, how much a company earns relative to sales generated, and the amount earned relative to the value of the firm's assets and equity.
- *Liquidity ratios* indicate how quickly and easily a company can obtain cash for its needs.
- *Debt ratios* measure how much a company owes to others.
- *Asset activity ratios* measure how efficiently a company uses its assets.

---

[3]"Euro Disney Reports Profit for '95, but the Future Remains Cloudy," *The New York Times* (November 16, 1995): D7.

- *Market value ratios* measure how the market value of the company's stock compares to its accounting values.

Calculating the Ratios

We will use the financial statements for the Acme Corporation presented in chapter 4 as the basis for our ratio analysis. Figure 5-1 shows Acme Corporation's income statement for 1999, and Figure 5-2 shows its December 31, 1999 balance sheet.

Now let's analyze Acme Corporation's financial health by calculating its profitability, liquidity, debt, asset utilization, and market value ratios.

Profitability Ratios   Profitability ratios measure how the firm's returns compare to its sales, asset investments, and equity. Stockholders have a special interest in the profitability ratios because profit ultimately leads to cash flow, a primary source of value for a firm. Managers, acting on behalf of stockholders, also pay close attention to profitability ratios to ensure that the managers preserve the firm's value.

We will discuss five profitability ratios: gross profit margin, operating profit margin, net profit margin, return on assets, and return on equity. Some of the profitability ratios use figures from two different financial statements.

Gross Profit Margin.   The *gross profit margin* measures how much profit remains out of each sales dollar after the cost of the goods sold is subtracted. The ratio formula follows:

$$\text{Gross Profit Margin} = \frac{\text{Gross Profit}}{\text{Sales}}$$

This ratio shows how well a firm generates revenue compared to its costs of goods sold. The higher the ratio, the better the cost controls compared to the sales revenues.

Net Sales	$15,000,000
Cost of Goods Sold	5,000,000
Gross Profit	10,000,000
Selling and Administrative Expenses	800,000
Depreciation	2,000,000
Operating Income (EBIT)	7,200,000
Interest Expense	1,710,000
Income before Taxes	5,490,000
Income Taxes (40%)	2,196,000
Net Income	$ 3,294,000
Preferred Dividends	110,000
Earnings Available to Common Stockholders	$ 3,184,000
Earnings per Share (3 million shares)	$1.06
Common Dividends Paid	400,000
Increase in Retained Earnings	$ 2,784,000

Figure 5-1   Acme Corporation Income Statement for the Year 1999

Figure 5-2 Acme Corporation Balance Sheet as of December 31, 1999

Assets:	
Cash	$10,000,000
Marketable Securities	8,000,000
Accounts Receivable	1,000,000
Inventory	10,000,000
Prepaid Expenses	1,000,000
Total Current Assets	30,000,000
Fixed Assets, Gross	28,000,000
Less Accumulated Depreciation	(8,000,000)
Fixed Assets, Net	20,000,000
Total Assets	$50,000,000
Liabilities and Equity:	
Accounts Payable	$ 4,000,000
Notes Payable	3,000,000
Accrued Expenses	2,000,000
Total Current Liabilities	9,000,000
Long-Term Debt	15,000,000
Total Liabilities	24,000,000
Preferred Stock	1,000,000
Common Stock (3 million shares)	3,000,000
Capital in Excess of Par	12,000,000
Retained Earnings	10,000,000
Total Equity	26,000,000
Total Liabilities and Equity	$50,000,000

To find the gross profit margin ratio for Acme, look at Figure 5-1, Acme's income statement. We see that Acme's gross profit for the year was $10 million and its sales revenue was $15 million. Dividing $10 million by $15 million yields Acme Corporation's gross profit margin of .67 or 67 percent. That ratio shows that Acme's cost of products and services sold was 33 percent of sales revenue, leaving the company with 67 percent of sales revenue to use for other purposes.

Operating Profit Margin. The *operating profit margin ratio* measures the cost of goods sold, as reflected in the gross profit margin ratio, as well as all other operating expenses. This ratio is calculated by dividing earnings before interest and taxes (EBIT or operating income) by sales revenue.

$$\text{Operating Profit Margin} = \frac{\text{Earnings Before Interest and Taxes}}{\text{Sales}}$$

$$= \frac{7,200,000}{15,000,000} = .48 = 48\%$$

Acme's EBIT, as shown on its income statement (see Figure 5-1), is $7,200,000. Dividing $7.2 million by its sales revenue of $15 million gives an operating profit margin of 48 percent (7,200,000 ÷ 15,000,000 = .48 or 48%). Acme's operating profit margin indicates that 48 percent of its sales revenues remain after subtracting all operating expenses.

Net Profit Margin. The *net profit margin* measures how much profit out of each sales dollar is left after *all* expenses are subtracted—that is, after all operating, interest, and tax expenses are subtracted. It is computed by dividing net income by sales revenue. Acme's net income for the year was $3.294 million. Dividing $3.294 million by $15 million in sales yields a 22 percent net profit margin. Here's the computation:

$$\text{Net Profit Margin} = \frac{\text{Net Income}}{\text{Sales}} = \frac{3{,}294{,}000}{15{,}000{,}000} = .22 \text{ or } 22\%$$

Net income, and the net profit margin ratio, are often referred to as "bottom-line" measures. The net profit margin includes adjustments for nonoperating expenses, such as interest and taxes, and operating expenses. We see that in 1999 Acme Corporation had 22 percent of each sales dollar remaining after all expenses were paid.

Return on Assets. The *return on assets* (ROA) ratio indicates how much income each dollar of assets produces on average. It shows whether the business is employing its assets effectively. The ROA ratio is calculated by dividing net income by the total assets of the firm. For Acme Corporation, we calculate this ratio by dividing $3.294 million in net income (see Figure 5-1, Acme income statement) by $50 million of total assets (see Figure 5-2, Acme balance sheet), for a return on assets (ROA) of 6.6 percent. Here's the calculation:

$$\text{Return on Assets} = \frac{\text{Net Income}}{\text{Total Assets}} = \frac{\$3{,}294{,}000}{\$50{,}000{,}000} = .066 \text{ or } 6.6\%$$

In 1999 each dollar of Acme Corporation's assets produced, on average, income of $.066. Although this return on assets figure may seem low, it is not unusual for certain types of companies, such as commercial banks, to have low ROA ratios.

Return on Equity. The *return on equity* (ROE) ratio measures the average return on the firm's capital contributions from its owners (for a corporation, that means the contributions of common stockholders). It indicates how many dollars of income were produced for each dollar invested by the common stockholders.

ROE is calculated by dividing net income by common stockholders' equity. To calculate ROE for Acme Corporation, divide $3.294 million in net income by $25 million in total common stockholders' equity ($26 million total equity − $1 million preferred stock. See Figure 5-2, Acme balance statement). Acme's ROE is 13.2 percent, shown as follows:

$$\text{Return on Equity} = \frac{\text{Net Income}}{\text{Common Equity}} = \frac{3{,}294{,}000}{25{,}000{,}000} = .132 \text{ or } 13.2\%$$

The ROE figure shows that Acme Corporation returned, on average, 13.2 percent for every dollar that common stockholders invested in the firm.

Mixing Numbers from Income Statements and Balance Sheets. When financial managers calculate the gross profit margin, operating profit margin, and net profit margin ratios they use only income statement variables. In contrast, analysts use both income statement and balance sheet variables to find the return on assets and return on equity ratios. A **mixed ratio** is a ratio that uses both income statement and balance sheet variables as inputs.

Take Note
Do not confuse the ROE ratio with the return earned by the individual common stockholders on their common stock investment. The changes in the market price of the stock and dividends received determine the total return on an individual's common stock investment.

Because income statement variables show values over a period of time and balance sheet variables show values for one moment in time, using mixed ratios poses the question of how to deal with the different time dimensions. For example, should the analyst select balance sheet variable values from the beginning of the year, the end of the year, or the midpoint of the year? If there was a large change in the balance sheet account during the year, the choice could make a big difference. Consider the following situation:

Total Assets Jan 1, 1999	$1,000,000
Total Assets Dec 31, 1999	2,000,000
Net Income in 1999	100,000

Return on assets based on Jan 1 balance sheet:

$$100{,}000/1{,}000{,}000 = .10, \text{ or } 10\%$$

Return on assets based on Dec 31 balance sheet:

$$100{,}000/2{,}000{,}000 = .05, \text{ or } 5\%$$

Which figure is correct? There is no black-and-white answer to this problem. Some analysts add the beginning of the year balance sheet figure to the end of the year figure and divide by two to get an average figure.

Logic and common sense suggest that analysts should pick figures that best match the returns to the assets or to the equity. Say that Acme purchased a large amount of assets early in the year. The middle or end of year balance sheet figures would probably match the returns to the assets more effectively than beginning of the year figures because assets can only affect profit if they have been used. For simplicity, we used end of year balance sheet figures to calculate Acme's mixed profitability ratios.

Liquidity Ratios Liquidity ratios measure the ability of a firm to meet its short-term obligations. These ratios are important because failure to pay such obligations can lead to bankruptcy. Bankers and other lenders use liquidity ratios to see whether to extend short-term credit to a firm. Generally, the higher the liquidity ratio, the more able a firm is to pay its short-term obligations. Stockholders, however, use liquidity ratios to see how the firm has invested in assets. Too much investment in current—as compared to long-term—assets indicates inefficiency.

The two main liquidity ratios are the current ratio and the quick ratio.

The Current Ratio. The *current ratio* compares all the current assets of the firm (cash and other assets that can be easily converted to cash) to all the company's current liabilities (liabilities that must be paid with cash soon). At the end of 1999, Acme Corporation's current assets were $30 million and its current liabilities were $9 million. Dividing Acme's current assets by its current liabilities, as follows, we see that:

$$\text{Current Ratio} = \frac{\text{Current Assets}}{\text{Current Liabilities}} = \frac{\$30{,}000{,}000}{\$9{,}000{,}000} = 3.33$$

Acme's current ratio value, then, is 3.33. The ratio result shows that Acme has $3.33 of current assets for every dollar of current liabilities, indicating that Acme could pay all its short-term debts by liquidating about a third of its current assets.

The Quick Ratio. The *quick ratio* is similar to the current ratio but is a more rigorous measure of liquidity because it excludes inventory from current assets. To calculate the quick ratio, then, divide current assets less inventory by current liabilities.

$$\text{Quick Ratio} = \frac{\text{Current Assets Less Inventory}}{\text{Current Liabilities}}$$

This conservative measure of a firm's liquidity may be useful for some businesses. To illustrate, suppose a computer retail store had a large inventory of personal computers with out-of-date Intel® 486 microprocessors. The computer store would have a tough time selling its inventory for much money.

At the end of 1999, the balance sheet figures show that Acme Corporation's current assets less inventory are worth $20 million ($30,000,000 − $10,000,000). Its current liabilities are $9 million. Dividing $20 million by $9 million, we see that its quick ratio is 2.22. A quick ratio of 2.22 means that Acme could pay off 222 percent of its current liabilities by liquidating its current assets, excluding inventory.

If Acme Corporation's inventory is hard to liquidate, the quick ratio is more important. If the company being analyzed had very liquid inventory, such as a government securities dealer, the quick ratio would not be a useful analysis tool compared to the current ratio.

Debt Ratios The financial analyst uses debt ratios to assess the relative size of a firm's debt load and the firm's ability pay off the debt. The three primary debt ratios are the debt to total assets, debt to equity, and times interest earned ratios.

Current and potential lenders of long-term funds, such as banks and bondholders, are interested in debt ratios. When a business's debt ratios increase significantly, bondholder and lender risk increases because more creditors compete for that firm's resources if the company runs into financial trouble. Stockholders are also concerned with the amount of debt a business has because bondholders are paid before stockholders.

The optimal debt ratio depends on many factors, including the type of business and the amount of risk lenders and stockholders will tolerate. Generally, a profitable firm in a stable business can handle more debt—and a higher debt ratio—than a growth firm in a volatile business.

Debt to Total Assets. The *debt to total assets* ratio measures the percentage of the firm's assets that is financed with debt. Acme Corporation's total debt at the end of 1999 was $25 million. This includes the $1 million in preferred stock, which for this calculation we've classified as debt rather than equity. Its total assets were $50 million. The calculations for the debt to total assets ratio follow:

$$\text{Debt to Total Assets} = \frac{\text{Total Debt}}{\text{Total Assets}} = \frac{\$25,000,000}{\$50,000,000} = .50 \text{ or } 50\%$$

Acme's debt to total assets ratio value is 50 percent, indicating that the other 50 percent of financing came from equity investors (the common stockholders).

Debt to Equity. The *debt to equity* ratio is the percentage of debt relative to the amount of common equity of the firm. At the end of 1999, Acme's amount of total debt was $25 million and the amount of common equity was $25 million. Here is the calculation of Acme's debt to equity ratio:

$$\text{Debt to Equity} = \frac{\text{Total Debt}}{\text{Common Equity}} = \frac{\$25,000,000}{\$25,000,000} = 1.00$$

Acme's debt to equity ratio of 1.00 shows that the firm had the same amount of debt as it had equity. Once again, preferred stock is treated here as debt rather than equity. Preferred stockholders are not owners and are not entitled to the residual income of the firm.

Times Interest Earned. The times interest earned ratio is often used to assess a company's ability to service the interest on its debt with operating income from the current period. The *times interest earned* ratio is equal to earnings before interest and taxes (EBIT) divided by interest expense. Acme Corporation has EBIT of $7.2 million and interest expense of $1.71 million for 1999. Acme's times interest earned ratio is as follows:

$$\text{Times Interest Earned} = \frac{\text{EBIT}}{\text{Interest Expense}} = \frac{\$7,200,000}{\$1,710,000} = 4.21$$

Acme's times interest earned ratio value of 4.21 means that the company earned $4.21 of operating income (EBIT) for each $1 of interest expense incurred during that year.

A high times interest earned ratio suggests that the company will have ample operating income to cover its interest expense. A low ratio signals that the company may have insufficient operating income to pay interest as it becomes due. If so, the business might need to liquidate assets, or raise new debt or equity funds to pay the interest due. Recall, however, that operating income is not the same as cash flow. Operating income figures do not show the amount of *cash* available to pay interest. Because interest payments are made with cash, the times interest earned ratio is only a rough measure of a firm's ability to pay interest with current funds.

Asset Activity Ratios Financial analysts use asset activity ratios to measure how efficiently a firm uses its assets. They analyze specific assets and classes of assets. The three asset activity ratios we'll examine here are the average collection period (for accounts receivable), the inventory turnover, and the total asset turnover ratios.

Average Collection Period. The *average collection period* ratio measures how many days, on average, the company's credit customers take to pay their accounts. Managers, especially credit managers, use this ratio to decide to whom the firm should extend credit. Slow payers are not welcome customers.

To calculate the average collection period, divide accounts receivable by the company's average credit sales per day. (This, in turn, is the company's annual credit sales divided by the number of days in a year, 365.)

$$\text{Average Collection Period} = \frac{\text{Accounts Receivable}}{\text{Average Daily Credit Sales}}$$

Acme Corporation had $1 million in accounts receivable and average daily credit sales of $41,095.89 (i.e., $15 million total credit sales divided by 365 days in one year). Dividing $1 million by $41,095.89 gives a value of 24.33. The ratio shows that in 1999 Acme Corporation's credit customers took an average of 24.33 days to pay their account balances.

Notice that, in calculating the ratio, we used Acme Corporation's total sales figure for 1999 in the denominator, assuming that all of Acme's sales for the year were made on credit. We made no attempt to break down Acme's sales into cash sales and credit sales. Financial analysts usually calculate this ratio using the total sales figure when they do not have the credit sales only figure.

Inventory Turnover. The *inventory turnover* ratio tells us how efficiently the firm converts inventory to sales. If the company has inventory that sells well, the ratio value will be high. If the inventory does not sell well due to lack of demand or if there is excess inventory, the ratio value will be low.

The inventory turnover formula follows:

$$\text{Inventory Turnover} = \frac{\text{Sales}}{\text{Inventory}}$$

Acme Corporation had sales of $15 million and inventory of $10 million in 1999. Dividing $15 million by $10 million, we see that the inventory turnover value is 1.5. This number means that in 1999 Acme "turned" its inventory into sales 1.5 times during the year.[4]

Total Asset Turnover. The *total asset turnover* ratio measures how efficiently a firm utilizes its assets. Stockholders, bondholders, and managers know that the more efficiently the firm operates, the better the returns.

If a company has many assets that do not help generate sales (such as fancy offices and corporate jets for senior management), then the total asset turnover ratio will be relatively low. A company that has a high asset utilization ratio suggests that its assets help promote sales revenue.

To calculate the asset turnover ratio for Acme, divide sales by total assets as follows:

$$\text{Total Asset Turnover} = \frac{\text{Sales}}{\text{Total Assets}}$$

The 1999 total asset turnover ratio for Acme Corporation is its sales of $15 million divided by its total assets of $50 million. The result is .30, indicating that Acme's sales were .30 of its assets. Put another way, the dollar amount of sales was 30 percent of the dollar amount of its assets.

Market Value Ratios The ratios examined so far rely on financial statement figures. But market value ratios mainly rely on financial marketplace data, such as the market price of a company's common stock. Market value ratios measure the market's perception of the future earning power of a company, as reflected in the stock share price. The two market value ratios we discuss are the price to earnings ratio and the market to book value ratio.

Price to Earnings Ratio. The price to earnings (P/E) ratio is defined as:

$$\text{P/E} = \frac{\text{Market Price per Share of Common Stock}}{\text{Earnings per Share}}$$

To calculate earnings per share (EPS), we divide earnings available to common stockholders by the number of shares of common stock outstanding.

Investors and managers use the P/E ratio to gauge the future prospects of a company. The ratio measures how much investors are willing to pay for claim to one dollar of the earnings per share of the firm. The more investors are willing to pay over the value of EPS for the stock, the more confidence they are displaying about

---

[4] Many financial analysts define the inventory turnover ratio using *cost of goods sold* instead of *sales* in the numerator. They use cost of goods sold because sales is defined in terms of *sales price* and inventory is defined in terms of *cost*. We will use *sales* in the numerator of the inventory turnover ratio to be consistent with the other turnover ratios.

the firm's future growth—that is, the higher the P/E ratio, the higher are investors' growth expectations. Consider the following marketplace data for Acme:

Current market price of Acme Corporation's stock:	$20.00
1999 EPS:	$1.06

$$\text{P/E Ratio} = \frac{\text{Market Price per Share}}{\text{Earnings per Share}} = \frac{\$20}{\$1.06} = 18.87$$

We see that the $20 per share market price of Acme Corporation's common stock is 18.87 times the level of its 1999 earnings per share ($1.06 EPS). The 18.87 result indicates that stock traders predict that Acme has average future earnings potential. It would take just under 19 years, at Acme's 1999 earnings rate, for the company to earn net profits of $20 per share, the amount an investor would pay today to buy this stock.

Market to Book Value. The *market to book value* (M/B) ratio is the market price per share of a company's common stock divided by the accounting book value per share (BPS) ratio. The book value per share ratio is the amount of common stock equity on the firm's balance sheet divided by the number of common shares outstanding.

The book value per share is a proxy for the amount remaining per share after selling the firm's assets for their balance sheet values, and paying the debt owed to all creditors and preferred stockholders. We calculate Acme's BPS ratio, assuming the following information:

Total Common Stock Equity at Year-End 1999:	$25,000,000
Number of Common Shares Outstanding:	3,000,000

$$\text{BPS} = \frac{\text{Common Stock Equity}}{\text{Number of Common Shares Outstanding}} = \frac{25{,}000{,}000}{3{,}000{,}000} = \$8.33$$

Now that we know the book value per share of Acme's stock is $8.33, we can find the market to book value ratio as follows:

$$\text{Market to Book Value Ratio} = \frac{\text{Market Price per Share}}{\text{Book Value per Share}} = \frac{\$20}{\$8.33} = 2.4$$

We see that Acme's M/B ratio is 2.4. That value indicates that the market price of Acme common stock ($20) is 2.4 times its book value per share ($8.33).

When the market price per share of stock is greater than the book value per share, analysts often conclude that the market believes the company's future earnings are worth more than the firm's liquidation value. The value of the firm's future earnings minus the liquidation value is the **going concern value** of the firm. The higher the M/B ratio, when it is greater than 1, the greater the going concern value of the company seems to be. In our case Acme seems to have positive going concern value.

Companies that have a market to book value of less than 1 are sometimes considered to be "worth more dead than alive." Such an M/B ratio suggests that if the company liquidated and paid off all creditors and preferred stockholders, it would have more left over for the common stockholders than what the common stock could be sold for in the marketplace.

The M/B ratio is useful, but it is only a rough approximation of how liquidation and going concern values compare. This is because the M/B ratio uses an accounting-based book value. The actual liquidation value of a firm is likely to be different than the book value. For instance, the assets of the firm may be worth more or less than the value at which they are currently carried on the company's balance sheet. In addition, the current market price of the company's bonds and preferred stock may also differ from the accounting value of these claims.

Economic Value Added and Market Value Added. Two new financial indicators that have become popular are *economic value added* (EVA) and *market value added* (MVA). These indicators were developed by Stern Stewart & Company, a consulting firm in New York City. EVA is a measure of the amount of profit remaining after accounting for the return expected by the firm's investors, whereas MVA compares the firm's current value to the amount the current owners paid for it. According to Stern Stewart, the use of the EVA and MVA indicators can help add value to a company because they help managers focus on rewards to stockholders instead of traditional accounting measures.[5] In the following paragraphs we discuss EVA and MVA individually.

Economic Value Added (EVA). As we mentioned previously, EVA is a measure of the amount of profit remaining after accounting for the return expected by the firm's investors. As such, EVA is said to be an "estimate of true economic profit, or the amount by which earnings exceed or fall short of the required minimum rate of return investors could get by investing in other securities of comparable risk."[6] The formula to calculate EVA is as follows:

$$EVA = EBIT(1 - TR) - (IC \times Ka)$$

where EBIT = earnings before interest and taxes (i.e., operating income)
  TR = the effective or average income tax rate
  IC = invested capital (explained later)
  Ka = investors' required rate of return on their investment (explained later)

Invested capital (IC) is the total amount of capital invested in the company. It is the sum of the *market values* of the firm's equity, preferred stock, and debt capital. Ka is the weighted average of the rates of return expected by the suppliers of the firm's capital, sometimes called the weighted average cost of capital, or WACC.

To illustrate how EVA is calculated, assume Acme's common stock is currently selling for $20 a share, and the weighted average return expected by investors (Ka) is 12 percent. Also assume that the book values of debt and preferred stock on Acme's balance sheet are the same as the market values.[7] Also recall from

The cost of a firm's capital is discussed fully in chapter 12.

---

[5] ⟨http://www.sternstewart.com⟩.
[6] Ibid.
[7] This assumption is frequently made in financial analysis to ease the difficulties of locating current market prices for debt and preferred securities. Because prices of debt and preferred securities do not tend to fluctuate widely, the assumption does not generally introduce an excessive amount of error into the EVA calculation.

Figures 5-1 and 5-2 that Acme's EBIT for 1999 is $7,200,000, its effective income tax rate is 40 percent, and there are 3 million shares of common stock outstanding.

The last term we need before calculating Acme's EVA is invested capital (IC). Remember it is the sum of the *market values* of the firm's equity, preferred stock, and debt capital. Acme's IC is found as follows:

$$\begin{aligned}
\text{Market Value of Common Equity} &= 3,000,000 \text{ shares} \times \$20 \\
&= \$60,000,000 \\
\text{Market Value of Preferred Stock} &= \text{Book Value} \\
&= \$1,000,000 \\
\text{Market Value of Debt Capital} &= \text{Book Value} \\
&= \text{Notes Payable} + \text{Long-Term Debt}^8 \\
&= \$3,000,000 + \$15,000,000 \\
&= \$18,000,000 \\
\text{Total Invested Capital (IC)} &= \$60,000,000 \\
&+ 1,000,000 \\
&+ \underline{18,000,000} \\
&\$79,000,000
\end{aligned}$$

Now we have all the amounts necessary to solve the EVA equation for Acme:

$$\text{EVA} = \text{EBIT}(1 - \text{TR}) - (\text{IC} \times \text{Ka})$$

For Acme in 1999:

$$\begin{aligned}
\text{EVA} &= \$7,200,000\,(1 - .40) - (\$79,000,000 \times .12) \\
&= \$4,320,000 - \$9,480,000 \\
&= \$(5,160,000)
\end{aligned}$$

Acme's EVA for 1999 is negative, indicating the company did not earn a sufficient amount during the year to provide the return expected by all those who contributed capital to the firm. Even though Acme had $7,200,000 of operating income in 1999, and $3,294,000 of net income, it was not enough to provide the 12 percent return expected by Acme's creditors and stockholders.

Does the negative EVA result for 1999 indicate that Acme is in trouble? Not at all. Remember the negative result is only for one year, whereas it is the trend over the long term that counts. The negative result for this year could be due to any number of factors, all of which might be approved of by the creditors and stockholders. As long as Acme's average EVA over time is positive, occasional negative years are not cause for alarm.

Market Value Added (MVA). Market value added (MVA) is the market value of the firm, debt plus equity, minus the total amount of capital invested in the firm. MVA is similar to the market to book ratio (M/B). MVA focuses on total market value and total invested capital whereas M/B focuses on the per share stock price and invested equity capital. The two measures are highly correlated.

---

[8] Take note that total debt capital is *not* the same as total liabilities. Liabilities that are spontaneously generated, such as accounts payable and accrued expenses, are not generally included in the definition of debt capital. True debt capital is created when a specified amount of money is lent to the firm at a specified interest rate.

Companies that consistently have high EVAs would normally have a positive MVA. If a company consistently has negative EVAs, it should have a negative MVA too.

In this section we examined the key profitability, liquidity, debt, asset activity, and market value ratios. The value of each ratio tells part of the story about the financial health of the firm. Next we explore relationships among ratios.

Relationships among Ratios: The Du Pont System As we discussed earlier, ratios may be used to compare one ratio to another related ratio. Financial analysts compare related ratios to see what specific activities add to or detract from a firm's performance.

The Du Pont system of ratio analysis is named for the company whose managers developed the general system. It first examines the relationships between total revenues relative to sales, and sales relative to total assets. The product of the net profit margin and the total asset turnover is the return on assets (or ROA). This equation, known as the Du Pont equation, follows:

$$\text{Du Pont Equation}$$
$$\text{Return on Assets} = \text{Net Profit Margin} \times \text{Total Asset Turnover} \quad (5\text{-}1)$$
$$\frac{\text{Net Income}}{\text{Total Assets}} = \frac{\text{Net Income}}{\text{Sales}} \times \frac{\text{Sales}}{\text{Total Assets}}$$

Sales, on the right side of the equation, appears in the denominator of the net profit margin and in the numerator of the total asset turnover. These two equal sales figures would cancel each other out if the equation were simplified, leaving net income over total assets on the right. This, of course, equals net income over total assets, which is on the left side of the equal sign, indicating that the equation is valid.

This version of the Du Pont equation helps us analyze factors that contribute to a firm's return on assets. For example, we already know from our basic ratio analysis that Acme Corporation's return on assets for 1999 was 6.59 percent. Now suppose you wanted to know how much of that 6.59 percent was the result of Acme's net profit margin for 1999, and how much was the result of the activity of Acme's assets in 1999. Equation 5-1, the Du Pont equation, provides the following answer:

$$\text{Return on Assets} = \text{Net Profit Margin} \times \text{Total Asset Turnover}$$
$$\text{Return on Assets} = \frac{\text{Net Income}}{\text{Sales}} \times \frac{\text{Sales}}{\text{Total Assets}}$$

For Acme Corporation:

$$.0659, \text{ or } 6.59\% = \frac{\$3{,}294{,}000}{\$15{,}000{,}000} \times \frac{\$15{,}000{,}000}{\$50{,}000{,}000}$$
$$.0659 = .2196 \times .3$$
$$\text{or}$$
$$6.59\% = 21.96\% \times .3$$

Acme Corporation, we see, has a fairly healthy net profit margin, 21.96 percent, but its total asset turnover is only three-tenths its sales. The .3 total asset turnover has the effect of dividing the 21.96 percent net profit margin by 3.33, such that ROA is only 6.59 percent.

We might see a low total asset turnover and high net profit margin in a jewelry store, where few items are sold each day but high profit is made on each item sold. A grocery store, however, would have a low net profit margin and a high total asset turnover because many items are sold each day but little profit is made on each dollar of sales.

Another version of the Du Pont equation, called the Modified Du Pont equation, measures how the return on equity (ROE) is affected by net profit margin, asset activity, and debt financing. As shown in Equation 5-2, in the modified Du Pont equation, ROE is the product of net profit margin, total asset turnover, and the **equity multiplier** (the ratio of total assets to common equity).

$$\text{Modified Du Pont Equation}$$

$$\text{ROE} = \text{Net Profit Margin} \times \text{Total Asset Turnover} \times \text{Equity Multiplier} \tag{5-2}$$

$$\frac{\text{Net Income}}{\text{Equity}} = \frac{\text{Net Income}}{\text{Sales}} \times \frac{\text{Sales}}{\text{Total Assets}} \times \frac{\text{Total Assets}}{\text{Equity}}[9]$$

for Acme Corporation:

$$.1318 = 13.18\% = \frac{\$3{,}294{,}000}{\$15{,}000{,}000} \times \frac{\$15{,}000{,}000}{\$50{,}000{,}000} \times \frac{\$50{,}000{,}000}{\$25{,}000{,}000}$$

$$.1318 = .2196 \times .3 \times 2$$
$$13.18\% = 21.96\% \times .3 \times 2$$

Examining the preceding equation, we see that Acme's net profit margin of 21.96 percent is greater than its 13.18 percent ROE. However, Acme's low productivity of assets ($.30 in sales for every dollar of assets employed) reduces the effect of the profit margin—21.96% × .3 = 6.59%. If no other factors were present, Acme's ROE would be 6.59 percent.

Now the equity multiplier comes into play. The equity multiplier indicates the amount of financial leverage a firm has. A firm that uses only equity to finance its assets should have an equity multiplier that equals 1.0. To arrive at this conclusion, recall the basic accounting formula—total assets = liabilities + equity. If a firm had no debt on its balance sheet, its liabilities would equal zero, so equity would equal total assets. If equity equals total assets, then the equity multiplier would be 1. Multiplying 1 times any other number has no effect, so in such a situation ROE would depend solely on net profit margin and total asset turnover.

If a firm does have debt on its balance sheet (as Acme does), it will have greater assets than equity and the equity multiplier will be greater than 1. This produces a multiplier effect that drives ROE higher (assuming net income is positive) than can otherwise be accounted for by net profit margin and asset turnover.[10]

---

[9]Notice that sales and total assets appear in both a numerator and a denominator in the right side of the equation and would cancel out if the equation were simplified, leaving net income over equity on both the right and the left of the equal sign.

$$\frac{\text{Net Income}}{\text{Equity}} = \frac{\text{Net Income}}{\cancel{\text{Sales}}} \times \frac{\cancel{\text{Sales}}}{\cancel{\text{Total Assets}}} \times \frac{\cancel{\text{Total Assets}}}{\text{Equity}}$$

$$\frac{\text{Net Income}}{\text{Equity}} = \frac{\text{Net Income}}{\text{Equity}}$$

[10]We'll discuss leverage in more detail in chapter 13.

FINANCE AT WORK

Lee Anne Schuster, Kitchell Contractors, Marketing

Lee Anne Schuster

Lee Anne Schuster is not rich, but she owns part of a major construction company. She is an employee-owner of Kitchell Contractors, one of the largest building contractors in the West. Lee Anne is not a financial expert, as she is the first to admit. But in her job as marketing coordinator for Kitchell, she must have a firm grasp of financial concepts because when Lee Anne submits a proposal for a new building project to customers, they want to see Kitchell's financial position. Plus, as an employee-owner, Lee Anne carefully reads the financial statements to understand her company's financial health.

Q. When you look at your company's financial statements, what do you look at?

As an employee-owner, I have a strong interest in working smarter, to improve our company's bottom line. I always start with the profitability ratios. Those ratios combine many factors and reveal the bottom line. By looking at the net profit margin—what's left of revenues after all expenses and taxes have been taken out—I can get a quick snapshot of how the company is doing.

I also check to see our company's debt load. If we have too high a debt to equity ratio, we may have a hard time raising cash and staying ahead of our interest payments later. I prepare many reports for banks as part of my job duties, so I've learned that creditors also look at our debt ratios carefully, especially the debt to equity ratio.

Q. What else do your creditors look at?

They look at our current assets versus our current liabilities, the current ratio. They want to see how liquid the company is before they extend credit to us. For instance, say that the company hits a bad stretch. Lenders want to know that we can pay off our liabilities and still survive. If the current ratio is 1:1, the company's chances do not appear secure. If the ratio is higher, say it is two times the current assets compared to current liabilities, then creditors view Kitchell as a safer business to lend to in the event that we have a downturn.

Q. Kitchell is not a publicly traded company. How do you value your stock?

Because our stock is not traded publicly, it is tough to assign a value to it. Our stock is based on the value of the company, so after I review the profitability and debt ratios, I examine the asset activity ratios. The value of our assets creates value for the business. I try to find out how well the assets are performing over time to see if we're investing in assets wisely. The more our assets help generate sales, the brighter our future will be.

Q. How else has reading financial statements helped you?

I analyze the financial statements of my customers to see how well they are doing. If a customer has poor profitability or liquidity, debt ratios that are too high, or assets that are underperforming, our firm pays attention to those red flags before we agree to work with that customer. As an employee-owner, I appreciate that we don't sign up every customer that walks through our door. The risk of not getting paid may simply be too great.

Source: Interview with Lee Anne Schuster.

Acme's equity multiplier of 2 indicates that Acme has assets that are 2 times its equity. This has the effect (called the **leverage effect**) of boosting Acme's return on equity from 6.59 percent to 13.18 percent. The leverage effect, caused by debt of $41 million shown on Acme's balance sheet, significantly alters Acme's ROE.

In this section we reviewed basic ratios, and analyzed relationships of one ratio to another to assess the firm's financial condition. Next we will investigate how ratio analysis can be used to compare trends in a firm's performance, and to compare the firm's performance to other firms in the same industry.

TREND ANALYSIS AND INDUSTRY COMPARISONS

Ratios are used to compare a firm's past and present performance and its industry performance. In this section we will examine trend analysis and industry comparison. Comparing a ratio for one year with the same ratio for other years is known as **trend analysis.** Comparing a ratio for one company with the same ratio for other companies in the same industry is **industry comparison.**

Trend Analysis

Trend analysis helps financial managers and analysts see whether a company's current financial situation is improving or deteriorating. To prepare a trend analysis, compute the ratio values for several time periods (usually years) and compare them. Table 5-1 shows a five-year trend for Acme Corporation's ROA.

As Table 5-1 shows, Acme Corporation's ROA has risen steadily over this five-year period, with the largest growth occurring between 1998 and 1999. Overall, the trend analysis indicates that Acme's 1999 ROA of 6.59 percent is positive, compared to earlier years.

Usually, analysts plot ratio value trends on a graph to provide a picture of the results. Figure 5-3 is a graph of Acme's 1995–1999 ROA ratios.

TABLE 5-1

Acme Corporation Five-Year Return on Assets (ROA) Trend Analysis

	1995	1996	1997	1998	1999
ROA	−1.8%	2.2%	2.6%	4.1%	6.6%

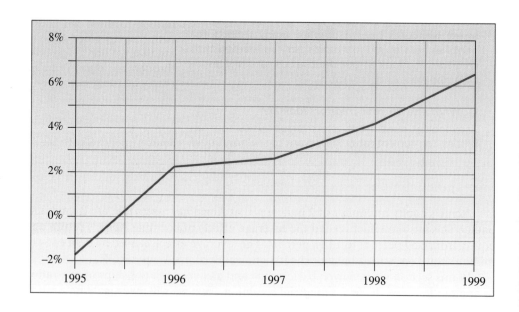

Figure 5-3 **Acme Corporation Five-Year Trend in ROA**

This figure shows the trend in the return on assets ratio from 1995 to 1999 for Acme Corporation

TABLE 5-2

Acme Corporation Cross-Sectional Analysis of ROA 1999

Company	ROA
Acme Corporation	6.6%
Company A	1.0%
Company B	7.1%
Company C	0.9%
Industry Average (Acme + A + B + C) ÷ 4 = 3.9	3.9%

The five-year upward trend in ROA, depicted in Figure 5-3, indicates that Acme Corporation increased the amount of profit it generated from its assets.

Trend analysis is an invaluable part of ratio analysis. It helps management spot a deteriorating condition and take corrective action, or identify the company's strengths. By assessing the firm's strengths and weaknesses, and the pace of change in a strength or weakness, management can plan effectively for the future.

Industry Comparisons

Another way to judge whether a firm's ratio is too high or too low is to compare it to the ratios of other firms in the industry (this is sometimes called **cross-sectional analysis**). This type of comparison pinpoints deviations from the norm that may indicate problems.

Table 5-2 shows a comparison between Acme Corporation's ROA ratio and the average ROA in Acme Corporation's industry for 1999.

Table 5-2 shows that, compared to other firms in Acme's industry, Acme achieved an above-average ROA in 1999. Only Company *B* managed to do better than Acme.

Cross-sectional analysis allows analysts to put the value of a firm's ratio in the context of its industry. For example, Acme's ROA of 6.6% is higher than average for its industry, and thus Acme would be looked upon favorably. In another industry, however, the average ROA might be 10 percent, causing Acme's 6.6% to appear much too low. Whether a ratio value is good or bad depends on the characteristics of the industry. By putting the ratio in context, analysts compare apples to apples and not apples to oranges.

Summary Analysis: Trend and Industry Comparisons Together

A complete ratio analysis of a company combines both trend analysis and industry comparisons. Table 5-3 shows all the ratios presented in this chapter for Acme Corporation for the years 1995 through 1999 along with the industry averages for those ratios. (The industry averages are labeled IND in the table.)

First, let's review Acme's profitability ratios compared to the industry average for 1995 to 1999. In 1995 and 1996, Acme Corporation had negative net income. This gave a negative value to the net profit margin, return on assets, and return on equity for each of these years (because net income is the numerator for each ratio). There was steady improvement, however, in the profit ratios from 1995 to 1999.

Acme Corporation had lower gross profit, operating profit, and net profit margins than the industry norm for that five-year time period except for the 1999 operating and net profit margins. For 1995 and 1996, Acme also had a lower return

TABLE 5-3 Five-Year Ratio Analysis for Acme Corporation

RATIOS		1995	1996	1997	1998	1999
Profitability Ratios						
Gross Profit Margin	ACME	36.2%	38.9%	42.8%	58.9%	67.0%
(Gross Profit ÷ Sales)	IND	55.7%	58.9%	62.2%	66.0%	68.0%
Operating Profit Margin	ACME	14.3%	16.5%	18.6%	28.9%	48.0%
(EBIT ÷ Sales)	IND	34.2%	35.1%	37.5%	40.0%	42.0%
Net Profit Margin	ACME	−8.5%	−5.8%	3.4%	7.8%	22.0%
(Net Income ÷ Sales)	IND	4.3%	6.4%	10.2%	11.5%	13.4%
Return on Assets	ACME	−1.8%	−2.2%	2.6%	4.1%	6.6%
(Net Income ÷ Total Assets)	IND	1.2%	1.8%	2.2%	2.5%	2.1%
Return on Equity	ACME	−14.6%	−7.5%	2.8%	8.3%	13.2%
(Net Income ÷ Equity)	IND	3.9%	4.4%	5.1%	5.6%	7.8%
Liquidity Ratios						
Current Ratio	ACME	2.15	2.22	2.30	2.48	3.33
(Current Assets ÷ Current Liabilities)	IND	1.98	2.00	2.10	2.15	2.20
Quick Ratio	ACME	.98	1.22	1.30	1.30	2.22
(Current Assets Less Inventory ÷ Current Liabilities)	IND	1.10	1.00	1.10	1.12	1.15
Debt Ratios						
Debt to Total Assets	ACME	81%	81%	82%	81.5%	50%
(Total Debt ÷ Total Assets)	IND	62%	59%	57%	58%	60%
Debt to Equity	ACME	426%	426%	456%	442%	100%
(Total Debt ÷ Common Equity)	IND	163%	144%	133%	138%	150%
Times Interest Earned	ACME	1.10	1.20	1.20	1.37	4.21
(EBIT ÷ Interest Expense)	IND	3.70	3.80	4.00	4.20	4.25
Asset Activity Ratios						
Average Collection Period (Days)	ACME	33.80	31.50	30.10	28.40	24.33
(Accounts Receivable ÷ Average Daily Credit Sales)	IND	40.20	39.80	38.40	37.30	40.00
Inventory Turnover	ACME	.38	.49	.80	.51	1.5
(Sales ÷ Inventory)	IND	.62	.66	.68	.72	.70
Total Asset Turnover	ACME	.33	.24	.15	.19	.30
(Sales ÷ Total Assets)	IND	.22	.20	.12	.15	.16
Market Value Ratios						
P/E Ratio	ACME	—	—	80.0	36.0	18.87
(Market Price per Share ÷ EPS)	IND	15.0	17.0	19.0	15.0	16.0
Market to Book Value	ACME	1.3	1.6	1.8	2.0	2.4
(Market Price ÷ Book Value)	IND	2.1	2.2	1.9	2.0	2.0

on assets ratio than the industry average. As the summary analysis shows, the 1997–1999 ROAs are the result of higher asset turnover ratios.

The return on equity figures paint a telling story over this five-year period. From 1995 to 1997, Acme Corporation had a much lower return on equity than did the average firm in its industry. In the years 1995 and 1996, these figures had significant negative values, whereas the industry norm was positive. In 1998 and 1999, however, Acme Corporation had a much higher return on equity than the average firm in its industry.

Acme had a high debt load until 1999. Look at the debt to asset and debt to equity ratio values. The debt to total asset ratio is consistently above 80 percent, whereas the industry norm for this ratio is 62 percent or less from 1995 to 1998. A high debt load magnifies the changes in the return on equity ratio values.

The times interest earned ratio shows that Acme Corporation barely covered its interest expense with its operating income. The value of this ratio was slightly more than 1 except for 1999, when it jumped to 4.21.

Next, we examine the liquidity ratios: the current and quick ratios. The current ratio has been approximately 2 each year except for 1999, when it was 3.33. Having two times or more the amount of current assets as current liabilities is a good target for most companies. Because the industry norm for the current ratio was below the value Acme Corporation had each of these years, Acme had a comparatively good liquidity position.

The quick ratio stayed near the industry norm throughout this period until it spiked to 2.22 in 1999. This means that when inventory is subtracted from total current assets, Acme Corporation's liquidity looked steady. Again, however, 1999's value (2.22 for Acme versus 1.15 for the industry norm) suggests that management watch liquidity in 2000.

Now let's look at the asset activity ratios: average collection period, inventory turnover, and asset turnover. The average collection period was significantly lower for Acme than for the average firm in its industry. It appears that Acme did a better than average job of collecting its accounts receivable.

The inventory turnover ratio was erratic over this five-year period. The fluctuations suggest that Acme did not match its inventory to its demand for products. The numbers suggest that Acme's managers should have studied its inventory control policies to look for ways to match demand and inventory more closely. There was a big increase in 1999. More about this in chapter 19.

The total asset turnover ratio was consistently higher than the industry norm. This helped the return on assets ratio during the years when net income was positive as described earlier.

Finally, we turn to the market value ratios. Acme had no meaningful P/E ratios for 1995 and 1996 because net income, and, therefore, EPS, were negative. The P/E ratio of 80 in 1997 shows investors had high expectations about Acme's future growth, but these expectations moderated in the next two years as the company matured. The market to book value ratio shows an upward trend over the five-year period showing that investors increasingly valued Acme's future earnings potential above the company's asset liquidation value.

We have just finished a complete ratio analysis of Acme Corporation, including examinations of the company's profitability, liquidity, debt management, asset activity, and market value ratios. To conduct the analysis, we combined trend and industry analysis so that we could see how Acme performed over time and how it performed relative to its industry. Managers inside the company can use the results of the analysis to support proposed changes in operations or organization; and creditors and investors outside the company can use the results to support decisions about lending money to the company or buying its stock.

LOCATING INFORMATION ABOUT FINANCIAL RATIOS

Ratio analysis involves a fair amount of research. Before analysts can calculate all the ratios, they must locate the underlying, raw financial data. Analysts can gather

TABLE 5-4   Sources of Financial Information

INFORMATION	MEDIUM	SOURCE
Business news, articles, market data, stock, bond, mutual fund price quotes	Newspapers	*The Wall Street Journal* *Barron's* *USA Today*
Business news, articles	Magazines	*Forbes* *Fortune* *Business Week* *Money Magazine*
Data on the economy, industries; many financial statistics (interest rates, inflation, etc.)	Bound publications	*US Industrial Outlook* *Standard & Poor's Statistical Surveys* *Standard & Poor's Industry Surveys* *Federal Reserve Bulletin* *World Almanac* *Statistical Abstract of the United States* *Business Conditions Digest* (contains leading, lagging, and coincident indicators of the economy) *Economic Report of the President*
Summary data about industries, companies; advice on industries, stocks; analysis and forecasts	Investment advisory publications	*Value-Line Investment Survey* (each company report appears on one page) *Standard & Poor's Outlook*
Data on companies and industries	Computer databases	*Compustat PC Plus CD-ROM* *Value Screen*
Company performance; corporate financial data		Annual reports from the company (can often be obtained by phone from the company) *Standard & Poor's Stock Reports* (versions for the NYSE, AMEX, and OTC markets) *Standard & Poor's Corporation Records* (contains in-depth reports about companies, including financial statement data) *Moody's Handbook of Common Stocks* (similar to the *S&P Stock Reports*) *Moody's Industrial Manual, Moody's Bank & Finance Manual, Moody's OTC Manual, Moody's Public Utility Manual, Moody's Transportation Manual, Moody's International Manual* (all these *Moody's Manuals* contain in-depth reports on companies)
Information about bonds	Bound publications	*Moody's Bond Record* *Moody's Bond Survey*
Information about mutual funds	Bound publications	*Morningstar Mutual Funds* (similar to *Value-Line* but for mutual funds) *Weisenberger's Management Results* *Weisenberger's Current Performance & Dividend Record* (similar to *Moody's Manuals* but covers mutual funds)
Variety of business and financial news and information (some require paid subscriptions)	Web	⟨http://www.bloomberg.com⟩ ⟨http://www.compustat.com⟩ ⟨http://www.valueline.com⟩ ⟨http://www.morningstar.net⟩ ⟨http://www.wsj.com⟩ ⟨http://www.cnnfn.com⟩ ⟨http://www.yahoo.com/Business_and_Economy/Finance_and_Investment⟩ ⟨http://www.sec.gov/edgarhp.htm⟩

information about publicly traded corporations at most libraries, on CD-ROM databases, and on the Internet.

A number of organizations publish financial data about companies and industries. Many publications contain ratios that are already calculated. Table 5-4 (page 104) contains a list of publications that financial analysts find useful when they are researching companies and industries. Many of them are available at local libraries.

What's Next

In this chapter we learned how to calculate and apply financial ratios to analyze the financial condition of the firm. In chapter 6 we will see how to use analyses to forecast and plan for the company's future.

Summary

1. **Explain how financial ratio analysis helps assess the health of a company.**

Just as doctors assess a patient's health, financial analysts assess the financial health of a firm. One of the most powerful assessment tools is financial ratio analysis. Financial ratios are comparative measures that allow analysts to interpret raw accounting data and identify strengths and weaknesses of the firm.

2. **Compute profitability, liquidity, debt, asset activity, and market value ratios.**

Profitability, liquidity, debt, asset activity, and market value ratios show different aspects of a firm's financial performance. Profitability, liquidity, debt, and asset activity ratios use information from a firm's income statement or balance sheet to compute the ratios. Market value ratios use market and financial statement information.

Profitability ratios measure how the firm's returns compare to its sales, asset investments, and equity. Liquidity ratios measure the ability of a firm to meet its short-term obligations. Debt ratios measure the firm's debt financing and its ability to pay off its debt. Asset activity ratios measure how efficiently a firm uses its assets. Finally, market value ratios measure the market's perception about the future earning power of a business.

The Du Pont system analyzes the sources of ROA and ROE. Two versions of the Du Pont equation were covered in this chapter. The first analyzes the contributions of net profit margin and total asset turnover to ROA. The second version analyzes how the influences of net profit margin, total asset turnover, and leverage affect ROE.

3. **Compare financial information over time and among companies.**

Trend analysis compares the past and present financial ratios to see how a firm has performed over time. Industry analysis compares a firm's ratios with the ratios of companies in the same industry. Summary analysis, one of the most useful financial analysis tools, combines trend and industry analysis to measure how a company performed over time in the context of the industry.

4. **Locate ratio value data for specific companies and industries.**

A number of organizations publish financial data about companies and industries. Many publications contain ratios that are already calculated. Table 5-4 contains a list of publications that financial analysts find useful when they are researching companies and industries.

Companion Website
content by Active Learning Technologies

You can access your FREE PHLIP/CW On–Line Study Guide, Current Events, and Spreadsheet Problem templates, which may correspond to this chapter either through the Prentice Hall Finance Center CD–ROM or by directly going to: **http://www.prenhall.com/gallagher.**

PHLIP/CW — *The Prentice Hall Learning on the Internet Partnership–Companion Web Site*

Equations Introduced in This Chapter

Profitability Ratios:

$$\text{Gross Profit Margin} = \frac{\text{Gross Profit}}{\text{Sales}}$$

$$\text{Operating Profit Margin} = \frac{\text{Earnings Before Interest and Taxes}}{\text{Sales}}$$

$$\text{Net Profit Margin} = \frac{\text{Net Income}}{\text{Sales}}$$

$$\text{Return on Assets} = \frac{\text{Net Income}}{\text{Total Assets}}$$

$$\text{Return on Equity} = \frac{\text{Net Income}}{\text{Common Equity}}$$

Liquidity Ratios:

$$\text{Current Ratio} = \frac{\text{Current Assets}}{\text{Current Liabilities}}$$

$$\text{Quick Ratio} = \frac{\text{Current Assets Less Inventory}}{\text{Current Liabilities}}$$

Debt Ratios:

$$\text{Debt to Total Assets Ratio} = \frac{\text{Total Debt}}{\text{Total Assets}}$$

$$\text{Debt to Equity} = \frac{\text{Total Debt}}{\text{Equity}}$$

$$\text{Times Interest Earned} = \frac{\text{Earnings Before Interest and Taxes}}{\text{Interest Expense}}$$

Asset Activity Ratios:

$$\text{Average Collection Period} = \frac{\text{Accounts Receivable}}{\text{Average Daily Credit Sales}}$$

$$\text{Inventory Turnover} = \frac{\text{Sales}}{\text{Inventory}}$$

$$\text{Total Asset Turnover} = \frac{\text{Sales}}{\text{Total Assets}}$$

Market Value Ratios:

$$\text{P/E Ratio} = \frac{\text{Market Price per Share}}{\text{Earnings per Share}}$$

$$\text{Market to Book Ratio} = \frac{\text{Market Price per Share}}{\text{Book Value per Share}}$$

Equation 5-1. The Du Pont Formula:

$$\text{Return on Assets} = \text{Net Profit Margin} \times \text{Total Asset Turnover}$$

$$\frac{\text{Net Income}}{\text{Total Assets}} = \frac{\text{Net Income}}{\text{Sales}} \times \frac{\text{Sales}}{\text{Total Assets}}$$

Equation 5-2. The Modified Du Pont Formula:

$$\text{ROE} = \text{Net Profit Margin} \times \text{Total Asset Turnover} \times \text{Equity Multiplier}$$

$$\frac{\text{Net Income}}{\text{Equity}} = \frac{\text{Net Income}}{\text{Sales}} \times \frac{\text{Sales}}{\text{Total Assets}} \times \frac{\text{Total Assets}}{\text{Equity}}$$

Self-Test

ST-1. De Marco Corporation has total assets of $5 million and an asset turnover ratio of 4. If net income is $2 million, what is the value of the net profit margin?

ST-2. Francisco Company has current assets of $50,000. Total assets are $200,000; and long-term liabilities, preferred stock, and common stock collectively total $180,000. What is the value of the current ratio?

ST-3. If one-half the current assets in ST-2 consist of inventory, what is the value of the quick ratio?

ST-4. Sheth Corporation has a return on assets ratio of 6 percent. If the debt to total assets ratio is .5, what is the firm's return on equity?

ST-5. Mitra Company has a quick ratio value of 1.5. It has total current assets of $100,000 and total current liabilities of $25,000. If sales are $200,000, what is the value of the inventory turnover ratio?

ST-6. Yates Corporation has total assets of $500,000. Its equity is $200,000. What is the company's debt to total asset ratio?

ST-7. Pendell Company has total sales of $4 million. One-fourth of these are credit sales. The amount of accounts receivable is $100,000. What is the average collection period for the company? Use a 365-day year.

Review Questions

1. What is a financial ratio?
2. Why do analysts calculate financial ratios?
3. Which ratios would a banker be most interested in when considering whether to approve an application for a short-term business loan? Explain.
4. In which ratios would a potential long-term bond investor be most interested? Explain.
5. Under what circumstances would market to book value ratios be misleading? Explain.
6. Why would an analyst use the Modified Du Pont system to calculate ROE when ROE may be calculated more simply? Explain.
7. Why are trend analysis and industry comparison important to financial ratio analysis?

Build Your Communication Skills

CS-1. Research a publicly traded company that has a presence in your community. Assess the financial health of this company in the areas of profitability, liquidity, debt, and asset activity. Write a report of your findings. Include in your report a discussion of the strengths and weaknesses of the company, key trends, and how the company's ratios compare to other companies in its industry.

CS-2. You have just been given a job as a loan officer. It is your job to evaluate business loan applications. Your boss would like you to prepare a new set of guidelines to be used by the bank to evaluate loan requests, leading to the approval or denial decision.

Prepare a loan application packet. Include the specific quantitative and qualitative information you would want an applicant for a loan to provide to you. Explain in a brief report, oral or written, how you would use the requested information to decide whether a loan should be approved.

Problems

Problems 5-1 to 5-5 refer to the consolidated income statement and consolidated balance sheet of Pinewood Company and Subsidiaries that follow.

(for problems 5-1 through 5-5)

Pinewood Company and Subsidiaries Income Statement for the Year 1999 ('000 dollars)

Sales	$94,001
Cost of Goods Sold	46,623
Gross Profit	47,378
Selling and Administrative Expenses	28,685
Depreciation and R&D Expense (both tax deductible)	5,752
EBIT or Operating Income	12,941
Interest Expense	48
Interest Income	427
Earning Before Taxes (EBT)	13,320
Income Taxes	4,700
Net Income (NI)	8,620
Preferred Stock Dividends	—
Earnings Available to Common Stockholders	8,620
Earnings per Share	.76

Pinewood Company and Subsidiaries Balance Sheet as of End of the Year 1999 ('000 dollars)

Assets:
Cash	$ 5,534
Accounts Receivable (gross)	14,956
Less: Allowance for Bad Debts	211
Accounts Receivable (net)	14,745
Inventory	10,733
Marketable Securities	952
Prepaid Expenses	3,234
Plant and Equipment (gross)	57,340
Less: Accumulated Depreciation	29,080
Plant and Equipment (net)	28,260
Land	1,010
Long-Term Investments	2,503
Total Assets	66,971

Liabilities:
Accounts Payable	3,253
Notes Payable	—
Accrued Expenses	6,821
Bonds Payable	2,389

Stockholders' Equity:
Common Stock	8,549
Retained Earnings	45,959
Total Liabilities and Equity	66,971

5-1. Calculate the following profitability ratios for 1999. *Profit Ratios*

  a. Gross profit margin
  b. Operating profit margin
  c. Net profit margin
  d. Return on assets
  e. Return on equity

Comment on net profit margin and return on assets ratios if the industry average for these two ratios are 5 percent and 14 percent respectively.

5-2. Calculate the following liquidity ratios for the end of 1999. *Liquidity Ratios*

  a. Current ratio
  b. Quick ratio

Comment on the company's ability to pay off short-term debts.

5-3. Calculate the following debt ratios for the end of 1999. *Debt Ratios*

  a. Debt to total assets
  b. Times interest earned

Would a banker agree to extend a loan to Pinewood? Explain.

5-4. Calculate the following asset activity ratios for the end of 1999. *Asset Activity Ratios*

  a. Average collection period
  b. Inventory turnover
  c. Total asset turnover

Comment on Pinewood's asset utilization.

5-5. Construct and solve Pinewood's Modified Du Pont equation for 1999. Use the end of 1999 asset figures. Comment on the company's sources of ROE. *Modified Du Pont Equation*

Du Pont Equation

5-6. The following financial data relate to ABC Textile Company's business in 1999.

Sales	=	$1,000,000
Net Income	=	80,000
Total Assets	=	500,000
Debt to Total Assets Ratio	=	0.5 or 50%

a. Construct and solve the Du Pont and Modified Du Pont equations for ABC.
b. What would be the value of the ROE ratio if the debt to total asset ratio were 70 percent?
c. What would be the value of the ROE ratio if the debt to total asset ratio were 90 percent?
d. What would be the value of the ROE ratio if the debt to total asset ratio were 10 percent?

Financial Relationships

5-7. From the values of the different ratios that follow, calculate the missing balance sheet items and complete the balance sheet.

Sales	$100,000
Average Collection Period	55 days
Inventory Turnover	15
Debt to Assets Ratio	.4 or 40%
Current Ratio	3
Total Asset Turnover	1.6
Fixed Asset Turnover	2.9

Assets		Liabilities + Equity	
Cash	$6,000	Accounts Payable	6,000
Accounts Receivable	_____	Notes Payable	_____
Inventory	_____	Accrued Expenses	600
Prepaid Expenses	_____	Total Current Liabilities	_____
Total Current Assets	_____	Bonds Payable	_____
Fixed Assets	_____	Common Stock	16,000
		Retained Earnings	_____
Total Assets	_____	Total Liabilities + Equity	_____

Financial Relationships

5-8. Given the partial financial statement information from La Strada Corporation, a circus equipment supplier, calculate the return on equity ratio.

Total Assets:	$10,000
Total Liabilities:	6,000
Total Sales:	5,000
Net Profit Margin:	10%

Liquidity Ratios

5-9. What is the current ratio of Ah, Wilderness! Corporation, given the following information from its end of 1999 balance sheet?

Current Assets:	$ 5,000
Long-Term Liabilities:	18,000
Total Liabilities:	20,000
Total Equity:	30,000

Part Two   Essential Concepts in Finance

5-10. Rocinante, Inc. manufactures windmills. What is Rocinante's total asset turnover if its return on assets is 12 percent and its net profit margin is 4 percent? *Du Pont Equation*

Use the following information to answer questions 5-11 to 5-17.
In 1999 Iron Jay opened a small sporting goods retail store called Iron Jay's Sports Stuff (IJSS). It immediately became very popular and growth was only limited by the amount of capital Jay could generate through profits and loans. Jay's financial manager advised him to incorporate. His manager said that by selling stock, Jay would have the necessary capital to expand his business at an accelerated pace. Answer the following questions relating to Iron Jay's Sports Stuff.

5-11. The management team at IJSS is looking toward the future. They want to maintain a gross profit margin of 50 percent. If the estimate for net sales in 2000 is $5 million, how much gross profit will be necessary in 2000 to maintain this ratio? *Profitability Ratios*

5-12. Using the data in 5-11, if the management team estimated $200,000 in selling and administration expenses and $50,000 in depreciation expenses for 2000, with net sales of $5 million, what operating profit margin can they expect? *Profitability Ratios*

5-13. What must net income be in 2000 if IJSS also wants to maintain a net profit margin of 20 percent on net sales of $5 million? *Profitability Ratios*

5-14. What will IJSS's return on assets be if its total assets at the end of 2000 are estimated to be $20 million? Net sales are $5 million and the net profit margin is 20 percent in that year. *Financial Relationships*

5-15. IJSS management knows the astute owners of IJSS stock will sell their stock if the return on stockholders' equity investment (return on equity ratio) drops below 10 percent. Total stockholders' equity for the end of 2000 is estimated to be $15 million. How much net income will IJSS need in 2000 to fulfill the stockholders' expectation of the return on equity ratio of 10 percent? *Profitability Ratios*

5-16. Of the $20 million in total assets estimated for the end of 2000, only $2 million will be classified as noncurrent assets. If current liabilities are $4 million, what will IJSS's current ratio be? *Liquidity Ratios*

5-17. Inventory on the balance sheet for the end of 2000 is expected to be $3 million. With total assets of $20 million, noncurrent assets of $2 million, and current liabilities of $4 million, what will be the value of IJSS's quick ratio? *Liquidity Ratios*

5-18. Given $20 million in total assets, $14 million in total stockholders' equity, and a debt to total asset ratio of 30 percent for Folson Corporation, what will be the debt to equity ratio? *Debt Ratios*

5-19. If total assets are $20 million, noncurrent assets are $2 million, inventory is $3 million, and sales are $5 million for Toronto Brewing Company, what is the inventory turnover ratio? *Asset Activity Ratios*

5-20. If the net profit margin of Dobie's Dog Hotel is maintained at 20 percent and total asset turnover ratio is .25, calculate return on assets. *Du Pont Equation*

5-21. The following data are from Saratoga Farms, Inc. 1999 financial statements. *Du Pont Equation*

Sales	$2,000,000
Net Income	200,000
Total Assets	1,000,000
Debt to Total Asset Ratio	60%

a. Construct and solve the Du Pont and Modified Du Pont equations for Saratoga Farms.
b. What will be the impact on ROE if the debt to total asset ratio were 80 percent?
c. What will be the impact on ROE if the debt to total asset ratio were 20 percent?

Various Ratios

5-22. The following financial information is from two successful retail operations in Niagara Falls. Rose and George Loomis own Notoriously Niagara, a lavish jewelry store that caters to the "personal jet-set" crowd. The other store, Niagara's Notions, is a big hit with the typical tourist. Polly and Ray Cutler, the owners, specialize in inexpensive souvenirs such as postcards, mugs, and T-shirts.

Notoriously Niagara		Niagara's Notions	
Sales	$500,000	Sales	$500,000
Net Income	100,000	Net Income	10,000
Assets	5,000,000	Assets	500,000

a. Calculate the net profit margin for each store.
b. Calculate the total asset turnover for each store.
c. Combine the preceding equations to calculate the return on assets for each store.
d. Why would you expect Notoriously Niagara's net profit margin to be higher than Niagara's Notions considering both stores had annual sales of $500,000 and the same figure for return on assets?

Various Ratios

5-23. Thunder Alley Corporation supplies parts for Indianapolis-type race cars. Current market price per share of Thunder Alley's common stock is $40. The latest annual report showed net income of $2,250,000 and total common stock equity of $15 million. The report also listed 1,750,000 shares of common stock outstanding. No common stock dividends are paid.

a. Calculate Thunder Alley's earnings per share (EPS).
b. Calculate Thunder Alley's price to earnings (P/E) ratio.
c. Calculate Thunder Alley's book value per share.
d. What is Thunder Alley's market to book ratio?
e. Based on this information, does the market believe that the future earning power of Thunder Alley justifies a higher value than could be obtained by liquidating the firm? Why or why not?

Industry Comparisons

5-24. Carrie White, the new financial analyst of Golden Products, Inc., has been given the task of reviewing the performance of her company over three recent years against the following industry information (figures in $000):

Year	Net Income	Current Assets	Current Liabilities	Total Assets	Total Liabilities	Sales
1996	$400	$500	$530	$3,800	$2,600	$4,000
1997	425	520	510	3,900	2,500	4,500
1998	440	550	510	4,000	2,400	4,700

The industry averages are:

NI/Sales	Current Ratio	Total Assets Turnover
9.42%	1.13	2.00

Should Carrie be critical of her company's performance?

5-25. Johnny Hooker, another financial analyst of Golden Products, Inc., is working with the same yearly figures shown in 5-24, but is trying to compare the performance trend using another set of industry averages:

Industry Comparisons

The industry averages are:

Fixed Asset Turnover	Return on Assets	Debt to Equity Ratio	Return on Equity
1.33	11.00%	1.8	26%

Should Johnny be appreciative of his company's performance?

5-26. Vernon Pinkby, the financial analyst reporting to the chief financial officer of Alufab Aluminum Company, is comparing the performance of the company's four separate divisions based on profit margin and return on assets. The relevant figures follow (figures in $000):

Profitability Ratios

	Mining	Smelting	Rolling	Extrusion
Net Income	$ 500	$ 2,600	$ 7,000	$ 2,500
Sales	15,000	30,000	60,000	25,000
Total Assets	12,000	25,000	39,000	18,000

a. Compare profit margin ratios of the four divisions.
b. Compare return on assets ratios of the four divisions.
c. Compute profit margin of the entire company.
d. Compute return on assets of the entire company.

5-27. George Hanson, the financial analyst reporting to the chief financial officer of FabAl Aluminum Company, is comparing the performance of the company's three separate divisions based on profit margin and return on assets. The relevant figures follow (figures in $000):

Challenge Problem

	Mining	Smelting	Rolling
Net Income	$ 600	$ 3,120	$ 8,400
Sales	18,000	36,000	72,000
Total Assets	14,400	30,000	46,800

a. Assuming that it is possible to sell the division with the lowest return on assets at its book value and invest the proceeds in the division with the highest return on assets without changing its value, what will be the return on assets of the entire company?
b. Compute the profit margin of the entire company before and after the proposed sale and transfer of assets assuming that the additional assets will generate additional sales in the same proportion.

5-28. From the values of the different ratios given, calculate the missing balance sheet and income statement items of National Glass Company.

Challenge Problem

Average Collection Period	48.67 days
Inventory Turnover	9×
Debt to Asset Ratio	.4 or 40%
Current Ratio	1.6250
Total Asset Turnover	1.5
Fixed Asset Turnover	2.647
Return on Equity	0.1933 or 19.33%
Return on Assets	0.116 or 11.6%
Operating Profit Margin	13.33%
Gross Profit Margin	48.89%

National Glass Company
Income Statement
for the Year 1999 (000 dollars)

Sales	$45,000
Cost of Goods Sold	_____
Gross Profit	_____
Selling and Administrative Expenses	_____
Depreciation	3,000
Operating Income (EBIT)	_____
Interest Expense	_____
Earnings Before Taxes (EBT)	_____
Income Taxes (@T = 40%)	2,320
Net Income (NI)	_____
Preferred Stock Dividends	0
Earning Available to Common Stockholders	_____

National Glass Company
Balance Sheet
as of End of the Year 1999

	1999
Assets:	$
Cash	_____
Accounts Receivable (gross)	_____
Inventory	_____
Plant and Equipment (net)	_____
Land	1,000
Liabilities:	
Accounts Payable	2,000
Notes Payable	_____
Accrued Expenses	3,000
Bonds Payable	_____
Stockholders' Equity:	
Common Stock	4,000
Retained Earnings	_____

Comprehensive Problem

5-29. Kingston Tools Company (KTC) manufactures various types of high-quality punching and deep-drawing press tools for kitchen appliance manufacturers. Homer Smith, the finance manager of KTC, has submitted a justification to support the application for a short-term loan from the Queensville Interstate Bank (QIB) to finance increased sales. The consolidated income statement and balance sheet of KTC, submitted with the justification to QIB, are on the following page.

Kingston Tools Company
Income Statement for the Years 1999 and 2000 (000 dollars)

	1999	2000
Sales	$40,909	$45,000
Cost of Goods Sold	20,909	23,000
Gross Profit	20,000	22,000
Selling and Administrative Expenses	11,818	13,000
Depreciation	2,000	3,000
Operating Income (EBIT)	6,182	6,000
Interest Expense	400	412
Earnings before Taxes (EBT)	5,782	5,588
Income Taxes (@ 40%)	2,313	2,235
Net Income (NI)	3,469	3,353
Preferred Stock Dividends	—	—
Earnings Available to Common Stockholders	3,469	3,353
Dividends Paid (@ 21.86%)	758	733

Kingston Tools Company
Balance Sheet
as of End of the Year 1999 and 2000 (000 dollars)

	1999	2000
Assets:		
Cash	$ 2,000	$ 1,800
Accounts Receivable (net)	6,000	7,600
Inventory	5,000	5,220
Plant and Equipment (gross)	26,000	31,000
Less: Accumulated Depreciation	10,000	13,000
Plant and Equipment (net)	16,000	18,000
Land	1,000	1,000
Liabilities:		
Accounts Payable	2,000	2,600
Notes Payable	3,000	3,300
Accrued Expenses	3,000	3,100
Bonds Payable	4,000	4,000
Stockholders' Equity:		
Common Stock	4,000	4,000
Retained Earnings	14,000	16,620

You are the loan officer at QIB responsible for determining whether KTC's business is strong enough to be able to repay the loan. To do so, accomplish the following:

a. Calculate the following ratios for 1999 and 2000, compare with the industry averages shown in parentheses, and indicate if the company is doing better or worse than the industry and whether the performance is improving or deteriorating in 2000 as compared to 1999.
  (i) Gross profit margin (50 percent)
  (ii) Operating profit margin (15 percent)
  (iii) Net profit margin (8 percent)
  (iv) Return on assets (10 percent)
  (v) Return on equity (20 percent)
  (vi) Current ratio (1.5)
  (vii) Quick ratio (1.0)

  (viii) Debt to total asset ratio (0.5)
  (ix) Times interest earned (25)
  (x) Average collection period (45 days)
  (xi) Inventory turnover (8)
  (xii) Total asset turnover (1.6)
b. Discuss the financial strengths and weaknesses of KTC.
c. Determine the sources and uses of funds and prepare a statement of cash flows for 2000.
d. Compare and comment on the financial condition as evident from the ratio analysis and the cash flow statement.
e. Which ratios should you analyze more critically before recommending granting of the loan and what is your recommendation?

Answers to Self-Test

ST-1. Sales ÷ $5,000,000 = 4 ∴ sales = $20,000,000
$2,000,000 net income ÷ $20,000,000 sales = .1 = 10% net profit margin

ST-2. Current liabilities = $200,000 total assets − $180,000 LTD, PS, & CS = $20,000
$50,000 current assets ÷ $20,000 current liabilities = 2.5 current ratio

ST-3. Current assets − inventory = $50,000 − (.5 × $50,000) = $25,000
$25,000 ÷ $20,000 current liabilities = 1.25 quick ratio

ST-4. Debt ÷ assets = .5 ∴ equity ÷ assets = .5 ∴ assets ÷ equity = 1 ÷ .5 = 2
ROE = ROA × (A/E)
   = .06 × 2
   = .12 = 12%

ST-5. ($100,000 current assets − inventory) ÷ $25,000 = 1.5 quick ratio ∴ inventory = $62,500
$200,000 sales ÷ $62,500 inventory = 3.2 inventory turnover ratio

ST-6. Debt = $500,000 assets − $200,000 equity = $300,000
$300,000 debt ÷ $500,000 assets = .6 = 60% debt to total asset ratio

ST-7. Credit sales = $4,000,000 ÷ 4 = $1,000,000
Average collection period = accounts receivable ÷ average daily credit sales = $100,000 accounts receivable ÷ ($1,000,000 annual credit sales ÷ 365 days per year) = 36.5 days

CHAPTER 6

FORECASTING FOR FINANCIAL PLANNING

CHAPTER OBJECTIVES

After reading this chapter, you should be able to:

- Explain why forecasting is vital to business success.
- Describe the financial statement forecasting process.
- Prepare *pro forma* (projected) financial statements.
- Explain the importance of analyzing forecasts.

An economist's guess is liable to be as good as anybody else's.
—Will Rogers

DELTA AIRLINES: DELTA SPONSORS ECONOMIC FORECASTING CENTER AT GEORGIA STATE UNIVERSITY

Delta Air Lines today announced corporate sponsorship of the Economic Forecasting Center (EFC) at Georgia State University. The nationally recognized center, founded by renowned economist Dr. Donald Ratajczak, has been consistently ranked among the top economic forecasting centers in the country.

"Delta is pleased to join the ranks of the Economic Forecasting Center's other prestigious corporate sponsors," said Paul Matsen, Delta's senior vice president for corporate planning and information technologies. "The decision to enter into this important partnership was closely linked to our recognition of the importance of using solid economic research in Delta's strategic planning efforts. The Economic Forecasting Center's reputation is unparalleled in terms of providing high quality economic research and timely economic forecasts."

As a corporate sponsor of the Economic Forecasting Center at Georgia State University, Delta joins more than 25 other major corporations, including the Coca-Cola Company, BellSouth Corporation, and NationsBank Corporation, whose funding supports ongoing economic research efforts.

"By monitoring transitions in the economy, successful organizations can anticipate changes and react to them accordingly," said Ratajczak. "Most of our sponsors have found that use of our research allows them to maximize opportunities and minimize threats from the economic environment. Simply put, corporations today know that an accurate, dependable forecast is vital for any long or short-term strategic planning. We're proud to have Delta Air Lines join us in our efforts to provide the highest quality economic research available in the world."

"This sponsorship reflects the airline's ongoing commitment to the Atlanta community," said Matsen. "It extends an already strong and supportive relationship with Georgia State University that was initiated with the founding of the University's Beebe Institute. Delta's financial support will help provide the center with the resources required to instill excellence in tomorrow's economists. We're looking ahead to a long and successful partnership with this renowned academic center for economic research."

Delta has served as the major sponsor of Georgia State University's Beebe Institute since 1987. The Beebe Institute provides one of the nation's only academic centers dedicated to the study of personnel and human resource management.

The Economic Forecasting Center was established in 1973 in the College of Business Administration at Georgia State University. The center conducts research and develops forecasts of the national, regional, and local economies. It works closely with the business community and provides organizations with important information about economic conditions that may affect their strategic planning and decision making.

Delta Air Lines carries more passengers worldwide than any other airline. Delta, Delta Express, the Delta Shuttle, the Delta Connection carriers and Delta's Worldwide Partners operate 4,881 flights each day to 324 cities in 45 countries.

Source: M2 Press WIRE © 1998 M2 Communications. Used by permission of M2 Comunications Ltd. M2 Press WIRE, ⟨http://www.presswire.net⟩.

Chapter Overview

A business owner who wants to run a successful business must be able to answer many questions about the future, including the following:

- How much profit will your business make?
- How much demand will there be for your product or service?
- How much will it cost to produce your product or offer your service?
- How much money will you need to borrow, when, and how will you pay it back?

Business people must estimate the future all the time. The task of estimating future business events is daunting and darn near impossible, but in business it is necessary. Without some idea of what is going to happen in the future, it is impossible to plan and to succeed in business.

We will look first at the forecasting task, discuss why it is important, and explain what forecasting approaches business people use. Then, step by step, we will build a set of projected financial statements. We will conclude with a discussion of how to analyze forecasts to plan for the financial success of the company.

Why Forecasting Is Important

Every day you make decisions based on forecasts. When you go shopping, for example, you decide how much money to spend based on your forecast of how much money you need for other reasons. When you plan trips, you decide how much money to take along based on your forecast of the trip's expenses. You choose what to wear in the morning based on the weather forecaster's prediction of good or bad weather.

The situation is similar in business—particularly in finance. Financial decisions are based on forecasts of situations a business expects to confront in the future. Businesses develop new products, set production quotas, and select financing sources based on forecasts about the future economic environment and the firm's condition. If economists predict interest rates will be relatively high, for example, a firm may plan to limit borrowing and defer expansion plans.

Forecasting in business is especially important because failing to anticipate future trends can be devastating. The following business examples show why:

- Sony Corporation's devotion to beta-formatted videotapes caused the firm to miss out on millions of dollars in sales of VCRs formatted for VHS.
- Cray Computer Corporation's underestimate of the time and cost to develop new supercomputers, coupled with a drop in demand for such machines, caused the company to go bankrupt in 1995.
- Coca-Cola's introduction of New Coke in 1985 cost millions of dollars and failed miserably. Coca-Cola suffered another marketing flop in 1988 with the development of BreakMate, a small microwave-oven-sized device designed to replace soda vending machines in small offices.[1]

Undoubtedly, each of these firms wishes that their forecasts had been more accurate.

Firms often spend a large amount of time, effort, and money to obtain accurate forecasts. Let's take a look at some of the ways forecasters approach the forecasting task.

Forecasting Approaches

Forecasting simply means making projections about what we think will happen in the future. To make these projections, forecasters use different approaches depending on the situation. Financial managers concentrate on three general approaches: *experience, probability,* and *correlation.*

Experience Sometimes we think things will happen a certain way in the future because they happened that way in the past. For instance, if it has always taken you

---

[1]Michael A. Cusamo, et al., "Strategic Maneuvering and Mass-Market Dynamics: The Triumph of VHS over Beta," *Business History Review* 66, No. 1 (Spring 1992): 51–94; "How the Numbers Crunched Cray Computer," *Business Week* (April 10, 1995): retrieved from Internet; Pat Selig, "All Over the Map," *Sales & Marketing Management* 141, No. 4 (March 1989): 58–64.

15 minutes to drive to the grocery store, then you will probably assume that it will take you about 15 minutes the next time you drive to the store. Similarly, financial managers often assume sales, expenses, or earnings will grow at certain rates in the future because that is how the rates grew in the past.

Probability Sometimes we think things will happen a certain way in the future because the laws of probability indicate that it will be so. For example, let's say insurance company statisticians calculate that male drivers under 25 years of age during a one-year period will have a .25 probability of having an accident. That company insures 10,000 male drivers under 25, so the company's financial manager—based on probability figures—would forecast that 2,500 of the firm's under-25 male policyholders (.25 × 10,000 = 2,500) will have an accident during the coming year. Financial managers use probabilities to forecast the number of customers who won't pay their bills, the number of rejects in a production process, and so on.

Correlation Sometimes we think things will happen a certain way in the future because there is a high correlation between the behavior of one item and the behavior of another item that we know more about. For example, if you know that most people buy new cars right after they graduate from college, you could forecast new car sales to recent graduates fairly accurately. You would base your plans on the amount of correlation between the number of people graduating and new car sales. In a similar manner, financial managers forecast more sales for snow tires in the winter and more sales of lawn mowers in the summer.

Why Forecasts Are Sometimes Wrong

In general, forecasting the future is based on what has happened in the past (either past experience, past probability, or past correlation). However, just because something has occurred in the past does not *ensure* that it will happen the same way in the future, which is why forecasts are sometimes spectacularly wrong. No one can forecast the future with absolute precision.

To illustrate, in 1987, Dr. Ravi Batra wrote in *The Great Depression of 1990,* that "at the end of 1989 or in the first half of 1990, the stock market will crash and will be followed by an abysmal decline in business activity and a sharply higher rate of unemployment." He added, "We can actually pinpoint 1990 as the year of the world's greatest depression."[2] Fortunately, Dr. Batra's forecast did not come true.

The approaches to forecasting range from being quite simple to complex and sophisticated, depending on what is being forecast and how much the business needs to rely on an accurate forecast. In the sections that follow, we will examine some of the approaches that business people use to predict future sales, and the way that finance people predict their firms' future.

FORECASTING SALES

Producing a sales forecast is not purely a financial task. Estimates of future sales depend on the demand for the firm's products and the strength of the competition. Sales and marketing personnel usually provide assessments of demand and the com-

---

[2]Dr. Ravi Batra, *The Great Depression of 1990* (New York: Simon & Schuster, 1987): 133, 140.

petition. Production personnel usually provide estimates of manufacturing capacity and other production constraints. Top management will make strategic decisions affecting the firm as a whole. Sales forecasting, then, is a group effort. Financial managers coordinate, collect, and analyze the sales forecasting information. Figure 6-1 shows a diagram of the process.

Sometimes financial analysts make a quick estimate of a company's future sales by extending the trend of past sales. Figure 6-2 illustrates this technique, with the sales record of Esoteric Enterprises, Inc., for the years 1994 to 1999.

The graph in Figure 6-2 shows that Esoteric's sales have been somewhat constant during the five-year period. A forecaster, by extending the past trend, would estimate that Esoteric's sales in the next year are likely to be about $222,000. The technique of extending a past trend works well when the past trend is discernible and no outside factors, such as new competition, act to change the trend.

Figure 6-1 **The Sales Forecasting Process**

This chart shows how a company's sales forecast is developed from many different sources. Marketing data, company goals, production capabilities, and accounting data are analyzed, weighed, and combined to produce the final sales estimate.

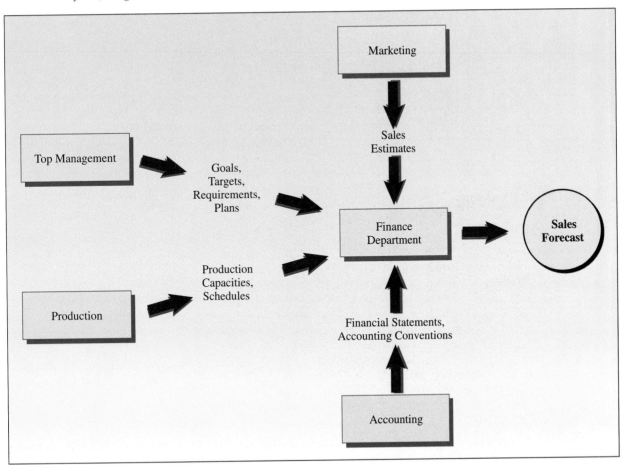

Figure 6-2 Sales Record, Esoteric Enterprises

This chart shows the sales record for Esoteric Enterprises, Inc., during 1994 to 1999. Sales growth has been fairly constant during the five-year period. By extending the sales trend, an analyst might estimate Esoteric's sales in 2000 to be about $222,000.

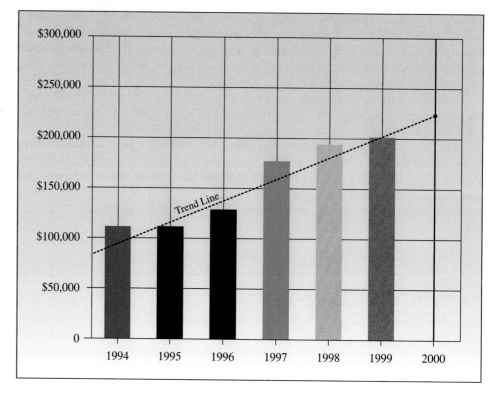

Forecasting Financial Statements

After the sales forecast is complete, financial managers determine whether the firm can support the sales forecast financially. They do this by extending the firm's financial statements into future time periods. These forecasted financial statements are commonly referred to as **pro forma financial statements**.[3] *Pro forma* financial statements show what the firm will look like if the sales forecasts are indeed realized and management's plans carried out. By analyzing the projected financial statements, managers can tell if funds needed to make purchases will be available, if the firm will be able to make loan payments on schedule, if external financing will be required, what return on investment the stockholders can expect, and so on.

Take Note

Developing a cash budget and a capital budget is a complex matter. We discuss cash budgets in chapter 18, and capital budgets in chapters 10 and 11.

Budgets

Financial managers use a variety of budgets to help produce *pro forma* financial statements. Budgets contain estimates of future receipts and expenditures for various activities. Financial managers produce *pro forma* statements that assume the budget figures will, in fact, occur.

Two budgets are particularly important to forecasters. These are the *cash budget* and the *capital budget*. The **cash budget** shows the projected flow of cash

---

[3] *Pro forma* is a Latin term meaning "as a matter of form." *Pro forma* financial statements show what the business will look like (its form) if expected events take place.

122   Part Two   Essential Concepts in Finance

in and out of the firm for specified time periods. The **capital budget** shows planned expenditures for major asset acquisitions. In the sections that follow, we will see how forecasters incorporate data from these budgets into *pro forma* financial statements.

Producing *Pro Forma* Financial Statements

Now that we have reviewed the sales forecast and the budgets from which data are drawn to produce *pro forma* financial statements, we can discuss how to produce *pro forma* financial statements. In the following sections, we explore the step-by-step process of creating both a *pro forma* income statement and a *pro forma* balance sheet for Esoteric Enterprises, Inc.

Esoteric Enterprises makes one product: a rechargeable lithium battery used in industrial facilities throughout the United States to power emergency lights. The company's income statement and balance sheet for the year 1999 are shown in the first column of Figures 6-3a and 6-3b. We will create a *pro forma* income statement and balance sheet for 2000 by filling in the blanks in the second column of Figures 6-3a and 6-3b.

For convenience, assume that we are preparing this forecast on January 1, 2000. (Being dedicated, we have come in on New Year's Day to do this.)

Choosing the Forecasting Basis Before we make *pro forma* forecasts of each income statement and balance sheet line item, let's consider our procedure. Unfortunately, no universal procedure applies to all line items on *pro forma* financial statements because forecasters choose values for a variety of reasons. The three main reasons are: (1) Management specifies a target goal, (2) the value is taken from either the cash or capital budgets, or (3) the value is an extension of a past trend. If management does not specify a target value, and if no value from a budget is available,

	1999	FORECAST FOR 2000
Net Sales	$201,734	_____
Cost of Goods Sold	107,280	_____
Gross Profit	94,454	_____
Operating Expenses:		
Selling and Marketing Expenses	32,392	_____
General and Administrative Expense	10,837	_____
Depreciation	4,500	_____
Operating Income	46,725	_____
Interest Expense	2,971	_____
Before-Tax Income	43,754	_____
Income Taxes (tax rate = 40%)	17,502	_____
Net Income	$26,252	_____
Dividends Paid	23,627	_____
Addition to Retained Earnings	2,625	_____

Figure 6-3a **Esoteric Enterprises Income Statements**

The first column in this figure shows Esoteric Enterprises' income statement for 1999. The pro forma forecast for 2000 will be inserted in the second column.

Figure 6-3b Esoteric Enterprises Balance Sheets

The first column in this figure shows Esoteric Enterprises' balance sheet as of December 31, 1999. The pro forma *forecast for December 31, 2000, will be inserted in the second column.*

	ACTUAL DEC. 31 1999	FORECAST FOR DEC. 31 2000
Assets:		
Current Assets:		
Cash and Marketable Securities	$ 65,313	_____
Accounts Receivable	13,035	_____
Inventory	21,453	_____
Total Current Assets	99,801	_____
Property, Plant, and Equipment (gross)	133,369	_____
Less Accumulated Depreciation	(40,386)	_____
Property, Plant, and Equipment (net)	92,983	_____
Total Assets	$192,784	_____
Liabilities and Equity:		
Current Liabilities:		
Accounts Payable	$ 4,733	_____
Notes Payable	302	_____
Total Current Liabilities	5,035	_____
Long-Term Debt	37,142	_____
Total Liabilities	42,177	_____
Common Stock		
(35 million shares, $1.00 par value)	35,000	_____
Capital in Excess of Par	32,100	_____
Retained Earnings	83,507	_____
Total Stockholders' Equity	150,607	_____
Total Liabilities and Equity	$192,784	_____

then forecasters must evaluate how the item has behaved in the past and estimate a future value based on experience, probability, or correlation, as discussed earlier in the chapter.

As we consider each financial statement item on Esoteric's *pro forma* statements, we will determine its value by seeing whether management has a target goal, and whether a budget sets the value for the item. If not, we will extend the trend based on experience, probability, and correlation.

Let's begin with Esoteric Industries' *pro forma* income statement for 2000.

The *Pro Forma* Income Statement To prepare the *pro forma* income statement, we project the values of the following items: sales, costs and expenses associated with sales, general and administrative expenses, depreciation, interest, taxes, dividends paid, and addition to retained earnings. We will examine how to project each value next.

The Sales Projection. At the top of our *pro forma* income statement for 2000 (shown in Figure 6-3a), we need a figure for sales. Assume that our analysis of marketing, production, finance, and management information results in a sales forecast of $221,907. Enter this figure in the "Forecast for 2000" column of Figure 6-3a, as shown:

	FORECAST FOR 2000
Net Sales	$221,907

Cost of Goods Sold (COGS) and Selling and Marketing Expenses. After sales, the next two items are cost of goods sold (COGS) and selling and marketing expenses. We do not have a management target or budget figure for these expenses, so we will forecast them based on past experience. For Esoteric Enterprises, experience suggests that over time both these items have remained a constant percentage of sales. That is, over the years, COGS has been about 53 percent of sales and selling and marketing expenses have been about 16 percent of sales. So we conclude that in 2000 these items will be 53 percent and 16 percent of sales, respectively, shown as follows:

	FORECAST FOR 2000
Cost of Goods Sold	$221,907 × .53 = $117,611
Selling and Marketing Expenses	$221,907 × .16 = $35,505

General and Administrative Expenses. General and administrative expenses are closely related to the size of Esoteric's manufacturing plant. For our 2000 forecast, we assume that Esoteric's property, plant, and equipment will not change for 2000. This means that our projected value for general and administrative expenses is $10,837, the same value as in the previous year:

	FORECAST FOR 2000
General and Administrative Expenses	$10,837

Depreciation Expense. For our depreciation expenses forecast, let us say that Esoteric Enterprises' capital budget does not include the purchase of any additional property, plant, or equipment, and that no equipment is near the end of its projected useful life. The projected depreciation expense, then, will be $4,500, the same value as it was for 1999.

For simplicity we use the straight-line depreciation method instead of MACRS to determine depreciation expense for 2000. As a result, we obtain a constant depreciation expense value as long as no equipment is replaced.

	FORECAST FOR 2000
Depreciation Expense	$4,500

Interest Expense. The amount of interest to be paid in 2000 depends on the amount of debt outstanding in that year. That hasn't been determined yet because it is part of the balance sheet forecast. At this point we have no information indicating new debt will be obtained or old debt paid off, so we will project that interest expense in 2000 will be $2,971, the same as its 1999 value.

	FORECAST FOR 2000
Interest Expense	$2,971

Figure 6-4 Esoteric Enterprises *Pro Forma* Income Statement for 2000

This figure shows Esoteric Enterprises' anticipated 2000 values for each income statement line item. Each forecast value was calculated separately, according to the forecasting assumptions given.

	FORECAST FOR 2000
Net Sales	$221,907
Cost of Goods Sold	117,611
Gross Profit	104,296
Operating Expenses:	
Selling and Marketing Expenses	35,505
General and Administrative Expenses	10,837
Depreciation	4,500
Operating Income	53,454
Interest Expense	2,971
Before-Tax Income	50,483
Income Taxes (tax rate = 40%)	20,193
Net Income	$30,290
Dividends Paid	27,261
Addition to Retained Earnings	3,029

As we see in Figure 6-4, Esoteric's before-tax income is expected to be $50,483.

Income Taxes. Esoteric's effective 1999 tax rate, shown in Figure 6-3a, is 40 percent.[4] We assume no changes in the tax law for 2000, so to obtain income tax expense for 2000, multiply the 2000 before-tax income by 40 percent, as follows:

	FORECAST FOR 2000
Income Tax Expense	$50,483 × .40 = $20,193

Dividends Paid and Additions to Retained Earnings. Esoteric's management plans to continue the current dividend policy of paying 90 percent of net income in dividends and retaining 10 percent in 2000. We forecast net income in 2000 as $30,290 (see Figure 6-4), so dividends paid and the addition to retained earnings will be as follows:

	FORECAST FOR 2000
Dividends Paid	$30,290 × .90 = $27,261
Addition to Retained Earnings	$30,290 × .10 = $3,029

This completes our *pro forma* income statement. The results are summarized in Figure 6-4. Now let's turn to the *pro forma* balance sheet.

---

[4]We assume that Esoteric pays 35 percent of its income to the federal government and 5 percent to the state. Caution: When you are forecasting, be sure to check the latest tax rate schedule set by Congress.

The *Pro Forma* Balance Sheet   Now we will create the *pro forma* balance sheet for 2000 (December 31) by examining each individual line item account. If no target value is specified by management, and if no value from a budget is available, then we will evaluate the item's past performance and estimate its future value based on experience, probability, or correlation.

Cash and Marketable Securities.   The forecast value for cash and marketable securities is normally drawn from the company's *cash budget,* as discussed earlier. Let's assume that financial managers at Esoteric have prepared a cash budget that predicts the amount of cash on hand at the end of 2000 will be $71,853.

Accounts Receivable and Inventory.   Experience has shown that the accounts receivable and inventory accounts tend to remain the same percentage of sales, similar to the cost of goods sold and selling and marketing expenses on the income statement. In the past, accounts receivable has been 6 percent of sales and inventory has been 11 percent of sales. Therefore, we will assume that these items will be 6 percent and 11 percent of 2000 sales, respectively, at the end of 2000:

	FORECAST FOR 2000
Accounts Receivable	$221,907 × .06 = $13,314
Inventory	$221,907 × .11 = $24,410

Property, Plant, and Equipment.   Esoteric's capital budget does not include any provision for purchasing production equipment, buildings, or land. In our income statement forecast, we assumed that in 2000 Esoteric Enterprises will not need any additional equipment, no equipment will be disposed of, and no equipment will reach the end of its useful life. Property, plant, and equipment gross at the end of 2000, then, will be the same as its end of 1999 value of $133,369. Property, plant, and equipment net will be the 1999 gross value less the additional depreciation expense ($4,500) accumulated during 2000. Here are the calculations:

		FORECAST FOR 2000
		12/31/00
Property, Plant, and Equipment (gross)	(Same as 12/31/99)	$133,369
Less		—
Accumulated Depreciation		$40,386 + $4,500 = $44,886
[end of 1999 accumulated depreciation ($40,386) plus 2000 depreciation expense ($4,500)]		
Property, Plant, and Equipment (net)		$88,483

Accounts Payable.   Experience has shown that, like accounts receivable and inventory, accounts payable tends to remain the same percentage of sales. In the past accounts payable has been about 2 percent of sales. Therefore, we will assume that accounts payable at the end of 2000 will be 2 percent of 2000 sales. Here are the calculations:

	FORECAST FOR 2000
	12/31/00
Accounts Payable	$221,907 × .02 = $4,438

Notes Payable. We assume based on experience and management policy that any notes outstanding at the end of a year will be paid off by the end of the following year, resulting in a zero balance in the notes payable account. Accordingly, Esoteric's notes payable value for the end of 2000 will be $0, shown as follows:

	FORECAST FOR 2000
	12/31/00
Notes Payable	$0

Long-Term Debt. We will assume that no principal payments on Esoteric's long-term debt are due in 2000, and no new debt financing arrangements have been made yet. Therefore, the long-term debt value at the end of 2000 will be the same as the end of 1999 value, $37,142.

	FORECAST FOR 2000
	12/31/00
Long-Term Debt (Same as 1999 value)	$37,142

Common Stock and Capital in Excess of Par. Esoteric's management has no plans to issue or to buy back stock in 2000. The common stock and capital in excess of par values, then, will remain the same at the end of 2000 as they were at the end of 1999, $35,000 and $32,100, respectively. The forecast follows:

	FORECAST FOR 2000
	12/31/00
Common Stock (same as 1999 value)	$35,000
Capital in Excess of Par (same as 1999 value)	$32,100

We discussed retained earnings in chapter 4, on page 68.

Retained Earnings. As discussed in chapter 4, the retained earnings account represents the sum of all net income not paid out in the form of dividends to stockholders.[5]

---

[5]Esoteric has no preferred stock and, therefore, no preferred stock dividends paid. This means that its net income is the same as its earnings available to common stockholders discussed in chapters 4 and 5.

At the end of 2000, the retained earnings value will be the total of the end of 1999 figure ($83,507) plus the 2000 addition to retained earnings ($3,029), as shown on the income statement forecast.

	FORECAST FOR 2000
	12/31/00
Retained Earnings	$83,507 + $3,029 = $86,536

This completes our *pro forma* balance sheet. Figure 6-5 summarizes the results.

Now both our *pro forma* financial statements for Esoteric Enterprises are complete. Figures 6-6a and 6-6b contain the complete 2000 forecast, including source notes explaining the reasons for each forecasted item's value.

Additional Funds Needed When the *pro forma* balance sheet is completed, total assets and total liabilities and equity will rarely match. Our forecast in Figure 6-6b—in which total assets are forecast to be $198,060, but total liabilities and equity are forecast to be only $195,216—is typical. The discrepancy between forecasted assets and forecasted liabilities and equity ($2,844 in our example) results

	FORECAST FOR 12/31/00
Assets:	
Current Assets:	
Cash and Marketable Securities	$ 71,853
Accounts Receivable	13,314
Inventory	24,410
Total Current Assets	109,577
Property, Plant, and Equipment (gross)	133,369
Less Accumulated Depreciation	(44,886)
Property, Plant, and Equipment (net)	88,483
Total Assets	$198,060
Liabilities and Equity:	
Current Liabilities:	
Accounts Payable	$ 4,438
Notes Payable	0
Total Current Liabilities	4,438
Long-Term Debt	37,142
Total Liabilities	$ 41,580
Common Stock	35,000
Capital in Excess of Par	32,100
Retained Earnings	86,536
Total Stockholders' Equity	153,616
Total Liabilities and Equity	$195,216

Figure 6-5 Esoteric Enterprises *Pro Forma* Balance Sheet for December 31, 2000

This figure shows Esoteric Enterprises' projected end of 2000 values for each balance sheet line item. Each forecasted value was calculated separately by assessing management goals, budget figures, or past trends.

Figure 6-6a Esoteric Enterprises Income Statements

The first column in this figure shows Esoteric Enterprises' income statement for 1999. The second column shows the pro forma forecast for 2000. The last column contains notes on where each line item value was obtained.

	1999	FORECAST FOR 2000	SOURCE NOTES
Net Sales	$201,734	$221,907	Sales forecast
Cost of Goods Sold	107,280	117,611	53% of sales
Gross Profit	94,454	104,296	
Operating Expenses:			
Selling and Marketing Expenses	32,392	35,505	16% of sales
General and Administrative Expenses	10,837	10,837	Keep same
Depreciation	4,500	4,500	Keep same
Operating Income	46,725	53,454	
Interest Expense	2,971	2,971	Keep same
Before-Tax Income	43,754	50,483	
Income Taxes (tax rate = 40%)	17,502	20,193	Same tax rate
Net Income	$26,252	$30,290	
Dividends Paid	$23,627	$27,261	Same payout policy (90%)
Addition to Retained Earnings	$2,625	$3,029	Net income − dividends paid

Raising funds in the financial marketplace is covered in chapter 2.

when either too little or too much financing is projected for the amount of asset growth expected. The discrepancy is called **additional funds needed** (AFN) when forecasted assets exceed forecasted liabilities and equity. It is called **excess financing** when forecasted liabilities and equity exceed forecasted assets. Our forecast indicates that $2,844 of additional funds are needed to support Esoteric's needed asset growth.

The determination of additional funds needed is one of the most important reasons for producing *pro forma* financial statements. Armed with the knowledge of how much additional external funding is needed, financial managers can make the necessary financing arrangements in the financial markets before a crisis occurs. Esoteric only needs a small amount, $2,844, so the company would probably obtain the funds from a line of credit with its bank. When large amounts are required, other funding sources include a new bond or stock issue.

A Note on Interest Expense. According to Esoteric's *pro forma* financial statements, the company needs $2,844 of new external financing in 2000. If it borrows the money (as we implied it would), then Esoteric will incur new interest charges that were not included in the original *pro forma* income statement. To be accurate, forecasters should revise the *pro forma* income statement to include the new interest. However, if they make the revision, it will reduce 2000's net income—which, in turn, will reduce 2000's retained earnings on the balance sheet forecast. This will change the total liabilities and equity figure for 2000, throwing the balance sheet out of balance again and changing the amount of additional funds needed!

	ACTUAL 12/31/99	FORECAST 12/31/00	SOURCE NOTES
Assets			
Current Assets:			
Cash and Marketable Securities	$ 65,313	$ 71,853	Cash budget
Accounts Receivable	13,035	13,314	6% of sales
Inventory	21,453	24,410	11% of sales
Total Current Assets	99,801	109,577	
Property, Plant, and Equipment (gross)	133,369	133,369	Capital budget (keep same)
Less Accumulated Depreciation	(40,386)	(44,886)	1999 plus 2000 depreciation expense
Property, Plant, and Equipment (net)	92,983	88,483	
Total Assets	$192,784	$198,060	
Liabilities and Equity			
Current Liabilities:			
Accounts Payable	$ 4,733	$ 4,438	2% of sales
Notes Payable	302	0	Pay off
Total Current Liabilities	5,035	4,438	
Long-Term Debt	37,142	37,142	Keep same
Total Liabilities	42,177	41,580	
Common Stock			
(35 million shares, $1.00 par value)	35,000	35,000	Keep same
Additional Paid-In Capital	32,100	32,100	Keep same
Retained Earnings	83,507	86,536	1999 end of year retained earnings + 2000 addition to retained earnings
Total Stockholders' Equity	150,607	153,636	
Total Liabilities and Equity	$192,784	$195,216	2000 AFN = $2,844

Figure 6-6b Esoteric Enterprises Balance Sheets

The first column in this figure shows Esoteric Enterprises' balance sheet for December 31, 1999. The second column shows the pro forma forecast for December 31, 2000. The last column contains notes on where each line item value was obtained.

In forecasting circles, this is known as the *balancing problem*. If the forecast is done on an electronic spreadsheet, it is not difficult to recast the financial statements several times over until the additional amount of interest expense becomes negligible. In this chapter, however, to avoid repeating the forecast over and over, we will simply stay with our original interest expense figure.

Analyzing Forecasts for Financial Planning

The most important forecasting task begins after the *pro forma* financial statements are complete. At that time, financial managers must analyze the forecast to determine:

1. What current trends suggest will happen to the firm in the future
2. What effect management's current plans and budgets will have on the firm
3. What actions to take to avoid problems revealed in the *pro forma* statements

FINANCE AT WORK

SALES·RETAIL·SPORTS·MEDIA TECHNOLOGY·PUBLIC RELATIONS·PRODUCTION·EXPORTS

Keith Ender, Customer Service Representative, James River Paper Company

Keith Ender is customer service manager for the James River Paper Company, manufacturer of Brawny paper towels, Northern bathroom tissue, Dixie paper cups, and a variety of other products. Keith helps customers who buy in large quantities. For example, if a big grocery chain orders a huge amount of Brawny paper towels, Keith's department handles every aspect of the sale, from the initial order to the final payment.

According to Keith, forecasting is vital to the success of James River. He points out, however, that it is an inexact science. Internal and external factors can throw off even the best forecast.

Q. Just how important is forecasting to your business?

Our success is tied to accurate forecasts. Each department, including my own, makes assumptions about what will happen in the future by forecasting its needs and how it will perform. To help the company produce accurate financial forecasts, though, we must start with accurate sales forecasts. My department, for instance, may forecast how much product it will sell if given the resources to move the maximum amount of product.

To help with the sales forecast, we could eyeball sales trends and say, "We're going to sell this amount of Brawny paper towels, and this number of Dixie cups." Now, let's say that halfway through the year we realize that our sales team will sell fewer paper towels and more Dixie cups than forecasted. A chain reaction starts. Production will be affected because they planned, based on our sales forecast, to make more paper towels and less cups than we need. Production may not have the capacity, money, or materials to adjust to the need for more cups immediately. If we can't meet consumer demand, or have too much inventory, our projected revenues are affected. We lose sales revenue and so do our customers. In fact, a customer might withhold payments they owe us, claiming we caused them to lose business because they didn't have enough cups to meet their customers' demand.

In short, there are a lot of negative repercussions to the forecast if our sales projections are not on track.

Q. What kind of things can throw off your forecast?

Sometimes special promotions can affect our forecasts. We must always try to accurately cover the cost of sales and make the best net profit margin possible. But that doesn't always happen. Say we decide to push Brawny paper towels. To help, marketing offers customers a special promotion fee to sell the towels. Sometimes the promotion works too well, and we end up selling more towels than forecasted at that special promotion price. Now, we may be selling like mad, but our profitability is dropping, because we're basically selling the towels for a lower price. If we hadn't offered the promotion, more towels might have been sold for a higher profit to other customers.

Another thing that can throw off forecasts is the price of raw materials. These prices can fluctuate wildly. For example, the cost of paper, which we are totally dependent on, recently climbed to record highs that we didn't expect. As a result, our cost of goods sold will be much higher than our forecasted cost. The increased cost will reduce our profitability significantly.

Q. How can competition affect your forecasting?

Sometimes it's tough to forecast revenues accurately because we don't know what our competitors are going to do. They could decide to increase their promotional spending and take a short-term decrease in revenues if, in the long run, they will benefit. Say that a large, diversified competitor wants to increase its market share of the paper towel market, so it runs special promotions on its paper towels. Because it is such a large company, it may lose money for a while, but it will be worth it if its paper towels gain market share. And again, because the company is large and diversified, it can make up for the paper towel promotion if the profit it makes selling soap or other products can cover the cost.

The point is, if their promotion works, they may take away some of our market share. Once market share is lost, it is tough to regain. To maintain our market share, we often have to react quickly by lowering the price of our paper towels to compete—but that lower price shoots the heck out of our profit forecast.

FINANCE AT WORK—Continued

Q. How important is understanding finance to your career?

Very important. Although I don't have a finance background—I was a journalism major in college—I have many financial job responsibilities. I budget for our department, I handle incoming cash, and I help with sales forecasts. I review my department's financial numbers all the time.

Q. Do you have any advice for business students?

Learn as much about finance as you can, no matter what field you choose. In college, I was always intimidated by any course that dealt with numbers. I wish I hadn't been so intimidated because as I progressed in my work career, I learned I couldn't avoid numbers. The more I progressed, the more familiar I had to be with finance because finance is the language of business.

Source: Interview with Keith Ender.

When analyzing the *pro forma* statements, financial managers often see signs of emerging positive or negative conditions. If forecasters discover positive indicators, they will recommend that management continue its current plans. If forecasters see negative indicators, they will recommend corrective action.

To illustrate, let's see how Esoteric Enterprises' financial managers would analyze the company's *pro forma* financial statements and plan for the future.

Scanning the first column in Figure 6-6a, we calculate Esoteric Enterprises' 1999 net profit margin (net income divided by sales) as follows:

$$\text{Current Net Profit Margin} = \$26{,}252 \,/\, \$201{,}734 = .1301, \text{ or } 13.01\%$$

Using the figures from the *pro forma* forecast (see Figure 6-6a, column 2), Esoteric's forecasted net profit margin in 2000 is as follows:

$$\text{Forecasted Net Profit Margin} = \$30{,}290 \,/\, \$221{,}907 = .1365, \text{ or } 13.65\%$$

The expected increase in the net profit margin from 13.01 percent to 13.65 percent is a desirable trend, so Esoteric's financial managers will probably recommend that the business maintain its current course of action.

However, if the analysis had shown a projected decline in the net profit margin to 11 percent, the financial managers would try to determine the cause of this decline (perhaps administrative expenses are too high or asset productivity is too low). Once the financial managers find the cause, they would recommend appropriate corrective action. Once the company adjusts its plans to correct the problem, the financial managers would prepare a new set of *pro forma* financial statements that reflect the changes.

We presented a brief example to illustrate the process of analyzing *pro forma* financial statements. A complete analysis would involve calculating profitability ratios, asset productivity ratios, liquidity ratios, and debt management ratios, as described in chapter 5.

What's Next

In this chapter we described forecasting and prepared *pro forma* statements. In chapter 7 we turn to the risk/return relationship, one of the key concepts of finance.

Summary

1. **Explain the need for forecasting.**

Business planning is based on forecasts of the company's future financial performance. Without forecasting, a business cannot succeed. Incorrect forecasts can be costly—so costly, in some cases, that the mistakes lead to failure.

2. **Describe the financial statement forecasting process.**

Forecasting means making assumptions about what will happen in the future. The three main approaches to making these assumptions are:

- *Experience.* We assume things will happen a certain way in the future because they have happened that way in the past.
- *Probability.* We assume things will happen a certain way in the future because the laws of probability indicate that it will be so.
- *Correlation.* We assume things will happen a certain way in the future because of a high correlation between the thing we are interested in and another thing we know more about.

Financial managers use the sales forecast, a variety of budgets, and past trend information to produce financial statements for periods in the future. These projected financial statements are *pro forma* financial statements. *Pro forma* financial statements show what assets, liabilities, and equity a firm is expected to have in the future.

3. **Producing pro forma *financial statements*.**

Pro forma financial statements are based on a company's current financial statements. The forecasted value of each current financial statement line item is determined by a target specified by management, a value extracted from a budget, or an extension of a past trend. In *pro forma* financial statement preparation, no general rule can be applied universally to all line items. Instead, each item must be examined individually. If no target value is specified by management, and if no value from a budget is available, then forecasters must evaluate the past performance of the account and estimate a future value based on experience, probability, or correlation.

On the *pro forma* balance sheet the forecasted values for total assets and total liabilities and equity rarely match. When forecasted assets exceed forecasted liabilities and equity, the difference is called additional funds needed (AFN). When forecasted liabilities and equity exceed forecasted assets, the difference is called excess financing. Additional funds needed is the additional external financing required to support projected asset growth. Excess financing means that too much funding has been set aside for expected asset growth.

4. **Explain the importance of analyzing forecasts.**

Once the *pro forma* financial statements are complete, financial managers must analyze them to determine if the company should continue with its current plans (as in the case of *pro forma* statements that show a growth in revenues), or if plans need to be modified to avoid problems in the future. Financial managers analyze the *pro forma* statements by using the ratio analysis techniques described in chapter 5.

Companion Website
content by Active Learning Technologies

You can access your FREE PHLIP/CW On–Line Study Guide, Current Events, and Spreadsheet Problems templates, which may correspond to this chapter either through the Prentice Hall Finance Center CD–ROM or by directly going to: **http://www.prenhall.com/gallagher.**

PHLIP/CW — *The Prentice Hall Learning on the Internet Partnership–Companion Web Site*

Self-Test

ST-1. For the last five years, cost of goods sold (COGS) for the Heaven's Gate Corporation has averaged 60 percent of sales. This trend is expected to continue for the foreseeable future. If sales for 2000 are expected to be $2.5 million, what would be the forecast 2000 value for COGS?

ST-2. In 1999 the Ishtar Corporation had $180,000 in retained earnings. During 2000 the company expects net income to be $750,000. What will the value of retained earnings be on the company's *pro forma* balance sheet for December 31, 2000, if the company continues its policy of paying 50 percent of net income in dividends?

ST-3. The Far and Away Irish Import Company's *pro forma* balance sheet for December 31, 2000, indicates that total assets will be $8,420,000, but total liabilities and equity will be only $7,685,000. What should Far and Away do to resolve the discrepancy between assets and liabilities?

ST-4. Refer to the *pro forma* financial statements for Esoteric Enterprises, Figures 6-6a and 6-6b. Calculate Esoteric's return on equity (ROE) ratio for 1999 and 2000. Comment on the results.

Review Questions

1. Why do businesses spend time, effort, and money to produce forecasts?
2. What is the primary assumption behind the experience approach to forecasting?
3. Describe the sales forecasting process.
4. Explain how the cash budget and the capital budget relate to *pro forma* financial statement preparation.
5. Explain how management goals are integrated into *pro forma* financial statements.
6. Explain the significance of the term *additional funds needed*.
7. What do financial managers look for when they analyze *pro forma* financial statements?
8. What action(s) should be taken if analysis of *pro forma* financial statements reveals positive trends? Negative trends?

Build Your Communication Skills

CS-1. Refer to the current and *pro forma* financial statements for Esoteric Enterprises in Figures 6-6a and 6-6b. Analyze the financial statements using the techniques in chapter 5 and prepare a report of Esoteric's strengths and weaknesses.

CS-2. Form small groups of four to six. Based on each group member's assessment of Esoteric's strengths and weaknesses, discuss whether Esoteric should change its business plans and how. Once the group has prepared a strategy, select a spokesperson to report the group's conclusions to the class.

Problems

Sales Forecasts

6-1. Miniver Corporation grows flowers and sells them to major U.S. retail flower shops. Mrs. Miniver has asked you to prepare a forecast of expected future sales. The chart below shows the Miniver Corporation's sales record for the last six years. Make an estimate of 2000 sales by extending the trend. Justify your estimate to Mrs. Miniver.

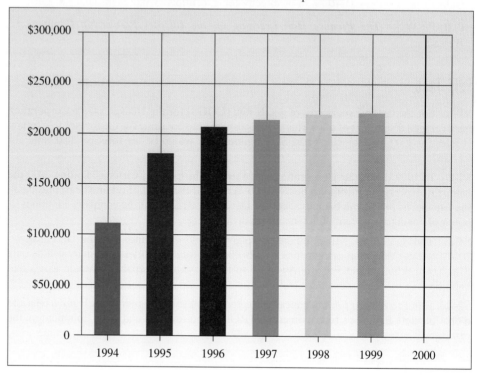

Additional Funds Needed

6-2. Complete the following *pro forma* financial statements. Use the forecasting assumptions in the far right-hand column.

	This Year	Next Year	Forecasting Assumption
Sales	100	_____	Sales will grow 20%
Variable Costs	50	_____	Constant % of sales
Fixed Costs	40	_____	Remains same
Net Income	10	_____	
Dividends	5	_____	Keep 50% payout ratio
Current Assets	60	_____	Constant % of sales
Fixed Assets	100	_____	Remains same
Total Assets	160	_____	
Current Liabilities	20	_____	Constant % of sales
Long-Term Debt	20	_____	Remains same
Common Stock	20	_____	Remains same
Retained Earnings	100	_____	
Total Liabilities and Equity	160	_____	
		AFN = _____	

6-3. Jolly Joe's Pizza has just come out with a new pizza that Joe is sure will cause sales to double between 1999 and 2000. Using the following worksheet, complete Joe's *pro forma* financial forecast and answer the related questions.

Additional Funds Needed

You may assume that COGS, current assets, and current liabilities will maintain the same percentage of sales as in 1999. Furthermore, you may assume that no new fixed assets will be needed in 2000, and the current dividend policy will be continued in 2000.

Jolly Joe's Pizza, Inc.
Financial Status and Forecast

	1999	Estimate for 2000
Sales	$10,000	_____
COGS	4,000	_____
Gross Profit	6,000	_____
Fixed Expenses	3,000	_____
Before-Tax Profit	3,000	_____
Tax @ 33.33%	1,000	_____
Net Profit	$ 2,000	_____
Dividends	$0	_____
Current Assets	$25,000	_____
Net Fixed Assets	15,000	_____
Total Assets	$40,000	_____
Current Liabilities	$17,000	_____
Long-Term Debt	3,000	_____
Common Stock	7,000	_____
Retained Earnings	13,000	_____
Total Liabilities and Equity	$40,000	_____

Will Joe be able to get by without any additional funds needed in 2000? If not, how much will he need?

6-4. Jose Tine owns Sugar Cane Alley, a small candy shop in Aspen, Colorado. He would like to expand his business and open a second store in Vail. Mr. Tine does not have the capital to undertake this project and would like to borrow the money from the local bank. He knows the banker will need projected income statement data for his current store when considering his loan application. Net sales for 1999 were $90,000. Considering previous growth rates in his business and the anticipated increase in tourism, projected net sales for 2000 are $110,000. Answer the questions based on the following assumptions. Cost of goods sold and selling and marketing expenses will remain the same percentage of sales as in 1999. General and administration expenses will remain the same value as 1999 at $5,000. Mr. Tine uses the straight-line method of depreciation, so last year's depreciation expense figure of $2,000 can also be applied to 2000.

Pro Forma Income Statement

  a. 1999 cost of goods sold was $48,000. What is the forecasted value of cost of goods sold for 2000?
  b. What is the forecasted gross profit for 2000?
  c. Selling and marketing expenses for 1999 were $13,000. What is the forecasted value for 2000?
  d. Calculate the forecasted operating income for 2000.
  e. Assume the interest expense for 2000 will be $800 and the tax rate is 30 percent. Calculate earnings before taxes (EBT) and net income expected for 2000 to the nearest dollar.
  f. If $10,000 is distributed in dividends in 2000, what will be 2000's addition to retained earnings?

Chapter 6   *Forecasting for Financial Planning*   **137**

Pro Forma Balance Sheet

6-5. After completing the *pro forma* income statement in problem 6-4, Mr. Tine now realizes he should also complete a *pro forma* balance sheet. Net sales in 1999 were $90,000 and his forecasted sales for 2000 are $110,000. All of Sugar Cane Alley's current assets will remain the same percentage of sales as they were in 1999. Mr. Tine does not plan to buy or sell any equipment, so his gross property and equipment amount will remain the same as 1999. In the liabilities and equity section, only accounts payable will remain the same percentage of sales as in 1999. Except for retained earnings, the other accounts are expected to remain the same value as 1999. The following balances were taken from Sugar Cane Alley's end of 1999 balance sheet:

Cash	$10,000
Accounts Receivable	2,220
Inventory	8,000
Property and Equipment (gross)	25,000
Accumulated Depreciation	4,000
Property and Equipment (net)	21,000
Accounts Payable	1,380
Long-Term Notes Payable	8,000
Retained Earnings	5,000
Common Stock	26,840

a. Calculate the forecasted end of 2000 values for each of the current asset accounts.
b. Depreciation expense for 2000 is estimated to be $2,000. Calculate the estimated total assets for the end of 2000.
c. Forecast the accounts payable for the end of 2000.
d. What will total liabilities be at the end of 2000?
e. Assuming the forecasted net income for 2000 is $19,351 and cash dividends paid equal $10,000, what total will be forecasted for the end of 2000 total liabilities and equity?
f. Based on these calculations of the *pro forma* balance sheet, are additional funds needed?
g. Net income for 1999 was $14,840. What was Sugar Cane Alley's net profit margin for 1999? The forecasted net income for 2000 is $19,357. What is Sugar Cane Alley's forecasted 2000 net profit margin?

Pro Forma Balance Sheet

6-6. The balance sheet of Free Enterprises, Inc., at the end of 1999 follows.

Free Enterprises, Inc.
Balance Sheet, December 31, 1999
(Thousands of Dollars)

Assets		Liabilities + Equity	
Cash	$ 4,000	Accounts Payable	$ 4,400
Accounts Receivable	10,000	Notes Payable	4,000
Inventory	13,000	Accrued Expenses	5,000
Prepaid Expenses	400	Total Current Liabilities	13,400
Total Current Assets	27,400	Bonds Payable	6,000
Fixed Assets	11,000	Common Equity	19,000
Total Assets	$38,400	Total Liabilities and Equity	$38,400

Net sales for 1999 were $85,000,000. The average annual sales growth rate is expected to be 10 percent every year for the next three years. Over the past several years the earnings before taxes (EBT) were 11 percent of net sales and are expected to remain the same over the next three years. The company's tax rate is 40 percent and it plans to maintain the dividend payout ratio (dividends paid/net income) at 60 percent.

Prepare a *pro forma* balance sheet for Free Enterprises, Inc., for December 31, 2000. Assume the accounts that are not a function of sales (fixed assets and bonds payable) remain the same as they were in 1999. Current assets and current liabilities will remain the same percentage of sales as in 1999. Assume the only change to common equity is for the addition to retained earnings.

6-7. Consider the current and *pro forma* financial statements that follow.

Forecasting Ratio Values

	1999	2000
Sales	200	220
Variable Costs	100	110
Fixed Costs	80	80
Net Income	20	30
Dividends	10	22
Current Assets	120	132
Fixed Assets	200	200
Total Assets	320	332
Current Liabilities	40	44
Long-Term Debt	40	40
Common Stock	40	40
Retained Earnings	200	208
Total Liabilities and Equity	320	332
		AFN = 0

Compute the following ratios for 1999 and 2000:

	1999	2000
Current Ratio	———	———
Debt to Assets Ratio	———	———
Sales to Assets Ratio	———	———
Net Profit Margin	———	———
Return on Assets	———	———
Return on Equity	———	———

Comment on any trends revealed by your ratio analysis.

6-8.

Challenge Problem

1. Develop a *pro forma* income statement and balance sheet for the Bright Future Corporation. The company's 1999 financial statements are shown on pages 140–141. Base your forecast on the financial statements and the following assumptions:

   - Sales growth is predicted to be 20 percent in 2000.
   - Cost of goods sold, selling and administrative expense, all current assets, accounts payable, and accrued expenses will remain the same percentage of sales they were in 1999.
   - Depreciation expense, interest expense, gross plant and equipment, notes payable, long-term debt, and equity accounts *other than retained earnings* in 2000 will be the same as in 1999.
   - The company's tax rate in 2000 will be 40 percent.
   - The same dollar amount of dividends will be paid to the preferred and common stockholders in 2000 as was paid in 1999.
   - Bad debt allowance in 2000 will be the same percentage of accounts receivable as it was in 1999.

Bright Future Corporation
Income Statement for the Year Ended 1999

Sales		$10,000,000
Cost of Goods Sold		− 4,000,000
Gross Profit		= 6,000,000
Selling and Administrative Expenses		− 800,000
Depreciation Expense		− 2,000,000
Earnings Before Interest and Taxes (EBIT, or Operating Income)		= 3,200,000
Interest Expense		− 1,350,000
Earnings Before Taxes (EBT)		= 1,850,000
Taxes (40%)		− 740,000
Net Income (NI)		= 1,110,000
Preferred Stock Dividends Paid		− 110,000
Earnings Available to Common Stockholders		= 1,000,000
Earnings per Share (EPS) (1 million shares) = $1.00		
Common Stock Dividends Paid		− 400,000
Addition to Retained Earnings		= 600,000

Bright Future Corporation
Balance Sheet December 31, 1999

ASSETS:

Current Assets:			
Cash		$ 9,000,000	
Marketable Securities		8,000,000	
Accounts Receivable (Gross)	$1,200,000		
Less Allowance for Bad Debts	200,000		
Accounts Receivable (Net)		1,000,000	
Inventory		20,000,000	
Prepaid Expenses		1,000,000	
Total Current Assets			$39,000,000
Fixed Assets:			
Plant and Equipment (Gross)	20,000,000		
Less Accumulated Depreciation	9,000,000		
Plant and Equipment (Net)		11,000,000	
Total Fixed Assets			11,000,000
TOTAL ASSETS			$50,000,000

LIABILITIES AND EQUITY:

Current Liabilities:			
Accounts Payable	$12,000,000		
Notes Payable	5,000,000		
Accrued Expenses	3,000,000		
Total Current Liabilities		$20,000,000	
Long-Term Debt:			
Bonds Payable (5%, due 2015)	20,000,000		
Total Long-Term Debt		20,000,000	
Total Liabilities			$40,000,000
Equity:			
Preferred Stock		1,000,000	
Common Stock (1 million shares, $1 par)		1,000,000	
Capital in Excess of Par		3,000,000	
Retained Earnings		5,000,000	
Total Equity			10,000,000
TOTAL LIABILITIES AND EQUITY			$50,000,000

2. **a.** Calculate Bright Future's additional funds needed, or excess financing. If additional funds are needed, add them to long-term debt to bring the balance sheet into balance. If excess financing is available, increase common stock dividends paid (and, therefore, decrease 2000 retained earnings) until the balance sheet is in balance.
   b. Calculate Bright Future's current ratio for the end of 1999 and 2000.
   c. Calculate Bright Future's total asset turnover and inventory turnover ratios for 2000.
   d. Calculate Bright Future's total debt to total assets ratio for 1999 and 2000.
   e. Calculate Bright Future's net profit margin, return on assets, and return on equity ratios for 1999 and 2000.

3. Comment on Bright Future's liquidity, asset productivity, debt management, and profitability based on the results of your ratio analysis in 2.b. through 2.e.

4. What recommendations would you provide to management based on your forecast and analysis?

Answers to Self-Test

ST-1. If COGS is expected to remain 60 percent of sales, then in 2000 COGS will be 60 percent of 2000's sales:

$$.60 \times \$2{,}500{,}000 = \$1{,}500{,}000$$

ST-2. If Ishtar earns $750,000 in 2000 and pays 50 percent of it to stockholders as dividends, then the other 50 percent, or $375,000 will be retained and added to the existing retained earnings account. Therefore, retained earnings on Ishtar's December 31, 2000, *pro forma* balance sheet will be:

$$\$180{,}000 + \$375{,}000 = \$555{,}000$$

ST-3. Far and Away's total liabilities and equity are forecast to be $735,000 less than total assets. This means Far and Away must arrange for $735,000 in additional financing to support expected asset growth. Possible sources of this financing include bank loans, a bond issue, a stock issue, or perhaps lowering the 2000 dividend (if any).

ST-4.

$$\text{ROE} = \frac{\text{Net Income}}{\text{Total Equity}}$$

For Esoteric in 1999 (from the 1999 financial statements):

$$\text{ROE} = \frac{\$26{,}252}{\$150{,}607} = .1743, \text{ or } 17.43\%$$

For Esoteric in 2000 (from the 2000 *pro forma* financial statements):

$$\text{ROE} = \frac{\$30{,}290}{\$153{,}636} = .1971, \text{ or } 19.71\%$$

Take Note

Remember that risk and return go hand in hand. The financial managers at Esoteric Enterprises must evaluate the risk associated with the ROE figures before concluding that current plans will lead to desirable results.

Note that this is a favorable forecast. The gain of over two percentage points in an already respectable ROE value should be particularly pleasing to the stockholders. Esoteric's financial managers should recommend that the company continue with its current plans, assuming that the increase in ROE does not signal an excessive increase in risk.

CHAPTER 7

RISK AND RETURN

CHAPTER OBJECTIVES

After reading this chapter, you should be able to:

- Define risk, risk aversion, and the risk–return relationship.

- Measure risk using the standard deviation and coefficient of variation.

- Identify the types of risk that business firms encounter.

- Explain methods of risk reduction.

- Describe how firms compensate for assuming risk.

- Discuss the capital asset pricing model (CAPM).

Believe me! The secret of reaping the greatest fruitfulness and the greatest enjoyment from life is to live dangerously!
—Friedrich Wilhelm Nietzsche

RISKY IN MORE WAYS THAN ONE

John Baldwin Long, 78, had an idea. It led to a risky new business venture. The meat business doesn't sound like an unusually risky business enterprise. It is, however, when you add high explosives.

Mr. Long found that you could put up to 500 pounds of boned beef in vacuum-packed plastic bags. These are lowered into a water-filled, stainless steel tank surrounded by concrete. A large explosive charge is then detonated. The blast tenderizes the beef.

Many consumers seek to avoid beef from animals treated with hormones or antibiotics. It is not clear how they will feel about beef that has been blown up. The explosion doesn't affect the outward appearance of the beef. There are surely risks, however, for the investors in this business, and perhaps some on the production line, too.

Source: "What a Blast," by Doug Donovan, from the July 27, 1998 issue of Forbes Magazine, July 25, 1998 ⟨http//www.forbes.com/forbes/98/0727/6202080a.htm⟩.

Chapter Overview

Business firms face risk in nearly everything they do. Assessing risk is one of the most important tasks financial managers perform. In this chapter we will discuss *risk, risk aversion,* and the *risk–return relationship*. We will measure risk using *standard deviation* and the *coefficient of variation*. We will identify types of risk and examine ways to reduce risk exposure or compensate for risk. Finally, we will see how the capital asset pricing model (CAPM) explains the risk–return relationship.

Risk

The world is a risky place. For instance, if you get out of bed in the morning and go to class, you run the risk of getting hit by a bus. If you stay in bed to minimize the chance of getting run over by a bus, you run the risk of getting coronary artery disease because of a lack of exercise. In everything we do—or don't do—there is a chance that something will happen that we didn't expect. **Risk** is the potential for unexpected events to occur.

Risk Aversion

Most people try to avoid risks if possible. Risk aversion doesn't mean that some people don't enjoy risky activities, such as skydiving, rock climbing, or automobile racing. In a financial setting, however, evidence shows that most people avoid risk where possible. Faced with financial alternatives that are equal except for their degree of risk, most people will choose the less risky alternative.

Risk aversion is the tendency to avoid additional risk. Risk-averse people will avoid risk if they can, unless they receive additional compensation for assuming that risk. In finance, the added compensation is a higher expected rate of return.

The Risk–Return Relationship

The relationship between risk and required rate of return is known as the **risk–return relationship.** It is a positive relationship because the more risk assumed, the higher the required rate of return most people will demand. It takes compensation to convince people to suffer.

Suppose, for instance, that you were offered a job in the Sahara Desert, working long hours for a boss everyone describes as a tyrant. You would surely be averse to the idea of taking such a job. But think about it: Is there any way you would take this job? What if you were told that your salary would be $1 million per year? This compensation might cause you to sign up immediately. Even though there is a high probability you would hate the job, you'd take that risk because of the high compensation.[1]

---

[1] We do not wish to suggest that people can be coaxed into doing anything they are averse to doing merely by offering them a sufficient amount of compensation. If people are asked to do something that offends their values, there may be no amount of compensation that could entice them.

Not everyone is risk averse, and among those who are, not all are equally risk averse. Some people would demand $2 million before taking the job in the Sahara Desert, whereas others would do it for a more modest salary.

People sometimes engage in risky financial activities, such as buying lottery tickets or gambling in casinos. This suggests that they like risk and will pay to experience it. Most people, however, view these activities as entertainment rather than financial investing. The entertainment value may be the excitement of being in a casino with all sorts of people, or being able to fantasize about spending the multimillion-dollar lotto jackpot. But in the financial markets, where people invest for the future, they almost always seek to avoid risk unless they are adequately compensated.

Risk aversion explains the positive risk–return relationship. It explains why risky junk bonds carry a higher market interest rate than essentially risk-free U.S. Treasury bonds. Hardly anyone would invest $5,000 in a risky junk bond if the interest rate on the bond were lower than that of a U.S. Treasury bond having the same maturity.

MEASURING RISK

We can never avoid risk entirely. That's why businesses must make sure that the anticipated return is sufficient to justify the degree of risk assumed. To do that, however, firms must first determine how much risk is present in a given financial situation. In other words, they must be able to answer the question, "How risky is it?"

Measuring risk quantitatively is a rather tall order. We all know when something feels risky, but we don't often quantify it. In business, risk measurement focuses on the degree of **uncertainty** present in a situation—the chance, or probability, of an unexpected outcome. The greater the probability of an unexpected outcome, the greater the degree of risk.

Using Standard Deviation to Measure Risk

In statistics, *distributions* are used to describe the many values variables may have. A company's sales in future years, for example, is a variable with many possible values. So the sales forecast may be described by a distribution of the possible sales values with different probabilities attached to each value. If this distribution is symmetric, its mean—the average of a set of values—would be the expected sales value. Similarly, possible returns on any investment can be described by a *probability distribution*—usually a graph, table, or formula that specifies the probability associated with each possible return the investment may generate. The mean of the distribution is the most likely, or expected, rate of return.

The graph in Figure 7-1 shows the distributions of forecast sales for two companies, Company Calm and Company Bold. Note how the distribution for Company Calm's possible sales values is clustered closely to the mean, and how the distribution of Company Bold's possible sales values is spread far above and far below the mean.

The narrowness or wideness of a distribution reflects the degree of uncertainty about the expected value of the variable in question (sales, in our example). The distributions in Figure 7-1 show, for instance, that although the most probable value of sales for both companies is $1,000, sales for Company Calm could vary between $600 and $1,400, whereas sales for Company Bold could vary between $200 and

Figure 7-1 Sales Forecast Distributions for Companies Calm and Bold

Possible future sales distributions for two companies. Calm has a relatively "tight" distribution and Bold has a relatively "wide" distribution. Note that sales for Company Bold has many more possible values than sales for Company Calm.[2]

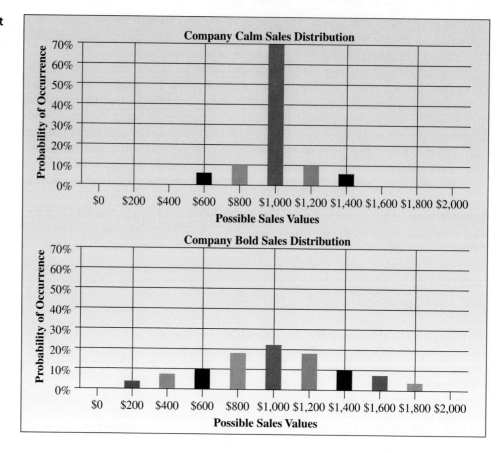

$1,800. Company Bold's relatively wide variations show that there is more uncertainty about its sales forecast than about Company Calm's sales forecast.

One way to measure risk is to compute the standard deviation of a variable's distribution of possible values. The **standard deviation** is a numerical indicator of how widely dispersed the possible values are around a mean. The more widely dispersed a distribution is, the larger the standard deviation, and the greater the probability that the value of a variable will be greatly different than the expected value. The standard deviation, then, indicates the likelihood that an outcome different from what is expected will occur.

Let's calculate the standard deviations of the sales forecast distributions for Companies Calm and Bold to illustrate how the standard deviation can measure the degree of uncertainty, or risk, that is present.

Calculating the Standard Deviation To calculate the standard deviation of the distribution of Company Calm's possible sales, we must first find the expected value, or mean, of the distribution using the following formula:

---

[2]These two distributions are *discrete*. If sales could take on any value within a given range, the distribution would be *continuous* and would be depicted by a curved line.

Formula for Expected Value, or Mean (μ)

$$\mu = \Sigma (V \times P) \qquad (7\text{-}1)$$

where: μ = the expected value, or mean
Σ = the sum of
V = the possible value for some variable
P = the probability of the value V occurring

Applying Equation 7-1, we can calculate the expected value, or mean, of Company Calm's forecasted sales. The following values for V and P are taken from Figure 7-1:

Calculating the Mean (μ) of Company Calm's Possible Future Sales Distribution

POSSIBLE SALES VALUE (V)	PROBABILITY OF OCCURRENCE (P)	V × P
$600	.05	30
$800	.10	80
$1,000	.70	700
$1,200	.10	120
$1,400	.05	70
	$\Sigma = 1.00$	$\Sigma = 1,000 = \mu$

Each possible sales value is multiplied by its respective probability. The probability values, taken from Figure 7-1, may be based on trends, industry ratios, experience, or other information sources. We add together the products of each value times its probability to find the mean of the possible sales distribution.

We now know that the mean of Company Calm's sales forecast distribution is $1,000. We are ready to calculate the standard deviation of the distribution using the following formula:

The Standard Deviation (σ) Formula

$$\sigma = \sqrt{\Sigma P(V - \mu)^2} \qquad (7\text{-}2)$$

where: σ = the standard deviation
Σ = the sum of
P = the probability of the value V occurring
V = the possible value for a variable
μ = the expected value

To calculate the standard deviation of Company Calm's sales distribution, we subtract the mean from each possible sales value, square that difference, and then multiply by the probability of that sales outcome. These differences squared, times their respective probabilities, are then added together, and the square root of this number is taken. The result is the standard deviation of the distribution of possible sales values.

Calculating the Standard Deviation (σ) of Company Calm's Possible Future Sales Distribution

Possible Sales Value (V)	Probability of Occurrence (P)	V − μ	(V − μ)²	P(V − μ)²
$600	.05	−400	160,000	8,000
$800	.10	−200	40,000	4,000
$1,000	.70	0	0	0
$1,200	.10	200	40,000	4,000
$1,400	.05	400	160,000	8,000
				Σ = 24,000
				$\sqrt{24,000} = 155 = \sigma$

This standard deviation result, 155, serves as the measure of the degree of risk present in Company Calm's sales forecast distribution.

Let's calculate the standard deviation of Company Bold's sales forecast distribution.

> **Take Note**
>
> In the following procedure, we combine two steps: (1) finding the mean of the distribution with Equation 7-1; and (2) calculating the standard deviation with Equation 7-2.

Mean (μ) and Standard Deviation (σ) of Company Bold's Possible Future Sales Distribution

Possible Sales Value (V)	Probability of Occurrence (P)	Mean Calculation V × P	V − μ	(V − μ)²	P(V − μ)²
$200	.04	8	−800	640,000	25,600
$400	.07	28	−600	360,000	25,200
$600	.10	60	−400	160,000	16,000
$800	.18	144	−200	40,000	7,200
$1,000	.22	220	0	0	0
$1,200	.18	216	200	40,000	7,200
$1,400	.10	140	400	160,000	16,000
$1,600	.07	112	600	360,000	25,200
$1,800	.04	72	800	640,000	25,600
		Σ = 1,000 = μ			Σ = 148,000
					$\sqrt{148,000} = 385 = \sigma$

As you can see, Company Bold's standard deviation of 385 is over twice that of Company Calm. This reflects the greater degree of risk in Company Bold's sales forecast.

Interpreting the Standard Deviation   Estimates of a company's possible sales, or a proposed project's future possible cash flows, can generally be thought of in terms of a *normal* probability distribution. The normal distribution is a special type of distribution. It allows us to make statements about how likely it is that the variable in question will be within in a certain range of the distribution.

Figure 7-2 shows a normal distribution of possible returns on an asset. The vertical axis measures probability density for this continuous distribution so that the area under the curve always sums to one. Statistics tells us that when a distribution is normal, there is about a 67 percent chance that the observed value

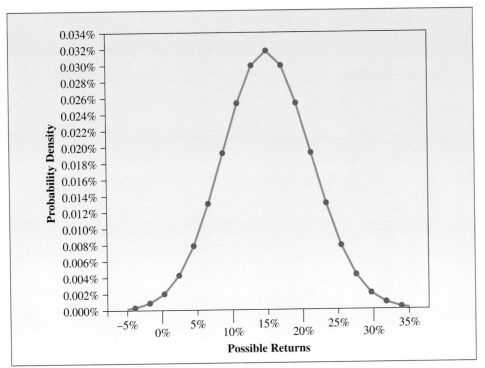

Figure 7-2 Normal Distribution

This normal probability distribution of possible returns has a mean, the expected rate of return, of 15%.

will be within one standard deviation of the mean. In the case of Company Calm, that means that there is a 67 percent probability that the actual sales will be $1,000 plus or minus $155 (between $845 and $1,155). For Company Bold it means there is a 67 percent probability that sales will be $1,000 plus or minus $385 (between $615 and $1,385).

Another characteristic of the normal distribution is that approximately 95 percent of the time, values observed will be within two standard deviations of the mean. For Company Calm this means that there is a 95 percent probability that sales will be $1,000 plus or minus $155 × 2, or $310 (between $690 and $1,310). For Company Bold it means that sales will be $1,000 plus or minus $385 × 2, or $770 (between $230 and $1,770). These relationships are shown graphically in Figure 7-3.

The greater the standard deviation value, the greater the uncertainty as to what the actual value of the variable in question will be. The greater the value of the standard deviation, the greater the possible deviations from the mean.

Using the Coefficient of Variation to Measure Risk

Whenever we want to compare the risk of investments that have different means, we use the *coefficient of variation*. We were safe in using the standard deviation to compare the riskiness of Company Calm's possible future sales distribution with that of Company Bold because the mean of the two distributions was the same ($1,000). Imagine, however, that Company Calm's sales were 10 times that of Company Bold. If that were the case and all other factors remained the same, then the standard deviation of Company Calm's possible future sales distribution would

Figure 7-3 The Degree of Risk Present in Company Calm's and Company Bold's Possible Future Sales Values as Measured by Standard Deviation

The standard deviation shows there is much more risk present in Company Bold's sales probability distribution than in Company Calm's. If the distributions are normal, then there is a 67% probability that Company Calm's sales will be between $845 and $1,155, and a 95% probability sales will be between $690 and $1,310. For Company Bold there is a 67% probability that sales will be between $615 and $1,385, and a 95% probability that sales will be between $230 and $1,770.

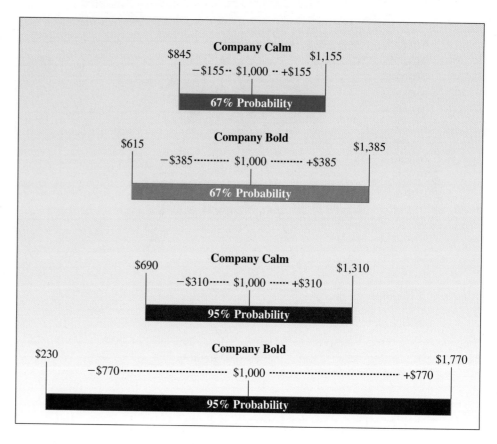

increase by a factor of 10, to $1,550. Company Calm's sales would appear to be much more risky than Company Bold's, whose standard deviation was only $385.

To compare the degree of risk among distributions of different sizes, we should use a statistic that measures *relative* riskiness. The **coefficient of variation** (CV) measures relative risk by relating the standard deviation to the mean. The formula follows:

$$\text{Coefficient of Variation (CV)}$$
$$CV = \frac{\text{Standard Deviation}}{\text{Mean}} \qquad (7\text{-}3)$$

The coefficient of variation represents the standard deviation's percentage of the mean. It provides a standardized measure of the degree of risk that can be used to compare alternatives.

To illustrate the use of the coefficient of variation, let's compare the relative risk depicted in Company Calm's and Company Bold's possible sales distributions. When we plug the figures into Equation 7-3, we see:

$$\text{Company Calm CV}_{sales} = \frac{\text{Standard Deviation}}{\text{Mean}} = \frac{155}{1{,}000} = .155, \text{ or } 15.5\%$$

$$\text{Company Bold CV}_{sales} = \frac{\text{Standard Deviation}}{\text{Mean}} = \frac{385}{1{,}000} = .385, \text{ or } 38.5\%$$

Company Bold's coefficient of variation of possible sales (38.5 percent) is more than twice that of Company Calm (15.5 percent). Furthermore, even if Company

Calm were 10 times the size of Company Bold—with a mean of its possible future sales of $10,000 and with a standard deviation of $1,550—it would not change the coefficient of variation. This would remain 1,550/10,000 = .155, or 15.5 percent. We use the coefficient of variation instead of the standard deviation to *compare* distributions that have means with different values because the CV adjusts for the difference, whereas the standard deviation does not.

The Types of Risks Firms Encounter

Risk refers to uncertainty—the chance that what you expect to happen *won't* happen. The forms of risk that businesses most often encounter are *business risk, financial risk,* and *portfolio risk.*

Business Risk

Business risk refers to the uncertainty a company has with regard to its operating income (also known as earnings before interest and taxes, or EBIT). The more uncertainty about a company's expected operating income, the more business risk the company has. For example, if we assume that grocery prices remain constant, the only grocery store in a small town probably has little business risk—the store owners can reliably predict how much their customers will buy each month. In contrast, a gold mining firm in Wyoming has a lot of business risk. Because the owners have no idea when, where, or how much gold they will strike, they can't predict how much they will earn in any period with any degree of certainty.

Measuring Business Risk The degree of uncertainty about operating income (and, therefore, the degree of business risk in the firm) depends on the volatility of operating income. If operating income is relatively constant, as in the grocery store example, then there is relatively little uncertainty associated with it. If operating income can take on many different values, as is the case with the gold mining firm, then there is a lot of uncertainty about it.

We can measure the variability of a company's operating income by calculating the standard deviation of the operating income forecast. A small standard deviation indicates little variability and, therefore, little uncertainty. A large standard deviation indicates a lot of variability and great uncertainty.

Some companies are large and others small. So to make comparisons among different firms, we must measure the risk by calculating the coefficient of variation of possible operating income values. The higher the coefficient of variation of possible operating income values, the greater the business risk of the firm.

Table 7-1 shows the expected value (μ), standard deviation (σ), and coefficient of variation (CV) of operating income for Company Calm and Company Bold, assuming that the expenses of both companies vary directly with sales (i.e., neither company has any fixed expenses).

The Influence of Sales Volatility Sales volatility affects business risk—the more volatile a company's sales, the more business risk the firm has. Indeed, when no fixed costs are present—as in the case of Company Calm and Company Bold—sales volatility is equivalent to operating income volatility. Table 7-1 shows that the coefficients of variation of Company Calm's and Company Bold's operating income are

TABLE 7-1

Expected Value (μ), Standard Deviation (σ), and Coefficient of Variation (CV) of Possible Operating Income Values for Companies Calm and Bold, Assuming All Expenses Are Variable

COMPANY CALM

	\multicolumn{5}{c}{Probability of Occurrence}
	5%	10%	70%	10%	5%
Sales	$600	$800	$1,000	$1,200	$1,400
Variable Expenses	516	688	860	1,032	1,204
Operating Income	84	112	140	168	196

μ of Possible Operating Income Values per Equation 7-1: $140
σ of Possible Operating Income Values per Equation 7-2: $21.69
CV of Possible Operating Income Values per Equation 7-3: 15.5%

COMPANY BOLD

	\multicolumn{9}{c}{Probability of Occurrence}
	4%	7%	10%	18%	22%	18%	10%	7%	4%
Sales	$200	$400	$600	$800	$1,000	$1,200	$1,400	$1,600	$1,800
Variable Expenses	172	344	516	688	860	1,032	1,204	1,376	1,548
Operating Income	28	56	84	112	140	168	196	224	252

μ of Possible Operating Income Values per Equation 7-1: $140
σ of Possible Operating Income Values per Equation 7-2: $53.86
CV of Possible Operating Income Values per Equation 7-3: 38.5%

This table shows the comparison of the standard deviation and coefficient of variation of possible operating income values for Company Calm (top) and Company Bold (bottom) **when all expenses are variable.** *The table indicates Company Bold's operating income is more than twice as volatile as the operating income of Company Calm.*

15.5 percent and 38.5 percent, respectively. Note that these coefficient numbers are exactly the same numbers as the two companies' coefficients of variation of expected sales.

The Influence of Fixed Operating Costs In Table 7-1 we assumed that all of Company Calm's and Company Bold's expenses varied proportionately with sales. We did this to illustrate how sales volatility affects operating income volatility. In the real world, of course, most companies have some fixed expenses as well, such as rent, insurance premiums, and the like. It turns out that fixed expenses magnify the effect of sales volatility on operating income volatility. In effect, fixed expenses intensify business risk. The tendency of fixed expenses to magnify business risk is called **operating leverage.** To see how this works, refer to Table 7-2, in which we assume that all of Company Calm's and Company Bold's expenses are fixed.

As Table 7-2 shows, the effect of replacing each company's variable expenses with fixed expenses increased the volatility of operating income considerably. The coefficient of variation of Company Calm's operating income jumped from 15.7 percent when all expenses were variable, to over 110 percent when all expenses were fixed. When all expenses are fixed, a 15.5% variation in sales is magnified to a 110.7 percent variation in operating income. A similar situation exists for Company Bold.

The greater the fixed expenses, the greater the change in operating income for a given change in sales. Capital-intensive companies, such as electric generating firms,

TABLE 7-2

Expected Value (μ), Standard Deviation (σ), and Coefficient of Variation (CV) of Possible Operating Income Values for Companies Calm and Bold, Assuming All Expenses Are Fixed

COMPANY CALM

	\multicolumn{5}{c}{Probability of Occurrence}				
	5%	10%	70%	10%	5%
Sales	$600	$800	$1,000	$1,200	$1,400
Fixed Expenses	860	860	860	860	860
Operating Income	(260)	(60)	140	340	540

μ of Possible Operating Income Values per Equation 7-1: $140
σ of Possible Operating Income Values per Equation 7-2: $155
CV of Possible Operating Income Values per Equation 7-3: 110.7%

COMPANY BOLD

	\multicolumn{9}{c}{Probability of Occurrence}								
	4%	7%	10%	18%	22%	18%	10%	7%	4%
Sales	$200	$400	$600	$800	$1,000	$1,200	$1,400	$1,600	$1,800
Fixed Expenses	860	860	860	860	860	860	860	860	860
Operating Income	(660)	(460)	(260)	(60)	140	340	540	740	940

μ of Possible Operating Income Values per Equation 7-1: $140
σ of Possible Operating Income Values per Equation 7-2: $385
CV of Possible Operating Income Values per Equation 7-3: 275.0%

Table 7-2 shows the comparison of the standard deviation and coefficient of variation of possible operating income values for Company Calm (top) and Company Bold (bottom) when all expenses are fixed. The table indicates that the volatility of both firms' operating income is increased by the presence of fixed expenses.

have high fixed expenses. Service companies, such as consulting firms, often have relatively low fixed expenses.

Financial Risk

When companies borrow money, they incur interest charges that appear as fixed expenses on their income statements. (For business loans, the entire amount borrowed normally remains outstanding until the end of the term of the loan. Interest on the unpaid balance, then, is a fixed amount that is paid each year until the loan matures.) Fixed interest charges act on a firm's net income in the same way that fixed operating expenses act on operating income—they increase volatility. The additional volatility of a firm's net income caused by the fixed interest expense is called **financial risk**, or **financial leverage**.

Measuring Financial Risk Financial risk is the additional volatility of net income caused by the presence of interest expense. So we measure financial risk by noting the difference between the volatility of net income when there is no interest expense and the volatility of net income when there is interest expense. To measure financial risk, we subtract the coefficient of variation of net income without interest expense from the coefficient of variation of net income with interest expense. The coefficient of variation of net income is the same as the coefficient of variation of

> **Take Note**
>
> For simplicity, Table 7-3 assumes that neither firm pays any income taxes. Income tax is not a fixed expense, so its presence would not change the volatility of net income.

operating income when no interest expense is present. Table 7-3 shows the calculation for Company Calm and Company Bold assuming (1) all variable operating expenses, and (2) $40 in interest expense.

Financial risk, which comes from borrowing money, compounds the effect of business risk and intensifies the volatility of net income. Fixed operating expenses increase the volatility of operating income and magnify business risk. In the same way, fixed financial expenses (such as interest on debt or a noncancellable lease expense) increase the volatility of net income and magnify financial risk.

TABLE 7-3

Expected Value (μ), Standard Deviation (σ), and Coefficient of Variation (CV) of Possible Net Income Values for Companies Calm and Bold

COMPANY CALM

	Probability of Occurrence				
	5%	10%	70%	10%	5%
Sales	$600	$800	$1,000	$1,200	$1,400
Variable Expenses	516	688	860	1,032	1,204
Operating Income	84	112	140	168	196
Interest Expense	40	40	40	40	40
Net Income	44	72	100	128	156

μ of Possible Net Income Values per Equation 7-1: $100
σ of Possible Net Income Values per Equation 7-2: $22
CV of Possible Net Income Values per Equation 7-3: 22.0%

Summary:
CV of Possible Net Income Values When Interest Expense Is Present: 22.0%
CV of Operating Income When Interest Expense Is *Not* Present
  (from Table 7-1): 15.5%
  Difference (financial risk): 6.5%

COMPANY BOLD

	Probability of Occurrence								
	4%	7%	10%	18%	22%	18%	10%	7%	4%
Sales	$200	$400	$600	$800	$1,000	$1,200	$1,400	$1,600	$1,800
Variable Expenses	172	344	516	688	860	1,032	1,204	1,376	1,548
Operating Income	28	56	84	112	140	168	196	224	252
Interest Expense	40	40	40	40	40	40	40	40	40
Net Income	(12)	16	44	72	100	128	156	184	212

μ of Possible Net Income Values per Equation 7-1: $100
σ of Possible Net Income Values per Equation 7-2: $54
CV of Possible Net Income Values per Equation 7-3: 54.0%

Summary:
CV of Possible Net Income Values When Interest Expense Is Present: 54.0%
CV of Operating Income When Interest Expense Is *Not* Present
  (from Table 7-1): 38.5%
  Difference (financial risk): 15.5%

*Table 7-3 illustrates the effect of financial risk. Compare the CVs of possible net income values in this table to those of operating income in Table 7-1. Note Company Calm's CV is increased by 6.5 percentage points and Company Bold's CV is increased by 15.5 percentage points as a result of the **presence of fixed interest expense**.*

Firms that have only equity financing have no financial risk because they have no debt on which to make fixed interest payments. Conversely, firms that operate primarily on borrowed money are exposed to a high degree of financial risk.

Portfolio Risk

A **portfolio** is any collection of assets managed as a group. Most large firms employ their assets in a number of different investments. Together, these make up the firm's portfolio of assets. Individual investors also have portfolios containing many different stocks or other investments.

Firms (and individuals for that matter) are interested in portfolio returns and the uncertainty associated with them. Investors want to know how much they can expect to get back from their portfolio compared to how much they invest (the portfolio's expected return) and what the chances are that they won't get that return (the portfolio's risk).

Investors who are dealing with an existing portfolio can usually estimate the expected return, or mean of the probability distribution of possible returns of the portfolio, and its standard deviation and coefficient of variation (as we did earlier in the chapter for the sales forecast distributions of Company Calm and Company Bold). Firms want to know whether adding another asset (or another portfolio) to an existing portfolio will change the overall risk of the firm.[3] To find the riskiness of the new combined portfolio, we must find the expected return and standard deviation of possible returns for the new portfolio. For example, suppose Company Calm merged with Company Bold to form a new firm called Company Cool. We can easily find the expected return of the new Company Cool portfolio, but calculating the standard deviation of the portfolio's possible return is a little more difficult.

For example, suppose the expected returns and standard deviations of possible returns of Company Calm's and Company Bold's asset portfolios are as follows:[4]

Take Note

One group of businesses is exposed to an extreme amount of financial risk because they operate almost entirely on borrowed money: banks and other financial institutions. Banks get almost all the money they use for loans from deposits—and deposits are liabilities on the bank's balance sheet. Banks must be careful to keep their revenues stable. Otherwise, fluctuations in revenues would cause losses that would drive the banks out of business. Now you know why the government regulates financial institutions so closely!

	COMPANY CALM	COMPANY BOLD
Expected Return E(R)	10%	12%
Standard Deviation σ	2%	4%

Also assume the new combined Company Cool is made up of 50 percent old Company Calm and 50 percent old Company Bold.

Finding the expected return of the combined Company Cool portfolio is easy. We simply calculate the weighted average expected return, $E(R_p)$ of the two-asset portfolio using the following formula:

Expected Rate of Return of a Portfolio, $E(R_p)$,
Comprised of Two Assets, a and b

$$E(R_p) = (w_a \times E(R_a)) + (w_b \times E(R_b)) \quad (7\text{-}4)$$

---

[3]This assumes that management is concerned with firm-specific risk and not stockholder risk.
[4]The mean (μ) of the probability distribution of possible returns is the expected value, E(V) of this distribution. Because we are referring to possible returns here, it is the expected return, E(R).

where: $E(R_p)$ = the expected rate of return of the portfolio composed of Asset a and Asset b
w_a = the weight of Asset a in the portfolio
$E(R_a)$ = the expected rate of return of Asset a
w_b = the weight of Asset b in the portfolio
$E(R_b)$ = the expected rate of return of Asset b

Let's call Company Calm "Asset a" and Company Bold "Asset b." Then, according to Equation 7-4, the expected rate of return of a portfolio containing 50 percent Company Calm and 50 percent Company Bold is:

$$\begin{aligned} E(R_p) &= (.50 \times .10) + (.50 \times .12) \\ &= (.05 + .06) \\ &= .11, \text{ or } 11\% \end{aligned}$$

Now let's turn to the standard deviation of possible returns of the new combined Company Cool portfolio. Determining the standard deviation of a portfolio requires special procedures. Why? Because gains from one asset in the portfolio may offset losses from another, lessening the overall degree of risk in the portfolio. Figure 7-4 shows how this works.

Figure 7-4 shows that even though the returns of each company vary, the timing of the variations is such that when one company's returns increase, the other's decrease. Therefore, the net change in the new combined Company Cool portfolio returns is very small—nearly zero. The weighted average of the standard deviations of returns of the two individual assets, then, does not result in the standard deviation of the portfolio containing both firms. The reduction in the fluctuations of the returns of Company Cool (the combination of Company Bold and Company Calm) is called the **diversification effect.**

Correlation How successfully diversification reduces risk depends on the degree of correlation between the two variables in question. **Correlation** indicates the de-

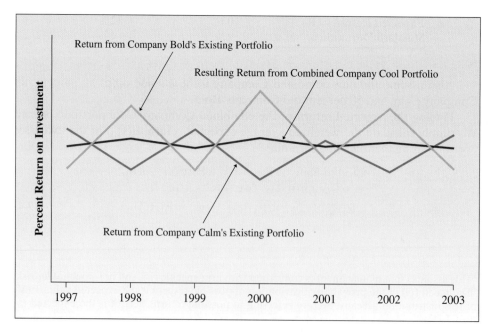

Figure 7-4 **The Variation in Returns over Time for Company Calm, Company Bold, and the Combined Company Cool**

Figure 7-4 shows how the returns of the portfolios of Companies Calm and Bold might vary over time. Notice that the fluctuations of each curve are such that gains in one almost completely offset losses in the other. The risk of the combined Company Cool portfolio is small due to the offsetting effects.

Source: DILBERT © 1996 United Features Syndicate. Reprinted by permission.

gree to which one variable is linearly related to another. Correlation is measured by the **correlation coefficient,** represented by the letter r. The correlation coefficient can take on values between +1.0 (perfect positive correlation) to −1.0 (perfect negative correlation). If two variables are perfectly positively correlated, it means they move together—that is, they change values proportionately in the same direction at the same time. If two variables are perfectly negatively correlated, it means that every positive change in one value is matched by a proportionate corresponding negative change in the other. In the case of Companies Calm and Bold in Figure 7-4, the assets are negatively correlated.

The closer r is to +1.0, the more the two variables will tend to move with each other at the same time. The closer r is to −1.0, the more the two variables will tend to move opposite each other at the same time. An r value of zero indicates that the variables' values aren't related at all. This is known as *statistical independence.*

In Figure 7-4 Company Calm and Company Bold had perfect negative correlation (r = −1.0). So the risk associated with each asset was nearly eliminated by combining the two assets into one portfolio. The risk would have been completely eliminated had the standard deviations of the two assets been equal.

Calculating the Correlation Coefficient. Determining the precise value of r between two variables can be extremely difficult. The process requires estimating the possible values that each variable could take, and their respective probabilities, simultaneously.

We can make a rough estimate of the degree of correlation between two variables by examining the nature of the assets involved. If one asset is, for instance, a firm's existing portfolio, and the other asset is a replacement piece of equipment, then the correlation between the returns of the two assets is probably close to +1.0. Why? Because there is no influence that would cause the returns of one asset to vary any differently than those of the other. A Coca-Cola® Bottling company expanding its capacity would be an example of a correlation of about +1.0.

What if a company planned to introduce a completely new product in a new market? In that case we might suspect that the correlation between the returns of the existing portfolio and the new product would be something significantly less than +1.0. Why? Because the cash flows of each asset would be due to different, and probably unrelated, factors. An example would be Disney buying the Anaheim Mighty Ducks National Hockey League team.

Any time the correlation coefficient of the returns of two assets is less than +1.0, then the standard deviation of the portfolio consisting of those assets will be less than the weighted average of the individual assets' standard deviations.

Calculating the Standard Deviation of a Two-Asset Portfolio To calculate the standard deviation of a portfolio, we must use a special formula that takes the diversification effect into account. Here is the formula for a portfolio containing two assets.[5] For convenience, they are labeled Asset a and Asset b:

Standard Deviation of the Returns of a Two-Asset Portfolio

$$\sigma_p = \sqrt{w_a^2\sigma_a^2 + w_b^2\sigma_b^2 + 2w_aw_br_{a,b}\sigma_a\sigma_b} \qquad (7\text{-}5)$$

where: σ_p = the standard deviation of the returns of the combined portfolio containing Asset a and Asset b
w_a = the weight of Asset a in the two-asset portfolio
σ_a = the standard deviation of the returns of Asset a
w_b = the weight of Asset b in the two-asset portfolio
σ_b = the standard deviation of the returns of Asset b
$r_{a,b}$ = the correlation coefficient of the returns of Asset a and Asset b

The formula may look scary, but don't panic. Once we know the values for each factor, we can solve the formula rather easily with a calculator. Let's use the formula to find the standard deviation of a portfolio composed of equal amounts invested in Company Calm and Company Bold (i.e., Company Cool).

To calculate the standard deviation of expected returns of the portfolio of Company Cool, we need to know that Company Cool's portfolio is composed of 50 percent Company Calm assets (w_a = .5) and 50 percent Company Bold Assets (w_b = .5). The standard deviation of Company Calm's expected returns is 2 percent (σ_a = .02) and the standard deviation of Company Bold's expected returns is 4 percent (σ_b = .04). To begin, assume the correlation coefficient (r) is −1.0, as shown in Figure 7-4.

Now we're ready to use Equation 7-5 to calculate the standard deviation of Company Cool's returns. For the purpose of the formula, assume Asset a is Company Calm and Asset b is Company Bold:

$$\begin{aligned}\sigma_p &= \sqrt{w_a^2\sigma_a^2 + w_b^2\sigma_b^2 + 2w_aw_br_{ab}\sigma_a\sigma_b}\\ &= \sqrt{(.50^2)(0.02^2) + (.50^2)(0.04^2) + (2)(.50)(.50)(-1.0)(.02)(.04)}\\ &= \sqrt{(.25)(.0004) + (.25)(.0016) - .0004}\\ &= \sqrt{.0001 + .0004 - .0004}\\ &= \sqrt{.0001}\\ &= .01, \text{ or } 1\%\end{aligned}$$

The diversification effect results in risk reduction. Why? Because we are combining two assets, Companies Calm and Bold, that have returns that are negatively correlated (r = −1.0). The standard deviation of the combined portfolio is much lower than that of either of the two individual companies (1 percent for Company Cool compared to 2 percent for Company Calm and 4 percent for Company Bold).

---

[5]You can adapt the formula to calculate the standard deviations of the returns of portfolios containing more than two assets, but doing so is complicated and usually unnecessary. Most of the time, you can view a firm's existing portfolio as one asset and a proposed addition to the portfolio as the second asset.

Nondiversifiable Risk   Unless the returns of one-half the assets in a portfolio are perfectly negatively correlated with the other half—which is extremely unlikely—some risk will remain after assets are combined into a portfolio. The degree of risk that remains is **nondiversifiable risk,** the part of a portfolio's total risk that can't be eliminated by diversifying.

Nondiversifiable risk is one of the characteristics of market risk because it is produced by factors that are shared, to a greater or lesser degree, by most assets in the market. These factors might include inflation and real gross domestic product changes. Figure 7-5 illustrates nondiversifiable risk.

In Figure 7-5 we assumed that the portfolio begins with one asset with possible returns having a probability distribution with a standard deviation of 10 percent. However, if the portfolio is divided equally between two assets, each with possible returns having a probability distribution with a standard deviation of 10 percent, and the correlation of the returns of the two assets is, say +.25, then the standard deviation of the returns or the portfolio drops to about 8 percent. If the portfolio is divided among greater numbers of stocks, the standard deviation of the portfolio will continue to fall—as long as the newly added stocks have returns that are less than perfectly positively correlated with those of the existing portfolio.

Note in Figure 7-5, however, that after about 20 assets have been included in the portfolio, adding more has little effect on the portfolio's standard deviation. Almost all the risk that can be eliminated by diversifying is gone. The remainder, about 5 percent in this example, represents the portfolio's nondiversifiable risk.

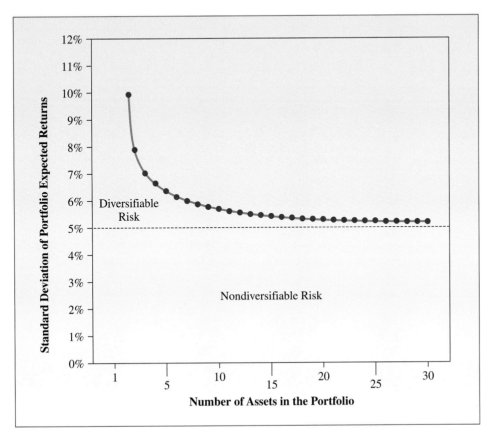

Figure 7-5   **The Relationship between the Number of Assets in a Portfolio and the Riskiness of the Portfolio**

The graph shows that as each new asset is added to a portfolio, the diversification effect causes the standard deviation of the portfolio to decrease. After about 20 assets have been added, however, the effect of adding further assets is slight. The remaining degree of risk is nondiversifiable risk.

Measuring Nondiversifiable Risk. Nondiversifiable risk is measured by a term called **beta** (β). The ultimate group of diversified assets, the market, has a beta of 1.0. The betas of portfolios, and individual assets, relate their returns to those of the overall stock market. Portfolios with betas higher than 1.0 are relatively more risky than the market. Portfolios with betas less than 1.0 are relatively less risky than the market. (Risk-free portfolios have a beta of zero.) The more the return of the portfolio in question fluctuates relative to the return of the overall market, the higher the beta, as shown graphically in Figure 7-6.

Figure 7-6 shows that returns of the overall market fluctuated between about 8 percent and 12 percent during the 10 periods that were measured. By definition, the market's beta is 1.0. The returns of the average-risk portfolio fluctuated exactly the same amount, so the beta of the average-risk portfolio is also 1.0. Returns of the low-risk portfolio fluctuated between 6 percent and 8 percent, half as much as the market. So the low-risk portfolio's beta is 0.5, only half that of the market. In contrast, returns of the high-risk portfolio fluctuated between 10 percent and 16 percent, one and a half times as much as the market. As a result, the high-risk portfolio's beta is 1.5, half again as high as the market.

Companies in low-risk, stable industries like public utilities will typically have low beta values because returns of their stock tend to be relatively stable. (When the economy goes into a recession, people generally continue to turn on their lights and use their refrigerators; and when the economy is booming, people do not splurge on additional electricity consumption.) Recreational boat companies, on the other hand, tend to have high beta values. That's because demand for recreational boats is

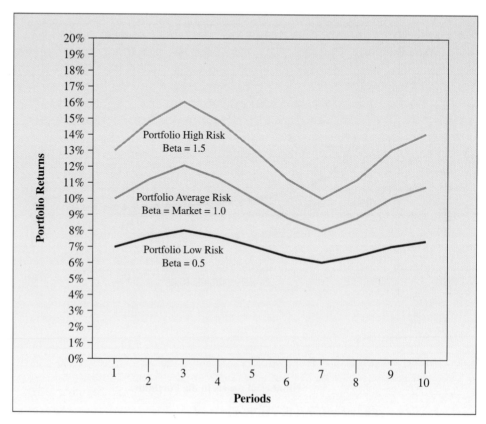

Figure 7-6 Portfolio Fluctuations and Beta

The relative fluctuation in returns for portfolios of different betas. The higher the beta, the more the portfolio's returns fluctuate relative to the returns of the overall market. The market itself has a beta of 1.0.

volatile. (When times are tough, people postpone the purchase of recreational boats. During good economic times, when people have extra cash in their pockets, sales of these boats take off.)

Dealing with Risk

Once companies determine the degree of risk present, what do they do about it? Suppose, for example, a firm determined that if a particular project were adopted, the standard deviation of possible returns of the firm's portfolio of assets would double. So what? How should a firm deal with the situation?

There are two broad classes of alternatives for dealing with risk. First, you might take some action to reduce the degree of risk present in the situation. Second (if the degree of risk can't be reduced), you may compensate for the degree of risk you are about to assume. We'll discuss these two classes of alternatives in the following sections.

Risk-Reduction Methods

One way companies can avoid risk, is simply to avoid risky situations entirely. Most of the time, however, refusing to get involved is an unsatisfactory business decision. Carried to its logical conclusion, this would mean that everyone would invest in risk-free assets only, and no products or services would be produced. Bill Gates, founder and CEO of Microsoft, didn't get rich by avoiding risks. To succeed, businesses must take risks.

If we assume that firms (and individuals) are willing to take some risk to achieve the higher expected returns that accompany that risk, then the task is to reduce the degree of risk as much as possible. The following three methods help to reduce risk: reducing sales volatility and fixed costs, insurance, and diversification.

Reducing Sales Volatility and Fixed Costs   Earlier in the chapter, we discussed how sales volatility and fixed costs contribute to a firm's business risk. Firms in volatile industries whose sales fluctuate widely are exposed to a high degree of business risk. That business risk is intensified even further if they have large amounts of fixed costs. Reducing the volatility of sales, and the amount of fixed costs a firm pays, then, will reduce risk.

Reducing Sales Volatility.   If a firm could smooth out its sales over time, then the fluctuation of its operating income (business risk) would also be reduced. Businesses try to stabilize sales in many ways. For example, retail ski equipment stores sell tennis equipment in the summer, summer vacation resorts offer winter specials, and movie theaters offer reduced prices for early shows to encourage more patronage during slow periods.

Insurance   Insurance is a time-honored way to spread risk among many participants and thus reduce the degree of risk borne by any one participant. Business firms insure themselves against many risks, such as flood, fire, and liability. However, one important risk—the risk that an investment might fail—is uninsurable. To reduce the risk of losing everything in one investment, firms turn to another risk-reduction technique, diversification.

Diversification is a hotly debated issue among financial theorists. Specifically, theorists question whether a firm provides value to its stockholders if it diversifies its asset portfolio to stabilize the firm's income. Many claim that individual stockholders can achieve diversification benefits more easily and cheaply than a firm, so firms that diversify actually do a disservice to their stockholders. What do you think?

Diversification Review Figure 7-4 and the discussion following the figure. We showed in that discussion how the standard deviation of returns of Company Calm (2 percent) and Company Bold (4 percent) could be reduced to 1 percent by combining the two firms into one portfolio. The diversification effect occurred because the returns of the two firms were not perfectly positively correlated. Any time firms invest in ventures whose returns are not perfectly positively correlated with the returns of their existing portfolios, they will experience diversification benefits.

Compensating for the Presence of Risk

In most cases it's not possible to avoid risk completely. Some risk usually remains even after firms use risk-reduction techniques. When firms assume risk to achieve an objective, they also take measures to receive compensation for assuming that risk. In the sections that follow, we discuss these compensation measures.

Adjusting the Required Rate of Return Most owners and financial managers are generally risk averse. So for a given expected rate of return, less risky investment projects are more desirable than more risky investment projects. The higher the expected rate of return, the more desirable the risky venture will appear. As we noted earlier in the chapter, the risk–return relationship is positive. That is, because of risk aversion, people demand a higher rate of return for taking on a higher-risk project.

In capital budgeting a rate of return to reflect risk is called a risk-adjusted discount rate. See chapter 10.

Although we know that the risk–return relationship is positive, an especially difficult question remains: How much return is appropriate for a given degree of risk? Say, for example, that a firm has all assets invested in a chain of convenience stores that provides a stable return on investment of about 6 percent a year. How much

Financial Management and You

Mutual Funds and Risk

Most people don't put all their eggs in one basket. Instead, they diversify to protect against risk.

Many investors achieve diversification benefits by buying shares in mutual funds. Mutual funds are companies that invest money on behalf of shareholders in many different securities. Mutual fund investors, therefore, achieve instant diversification with very little effort. Investing in too many funds, however, may lead to superfluous diversification.

To investigate the risks of a mutual fund, consider the following tips:

1. *Determine what type of fund it is.* Is the fund a growth, income, or growth and income fund? A growth fund will be more volatile, whereas a growth and income fund will be less so. An income fund will have steady income from dividends or bond interest to cushion the fund should stock prices fall. The type of fund you select often depends on your risk tolerance and how long you plan to invest your money.
2. *Learn whether the fund specializes.* If the fund invests in one industry only, you may want to invest in several funds to diversify or avoid that fund altogether.
3. *Check risk-rating services.* Mutual fund rating services—such as Morningstar Inc. of Chicago, Illinois—rate mutual funds according to risk. Mutual fund ratings are available in print or on-line.

Source: Timothy Middleton, "Trouble Ahead? Seven Funds Take Few Chances in the Long Run," *The New York Times* (November 12, 1995): F5.

more return should the firm require for investing some assets in a baseball team that may not provide steady returns[6]—8 percent? 10 percent? 25 percent? Unfortunately, no one knows for sure, but financial experts have researched the subject extensively.

One well-known model used to calculate the required rate of return of an investment is the **capital asset pricing model (CAPM).** We discuss CAPM next.

Relating Return and Risk: The Capital Asset Pricing Model Financial theorists William F. Sharpe, John Lintner, and Jan Mossin worked on the risk–return relationship and developed the capital asset pricing model, or CAPM. We can use this model to calculate the appropriate required rate of return for an investment project given its degree of risk as measured by beta (β).[7] The formula for CAPM is presented in Equation 7-6.

CAPM Formula

$$k_p = k_{RF} + (k_M - k_{RF}) \times \beta \qquad (7\text{-}6)$$

where: k_p = the required rate of return appropriate for the investment project
k_{RF} = the risk-free rate of return
k_M = the required rate of return on the overall market
β = the project's beta

The three components of the CAPM include the risk-free rate of return (k_{RF}), the market risk premium ($k_M - k_{RF}$), and the project's beta (β). The **risk-free rate of return (k_{RF})** is the rate of return that investors demand from a project that contains no risk. Risk-averse managers and owners will always demand at least this rate of return from any investment project.

The required rate of return on the overall market minus the risk-free rate ($k_M - k_{RF}$) represents the additional return demanded by investors for taking on the risk of investing in the market itself. The term is sometimes called the **market risk premium.** In the CAPM, the term for the market risk premium, ($k_M - k_{RF}$), can be viewed as the additional return over the risk-free rate that investors demand from an "average stock" or an "average-risk" investment project. The S&P 500 stock market index is often used as a proxy for the market.

As discussed earlier, a project's beta (β) represents a project's degree of risk relative to the overall stock market. In the CAPM, when the beta term is multiplied by the market risk premium term, ($k_M - k_{RF}$), the result is the additional return over the risk-free rate that investors demand from that individual project. Beta is the relevant risk measure according to the CAPM. High-risk (high-beta) projects have high required rates of return, and low-risk (low-beta) projects have low required rates of return.

Table 7-4 shows three examples of how the CAPM is used to determine the appropriate required rate of return for projects of different degrees of risk.

As we can see in Table 7-4, Project High Risk, with its beta of 1.5, has a required rate of return that is twice that of Project Low Risk, with its beta of 0.5. After all,

---

[6]Some major league baseball teams are losing money and others make a great deal. Television revenues differ greatly from team to team, as do ticket sales and salary expenses.
[7]See William Sharpe, "Capital Asset Prices: A Theory of Market Equilibrium," *Journal of Finance* (September 1964); John Lintner, "The Valuation of Risk Assets and the Selection of Risky Investments in Stock Portfolios and Capital Budgets," *Review of Economics and Statistics* (February 1965); and Jan Mossin, "Equilibrium in a Capital Asset Market," *Econometrica* (October 1966).

TABLE 7-4

Using the CAPM to Calculate Required Rates of Return for Investment Projects

Given:

The risk-free rate, $k_{RF} = 4\%$
The required rate of return on the market, $k_M = 12\%$
Project Low Risk's beta = 0.5
Project Average Risk's beta = 1.0
Project High Risk's beta = 1.5

Required rates of return on the project's per the CAPM:

$$\begin{aligned}
\text{Project Low Risk:} \quad k_p &= .04 + (.12 - .04) \times 0.5 \\
&= .04 + .04 \\
&= .08, \text{ or } 8\%
\end{aligned}$$

$$\begin{aligned}
\text{Project Average Risk:} \quad k_p &= .04 + (.12 - .04) \times 1.0 \\
&= .04 + .08 \\
&= .12, \text{ or } 12\%
\end{aligned}$$

$$\begin{aligned}
\text{Project High Risk:} \quad k_p &= .04 + (.12 - .04) \times 1.5 \\
&= .04 + .12 \\
&= .16, \text{ or } 16\%
\end{aligned}$$

Given that the risk-free rate of return is 4 percent and the required rate of return on the market is 12 percent, the CAPM indicates the appropriate required rate of return for a low-risk investment project with a beta of .5 is 8 percent. The appropriate required rate of return for an average-risk project is the same as that for the market, 12 percent, and the appropriate rate for a high-risk project with a beta of 1.5 is 16 percent.

Figure 7-7 CAPM and the Risk–Return Relationship

This graph illustrates the increasing return required for increasing risk as indicated by the CAPM beta. This graphical depiction of the risk–return relationship according to the CAPM is called the security market line (SML).

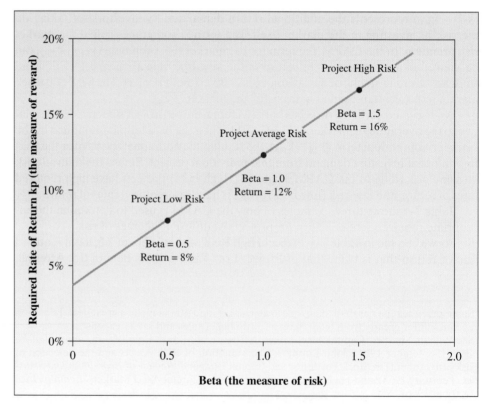

164   Part Two   Essential Concepts in Finance

CONTENT CHANGES

NEW TO THIS EDITION

◆ **Acetate overlays** are included in chapter 8, Time Value of Money, which demonstrate the sensitivity of present value and future value to changes in k and n values. These overlays help the student visually understand the time value of money concept by showing a visual step-by-step example.

◆ **Time-sensitive examples** have been thoroughly updated throughout the text with a number of companies that students can relate to, including Yahoo, Lucent Technologies, and GeoCities.

◆ **FinCoach Practice Exercise sections** are in every quantitative chapter. FinCoach is a Financial Math Practice Program that provides student tutoring. There are FinCoach icons integrated in the end-of-chapter exercises. This text includes FREE FinCoach software packaged with the Prentice Hall Finance Center CD-ROM in the back of the text.

◆ **PHLIP/CW icons** appear at the end of each chapter. PHLIP/CW (Prentice Hall Learning on the Internet Partnership – Companion Web site) is a Web-based enhancement designed specifically for this text. PHLIP is FREE to the student and includes access to current events, an on-line study guide, and a chat room that is directly tied to the text. A direct link to the PHLIP/CW Web site is located on the Prentice Hall Finance Center CD-ROM in the back of the text.

GENERAL CHANGES

◆ **Updated international coverage**, including a chapter on the financial troubles in Asia, the European Union and European Monetary Union, and the introduction to the euro, while weaving international coverage into all other chapters throughout the text.

◆ **Ethics boxes** identify many financial decisions with an ethical dimension. **Financial Management and You boxes** demonstrate how financial management concepts that were intended for use with a firm can be applied to personal financial decision making.

◆ The text and all supplements have been through a very thorough accuracy check. The authors themselves and several outside parties have reviewed the text and supplements extensively.

CONTENT CHANGES

shouldn't we ask for a higher rate of return if the risk is higher? Note also that Project Average Risk, which has the same beta as the market, 1.0, also has the same required rate of return as the market (12 percent). The risk–return relationship for these three projects is shown in Figure 7-7.

Remember that the beta term in the CAPM reflects only the nondiversifiable risk of an asset, not its diversifiable risk. Diversifiable risk is irrelevant because the diversity of each investor's portfolio essentially eliminates (or should eliminate) that risk. (After all, most investors are well diversified. They will not demand extra return for adding a security to their portfolios that contains diversifiable risk.) The return that well-diversified investors demand when they buy a security, as measured by the CAPM and beta, relates to the degree of nondiversifiable risk in the security.

What's Next

In this chapter we examined the risk–return relationship, types of risk, risk measurements, risk-reduction techniques, and the CAPM. In the next chapter, we will discuss the time value of money.

Summary

1. Define risk, risk aversion, and the risk–return relationship.

In everything you do, or don't do, there is a chance that something will happen that you didn't count on. Risk is the potential for unexpected events to occur.

Given two financial alternatives that are equal except for their degree of risk, most people will choose the less risky alternative because they are risk averse. Risk aversion is a common trait among almost all investors. Most investors avoid risk if they can, unless they are compensated for accepting risk. In an investment context, the additional compensation is a higher expected rate of return.

The risk–return relationship refers to the positive relationship between risk and the required rate of return. Due to risk aversion, the higher the risk, the more return investors expect.

2. Measure risk using the standard deviation and the coefficient of variation.

Risk is the chance, or probability, that outcomes other than what is expected will occur. This probability is reflected in the narrowness or width of the distribution of the possible values of the financial variable. In a distribution of variable values, the standard deviation is a number that indicates how widely dispersed the possible values are around the expected value. The more widely dispersed a distribution is, the larger the standard deviation, and the greater the probability that an actual value will be different than the expected value. The standard deviation, then, can be used to measure the likelihood that some outcome, substantially different than what is expected, will occur.

When the degree of risk in distributions of different sizes are compared, the coefficient of variation is a statistic used to measure *relative* riskiness. The coefficient of variation measures the standard deviation's percentage of the expected value. It relates the standard deviation to its mean to give a risk measure that is independent of the magnitude of the possible returns.

3. Identify the types of risk that business firms encounter.

Business risk is the risk that a company's operating income will differ from what is expected. The more volatile a company's operating income, the more business risk the firm contains. Business risk is a result of sales volatility, which translates into operating income volatility.

Business risk is increased by the presence of fixed costs, which magnify the effect on operating income of changes in sales.

Financial risk occurs when companies borrow money and incur interest charges that show up as fixed expenses on their income statements. Fixed interest charges act on a firm's net income the same way fixed operating expenses act on operating income—they increase volatility. The additional volatility of a firm's net income caused by the presence of fixed interest expense is called financial risk.

Portfolio risk is the chance that investors won't get the return they expect from a portfolio. Portfolio risk can be measured by the standard deviation of possible returns of a portfolio. It is affected by the correlation of returns of the assets making up the portfolio. The less correlated these returns are, the more gains on some assets offset losses on others, resulting in a reduction of the portfolio's risk. This phenomenon is known as the *diversification effect*. Nondiversifiable risk is risk that remains in a portfolio after all diversification benefits have been achieved. Nondiversifiable risk is measured by a term called beta (β). The market has a beta of 1.0. Portfolios with betas greater than 1.0 contain more nondiversifiable risk than the market, and portfolios with betas less than 1.0 contain less nondiversifiable risk than the market.

4. Explain methods of risk reduction.

Firms can reduce the degree of risk by taking steps to reduce the volatility of sales or their fixed costs. Firms also obtain insurance policies to protect against many risks, and diversify their asset portfolios to reduce the risk of income loss.

5. Describe how firms compensate for assuming risk.

Firms almost always demand a higher rate of return to compensate for assuming risk. The more risky a project, the higher the return firms demand.

6. Explain how the capital asset pricing model (CAPM) relates risk and return.

When investors adjust their required rates of return to compensate for risk, the question arises as to how much return is appropriate for a given degree of risk. The capital asset pricing model (CAPM) is a model that measures the required rate of return for an investment or project, given its degree of nondiversifiable risk as measured by beta (β).

You can access your FREE PHLIP/CW On–Line Study Guide, Current Events, and Spreadsheet Problem templates, which may correspond to this chapter either through the Prentice Hall Finance Center CD–ROM or by directly going to: **http://www.prenhall.com/gallagher.**

PHLIP/CW — *The Prentice Hall Learning on the Internet Partnership–Companion Web Site*

Equations Introduced in This Chapter

Equation 7-1. The Expected Value, or Mean (μ) of a Probability Distribution:

$$\mu = \Sigma (V \times P)$$

where: μ = the expected value, or mean
V = the possible value for some variable
P = the probability of the value V occurring

Equation 7-2. The Standard Deviation:

$$\sigma = \sqrt{\Sigma P(V - \mu)^2}$$

where: σ = the standard deviation
V = the value the variable can take
P = the probability of the value V occurring
μ = the expected value

Equation 7-3. The Coefficient of Variation of a Probability Distribution:

$$\text{Coefficient of Variation} = \frac{\text{Standard Deviation}}{\text{Mean}}$$

Equation 7-4. The Expected Rate of Return, $E(R_p)$, Comprised of Two Assets, a and b:

$$E(R_p) = [w_a \times E(R_a)] + [w_b \times E(R_b)]$$

where: w_a = the weight, or percentage of the new combined portfolio invested in Asset a
$E(R_a)$ = the expected return of Asset a
w_b = the weight, or percentage of the new combined portfolio invested in Asset b
$E(R_b)$ = the expected return of Asset b

Equation 7-5. The Standard Deviation of a Two-Asset Portfolio:

$$\sigma_p = \sqrt{w_a^2 \sigma_a^2 + w_b^2 \sigma_b^2 + 2 w_a w_b r_{a,b} \sigma_a \sigma_b}$$

where: σ_p = the standard deviation of the combined portfolio containing Assets a and b
w_a = the weight of Asset a in the portfolio
σ_a = the standard deviation of the returns of Asset a
$r_{a,b}$ = the correlation coefficient of the cash flows of Asset a and Asset b
w_b = the weight of Asset b in the portfolio
σ_b = the standard deviation of the returns of Asset b

Equation 7-6. The Capital Asset Pricing Model (CAPM):

$$k_p = k_{RF} + (k_M - k_{RF}) \times \beta$$

where: k_p = the required rate of return appropriate for the investment project
k_{RF} = the risk-free rate of return
k_M = the required rate of return on the overall market
β = the project's beta

Self-Test

ST-1. For Bryan Corporation, the mean of the distribution of next year's possible sales is $5 million. The standard deviation of this distribution is $400,000. Calculate the coefficient of variation (CV) for this distribution of possible sales.

ST-2. Investors in Hoeven Industries common stock have a .2 probability of earning a return of 4 percent, a .6 probability of earning a return of 10 percent, and a .2 probability of earning a return of 20 percent. What is the mean of this probability distribution (the expected rate of return)?

ST-3. What is the standard deviation for the Hoeven Industries common stock return probability distribution described in ST-2?

ST-4. The standard deviation of the possible returns of Boris Company common stock is .08, whereas the standard deviation of possible returns of Natasha Company common stock is .12. Calculate the standard deviation of a portfolio comprised of 40 percent Boris Company stock and 60 percent Natasha Company stock. The correlation coefficient of the returns of Boris Company stock relative to the returns of Natasha Company stock is $-.2$.

ST-5. The mean of the normal probability distribution of possible returns of Gidney and Cloyd Corporation common stock is 18 percent. The standard deviation is 3 percent. What is the range of possible values that you would be 95 percent sure would capture the return that will actually be earned on this stock?

ST-6. Dobie's Bagle Corporation common stock has a beta of 1.2. The market risk premium is 6 percent and the risk-free rate is 4 percent. What is the required rate of return on this stock according to the CAPM?

ST-7. Using the information provided in ST-6, what is the required rate of return on the common stock of Zack's Salt Corporation? This stock has a beta of .4.

ST-8. A portfolio of three stocks has an expected value of 14 percent. Stock A has an expected return of 6 percent and a weight of .25 in the portfolio. Stock B has an expected return of 10 percent and a weight of .5 in the portfolio. Stock C is the third stock in this portfolio. What is the expected rate of return of Stock C?

FinCoach Practice Exercises

To help improve your grade and master the mathematical skills covered in this chapter, open FinCoach on the Prentice Hall Finance Center CD-ROM and work practice problems in the following categories: 1) Portfolio Diversification and 2) CAPM.

Review Questions

1. What is risk aversion? If common stockholders are risk averse, how do you explain the fact that they often invest in very risky companies?
2. Explain the risk–return relationship.
3. Why is the coefficient of variation often a better risk measure when comparing different projects than the standard deviation?
4. What is the difference between business risk and financial risk?
5. Why does the riskiness of portfolios have to be looked at differently than the riskiness of individual assets?
6. What happens to the riskiness of a portfolio if assets with very low correlations (even negative correlations) are combined?
7. What does it mean when we say that the correlation coefficient for two variables is -1? What does it mean if this value were zero? What does it mean if it were $+1$?
8. What is nondiversifiable risk? How is it measured?
9. Compare diversifiable and nondiversifiable risk. Which do you think is more important to financial managers in business firms?
10. How do risk-averse investors compensate for risk when they take on investment projects?
11. Given that risk-averse investors demand more return for taking on more risk when they invest, how much more return is appropriate for, say, a share of common stock, than is appropriate for a Treasury bill?
12. Discuss risk from the perspective of the capital asset pricing model (CAPM).

Build Your Communication Skills

CS-1. Go to the library, use business magazines, computer databases, the Internet, or other sources that have financial information about businesses. (See chapter 5 for a list of specific resources.) Find three companies and compare their approaches to risk. Do the firms take a conservative or aggressive approach? Write a one- to two-page report, citing specific evidence of the risk-taking approach of each of the three companies you researched.

CS-2. Research three to five specific mutual funds. Then form small groups of four to six. Discuss whether the mutual funds each group member researched will help investors diversify the risk of their portfolio. Are some mutual funds better than others for an investor seeking good diversification? Prepare a list of mutual funds the group would select to diversify its risk and explain your choices. Present your recommendations to the class.

Problems

7-1. Manager Paul Smith believes an investment project will have the following yearly cash flows with the associated probabilities throughout its life of five years. Calculate the standard deviation and coefficient of variation of the cash flows.

Standard Deviation and Coefficient of Variation

Cash Flows($)	Probability of Occurrence
$10,000	.05
13,000	.10
16,000	.20
19,000	.30
22,000	.20
25,000	.10
28,000	.05

7-2. Milk-U, an agricultural consulting firm, has developed the following income statement forecast:

Measuring Risk

Milk-U Income Forecast (in 000's)

	Probability of Occurrence				
	2%	8%	80%	8%	2%
Sales	$500	$700	$1,200	$1,700	$1,900
Variable Expenses	250	350	600	850	950
Fixed Operating Expenses	250	250	250	250	250
Operating Income	0	100	350	600	700

a. Calculate the expected value of Milk-U's operating income.
b. Calculate the standard deviation of Milk-U's operating income.
c. Calculate the coefficient of variation of Milk-U's operating income.
d. Recalculate the expected value, standard deviation, and coefficient of variation of Milk-U's operating income if the company's sales forecast changed as follows:

	Probability of Occurrence				
	10%	15%	50%	15%	10%
Sales	$500	$700	$1,200	$1,700	$1,900

e. Comment on how Milk-U's degree of business risk changed as a result of the new sales forecast in part d.

7-3. As a new loan officer in the Bulwark Bank, you are comparing the financial riskiness of two firms. Selected information from *pro forma* statements for each firm follows.

Measuring Risk

Equity Eddie's Company
Net Income Forecast (in 000's)

	Probability of Occurrence				
	5%	10%	70%	10%	5%
Operating Income	$100	$200	$400	$600	$700
Interest Expense	0	0	0	0	0
Before-Tax Income	100	200	400	600	700
Taxes (28%)	28	56	112	168	196
Net Income	72	144	288	432	504

Barry Borrower's Company
Net Income Forecast (in 000's)

	Probability of Occurrence				
	5%	10%	70%	10%	5%
Operating Income	$110	$220	$440	$660	$770
Interest Expense	40	40	40	40	40
Before-Tax Income	70	180	400	620	730
Taxes (28%)	19.6	50.4	112	173.6	204.4
Net Income	50.4	129.6	288	446.4	525.6

a. Calculate the expected value of Equity Eddie's and Barry Borrower's net income.
b. Calculate the standard deviation of Equity Eddie's and Barry Borrower's net income.
c. Calculate the coefficient of variation of Equity Eddie's and Barry Borrower's net income.
d. Compare Equity Eddie's and Barry Borrower's degree of financial risk.

Standard Deviation and Coefficient of Variation

7-4. George Taylor, owner of a toy manufacturing company, is considering the addition of a new product line. Marketing research shows that gorilla action figures will be the next fad for the six- to ten-year-old age group. This new product line of gorilla-like action figures and their high-tech vehicles will be called Go-Rilla. George estimates that the most likely yearly incremental cash flow will be $26,000. There is some uncertainty about this value because George's company has never before made a product similar to the Go-Rilla. He has estimated the potential cash flows for the new product line along with their associated probabilities of occurrence. His estimates follow.

Go-Rilla Project

Cash Flows	Probability of Occurrence
$20,000	1%
$22,000	12%
$24,000	23%
$26,000	28%
$28,000	23%
$30,000	12%
$32,000	1%

a. Calculate the standard deviation of the estimated cash flows.
b. Calculate the coefficient of variation.
c. If George's other product lines have an average coefficient of variation of 12 percent, what can you say about the risk of the Go-Rilla Project relative to the average risk of the other product lines?

Portfolio Risk

7-5. Assume that a company has an existing portfolio A with an expected return of 9 percent and a standard deviation of 3 percent. The company is considering adding an asset B to its portfolio. Asset B's expected return is 12 percent with a standard deviation of 4 percent.

Also assume that the amount invested in A is $700,000 and the amount to be invested in B is $200,000. If the degree of correlation between returns from portfolio A and project B is zero, calculate:

  a. The standard deviation of the new combined portfolio and compare it with that of the existing portfolio.
  b. The coefficient of variation of the new combined portfolio and compare it with that of the existing portfolio.

7-6. A firm has an existing portfolio of projects with an expected return of 11 percent a year. The standard deviation of these returns is 4 percent. The existing portfolio's value is $820,000. As financial manager, you are considering the addition of a new project, PROJ1. PROJ1's expected return is 13 percent with a standard deviation of 5 percent. The initial cash outlay for PROJ1 is expected to be $194,000. *Challenge Problem*

  a. Calculate the coefficient of variation for the existing portfolio.
  b. Calculate the coefficient of variation for PROJ1.
  c. If PROJ1 is added to the existing portfolio, calculate the weight (proportion) of the existing portfolio in the combined portfolio.
  d. Calculate the weight (proportion) of PROJ1 in the combined portfolio.
  e. Assume the correlation coefficient of the cash flows of the existing portfolio and PROJ1 is zero. Calculate the standard deviation of the combined portfolio. Is the standard deviation of the combined portfolio higher or lower than the standard deviation of the existing portfolio?
  f. Calculate the coefficient of variation of the combined portfolio.
  g. If PROJ1 is added to the existing portfolio, will the firm's risk increase or decrease?

7-7. Assume the risk-free rate is 5 percent, the expected rate of return on the market is 15 percent, and the beta of your firm is 1.2. Given these conditions, what is the required rate of return on your company's stock per the capital asset pricing model? *CAPM*

7-8. Your firm has a beta of 1.5 and you are considering an investment project with a beta of 0.8. Answer the following questions assuming that short-term Treasury bills are currently yielding 5 percent and the expected return on the market is 15 percent. *Challenge Problem*

  a. What is the appropriate required rate of return for your company per the capital asset pricing model?
  b. What is the appropriate required rate of return for the investment project per the capital asset pricing model?
  c. If your firm invests 20 percent of its assets in the new investment project, what will be the beta of your firm after the project is adopted? (Hint: Compute the weighted average beta of the firm with the new asset using Equation 7-4.)

The following problems (7-9 to 7-13) relate to the expected business of Power Software Company (PSC) in the year 2000 (000's of dollars):

Power Software Company Forecasts

	Probability of Occurrence						
	2%	8%	20%	40%	20%	8%	2%
Sales	$800	$1,000	$1,400	$2,000	$2,600	$3,000	$3,200

7-9. Calculate the expected value, standard deviation, and coefficient of variation of sales revenue of PSC. *Measuring Risk*

7-10. Assume that PSC has no fixed expense but has a variable expense that is 60 percent of sales as follows: *Business Risk*

Power Software Company Forecasts

	Probability of Occurrence						
	2%	8%	20%	40%	20%	8%	2%
Sales	$800	$1,000	$1,400	$2,000	$2,600	$3,000	$3,200
Variable Expense	480	600	840	1,200	1,560	1,800	1,920

Calculate PSC's business risk (coefficient of variation of operating income).

Business Risk

7-11. Now assume that PSC has a fixed operating expense of $400,000, in addition to the variable expense of 60 percent of sales, shown as follows:

Power Software Company Forecasts

	Probability of Occurrence						
	2%	8%	20%	40%	20%	8%	2%
Sales	$800	$1,000	$1,400	$2,000	$2,600	$3,000	$3,200
Variable Expense	480	600	840	1,200	1,560	1,800	1,920
Fixed Operating Expense	400	400	400	400	400	400	400

Recalculate PSC's business risk (coefficient of variation of operating income). How does this figure compare with the business risk calculated with variable cost only?

Various Statistics and Financial Risk

7-12. Assume that PSC has a fixed interest expense of $60,000 on borrowed funds. Also assume that the applicable tax rate is 30 percent. What is the expected value, standard deviation, and coefficient of variation of PSC's net income? What is PSC's financial risk?

Business and Financial Risk

7-13. To reduce the various risks, PSC is planning to take suitable steps to reduce volatility of operating and net income. It has projected that fixed expenses and interest expenses can be reduced. The revised figures follow:

Power Software Company Forecasts

	Probability of Occurrence						
	1%	6%	13%	60%	13%	6%	1%
Sales	$800	$1,000	$1,400	$2,000	$2,600	$3,000	$3,200
Variable Expense	480	600	840	1,200	1,560	1,800	1,920
Fixed Operating Expense	250	250	250	250	250	250	250
Interest Expense	40	40	40	40	40	40	40

Recalculate PSC's business and financial risks and compare these figures with those calculated in problems 7-11 and 7-12. The tax rate is 30 percent.

Answers to Self-Test

ST-1. $CV = \sigma \div \mu = \$400{,}000 \div \$5{,}000{,}000 = .08 = 8\%$

ST-2. $\mu = (.2 \times .04) + (.6 \times .10) + (.2 \times .20) = .108 = 10.8\%$

ST-3. $\sigma = \{[.2 \times (.04 - .108)^2] + [.6 \times (.10 - .108)^2] + [.2 \times (.20 - .108)^2]\}^{.5}$
$= [(.2 \times .004624) + (.6 \times .000064) + (.2 \times .008464)]^{.5}$
$= .002656^{.5} = .0515 = 5.15\%$

ST-4. $\sigma_p = [(.4^2 \times .08^2) + (.6^2 + .12^2) + (2 \times .4 \times .6 \times -(.2) \times .08 \times .12)]^{.5}$
$= [(.16 \times .0064) + (.36 \times .0144) + (-.0009216)]^{.5}$
$= .0052864^{.5} = .0727 = 7.27\%$

ST-5. $.18 + (2 \times .03) = .24 = 24\%$
$.18 - (2 \times .03) = .12 = 12\%$
$\therefore$ We are 95% confident that the actual return will be between 12% and 24%.

ST-6. $k_s = .04 + (.06 \times 1.2) = .112 = 11.2\%$

ST-7. $k = .04 + (.06 \times .4) = .064 = 6.4\%$

ST-8. WT of Stock C must be .25 for the total of the weights to equal 1
$.14 = (.06 \times .25) + (.10 \times .5) + [E(R_C) \times .25]$
$.14 = .065 + [E(R_C) \times .25]$
$.075 = E(R_C) \times .25$
$E(R_C) = .30 = 30\%$

PART 2:
ESSENTIAL CONCEPTS IN FINANCE

CHAPTER 8

THE TIME VALUE OF MONEY

Time is money.
—Benjamin Franklin

COLLEGE QUIZ: PUT DOLLARS IN STOCKS OR IN SCHOLARS?

Want your kid to have a million dollars by the time he or she is 40? Forget college. Invest the tuition. Costs at some colleges are high enough that the tuition, if sunk into the stock market, would grow into $1 million in 22 years, and costs at less expensive schools are rising so rapidly that in a few years they too will reach this level. Conventional wisdom has it—and study after study confirms—that people who graduate from college earn far more than those who only finish high school.

A Census Bureau study a few years ago found that college graduates can expect to earn $1.421 million over their working lives (after allowing for inflation), whereas holders of high school diplomas could expect to pull in $821,000—a $600,000 advantage for the college grads. But with the cost of four years of undergraduate work nearing $120,000 for the most expensive private schools and $50,000 for top public ones, families are beginning to wonder whether even that earnings edge is worth it. And to today's high school seniors and their parents, weighing their offers of admission and (maybe) financial aid, the question has a particularly sharp edge.

Consider: If a family were capable of plunking down $120,000 for an Ivy League school but instead put that money into the stock market, where it earned 10.5 per-

CHAPTER OBJECTIVES

After reading this chapter, you should be able to:

- Explain the time value of money and its importance in the business world.

- Calculate the future value and present value of a single amount.

- Find the future and present values of an annuity.

- Solve time value of money problems with uneven cash flows.

- Solve for the interest rate, number or amount of payments, or the number of periods in a future or present value problem.

cent a year (roughly what the market has returned over the years), in 22 years, when today's 18-year-old turns 40, the money would have grown to slightly more than $1 million. That is not corrected for inflation, of course. But after making that correction, using today's rate of around 3 percent, the $120,000 would still be worth about $550,000. Leaving the money invested until retirement—47 years for an 18-year-old—would turn it into more than $13 million before inflation. Taking inflation into account cuts the number to $3.1 million, but that's still not a bad nest egg. Attending a less-expensive college, such as a state university, makes things a little better. An outlay of $48,000 invested in the market instead of college could be expected to grow to about $430,000, or $220,000 after inflation, over 22 years, and to $5.2 million, or a bit more than $1.2 million, over a 47-year working life. Taxes would whittle these amounts even further, but they also would reduce the graduate's earnings.

So what does this tell us? Pick the cheapest school you can find? Forget college and just go to work? Live at home until you're 40, or 65? Not really. For one thing, recent studies indicate that graduates of high-quality institutions do markedly better than those of less selective colleges, and price isn't always the best indicator of quality. "We found that investments in college quality were a good buy," said Dan Black, professor of economics at the University of Kentucky and co-author of a recent report on college values.

Black and his colleagues, Kermit Daniel of the University of Pennsylvania and Jeffrey Smith of the University of Western Ontario, looked at men age 22 to 30 who had attended colleges of varying quality, as measured by Scholastic Assessment Test scores, spending per student, faculty-student ratios and other measures. Graduates of better schools had higher earnings, and those from the top 20 percent of schools earned about 20 percent more than those from the bottom 20 percent, they found. Significantly, they found no correlation between earning power and having attended a private college—graduates of state schools did just as well, or just as poorly, as those from private schools of similar caliber. Thus, while the study makes it clear that "a high price for a not very good school" is likely to be a bad deal, Black said, "the real question is, if you have been admitted to the University of Virginia and you are going to pay [in-state] tuition, [and you've also been admitted to Princeton], is Princeton worth it? That's a much more difficult issue." Black suggested that families look at the choice as an investment, and think about how much more money, if any, the student might expect to earn as a graduate of one school than of another. Then figure how many years it will take to earn back the difference. After that, there are the "non-pecuniary amenities," such as location, living conditions and the like, said Black. "You have to decide if it's worth the extra money."

Then there's the ego question. As a parent, would you really rather tell the cocktail party crowd that your kid is at Princeton? And beyond ego, prestige may have real benefits. Attracting students from families that already are wealthy and successful "may actually be the way in which high-quality institutions are raising [graduates'] earnings. Whether the faculty is teaching better or whether the students are meeting the movers and shakers of the next generation" is difficult to sort out, Black said. It also may be that students at expensive schools try harder. Some studies have suggested that, though times and attitudes change.

John Boli, associate professor of sociology at Emory University, studied the class of 1981 from Stanford 10 years after graduation, and found them making money hand over fist. However, "these students are really assured that they are going to be able to do well. . . . They are coming from relatively intellectual or at least highly educated backgrounds," Boli said. But at the same time, he said, the 1980s were "a period when many students were very nakedly careerist-oriented." He also said there is an ironic possibility that the income of some of these graduates may decline as some of the women among them decide to leave the work world to raise children. In fact, "the only place we find that housewife phenomenon is in the upper-middle class" because this is the only group that can afford it, he said. A second question for families is the extensive discounting—schools prefer to call it financial aid—that goes on in higher education. State schools particularly are becoming very aggressive in trying to attract top high school students to their campuses, whether or not they can pay on their own.

In the most extreme case, a well-heeled student might have the option of going to a state school for free, thus getting a college education and still having the tuition money to invest. Athletic scholarships fall into the same category. Harvard has lost several prep school hockey players in recent years, when it offered little aid and the University of Michigan offered a "full ride." Finally, money isn't everything. Education is, after all, supposed to improve a student's mind and make him or her a better person. That might even be true. What it plainly does do, in an increasingly uncertain world, is give the student more resources to cope with change and to capitalize on opportunity. That, most likely, is what families are hoping to buy with all that money.

Source: Albert B. Crenshaw, "College Quiz," *The Washington Post* (April 28, 1996): H01.

Chapter Overview

A dollar in hand today is worth more than a promise of a dollar tomorrow. This is one of the basic principles of financial decision making. Time value analysis is a crucial part of financial decisions. It helps answer questions about how much money an investment will make over time and how much a firm must invest now to earn an expected payoff later.

In this chapter we will investigate why money has time value, as well as learn how to calculate the *future value* of cash invested today and the *present value* of cash to be received in the future. We will also discuss the present and future values of an *annuity*—a series of equal cash payments at regular time intervals. Finally, we will examine special time value of money problems, such as how to find the rate of return on an investment and how to deal with a series of uneven cash payments.

Why Money Has Time Value

The **time value of money** means that money you hold in your hand today is worth more than money you expect to receive in the future. Similarly, money you must pay out today is a greater burden than the same amount paid in the future.

In chapter 2 we learned that interest rates are positive in part because people prefer to consume now rather than later. Positive interest rates indicate, then, that

We discussed the real rate of interest and interest rate determinants in chapter 2, on pages 31–33.

money has time value. When one person lets another borrow money, the first person requires compensation in exchange for reducing current consumption. The person who borrows the money is willing to pay to increase current consumption. The cost paid by the borrower to the lender for reducing consumption, known as an *opportunity cost*, is the real rate of interest.

The real rate of interest reflects compensation for the **pure time value of money.** The real rate of interest does not include interest charged for expected inflation or the risk factors discussed in chapter 2. Recall from the interest rate discussion in chapter 2 that many factors—including the pure time value of money, inflation, default risk, liquidity risk, and maturity risk—determine market interest rates.

The required rate of return on an investment reflects the pure time value of money, an adjustment for expected inflation, and any risk premiums present.

Measuring the Time Value of Money

Financial managers adjust for the time value of money by calculating the *future value* and the *present value*. Future value and present value are mirror images of each other. **Future value** is the value of a starting amount at a future point in time, given the rate of growth per period and the number of periods until that future time. How much will $1,000 invested today at a 10 percent interest rate grow to in 15 years? **Present value** is the value of a future amount today, assuming a specific required interest rate for a number of periods until that future amount is realized. How much should we pay today to obtain a promised payment of $1,000 in 15 years if investing money today would yield a 10 percent rate of return per year?

The Future Value of a Single Amount

To calculate the future value of a single amount, we must first understand how money grows over time. Once money is invested, it earns an interest rate that compensates for the time value of money and, as we learned in chapter 2, for default risk, inflation, and other factors. Often, the interest earned on investments is **compound interest**—interest earned on interest *and* on the original principal. In contrast, *simple interest* is interest earned only on the original principal.

To illustrate compound interest, assume the financial manager of SaveCom decided to invest $100 of the firm's excess cash in an account that earns an annual interest rate of 5 percent. In year 1 SaveCom will earn $5 in interest, calculated as follows:

$$\begin{aligned}\text{Balance at End of One Year} &= \text{Principal} + \text{Interest} \\ &= \$100 + (100 \times .05) \\ &= \$100 \times (1 + .05) \\ &= \$100 \times 1.05 \\ &= \$105\end{aligned}$$

The total amount in the account at the end of year 1, then, is $105.

But look what happens in years 2 and 3. In year 2 SaveCom will earn 5 percent of *$105*. The $105 is the original principal of $100 *plus* the first year's interest—so the interest earned in year 2 is $5.25, rather than $5.00. The end of year 2 balance is $110.25—$100 in original principal and $10.25 in interest. In year 3 SaveCom will earn 5 percent of $110.25, or $5.51, for an ending balance of $115.76, shown as follows:

	Beginning Balance	×	**(1 + Interest Rate)**	**= Ending Balance**	**Interest**
Year 1	$100.00	×	1.05	= $105.00	$5.00
Year 2	$105.00	×	1.05	= $110.25	$5.25
Year 3	$110.25	×	1.05	= $115.76	$5.51

In our example SaveCom earned $5 in interest in year 1, $5.25 in interest in year 2 ($110.25 − $105.00), and $5.51 in year 3 ($115.76 − $110.25) because of the compounding effect. If the SaveCom deposit earned interest only on the original principal, rather than on the principal and interest, the balance in the account at the end of year 3 would be $115 ($100 + ($5 × 3) = $115). In our case the compounding effect accounts for an extra $.76 ($115.76 − $115.00 = .76).

The simplest way to find the balance at the end of year 3 is to multiply the original principal by 1 plus the interest rate per period, 1 + k, raised to the power of the

Source: *THE FAR SIDE* © 1985, 1992 FARWORKS, Inc. Dist. by *UNIVERSAL PRESS SYNDICATE.* Reprinted with permission. All rights reserved.

number of compounding periods, n.[1] Here's the formula for finding the future value—or ending balance—given the original principal, interest rate per period, and number of compounding periods:

Future Value for a Single Amount
Algebraic Method

$$FV = PV \times (1 + k)^n \tag{8-1a}$$

where: FV = Future Value, the ending amount
PV = Present Value, the starting amount, or original principal
k = Rate of interest per period (expressed as a decimal)
n = Number of time periods

In our SaveCom example, PV is the original deposit of $100, k is 5 percent, and n is 3. To solve for the ending balance, or FV, we apply Equation 8-1a as follows:

$$FV = PV \times (1 + k)^n = \$100 \times (1.05)^3 = \$100 \times 1.1576 = \$115.76$$

We may also solve for future value using a financial table. Financial tables are a compilation of values, known as interest factors, that represent a term, $(1 + k)^n$ in this case, in time value of money formulas. Table I in the Appendix at the end of the book is developed by calculating $(1 + k)^n$ for many combinations of k and n.

Table I is the **future value interest factor (FVIF)** table. The formula for future value using the FVIF table follows:

Future Value Formula for a Single Amount
Table Method

$$FV = PV \times (FVIF_{k,n}) \tag{8-1b}$$

where: FV = Future Value, the ending amount
PV = Present Value, the starting amount
$FVIF_{k,n}$ = Future Value Interest Factor given interest rate, k, and number of periods, n, from Table I

In our SaveCom example, in which $100 is deposited in an account at 5 percent interest for three years, the ending balance, or FV, according to Equation 8-1b, is as follows:

$$FV = PV \times (FVIF_{5\%, 3})$$
$$= \$100 \times 1.1576$$
$$= \$115.76$$

> **Take Note**
>
> *Because future value interest factors are rounded to four decimal places in Table I, you may get a slightly different solution compared to a problem solved by the algebraic method.*

To solve for FV using a financial calculator, we enter the numbers for PV, n, and k (k is depicted as I/Y on the TI BAII PLUS calculator; on other calculators it may be symbolized by i or I), and ask the calculator to compute FV. The keystrokes follow:[2]

---

[1] The compounding periods are usually years but not always. As you will see later in the chapter, compounding periods can be months, weeks, days, or any specified period of time.
[2] We will be using the keystroke patterns for the TI BAII PLUS financial calculator throughout this book. The separate reference card provided with this text gives keystroke sequences for other popular financial calculators.

Part Two   Essential Concepts in Finance

TI BAII PLUS FINANCIAL CALCULATOR SOLUTION

Step 1: First press `2nd` `CLR TVM`. This clears all the time value of money keys of all previously calculated or entered values.

Step 2: Press `2nd` `P/Y` 1 `ENTER`, `2nd` `QUIT`. This sets the calculator to the mode where one payment per year is expected, which is the assumption for the problems in this chapter.

Step 3: Input values for principal (PV), interest rate (k or I/Y on the calculator), and number of periods (n).

100 `+/−` `PV`   5 `I/Y`   3 `N`   `CPT` `FV`   Answer: 115.76

In the SaveCom example, we input −100 for the present value (PV), 3 for number of periods (N), and 5 for the interest rate per year (I/Y). Then we ask the calculator to compute the future value, FV. The result is $115.76. Our TI BAII PLUS is set to display two decimal places. You may choose a greater or lesser number if you wish.

We have learned three ways to calculate the future value of a single amount: the algebraic method, the financial table method, and the financial calculator method. In the next section, we see how future values are related to changes in the interest rate, k, and the number of time periods, n.

The Sensitivity of Future Values to Changes in Interest Rates or the Number of Compounding Periods

Future value has a positive relationship with the interest rate, k, and with the number of periods, n. That is, as the interest rate increases, future value increases. Similarly, as the number of periods increases, so does future value. In contrast, future value decreases with decreases in k and n values.

It is important to understand the sensitivity of future value to k and n because increases are exponential, as shown by the $(1 + k)^n$ term in the future value formula. Consider this: A business that invests $10,000 in a savings account at 5 percent for 20 years will have a future value of $26,532.98. If the interest rate is 8 percent for the same 20 years, the future value of the investment is $46,609.57. We see that the future value of the investment increases as k increases. Figure 8-1a on page 182b shows this graphically.

Now let's say that the business deposits $10,000 for 10 years at a 5 percent annual interest rate. The future value of that sum is $16,288.95. Another business deposits $10,000 for 20 years at the same 5 percent annual interest rate. The future value of that $10,000 investment is $26,532.98. Just as with the interest rate, the higher the number of periods, the higher the future value. Figure 8-1b on page 182b shows this graphically.

The Present Value of a Single Amount

Present value is today's dollar value of a specific future amount. With a bond, for instance, the issuer promises the investor future cash payments at specified points in time. With an investment in new plant or equipment, certain cash receipts are expected. When we calculate the present value of a future promised or expected cash payment, we discount it (mark it down in value) because it is worth *less* if it is to be received later rather than now. Similarly, future cash outflows are less burdensome than present cash outflows of the same amount. Future cash outflows are similarly discounted (made less negative). In present value analysis, then, the interest rate used in this discounting process is known as the **discount rate.** The discount rate is the required rate of return on an investment. It reflects the lost opportunity to spend or invest now (the opportunity cost) and the various risks assumed because we must wait for the funds.

Discounting is the inverse of compounding. Compound interest causes the value of a beginning amount to increase at an increasing rate. Discounting causes the present value of a future amount to decrease at an increasing rate.

To demonstrate, imagine the SaveCom financial manager needed to know how much to invest *now* to generate $115.76 in three years, given an interest rate of 5 percent. The calculation would look like this:

$$FV = PV \times (1 + k)^n$$
$$\$115.76 = PV \times 1.05^3$$
$$\$115.76 = PV \times 1.157625$$
$$PV = \$100.00$$

To solve present value problems, we modify the future value for a single amount equation by multiplying both sides by $1/(1 + k)^n$ to isolate PV on one side of the equal sign. The present value formula for a single amount follows:

The Present Value of a Single Amount Formula
Algebraic Method

$$PV = FV \times \frac{1}{(1 + k)^n} \tag{8-2a}$$

where: PV = Present Value, the starting amount
FV = Future Value, the ending amount
k = Discount rate of interest per period (expressed as a decimal)
n = Number of time periods

Applying this formula to the SaveCom example, in which its financial manager wanted to know how much the firm should pay today to receive $115.76 at the end of three years, assuming a 5 percent discount rate starting today, the following is the present value of the investment:

$$PV = FV \times \frac{1}{(1 + k)^n}$$
$$= \$115.76 \times \frac{1}{(1 + .05)^3}$$
$$= \$115.76 \times .86384$$
$$= \$100$$

SaveCom should be willing to pay $100 today to receive $115.76 three years from now at a 5 percent discount rate.

To solve for PV, we may also use the Present Value Interest Factor Table in Table II in the Appendix at the end of the book. A **present value interest factor,** or PVIF, is calculated and shown in Table II. It equals $1/(1 + k)^n$ for given combinations of k and n. The table method formula, Equation 8-2b, follows:

Present Value Formula for a Single Amount
Table Method

$$PV = FV \times (PVIF_{k,n}) \tag{8-2b}$$

where: PV = Present Value
FV = Future Value
$PVIF_{k,n}$ = Present Value Interest Factor given discount rate, k, and number of periods, n, from Table II

In our example, SaveCom's financial manager wanted to solve for the amount that must be invested today at a 5 percent interest rate to accumulate $115.76 within three years. Applying the present value table formula, we find the following solution:

$PV = FV \times (PVIF_{5\%,3})$
    $= \$115.76 \times .8638$ (from the PVIF table)
    $= \$99.99$ (slightly less than $100 due to the rounding to four places in the table)

The present value of $115.76, discounted back three years at a 5 percent discount rate, is $100.

To solve for present value using our financial calculator, enter the numbers for future value, FV, the number of periods, n, and the interest rate, k—symbolized as I/Y on the calculator—then hit the CPT (compute) and PV (present value) keys. The sequence follows:

TI BAII PLUS FINANCIAL CALCULATOR SOLUTION

Step 1: Press [2nd] [CLR TVM] to clear previous values.

Step 2: Press [2nd] [P/Y] 1 [ENTER] , [2nd] [QUIT] to ensure the calculator is in the mode for annual interest payments.

Step 3: Input the values for future value, the interest rate, and number of periods, and compute PV.

115.76 [FV]   5 [I/Y]   3 [N]   [CPT] [PV]   Answer: −100.00

The financial calculator result is displayed as a negative number to show that the present value sum is a cash outflow—that is, that sum will have to be invested to earn $115.76 in three years at a 5 percent annual interest rate.

We have examined how to find present value using the algebraic, table, and financial calculator methods. Next, we see how present value analysis is affected by the discount rate, k, and the number of compounding periods, n.

The Sensitivity of Present Values to Changes in k and n

In contrast to future value, present value is inversely related to k and n values. In other words, present value moves in the opposite direction of k and n. If k increases, present value decreases; if k decreases, present value increases. If n increases, present value decreases; if n decreases, present value increases.

Consider this: A business that expects a 5 percent annual return on its investment (k = 5%) should be willing to pay $3,768.89 today (the present value) for $10,000 to be received 20 years from now. If the expected annual return is 8 percent

Figure 8-1a Future Value at Different Interest Rates

Figure 8-1a shows the future value of $10,000 after 20 years at interest rates of 5% and 8%.

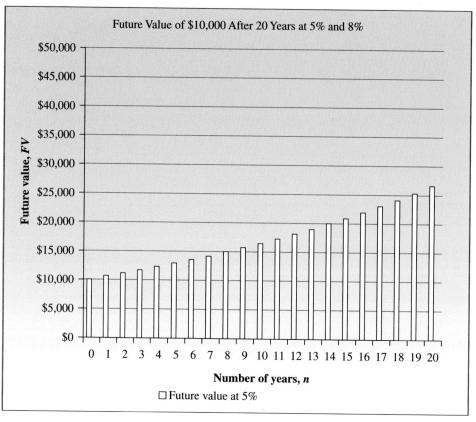

Figure 8-1b Future Value at Different Times

Figure 8-1b shows the future value of $10,000 after 10 years and 20 years at an interest rate of 5%.

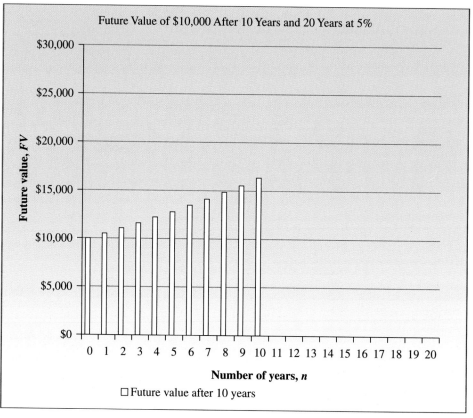

for the same 20 years, the present value of the investment is only $2,145.48. We see that the present value of the investment decreases as k increases. The way the present value of the $10,000 varies with changes in the required rate of return is shown graphically in Figure 8-2a on page 182d.

Now let's say that a business expects to receive $10,000 ten years from now. If its required rate of return for the investment is 5 percent annually, then it should be willing to pay $6,139 for the investment today (the present value is $6,139). If another business expects to receive $10,000 twenty years from now and it has the same 5 percent annual required rate of return, then it should be willing to pay $3,769 for the investment (the present value is $3,769). Just as with the interest rate, the greater the number of periods, the lower the present value. Figure 8-2b on page 182d shows how it works.

In this section we have learned how to find the future value and the present value of a single amount. Next, we will examine how to find the future value and present value of several amounts.

WORKING WITH ANNUITIES

Financial managers often need to assess a series of cash flows rather than just one. One common type of cash flow series is the **annuity**—a series of equal cash flows, spaced evenly over time.

Professional athletes often sign contracts that provide annuities for them after they retire, in addition to the signing bonus and regular salary they may receive during their playing years. Consumers can purchase annuities from insurance companies as a means of providing retirement income. The investor pays the insurance company a lump sum now in order to receive future payments of equal size at regularly spaced time intervals (usually monthly). Another example of an annuity is the interest on a bond. The interest payments are usually equal dollar amounts paid either annually or semiannually during the life of the bond.

Annuities are a significant part of many financial problems. You should learn to recognize annuities and determine their value, future or present. In this section we will explain how to calculate the future value and present value of annuities in which cash flows occur at the end of the specified time periods. Annuities in which the cash flows occur at the end of each of the specified time periods are known as *ordinary annuities*.

Future Value of an Ordinary Annuity

Financial managers often plan for the future. When they do, they often need to know how much money to save on a regular basis to accumulate a given amount of cash at a specified future time. The future value of an annuity is the amount that a given number of annuity payments, n, will grow to at a future date, for a given periodic interest rate, k.

For instance, suppose the SaveCom Company plans to invest $500 in a money market account at the end of each year for the next four years, beginning one year from today. The business expects to earn a 5 percent annual rate of return on its investment. How much money will SaveCom have in the account at the end of four years? The problem is illustrated in the timeline in Figure 8-3. The t values in the timeline represent the end of each time period. Thus, t_1 is the end of the first year, t_2 is the end of the second year, and so on. The symbol t_0 is *now,* the present point in time.

Figure 8-2a **Present Value at Different Interest Rates**

Figure 8-2a shows the present value of $10,000 to be received in 20 years at interest rates of 5% and 8%.

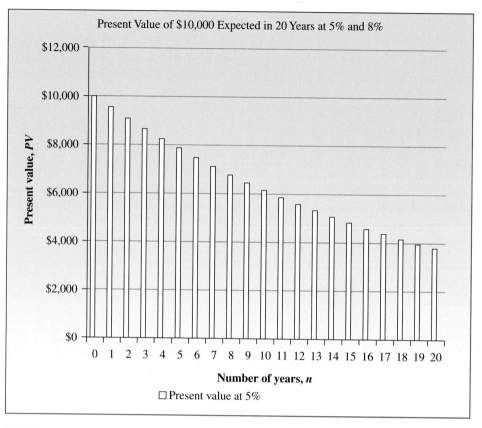

Figure 8-2b **Present Value at Different Times**

Figure 8-2b shows the present value of $10,000 to be received in 10 years and 20 years at an interest rate of 5%.

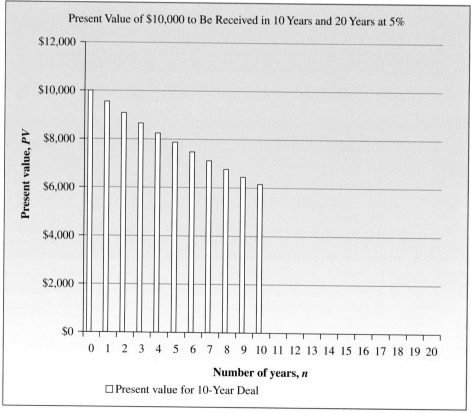

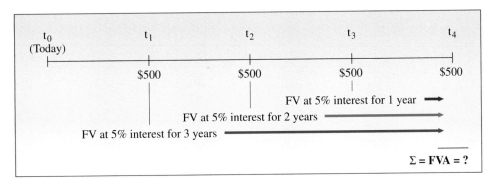

Figure 8-3 **SaveCom Annuity Timeline**

Because the $500 payments are each single amounts, we can solve this problem one step at a time. Looking at Figure 8-4, we see that the first step is to calculate the future value of t_1, t_2, t_3, and t_4 using the future value formula for a single amount. The next step is to add the four values together. The sum of those values is the annuity's future value.

The sum of the future values of the four single amounts is the annuity's future value, $2,155.05. However, the step-by-step process illustrated in Figure 8-4 is time consuming even in this simple example. Calculating the future value of a 20- or 30-year annuity, such as would be the case with many bonds, would take an enormous amount of time. Instead, we can calculate the future value of an annuity easily by using the following formula:

Future Value of an Annuity Formula
Algebraic Method

$$FVA = PMT \times \left[\frac{(1+k)^n - 1}{k} \right] \tag{8-3a}$$

where: FVA = Future Value of an Annuity
PMT = Amount of each annuity payment
k = Interest rate per time period
n = Number of annuity payments

The bracketed term in Equation 8-3a is the result of adding the corresponding $(1 + k)^n$ values from the future value of a single amount formulas, Equation 8-1a and 8-1b, for each of the individual annuity payments. This is the same as the sum of the FVIF values from Table I for each individual annuity payment. For example, the sum of the individual future value interest factors for each of the $500 payments in our example is as follows:

1.1576 (n = 3)
1.1025 (n = 2)
1.0500 (n = 1)
1.0000 (n = 0)
4.3101

Using Equation 8-3a in our SaveCom example, we solve for the future value of the annuity at 5 percent interest (k = 5%) with four $500 end-of-year payments (n = 4 and PMT = $500), as follows:

Figure 8-4 Future Value of the SaveCom Annuity

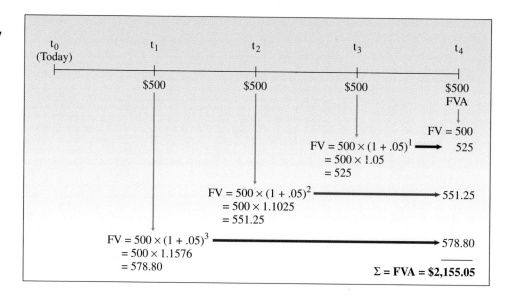

$$FVA = 500 \times \left[\frac{(1 + .05)^4 - 1}{.05}\right]$$

$$= 500 \times 4.3101$$

$$= \$2{,}155.05$$

Notice that 4.3101 is also the number we found by summing the values for the FVIF terms for the future value of a single amount computations. For a $500 annuity with a 5 percent interest rate and four annuity payments, we see that the future value of the SaveCom annuity is $2,155.05.

To find the future value of an annuity with the table method, we must find the **future value interest factor for an annuity** (FVIFA), found in Table III in the Appendix at the end of the book. The $FVIFA_{k,n}$ is the value of $[(1 + k)^n - 1] \div k$ for different combinations of k and n.

Future Value of an Annuity Formula
Table Method

$$FVA = PMT \times FVIFA_{k,n} \quad (8\text{-}3b)$$

where: FVA = Future Value of an Annuity
PMT = Amount of each annuity payment
$FVIFA_{k,n}$ = Future Value Interest Factor for an Annuity from Table III
k = Interest rate per period
n = Number of annuity payments

In our SaveCom example, then, we need to find the FVIFA for a discount rate of 5 percent with four annuity payments. Table III in the Appendix at the end of the book shows that the $FVIFA_{k=5\%, n=4}$ is 4.3101. Using the table method, we find the following future value of the SaveCom annuity:

$$FVA = 500 \times FVIFA_{5\%, 4}$$

$$= 500 \times 4.3101 \text{ (from the FVIFA table in Table III)}$$

$$= \$2{,}155.05$$

To find the future value of an annuity using a financial calculator, key in the values for the annuity payment (PMT), n, and k (remember that the notation for the interest rate on the TI BAII PLUS calculator is I/Y, not k). Then compute the future value of the annuity (FV on the calculator). For a series of four $500 end-of-year (ordinary annuity) payments where n = 4 and k = 5 percent, the computation is as follows:

TI BAII PLUS FINANCIAL CALCULATOR SOLUTION

Step 1: Press `2nd` `CLR TVM` to clear previous values.

Step 2: Press `2nd` `P/Y` 1 `ENTER` , `2nd` `BGN` `2nd` `SET` `2nd` `SET` repeat `2nd` `SET` until END shows in the display `2nd` `QUIT` to set the annual interest rate mode and to set the annuity payment to end of period mode.

Step 3: Input the values and compute.

0 `PV`   5 `I/Y`   4 `N`   500 `+/−` `PMT`   `CPT` `FV`   Answer: 2,155.06

In the financial calculator inputs, note that the payment is keyed in as a negative number to indicate that the payments are cash outflows—the payments flow out from the company into an investment.

The Present Value of an Ordinary Annuity

Because annuity payments are often promised (as with interest on a bond investment) or expected (as with cash inflows from an investment in new plant or equipment), it is important to know how much these investments are worth to us today. For example, assume that the financial manager of Buy4Later, Inc. learns of an annuity that promises to make four annual payments of $500, beginning one year from today. How much should the company be willing to pay to obtain that annuity? The answer is the present value of the annuity.

Because an annuity is nothing more than a series of single amounts, we could calculate the present value of an annuity with the present value formula for a single amount and sum the totals, but that would be a cumbersome process. Imagine calculating the present value of a 50-year annuity! We would have to find the present value for each of the 50 annuity payments and total them.

Fortunately, we can calculate the present value of an annuity in one step with the following formula:

The Present Value of an Annuity Formula
Algebraic Method

$$PVA = PMT \times \left[\frac{1 - \frac{1}{(1+k)^n}}{k} \right] \qquad (8\text{-}4a)$$

where: PVA = Present Value of an Annuity
PMT = Amount of each annuity payment
k = Discount rate per period
n = Number of annuity payments

> **Take Note**
>
> *Note that slight differences occur between the table method, algebraic method, and calculator solution. This is because our financial tables round interest factors to four decimal places, whereas the other methods generally use many more significant figures for greater accuracy.*

Using our example of a four-year ordinary annuity with payments of $500 per year and a 5 percent discount rate, we solve for the present value of the annuity as follows:

$$PVA = 500 \times \left[\frac{1 - \frac{1}{(1 + .05)^4}}{.05} \right]$$

$$= 500 \times 3.54595$$

$$= \$1{,}772.97$$

The present value of the four-year ordinary annuity with equal yearly payments of $500 at a 5 percent discount rate is 1,772.97.

We can also use the financial table for the **present value interest factor for an annuity (PVIFA)** to solve present value of annuity problems. The PVIFA table is found in Table IV in the Appendix at the end of the book. The formula for the table method follows:

The Present Value of an Annuity Formula
Table Method

$$PVA = PMT \times PVIFA_{k,n} \tag{8-4b}$$

where: PVA = Present Value of an Annuity
PMT = Amount of each annuity payment
$PVIFA_{k,n}$ = Present Value Interest Factor for an Annuity from Table IV
k = Discount rate per period
n = Number of annuity payments

Applying Equation 8-4b, we find that the present value of the four-year annuity with $500 equal payments and a 5 percent discount rate is as follows:[3]

$$PVA = 500 \times PVIFA_{5\%,4} = 500 \times 3.5460 = \$1{,}773.00$$

We may also solve for the present value of an annuity with a financial calculator. Simply key in the values for the payment, PMT, number of payment periods, n, and the interest rate, k—symbolized by I/Y on the TI BAII PLUS calculator—and ask the calculator to compute PVA (PV on the calculator). For the series of four $500 payments where n = 4 and k = 5% the computation follows:

TI BAII PLUS FINANCIAL CALCULATOR SOLUTION

Step 1: Press `2nd` `CLR TVM` to clear previous values.

Step 2: Press `2nd` `P/Y` 1 `ENTER` , `2nd` `BGN` `2nd` `SET` , repeat `2nd` `SET` until END shows in the display `2nd` `QUIT` to set to the annual interest rate mode and to set the annuity payment to end of period mode.

Step 3: Input the values and compute.

5 `I/Y`   4 `N`   500 `PMT`   `CPT` `PV`   Answer: −1,772.97

---

[3]The $.03 difference between the algebraic result and the table formula solution is due to differences in rounding.

ETHICAL CONNECTIONS

When a Million Isn't a Million. Taking a Chance on the Time Value of Money

You've won the lottery: $1 million. Congratulations! You're rich. No, wait a minute—maybe not.

"I wish they would stop *calling* us millionaires," whispers Cindy, who doesn't want to give out her last name. You see, Cindy, a recent lottery winner, has faced an awful truth: the time value of money.

In fact, most lottery winners wish the news media would tell the truth about those "millions." Those huge prizes that the media report are usually doled out over 20 years (25 in Colorado). The annual check for a million-dollar winner is $50,000—before taxes. They also take any student loans you haven't paid out of the check.

Michael Ondrish, a "millionaire" winner who thought he'd won a real, lump-sum million in 1982, tried suing the Arizona lottery for fraud and deception because lottery officials never mentioned the drawn-out payment scheme. He lost his case and received a good lesson in the time value of money.

A million dollars paid out over 20 years is an annuity. Like any other million-dollar annuity, it actually costs about $450,000 to buy because that is its present value. Depending on interest rates, taxes, and inflation, the value of the yearly check dwindles over time. Again, it's what financial experts call the time value of money. In fact, investors willing to buy those annuities from winners who need cash offer about 40 cents on the dollar.

But you won't see the time value of money in any lottery commercial. "You Could Win An Annuity Spread Over Twenty Years" just doesn't have the same appeal as "You've Won A Million." But look on the bright side. When you win the lottery, you'll know the truth. Especially after reading this chapter.

Questions to Consider

▶ Do you think it is ethical to advertise the lottery without explaining that the winnings are an annuity? Explain.

▶ Is it ethical to fail to disclose that the jackpot winnings are the future value of the jackpot, not the present value?

▶ Why do you think that Michael Ondrish lost his case against the Arizona lottery?

Source: Lois Gould "Ticket To Trouble," *New York Times Magazine* (April 23, 1995): 39.

The financial calculator present value result is displayed as a negative number to signal that the present value sum is a cash outflow—that is, $1,772.97 will have to be invested to earn a 5 percent annual rate of return on the four future annual annuity payments of $500 each to be received.

Future and Present Values of Annuities Due

Sometimes we must deal with annuities where the annuity payments occur at the *beginning* of each period. These are known as annuities *due*, in contrast to *ordinary* annuities in which the payments occurred at the end of each period, as described in the preceding section.

Annuities due are more likely to occur when doing future value of annuity (FVA) problems than when doing present value of annuity (PVA) problems. Today, for instance, you may start a retirement program, investing regular equal amounts each month or year. Calculating the amount you would accumulate when you reach retirement age would be a future value of an annuity due problem. Evaluating the present value of a promised or expected series of annuity payments that began today would be a present value of an annuity due problem. This is less common because car and mortgage payments almost always start at the end of the first period, making them ordinary annuities.

Whenever you run into an FVA or a PVA of an annuity due problem, the adjustment needed is the same in both cases. Use the FVA or PVA of an ordinary annuity formula shown earlier, then multiply your answer by (1 + k). We multiply the FVA or PVA formula by (1 + k) because annuities due have annuity payments earning interest one period sooner. So, higher FVA and PVA values result with an annuity due. The first payment occurs sooner in the case of a future value of an annuity due. In present value of annuity due problems, each annuity payment occurs one period sooner, so the payments are discounted less severely.

In our SaveCom example, the future value of a $500 ordinary annuity, with k = 5% and n = 4, was $2,155.06. If the $500 payments occurred at the *beginning* of each period instead of at the end, we would multiply $2,155.06 by 1.05 (1 + k = 1 + .05). The product, $2,262.81, is the future value of the annuity due. In our earlier Buy4Later, Inc. example, we found that the present value of a $500 ordinary annuity, with k = 5% and n = 4, was $1,772.97. If the $500 payments occurred at the *beginning* of each period instead of at the end, we would multiply $1,772.97 by 1.05 (1 + k = 1 + .05) and find that the present value of Buy4Later's annuity due is $1,861.62.

The financial calculator solutions for these annuity due problems are shown next.

Future Value of a Four-Year, $500 Annuity Due, k = 5%

TI BAII PLUS FINANCIAL CALCULATOR SOLUTION

Step 1: Press `2nd` `CLR TVM` to clear previous values.

Step 2: Press `2nd` `P/Y` 1 `ENTER`, `2nd` `BGN`, `2nd` `SET`. Repeat the `2nd` `SET` command until the display shows BGN, `2nd` `QUIT` to set to the annual interest rate mode and to set the annuity payment to beginning of period mode.

Step 3: Input the values for the annuity due and compute.

5 `I/Y`   4 `N`   500 `PMT`   `CPT` `FV`   Answer: −2,262,82

Present Value of a Four-Year, $500 Annuity Due, k = 5%

TI BAII PLUS FINANCIAL CALCULATOR SOLUTION

Step 1: Press `2nd` `CLR TVM`.

Step 2: Press `2nd` `P/Y` 1 `ENTER`, `2nd` `BGN`, `2nd` `SET`. Repeat the `2nd` `SET` command until the display shows BGN `2nd` `QUIT` to set to the annual interest rate mode and to set the annuity payment to beginning of period mode.

Step 3: Input the values for the annuity due and compute.

5 `I/Y`   4 `N`   500 `PMT`   `CPT` `PV`   Answer: −1,861.62

In this section we discussed ordinary annuities and annuities due and learned how to compute the present and future values of the annuities. Next, we will learn what a perpetuity is and how to solve for its present value.

Perpetuities

An annuity that goes on forever is called a perpetual annuity or a **perpetuity.** Perpetuities contain an infinite number of annuity payments. An example of a perpetuity is the dividends typically paid on a preferred stock issue.

The future value of perpetuities cannot be calculated, but the present value can be. We start with the present value of an annuity formula, Equation 8-3a.

$$PVA = PMT \times \left[\frac{1 - \frac{1}{(1+k)^n}}{k} \right]$$

Now imagine what happens in the equation as the number of payments (n) gets larger and larger. The $(1+k)^n$ term will get larger and larger. And as it does, it will cause the $1/(1+k)^n$ fraction to become smaller and smaller. As n approaches infinity, the $(1+k)^n$ term becomes infinitely large, and the $1/(1+k)^n$ term approaches zero. The entire formula reduces to the following equation:

Present Value of a Perpetuity Formula

$$PVP = PMT \times \left[\frac{1 - 0}{k} \right] \tag{8-5}$$

or

$$PVP = PMT \times \left(\frac{1}{k} \right)$$

where: PVP = Present value of a perpetuity
        k = Discount rate

Neither the table method nor the financial calculator can solve for the present value of a perpetuity. This is because the PVIFA table does not contain values for infinity and the financial calculator does not have an infinity key.

Suppose you had the opportunity to buy a share of preferred stock that pays $70 per year forever. If your required rate of return is 8 percent, what is the present value of the promised dividends to you? In other words, given your required rate of return, how much should you be willing to pay for the preferred stock? The answer, found by applying Equation 8-5, follows:

$$PVP = PMT \times \left(\frac{1}{k} \right) = \$70 \times \left(\frac{1}{.08} \right) = \$875$$

The present value of our preferred stock with a k of 8 percent and a payment of $70 per year forever is $875.

Present Value of an Investment with Uneven Cash Flows

Unlike annuities that have equal payments over time, many investments have payments that are unequal over time. That is, some investments have payments that vary over time. When the periodic payments vary, we say that the cash flow streams

TABLE 8-1

The Present Value of an Uneven Series of Cash Flows

t_0	$7,000,000	$7,000,000 \times \dfrac{1}{1.08^0}$ =	$7,000,000
t_1	$2,000,000	$2,000,000 \times \dfrac{1}{1.08^1}$ =	$1,851,851.85
t_2	$4,000,000	$4,000,000 \times \dfrac{1}{1.08^2}$ =	$3,429,355.28
t_3	$6,000,000	$6,000,000 \times \dfrac{1}{1.08^3}$ =	$4,762,993.45
t_4	$6,000,000	$6,000,000 \times \dfrac{1}{1.08^4}$ =	$4,410,179.12
		Sum of the PVs =	$21,454,379.70

are uneven. For instance, a professional athlete may sign a contract that provides for an immediate $7 million signing bonus, followed by a salary of $2 million in year 1, $4 million in year 2, then $6 million in years 3 and 4. What is the present value of the promised payments that total $25 million? Assume a discount rate of 8 percent. The present value calculations are shown in Table 8-1.

As we see from Table 8-1, we calculate the present value of an uneven series of cash flows by finding the present value of a single amount for each series and sum the totals.

We can also use a financial calculator to find the present value of this uneven series of cash flows. The worksheet mode of the TI BAII PLUS calculator is especially helpful in solving problems with uneven cash flows. The C display shows each cash payment following CF_0, the initial cash flow. The F display key indicates the frequency of that payment. The keystrokes follow.

TI BAII PLUS FINANCIAL CALCULATOR PV SOLUTION
Uneven Series of Cash Flows

KEYSTROKES	DISPLAY
CF	CF_0 = old contents
2nd CLR Work	CF_0 = 0.00
7000000 ENTER	7,000,000.00
↓ 2000000 ENTER	C01 = 2,000,000.00
↓	F01 = 1.00
↓ 4000000 ENTER	C02 = 4,000,000.00
↓	F02 = 1.00
↓ 6000000 ENTER	C03 = 6,000,000.00
↓ 2 ENTER	F03 = 2.00
NPV	I = 0.00
8 ENTER	I = 8.00
↓ CPT	NPV = 21,454,379.70

> **Take Note**
>
> *We used the NPV (net present value) key on our calculator to solve this problem. NPV will be discussed in chapter 10.*

188 Part Two Essential Concepts in Finance

We see from the calculator keystrokes that we are solving for the present value of a single amount for each payment in the series except for the last two payments, which are the same. The value of F03, the frequency of the third cash flow after the initial cash flow, was 2 instead of 1 because the $6 million payment occurred twice in the series (in years 3 and 4).

We have seen how to calculate the future value and present value of annuities, the present value of a perpetuity, and the present value of an investment with uneven cash flows. Now we turn to time value of money problems where we solve for k, n, or the annuity payment.

Special Time Value of Money Problems

Financial managers often face time value of money problems even when they know both the present value and the future value of an investment. In those cases financial managers may be asked to find out what return an investment made—that is, what the interest rate is on the investment. Still other times financial managers must find either the number of payment periods, or the amount of an annuity payment. In the next section, we will learn how to solve for k and n. We will also learn how to find the annuity payment (PMT).

Finding the Interest Rate

Financial managers frequently have to solve for the interest rate, k, when firms make a long-term investment. The method of solving for k depends on whether the investment is a single amount or an annuity.

Finding k of a Single-Amount Investment Financial managers may need to determine how much periodic return an investment generated over time. For example, imagine that you are head of the finance department of GrabLand, Inc. Say that GrabLand purchased a house on prime land 20 years ago for $40,000. Recently, GrabLand sold the property for $106,131. What average annual rate of return did the firm earn on its 20-year investment?

First, the future value—or ending amount—of the property is $106,131. The present value—the starting amount—is $40,000. The number of periods, n, is 20. Armed with those facts, you solve this problem using the future value of a single amount formula as follows:

$$FV = PV \times (FVIF_{k, n})$$
$$\$106{,}131 = \$40{,}000 \times (FVIF_{k=?, n=20})$$
$$\$106{,}131 \div \$40{,}000 = 2.6533 = (FVIF_{k=?, n=20})$$

Now find the FVIF value in Table I, shown in part on page 190. The whole table is in the Appendix at the end of the book. You know n = 20, so find the n = 20 row on the left-hand side of the table. You also know that the FVIF value is 2.6533, so move across the n = 20 row until you find (or come close to) the value 2.6533. You find the 2.6533 value in the k = 5% column. You discover, then, that GrabLand's property investment had an interest rate of 5 percent.

Future Value Interest Factors, Compounded at k Percent for n Periods, Part of Table I

	INTEREST RATE, k										
NUMBER OF PERIODS, n	**0%**	**1%**	**2%**	**3%**	**4%**	**5%**	**6%**	**7%**	**8%**	**9%**	**10%**
18	1.0000	1.1961	1.4282	1.7024	2.0258	2.4066	2.8543	3.3799	3.9960	4.7171	5.5599
19	1.0000	1.2081	1.4568	1.7535	2.1068	2.5270	3.0256	3.6165	4.3157	5.1417	6.1159
20	1.0000	1.2202	1.4859	1.8061	2.1911	2.6533	3.2071	3.8697	4.6610	5.6044	6.7275
25	1.0000	1.2824	1.6406	2.0938	2.6658	3.3864	4.2919	5.4274	6.8485	8.6231	10.8347

Solving for k using the FVIF table works well when the interest rate is a whole number, but it does not work well when the interest rate is not a whole number. To solve for the interest rate, k, we rewrite the future value of a single-amount formula, Equation 8-1a, to solve for k:

Solve for the Rate of Return, k

$$k = \left(\frac{FV}{PV}\right)^{\frac{1}{n}} - 1 \tag{8-6}$$

Let's use Equation 8-6 to find the average annual rate of return on GrabLand's house investment. Recall that the company bought it 20 years ago for $40,000 and sold it recently for $106,131. We solve for k applying Equation 8-6 as follows:

$$k = \left(\frac{FV}{PV}\right)^{\frac{1}{n}} - 1 = \left(\frac{\$106{,}131}{\$40{,}000}\right)^{\frac{1}{20}} - 1 = 2.653275^{.05} - 1 = 1.05 - 1$$
$$= .05 \text{ or } 5\%$$

Equation 8-6 will find any interest rate given a starting value, PV, an ending value, FV, and a number of compounding periods, n.

To solve for k with a financial calculator, key in all the other variables and ask the calculator to compute k (depicted as I/Y on your calculator). For GrabLand's house-buying example, the calculator solution follows:

TI BAII PLUS FINANCIAL CALCULATOR SOLUTION

Step 1: Press [2nd] [CLR TVM] .

Step 2: Press [2nd] [P/Y] 1 [ENTER] , [2nd] [QUIT] .

Step 3: Input the values and compute.

40000 [+/−] [PV]   106131 [FV]   20 [N]   [CPT] [I/Y]   Answer: 5.00

Remember when using the financial calculator to solve for the interest rate, you must enter cash outflows as a negative number. In our example, the $40,000 PV is entered as a negative number because GrabLand spent that amount to invest in the house.

Finding k for a PVA Problem Financial managers may need to find the interest rate for a PVA problem when they know the starting amount (PVA), n, and the annuity payment (PMT), but they do not know the interest rate, k. For example, suppose GrabLand wanted a 15-year, $100,000 amortized loan from a bank. An **amortized loan** is a loan that is paid off in equal amounts that include principal as well as interest.[4] GrabLand's payments will be $12,405.89 per year for 15 years. What interest rate is the bank charging on this loan?

To solve for k when the known values are PVA (the $100,000 loan proceeds), n (15), and PMT (the loan payments $12,405.89), we start with the present value of an annuity formula, Equation 8-3b, as follows:

Present Value of an Annuity Formula
Table Method

$$PVA = PMT \times (PVIFA_{k,n}) \tag{8-3b}$$

$$\$100{,}000 = \$12{,}405.89 \times (PVIFA_{k=?,\,n=15})$$

$$8.0607 = (PVIFA_{k=?,\,n=15})$$

Now refer to the PVIFA values in Table IV, shown in part as follows. You know n = 15, so find the n = 15 row on the left-hand side of the table. You have also determined that the PVIFA value is 8.0607 ($100,000/$12,405 = 8.0607), so move across the n = 15 row until you find (or come close to) the value of 8.0607. In our example, the location on the table where n = 15 and the PVIFA is 8.0607 is in the k = 9% column, so the interest rate on GrabLand's loan is 9 percent.

Present Value Interest Factors for an Annuity, Discounted at k Percent for n Periods, Part of Table IV

					DISCOUNT RATE, k						
NUMBER OF ANNUITY PAYMENTS, n	0%	1%	2%	3%	4%	5%	6%	7%	8%	9%	10%
13	13.0000	12.1337	11.3484	10.6350	9.9856	9.3936	8.8527	8.3577	7.9038	7.4869	7.1034
14	14.0000	13.0037	12.1062	11.2961	10.5631	9.8986	9.2950	8.7455	8.2442	7.7862	7.3667
15	15.0000	13.8651	12.8493	11.9379	11.1184	10.3797	9.7122	9.1079	8.5595	8.0607	7.6061
16	16.0000	14.7179	13.5777	12.5611	11.6523	10.8378	10.1059	9.4466	8.8514	8.3126	7.8237

To solve this problem with a financial calculator, key in all the variables but k, and ask the calculator to compute k (depicted as I/Y on the TI calculator) as follows:

TI BAII PLUS FINANCIAL CALCULATOR SOLUTION

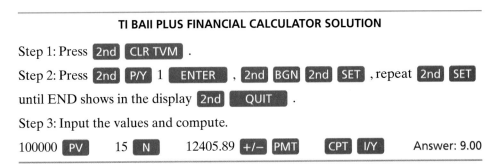

---

[4]Amortize comes from the Latin word *mortalis*, which means "death." You will kill off the entire loan after making the scheduled payments.

In this example the PMT was entered as a negative number to indicate that the loan payments are cash outflows, flowing away from the firm. The missing interest rate value was 9 percent, the interest rate on the loan.

Finding the Number of Periods

Suppose you found an investment that offered you a return of 6 percent per year. How long would it take you to double your money? In this problem you are looking for n, the number of compounding periods it will take for a starting amount, PV, to double in size (FV = 2 × PV).

To find n in our example, start with the formula for the future value of a single amount and solve for n as follows:

$$FV = PV \times (FVIF_{k, n})$$
$$2 \times PV = PV \times (FVIF_{k = 6\%, n = ?})$$
$$2.0 = (FVIF_{k = 6\%, n = ?})$$

Now refer to the FVIF values, shown as follows in part of Table I. You know k = 6%, so scan across the top row to find the k = 6% column. Knowing that the FVIF value is 2.0, move down the k = 6% column until you find (or come close to) the value 2.0. Note that it occurs in the row in which n = 12. Therefore, n in this problem, and the number of periods it would take for the value of an investment to double at 6 percent interest per period, is 12.

Future Value Interest Factors, Compounded at k Percent for n Periods, Part of Table I

	INTEREST RATE, k										
NUMBER OF PERIODS, n	0%	1%	2%	3%	4%	5%	6%	7%	8%	9%	10%
11	1.0000	1.1157	1.2434	1.3842	1.5395	1.7103	1.8983	2.1049	2.3316	2.5804	2.8531
12	1.0000	1.1268	1.2682	1.4258	1.6010	1.7959	2.0122	2.2522	2.5182	2.8127	3.1384
13	1.0000	1.1381	1.2936	1.4685	1.6651	1.8856	2.1329	2.4098	2.7196	3.0658	3.4523
14	1.0000	1.1495	1.3195	1.5126	1.7317	1.9799	2.2609	2.5785	2.9372	3.3417	3.7975

This problem can also be solved on a financial calculator quite quickly. Just key in all the known variables (PV, FV, and I/Y) and ask the calculator to compute n.

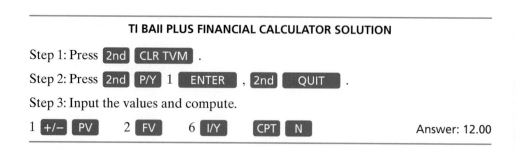

In our example n = 12 when $1 is paid out and $2 received with a rate of interest of 6 percent. That is, it takes 12 years to double your money at a 6 percent annual rate of interest.

Solving for the Payment

Lenders and financial managers frequently have to determine how much each payment—or installment—will need to be to repay an amortized loan. Suppose you are a business owner and you want to buy an office building for your company that costs $200,000. You have $50,000 for a down payment and the bank will lend you the $150,000 balance at a 6 percent annual interest rate. How much will the annual payments be if you obtain a 10-year amortized loan?

As we saw earlier, an amortized loan is repaid in equal payments over time. The period of time may vary. Let's assume in our example that your payments will occur annually, so that at the end of the 10-year period you will have paid off all interest and principal on the loan (FV = 0).

Because the payments are regular and equal in amount, this is an annuity problem. The present value of the annuity (PVA) is the $150,000 loan amount, the annual interest rate (k) is 6 percent, and n is 10 years. The payment amount (PMT) is the only unknown value.

Because all the variables but PMT are known, the problem can be solved by solving for PMT in the present value of an annuity formula as follows:

Present Value of an Annuity
Algebraic Method

$$PVA = PMT \times \left[\frac{1 - \frac{1}{(1+k)^n}}{k} \right] \quad (8\text{-}4a)$$

$$\$150{,}000 = PMT \times \left[\frac{1 - \frac{1}{(1.06)^{10}}}{.06} \right]$$

$$\$150{,}000 = PMT \times 7.36009$$

$$\frac{\$150{,}000}{7.36009} = PMT$$

$$\$20{,}380.19 = PMT$$

We see, then, that the payment for an annuity with a 6 percent interest rate, an n of 10, and a present value of $150,000 is $20,380.19.

We can also solve for PMT using the table formula as follows:

Present Value of an Annuity Formula
Table Method

$$PVA = PMT \times (PVIFA_{k, n}) \quad (8\text{-}4b)$$

$$\$150{,}000 = PMT \times (PVIFA_{6\%, \, 10 \, years})$$

$$\$150{,}000 = PMT \times 7.3601 \text{ (look up PVIFA, Table IV)}$$

$$\frac{\$150{,}000}{7.3601} = PMT$$

$$\$20{,}380.16 = PMT$$

The table formula shows that the payment for a loan with the present value of $150,000 at an annual interest rate of 6 percent and an n of 10 is $20,380.16.

With the financial calculator, simply key in all the variables but PMT and have the calculator compute PMT as follows:

TI BAII PLUS FINANCIAL CALCULATOR SOLUTION

Step 1: Press [2nd] [CLR TVM] to clear previous values.

Step 2: Press [2nd] [P/Y] 1 [ENTER], [2nd] [BGN], [2nd] [SET], repeat the [2nd] [SET] command until the display shows END [2nd] [QUIT] to set to the annual interest rate mode and to set the annuity payment to end of period mode.

Step 3: Input the values and compute the payment.

150000 [PV]   6 [I/Y]   10 [N]   [CPT] [PMT]   Answer: −20,380.19

The financial calculator will display the payment, $20,380.19, as a negative number because it is a cash outflow.

As each payment is made on an amortized loan, the interest due for that period is paid, along with a repayment of some of the principal that must also be "killed off." After the last payment is made, all the interest and principal on the loan have been paid. This step-by-step payment of the interest and principal payments owed is often shown in an *amortization table*. The amortization table for a four-year 10 percent annual interest rate loan of $1,000 with annual payments follows in Table 8-2. The annual payment of $315.47 is computed with our financial calculator.

TI BAII PLUS FINANCIAL CALCULATOR SOLUTION

Step 1: Press [2nd] [CLR TVM].

Step 2: [2nd] [P/Y] 1 [ENTER], [2nd] [BGN], [2nd] [SET], repeat the [2nd] [SET] command until the display shows END [2nd] [QUIT] to set to the annual interest rate mode and to set the annuity payment to end of period mode.

Step 3: Input all known values and compute the payment.

1000 [PV]   10 [I/Y]   4 [N]   [CPT] [PMT]   Answer: −315.47

We see from Table 8-2 how the $1,000 loan and interest is "killed off" a little each year until the balance at the end of year 4 is zero, with more interest paid early and more principal paid late.

We have seen how to solve for k, n, and PMT. Next, we will see how to solve time value of money problems in which interest is compounded more than once per year.

TABLE 8-2

Amortization Table $1,000 Loan, 10% Annual Interest Rate, 4 years

	COL. 1 Beginning Balance	COL. 2 Total Payment	COL. 3 COL. 1 × .10 Payment of Interest	COL. 4 COL. 2 − COL .3 Payment of Principal	COL. 5 COL. 1 − COL. 4 Ending Balance
YR1	$1,000.00	$315.47	$100.00	$215.47	$784.53
YR2	$784.53	$315.47	$78.45	$237.02	$547.51
YR3	$547.51	$315.47	$54.75	$260.72	$286.79
YR4	$286.79	$315.47	$28.68	$286.72	$0.00

COMPOUNDING MORE THAN ONCE PER YEAR

So far in this chapter, we have assumed that interest is compounded *annually*. However, there is nothing magical about annual compounding. Many investments pay interest that is compounded semiannually, quarterly, or even daily. Most banks, savings and loan associations, and credit unions, for example, compound interest on their deposits more frequently than annually.

Suppose you deposited $100 in a savings account that paid 12 percent interest, compounded *annually*. After one year you would have $112 in your account ($112 = $100 × 1.12^1).

Now, however, let's assume the bank used *semiannual* compounding. With semiannual compounding you would receive half a year's interest (6 percent) after six months. In the second half of the year, you would earn interest both on the interest earned in the first six months *and* on the original principal. The total interest earned during the year on a $100 investment at 12 percent annual interest would be:

$ 6.00 (interest for the first six months)
+ $ .36 (interest on the $6 interest during the second 6 months[5])
+ $ 6.00 (interest on the principal during the second six months)
= $12.36 total interest in year 1

At the end of the year, you will have a balance of $112.36 if the interest is compounded semiannually, compared to $112.00 with annual compounding—a difference of $.36.

Here's how to find answers to problems in which the compounding period is less than a year: Apply the relevant present value or future value equation, but adjust k and n so they reflect the actual compounding periods.

To demonstrate, let's look at our example of a $100 deposit in a savings account at 12 percent for one year with semiannual compounded interest. Because we want to find out what the future value of a single amount will be, we use that formula to solve for the future value of the account after one year. Next, we divide the *annual* interest rate, 12 percent, by two because interest will be compounded twice each year. Then, we multiply the number of *years* n (one in our case) by two because with semiannual interest there are two compounding periods in a year. The calculation follows:

---

[5]The $.36 was calculated by multiplying $6 by half of 12%. $6.00 × .06 = $.36.

$$FV = PV \times (1 + k/2)^{n \times 2}$$
$$= \$100 \times (1 + .12/2)^{1 \times 2}$$
$$= \$100 \times (1 + .06)^2$$
$$= \$100 \times 1.1236$$
$$= \$112.36$$

The future value of $100 after one year, earning 12 percent annual interest compounded semiannually, is $112.36.

To use the table method for finding the future value of a single amount, find the $FVIF_{k,n}$ in Table I in the Appendix at the end of the book. Then, divide the k by two and multiply the n by two as follows:

$$FV = PV \times (FVIF_{k/2, n \times 2})$$
$$= \$100 \times (FVIF_{12\%/2, 1 \times 2 \text{ periods}})$$
$$= \$100 \times (FVIF_{6\%, 2 \text{ periods}})$$
$$= \$100 \times 1.1236$$
$$= \$112.36$$

To solve the problem using a financial calculator, divide the k (represented as I/Y on the TI BAII PLUS calculator) by two and multiply the n by two. Next, key in the variables as follows:

TI BAII PLUS FINANCIAL CALCULATOR SOLUTION

Step 1: Press [2nd] [CLR TVM].

Step 2: Press [2nd] [P/Y] 1 [ENTER], [2nd] [QUIT].

Step 3: Input the values and compute.

100 [+/−] [PV]   6 [I/Y]   2 [N]   [CPT] [FV]   Answer: 112.36

The future value of $100 invested for two periods at 6 percent per period is $112.36.[6]

Other compounding rates, such as quarterly or monthly rates, can be found by modifying the applicable formula to adjust for the compounding periods. With a quarterly compounding period, then, annual k should be divided by four and annual n multiplied by four. For monthly compounding, annual k should be divided by twelve and annual n multiplied by twelve. Similar adjustments could be made for other compounding periods.

Annuity Compounding Periods Many annuity problems also involve compounding or discounting periods of less than a year. For instance, suppose you want to buy a car that costs $20,000. You have $5,000 for a down payment and plan to finance the remaining $15,000 at 6 percent annual interest for four years. What would your *monthly* loan payments be?

---

[6]Note that we "lied" to our TI BAII PLUS calculator. It asks us for the interest rate per year (I/Y). We gave it the semiannual interest rate of 6 percent, not the annual interest rate of 12 percent. Similarly, n was expressed as the number of semiannual periods, two in one year. As long as we are consistent in expressing the k and n values according to the number of compounding or discounting periods per year, the calculator will give us the correct answer.

First, change the stated annual rate of interest, 6 percent, to a monthly rate by dividing by 12 (6% / 12 = 1/2% or .005). Second, multiply the four-year period by 12 to obtain the number of months involved (4 × 12 = 48 months). Now solve for the annuity payment size using the annuity formula.

In our case, we apply the present value of an annuity formula as follows:

Present Value of an Annuity Algebraic Method

$$\text{PVA} = \text{PMT} \times \left[\frac{1 - \frac{1}{(1+k)^n}}{k} \right] \quad (8\text{-}3a)$$

$$\$15{,}000 = \text{PMT} \times \left[\frac{1 - \frac{1}{(1.005)^{48}}}{.005} \right]$$

$$\$15{,}000 = \text{PMT} \times 42.5803$$

$$\frac{\$15{,}000}{42.5803} = \text{PMT}$$

$$\$352.28 = \text{PMT}$$

The monthly payment on a $15,000 car loan with a 6 percent annual interest rate (.5 percent per month) for four years (48 months) is $352.28.

Solving this problem with the PVIFA table in Table IV in the Appendix at the end of the book would be difficult because the .5 percent interest rate is not listed in the PVIFA table. If the PVIFA were listed, we would apply the table formula, make the adjustments to reflect the monthly interest rate and the number of periods, and solve for the present value of the annuity.

On a financial calculator, we would first adjust the k and n to reflect the same time period—monthly, in our case—and then input the adjusted variables to solve the problem as follows:

TI BAII PLUS FINANCIAL CALCULATOR SOLUTION

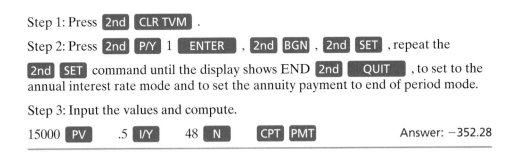

Note that once again we have lied to our TI BAII PLUS financial calculator. The interest rate we entered was not the 6 percent rate per year but rather the .5 percent rate per month. We entered not the number of years, four, but rather the number of months, 48. Because we were consistent in entering the k and n values in monthly terms, the calculator gave us the correct monthly payment of −352.28 (an outflow of $352.28 per month).

Continuous Compounding The effect of increasing the number of compounding periods per year is to increase the future value of the investment. The more frequently interest is compounded, the greater the future value. The smallest compounding period is used when we do **continuous compounding**—compounding that occurs every tiny unit of time (the smallest unit of time imaginable).

Recall our $100 deposit in an account at 12 percent for one year with annual compounding. At the end of year 1, our balance was $112. With semiannual compounding, the amount increased to $112.36.

When continuous compounding is involved, we cannot divide k by infinity and multiply n by infinity. Instead, we use the term e, which you may remember from your math class. We define e as follows:

$$e = \lim_{h \to \infty} \left[1 + \frac{1}{h}\right]^h \approx 2.71828$$

The value of e is the natural antilog of 1 and is approximately equal to 2.71828. This number is one of those like pi (approximately equal to 3.14159), which we can never express exactly, but can approximate. Using e, the formula for finding the future value of a given amount of money, PV, invested at annual rate, k, for n years, with continuous compounding, is as follows:

Future Value with Continuous Compounding

$$FV = PV \times e^{(k \times n)} \tag{8-7}$$

where k and n are expressed in annual terms

Applying Equation 8-7 to our example of a $100 deposit at 12 percent annual interest with continuous compounding, at the end of one year we would have the following balance:

$$FV = \$100 \times 2.71828^{(.12 \times 1)}$$
$$= \$112.75$$

The future value of $100, earning 12 percent annual interest compounded continuously, is $112.75.

As this section demonstrates, the compounding frequency can impact the value of an investment. Investors, then, should look carefully at the frequency of compounding. Is it annual, semiannual, quarterly, daily, or continuous? Other things being equal, the more frequently interest is compounded, the more interest the investment will earn.

What's Next

In this chapter we investigated the importance of the time value of money in financial decisions, and learned how to calculate present and future values for a single amount, for ordinary annuities, and for annuities due. We also learned how to solve special time value of money problems, such as finding the interest rate or the number of periods.

The skills acquired in this chapter will be applied in later chapters, as we evaluate proposed projects, bonds, and preferred and common stock. They will also be used when we estimate the rate of return expected by suppliers of capital. In the next chapter, we will turn to bond and stock valuation.

Summary

1. Explain the time value of money and its importance in the business world.

Money grows over time when it earns interest. Money expected or promised in the future is worth less than the same amount of money in hand today. This is because we lose the opportunity to earn interest when we have to wait to receive money. Similarly, money we owe is less burdensome if it is to be paid in the future rather than now. These concepts are at the heart of investment and valuation decisions of a firm.

2. Calculate the future value and present value of a single amount.

To calculate the future value and the present value of a single dollar amount, we may use the algebraic, table, or calculator methods. Future value and present value are mirror images of each other. They are compounding and discounting, respectively. With future value, increases in k and n result in an exponential increase in future value. Increases in k and n result in an exponential decrease in present value.

3. Find the future and present values of an annuity.

Annuities are a series of equal cash flows. An annuity that has payments that occur at the end of each period is an ordinary annuity. An annuity that has payments that occur at the beginning of each period is an annuity due. A perpetuity is a perpetual annuity.

To find the future and present values of an ordinary annuity, we may use the algebraic, table, or financial calculator method. To find the future and present values of an annuity due, multiply the applicable formula by $(1 + k)$ to reflect the earlier payment.

4. Solve time value of money problems with uneven cash flows.

To solve time value of money problems with uneven cash flows, we find the value of each payment (each single amount) in the cash flow series and total each single amount. Sometimes the series has several cash flows of the same amount. If so, calculate the present value of those cash flows as an annuity, and add the total to the sum of the present values of the single amounts to find the total present value of the uneven cash flow series.

5. Solve special time value of money problems, such as finding the interest rate, number or amount of payments, or number of periods in a future or present value problem.

To solve special time value of money problems, we use the present value and future value equations, and solve for the missing variable, such as the loan payment, k, or n. We may also solve for the present and future values of single amounts or annuities in which the interest rate, payments, and number of time periods are expressed in terms other than a year. The more often interest is compounded, the larger the future value.

You can access your FREE PHLIP/CW On–Line Study Guide, Current Events, and Spreadsheet Problems templates, which may correspond to this chapter either through the Prentice Hall Finance Center CD–ROM or by directly going to: **http://www.prenhall.com/gallagher.**

PHLIP/CW — *The Prentice Hall Learning on the Internet Partnership–Companion Web Site*

Equations Introduced in this Chapter

Equation 8-1a. Future Value of a Single Amount—Algebraic Method:

$$FV = PV \times (1 + k)^n$$

where: FV = Future Value, the ending amount
PV = Present Value, the starting amount, or original principal
k = Rate of interest per period (expressed as a decimal)
n = Number of time periods

Equation 8-1b. Future Value of a Single Amount—Table Method:

$$FV = PV \times (FVIF_{k,n})$$

where: FV = Future Value, the ending amount
PV = Present Value, the starting amount
$FVIF_{k,n}$ = Future Value Interest Factor given interest rate, k, and number of periods, n, from Table I

Equation 8-2a. Present Value of a Single Amount—Algebraic Method:

$$PV = FV \times \frac{1}{(1 + k)^n}$$

where: PV = Present Value, the starting amount
FV = Future Value, the ending amount
k = Discount rate of interest per period (expressed as a decimal)
n = Number of time periods

Equation 8-2b. Present Value of a Single Amount—Table Approach:

$$PV = FV \times (PVIF_{k,n})$$

where: PV = Present Value
FV = Future Value
$PVIF_{k,n}$ = Present Value Interest Factor given interest rate, k, and number of periods, n, from Table II

Equation 8-3a. Future Value of an Annuity—Algebraic Method:

$$FVA = PMT \times \left[\frac{(1 + k)^n - 1}{k} \right]$$

where: FVA = Future Value of an Annuity
PMT = Amount of each annuity payment
k = Interest rate per time period
n = Number of annuity payments

Equation 8-3b. Future Value of an Annuity—Table Approach:

$$FVA = PMT \times FVIFA_{k,n}$$

where: FVA = Future Value of an Annuity
PMT = Amount of each annuity payment
$FVIFA_{k,n}$ = Future Value Interest Factor for an Annuity from Table III
k = Interest rate per period
n = Number of annuity payments

Equation 8-4a. Present Value of an Annuity—Algebraic Method:

$$PVA = PMT \times \left[\frac{1 - \frac{1}{(1+k)^n}}{k} \right]$$

where: PVA = Present Value of an Annuity
PMT = Amount of each annuity payment
k = Discount rate per period
n = Number of annuity payments

Equation 8-4b. Present Value of an Annuity—Table Method:

$$PVA = PMT \times PVIFA_{k,n}$$

where: PVA = Present Value of an Annuity
PMT = Amount of each annuity payment
$PVIFA_{k,n}$ = Present Value Interest Factor for an Annuity from Table IV
k = Discount rate per time period
n = Number of annuity payments

Equation 8-5. Present Value of a Perpetuity:

$$PVP = PMT \times \left(\frac{1}{k} \right)$$

where: PVP = Present value of a perpetuity
PMT = Amount of each of the perpetual annuity payments
k = Discount rate

Equation 8-6. Rate of Return:

$$k = \left(\frac{FV}{PV} \right)^{\frac{1}{n}} - 1$$

where: k = Rate of return
FV = Future Value
PV = Present Value
n = Number of compounding periods

Equation 8-7. Future Value with Continuous Compounding:

$$FV = PV \times e^{(k \times n)}$$

where: FV = Future Value
PV = Present Value
e = Natural antilog of 1
k = Stated annual interest rate
n = Number of years

Self-Test

ST-1. Jed is investing $5,000 into an eight-year certificate of deposit (CD) that pays 6 percent annual interest with annual compounding. How much will he have when the CD matures?

ST-2. How much money would Jed have at maturity if he put his $5,000 into an eight-year CD that pays 6 percent annual interest compounded monthly?

ST-3. Heidi's grandmother died and provided in her will that Heidi will receive $100,000 from a trust when Heidi turns 21 years of age, 10 years from now. If the appropriate discount rate is 8 percent, what is the present value of this $100,000 to Heidi?

ST-4. Zack wants to buy a new Ford Mustang automobile. He will need to borrow $20,000 to go with his down payment in order to afford this car. If car loans are available at a 6 percent annual interest rate, what would Zack's monthly payment be on a four-year loan?

ST-5. Bridget invested $5,000 in a growth mutual fund and in 10 years her investment had grown to $15,529.24. What annual rate of return did Bridget earn over this 10-year period?

ST-6. If Tom invests $1,000 a year *beginning today* into a portfolio that earns a 10 percent return per year, how much will he have at the end of 10 years? (Hint: Recognize that this is an annuity due problem.)

FinCoach Practice Exercises

To help improve your grade and master the mathematical skills covered in this chapter, open FinCoach on the Prentice Hall Finance Center CD-ROM and work practice problems in the following categories: 1) Valuation of Single Cash Flows, 2) Valuation of Multiple Cash Flows, and 3) Valuation of Infinitely Many Cash Flows.

Review Questions

1. What is the time value of money?
2. Why does money have time value?
3. What is compound interest? Compare compound interest to discounting.
4. How is present value affected by a change in the discount rate?
5. What is an annuity?
6. Suppose you are planning to make regular contributions in equal payments to an investment fund for your retirement. Which formula would you use to figure out how much your investments will be worth at retirement time, given an assumed rate of return on your investments?
7. How does continuous compounding benefit an investor?
8. If you are doing PVA and FVA problems, what difference does it make if the annuities are ordinary annuities or annuities due?
9. Which formula would you use to solve for the payment required for a car loan if you know the interest rate, length of the loan, and the borrowed amount? Explain.

Build Your Communication Skills

CS-1. Obtain information from four different financial institutions about the terms of their basic savings accounts. Compare the interest rates paid and the frequency of the compounding. For each account, how much money would you have in 10 years if you deposited $100 today? Write a one- to two-page report of your findings.

CS-2. Interview a mortgage lender in your community. Write a brief report about the mortgage terms. Include in your discussion comments about what interest rate is charged on different types of loans, why rates differ, what fees are charged in addition to the interest and principal that mortgagees must pay to this lender, and describe the advantages and disadvantages of some of the loans offered.

Problems

8-1. What is the future value of $1,000 invested today if you earn 7 percent annual interest for 5 years? *Future Value*

8-2. Calculate the future value of $50,000 10 years from now if the annual interest rate is: *Future Value*

   a. 0 percent
   b. 5 percent
   c. 10 percent
   d. 20 percent

8-3. What is the present value of $20,000 to be received 10 years from now using a 12 percent annual discount rate? *Present Value*

8-4. Calculate the present value of $60,000 to be received 20 years from now at an annual discount rate of: *Present Value*

   a. 0 percent
   b. 5 percent
   c. 10 percent
   d. 20 percent

8-5. What is the present value of a $500 ten-year annual ordinary annuity at a 6 percent annual discount rate? *Present Value of an Annuity*

8-6. Calculate the present value of a $10,000 thirty-year annual ordinary annuity at an annual discount rate of: *Present Value of an Annuity*

   a. 0 percent
   b. 10 percent
   c. 20 percent
   d. 50 percent

8-7. What is the future value of a five-year annual ordinary annuity of $500, using a 9 percent interest rate? *Future Value of an Annuity*

8-8. Calculate the future value of a 12-year, $6,000 annual ordinary annuity, using a discount rate of: *Future Value of an Annuity*

   a. 0 percent
   b. 2 percent
   c. 10 percent
   d. 20 percent

8-9. Starting today, you invest $1,200 a year into your individual retirement account (IRA). If your IRA earns 12 percent a year, how much will be available at the end of 40 years? *Future Value of an Annuity Due*

8-10. If your required rate of return is 12 percent, how much will an investment that pays $80 a year at the beginning of each of the next 20 years be worth to you today? *Present Value of an Annuity Due*

8-11. You invested $50,000, and 10 years later the value of your investment has grown to $185,361. What is your compounded annual rate of return over this period? *Solving for k*

8-12. You invested $1,000 five years ago, and the value of your investment has *fallen* to $773.78. What is your compounded annual rate of return over this period? *Solving for k*

8-13. What is the present value of a $50 annual perpetual annuity using a discount rate of 8 percent? *Present Value of a Perpetuity*

8-14. What is the future value of $10, earning 8 percent annual interest, 200 years later? *Future Value*

8-15. Indicate the number of periods (n) and the interest rate per period (k) for each of the following: *Different Compounding Frequencies*

a. 10 years, 8 percent annual interest rate, compounded annually.
b. 10 years, 8 percent annual interest rate, compounded semiannually.
c. 10 years, 8 percent annual interest rate, compounded quarterly.
d. 10 years, 8 percent annual interest rate, compounded monthly.
e. 10 years, 8 percent annual interest rate, compounded daily.

Future Value

8-16. If you deposit $8,000 into your savings account today, what will your balance be after five years assuming the following independent scenarios?

a. 10 percent annual interest, compounded annually.
b. 10 percent annual interest, compounded semiannually.
c. 10 percent annual interest, compounded monthly.
d. 10 percent annual interest, compounded daily.
e. 10 percent annual interest, compounded continuously.

Present Value

8-17. What is the amount you have to invest today at 7 percent annual interest to be able to receive $10,000 after

a. 5 years?
b. 10 years?
c. 20 years?

Future Value

8-18. How much money would Ruby Carter need to deposit in her savings account at Great Western Bank today in order to have $16,850.58 in her account after five years? Assume she makes no other deposits or withdrawals and the bank guarantees an 11 percent annual interest rate, compounded annually.

Future Value

8-19. If you invest $20,000 today, how much will you receive after

a. 7 years at a 5 percent annual interest rate?
b. 10 years at a 7 percent annual interest rate?

Solving for k

8-20. The property you purchased 12 years ago for $40,000 is now worth $200,000. What is the compounded annual rate of return you have earned on this investment?

Solving for n

8-21. Amy Jolly deposited $1,000 in a savings account. The annual interest rate is 10 percent, compounded semiannually. How many years will it take for her money to grow to $2,653.30?

Present Value of an Annuity

8-22. Beginning a year from now, Bernardo O'Reilly will receive $20,000 a year from his pension fund. There will be 15 of these annual payments. What is the present value of these payments if a 6 percent annual interest rate is applied as the discount rate?

Future Value of an Annuity

8-23. If you invest $ 2,000 per year for the next 10 years at an 8 percent annual interest rate, beginning one year from today compounded annually, how much are you going to have at the end of the tenth year?

8-24. It is the beginning of the quarter and you intend to invest $300 into your retirement fund at the end of every quarter for the next 30 years. You are promised an annual interest rate of 8 percent, compounded quarterly.

Challenge Problem
Future Value of an Annuity

a. How much will you have after 30 years upon your retirement?
b. How long will your money last if you start withdrawing $6,000 at the end of every quarter after you retire?

Solve for the Payment

8-25. A $30,000 loan obtained today is to be repaid in equal annual installments over the next seven years starting at the end of this year. If the annual interest rate is 10 percent, compounded annually, how much is to be paid each year?

8-26. Allie Fox is moving to Central America. Before he packs up his wife and son he purchases an annuity that guarantees payments of $10,000 a year in perpetuity. How much did he pay if the annual interest rate is 12 percent?

Present Value of a Perpetuity

8-27. Matt and Christina Drayton deposited $500 into a savings account the day their daughter, Joey, was born. Their intention was to use this money to help pay for Joey's wedding expenses when and if she decided to get married. The account pays 5 percent annual interest with continuous compounding. Upon her return from a graduation trip to Hawaii, Joey surprises her parents with the sudden announcement of her planned marriage to John Prentice. The couple set the wedding date to coincide with Joey's twenty-third birthday. How much money will be in Joey's account on her wedding day?

Future Value

8-28. To finance your college education your bank has agreed to lend you $5,000 on January 1, 1998, and $5,000 on the first day of each year for the next three years. The annual interest rate of 10 percent is compounded yearly. You plan to graduate on December 31, 2001. What will be the amount you will owe the bank on that date?

Future Value

8-29. You plan to repay the loan in Problem 8-28 in 10 equal installments beginning at the beginning of each year starting in the year 2002. What amount will you have to pay each year?

Solving for the Payment

8-30. Joanne and Walter borrow $14,568.50 for a new car before they move to Stepford, Connecticut. They are required to repay the amortized loan with four annual payments of $5,000 each. What is the interest rate on their loan?

Solving for k

8-31. Norman Bates is planning for his eventual retirement from the motel business. He plans to make quarterly deposits of $1,000 into an IRA starting three months from today. The guaranteed annual interest rate is 8 percent, compounded quarterly. He plans to retire in 15 years.

  a. How much money will be in his retirement account when he retires?
  b. Norman also supports his mother. At Norman's retirement party, Mother tells him they will need $2,000 each month in order to pay for their living expenses. Using the preceding interest rate and the total account balance from part a, for how many years will Norman keep Mother happy by withdrawing $6,000 at the end of each quarter? It is very important that Norman keep Mother happy.

*Challenge Problem
Future Value of an Annuity*

8-32. Jack Torrance comes to you for financial advice. He hired the Redrum Weed-N-Whack Lawn Service to trim the hedges in his garden. Because of the large size of the project (the shrubs were really out of control), Redrum has given Jack a choice of four different payment options. Which of the following four options would you recommend that Jack choose? Why?

  Option 1. Pay $5,650 cash immediately.
  Option 2. Pay $6,750 cash in one lump sum two years from now.
  Option 3. Pay $800 at the end of each quarter for two years.
  Option 4. Pay $1,000 immediately plus $5,250 in one lump sum two years from now. Jack tells you he can earn 8 percent interest, compounded quarterly, on his money. You have no reason to question his assumption.

Challenge Problem

8-33. Sarah has $30,000 for a down payment on a house and wants to borrow $120,000 from a mortgage banker to purchase a $150,000 house. The mortgage loan is to be repaid in monthly installments over a 30-year period. The annual interest rate is 9 percent. How much will Sarah's monthly mortgage payments be?

Solving for the Payment

8-34. Slick has his heart set on a new Miata sportscar. He will need to borrow $18,000 to get the car he wants. The bank will loan Slick the $18,000 at an annual interest rate of 6 percent.

  a. How much would Slick's monthly car payments be for a four-year loan?
  b. How much would Slick's monthly car payments be if he obtains a six-year loan at the same interest rate?

Solving for Payment

Solving for Payment

Challenge Problem
Value of Missing Cash Flow

8-35. Assume the following set of cash flows:

Time 1	Time 2	Time 3	Time 4
$100	$150	?	$100

At a discount rate of 10%, the total present value of all the cash flows above, including the missing cash flow, is $320,743. Given these conditions, what is the value of the missing cash flow?

Answers to Self-Test

ST-1. $FV = \$5{,}000 \times 1.06^8 = \$7{,}969.24$ = Jed's balance when the eight-year CD matures

ST-2. $FV = \$5{,}000 \times (1 + .06/12)^{12 \times 8}$
$= \$5{,}000 \times 1.005^{96} = \$8{,}070.71$
= Jed's balance in 8 years with monthly compounding

ST-3. $PV = \$100{,}000 \times 1/1.08^{10} = \$46{,}319.35$
= the present value of Heidi's $100,000

ST-4. $\$20{,}000 = PMT \times [1-(1/1.005^{48})]/.005$
$\$20{,}000 = PMT \times 42.5803$
$PMT = \$469.70$ = Zack's car loan payment

ST-5. $\$5{,}000 \times FVIF_{k=?,\,n=10} = \$15{,}529.24$
$FVIF_{k=?,\,n=10} = 3.1058 \therefore k = 12\%$ = Bridget's annual rate of return on the mutual fund

ST-6. $FV = PMT\,(FVIFA_{k\%\,n}) \times (1 + k)$
$= \$1{,}000\,(FVIFA_{10\%\,10}) \times (1 + .10)$
$= \$1{,}000\,(15.9374) \times (1.10)$
$= \$17{,}531.14$

CHAPTER 9

BOND AND STOCK VALUATION

*Nowadays we know the price of everything and
the value of nothing.*
—Oscar Wilde

VALUING COMPANIES

Ms. Jennifer Owen is a senior financial analyst for BIA Consulting. As a business appraiser, she uses a variety of models to measure the value of a company's common stock. Several of these models are discussed in this chapter. Following are excerpts from an interview with Ms. Owen.

Q. What approaches do you use to value private companies?

A. Relevant data are much more available for a public company than for a private one. With a public company you can simply multiply the company's stock price by the number of shares outstanding to come up with the value of the company's equity and then add the value of the company's interest-bearing debt to get the enterprise value of the company. When valuing a private company, we consider three approaches to value: the cost approach, the cash flow approach, and the market approach. The cost approach considers the replacement cost of the assets and is normally used when an entity is worth more on a piecemeal basis than as an operating entity. The cash flow approach determines the value of a company by calculating the net present value of projected cash flows to be generated by the company. The market approach determines the value of a company by comparing the company to similar companies that are publicly traded.

CHAPTER OBJECTIVES

After reading this chapter, you should be able to:

- Explain the importance of bond and stock valuation.

- Discuss the concept of securities valuation.

- Compute the market value and the yield to maturity of a bond.

- Calculate the market value and expected yield of preferred stock.

- Compute the market value and expected yield of common stock.

Q. Do some valuation methods work better than others?

A. It depends on the information that is available. If there are only a few transactions involving publicly traded companies, or if there are only a couple of publicly traded companies to look at for comparisons, the cash flow approach would probably be used. If the future cash flows are difficult to forecast, then the market approach would be preferred. If feasible, we like to consider more than one approach when valuing a private company.

Q. It sounds as if "value is in the eye of the beholder." Is that true?

A. Generally speaking, this is true. In most cases we are contracted to estimate the fair market value of an entity. Sometimes we are asked to estimate that value from the perspective of a specified person or company. For example, a wireless telecommunications company might be more valuable to Company A than to Company B because this is the last market Company A needs to operate a nationwide network.

Q. What advice would you give to a finance student who is interested in entering the field of valuations?

A. An organization called the American Society of Appraisers (ASA) provides a lot of helpful information for those considering this career path. The organization has a certification program that establishes appraiser designations. An ASA designation is a good thing to have for career advancement.

Source: Interview with Ms. Jennifer Owen, September 1998.

Chapter Overview

In this chapter we will discuss how to value bonds and shares of stock in a dynamic marketplace. First, we will investigate the importance of valuation and introduce a general model that analysts and investors use to value securities. Then we will show how to adapt the model to bonds, preferred stock, and common stock. For common stock, we'll explore additional valuation techniques.

The Importance of Bond and Stock Valuation

As chapter 1 explained, the primary financial goal of financial managers is to maximize the market value of their firm. It follows, then, that financial managers need to assess the market value of their bonds and stock to gauge progress.

Accurate bond and stock valuation is also a concern when a corporation contemplates selling securities to raise long-term funds. Issuers want to raise the most money possible from selling securities. Issuers lose money if they undervalue their stock or bonds. Likewise, would-be purchasers are concerned about securities' value because they don't want to pay more than the securities are worth.

A General Valuation Model

The value of a security depends on its future earning power. To value a security, then, we consider three factors that affect future earnings:

- size of cash flows
- timing of cash flows
- risk

In chapter 7 we examined how risk factors affect an investor's required rate of return. In chapter 8 we learned that time value of money calculations can determine an investment's value, given the size and timing of the cash flows.

Financial managers and investors determine the value of a security by finding the present value of the security's future cash flows. We can calculate a bond's value by taking the sum of the present values of each of future cash flows from the bond's interest and principal payments. We can calculate a stock's value by taking the sum of the present values of future dividend cash flow payments.

Analysts and investors use a general valuation model to calculate the present value of future cash flows of a security. This model, the **discounted cash flow model (DCF)**, is a basic valuation model for a security that is expected to generate cash payments, such as dividends or interest and principal. The DCF equation is shown in Equation 9-1:

The Discounted Cash Flow Valuation Model

$$V_0 = \frac{CF_1}{(1+k)^1} + \frac{CF_2}{(1+k)^2} + \frac{CF_3}{(1+k)^3} \cdots + \cdots \frac{CF_n}{(1+k)^n} \quad (9\text{-}1)$$

where: V_0 = Present value of the anticipated cash flows from the security, its current value
$CF_{1, 2, 3, \text{ and } n}$ = Cash flows expected to be received one, two, three, and so on up to n periods in the future
k = Discount rate, the required rate of return

The discounted cash flow model, Equation 9-1, is derived from the present value equations in chapter 8, pages 181–183.

The DCF model values a security by calculating the sum of the present values of all expected future cash flows. Each cash flow is discounted back the appropriate number of periods according to its expected timing and required rate of return.

The discount rate in Equation 9-1 is the investor's required rate of return, which is a function of the risk of the investment. Recall from chapter 7 that the riskier the security, the higher the required rate of return.

The discounted cash flow model is easy to use if we know the cash flows and discount rate. For example, suppose you were considering purchasing a security that entitled you to receive payments of $100 in one year, another $100 in two years, and $1,000 in three years. If your required rate of return for securities of this type were 20 percent, then we would calculate the value of the security as follows:

$$V_0 = \frac{\$100}{(1+.20)^1} + \frac{\$100}{(1+.20)^2} + \frac{\$1,000}{(1+.20)^3}$$
$$= \$83.3333 + \$69.4444 + \$578.7037$$
$$= \$731.48$$

The total of the security's three future cash flows at a 20 percent required rate of return yields a present value of $731.48.

In the sections that follow, we'll adapt the discounted cash flow valuation model to apply to bonds and stocks.

We discussed bonds and bond terminology in chapter 2, pages 27–29.

BOND VALUATION

Remember from chapter 2 that a bond's cash flows are determined by the bond's coupon interest payments, face value, and maturity.

Because coupon interest payments occur at regular intervals throughout the life of the bond, those payments are an annuity. Instead of using several terms representing the individual cash flows from the future coupon interest payments (CF_1, CF_2, and so on), we adapt Equation 9-1 by using one term to show the annuity. The remaining term represents the future cash flow of the bond's face value, or principal, that is paid at maturity. Equation 9-2 shows the adapted valuation model:

The Bond Valuation Formula (Algebraic Version)

$$V_B = INT \times \left[\frac{1 - \frac{1}{(1 + k_d)^n}}{k_d} \right] + \frac{M}{(1 + k_d)^n} \quad (9\text{-}2)$$

where: V_B = Current market value of the bond
INT = Dollar amount of each periodic interest payment
n = Number of times the interest payment is received (which is also the number of periods until maturity)
M = Principal payment received at maturity
k_d = Required rate of return per period on the bond debt instrument

The table version of the bond valuation model is shown in Equation 9-3, as follows:

The Bond Valuation Formula (Table Version)

$$V_B = (INT \times PVIFA_{k,n}) + (M \times PVIF_{k,n}) \quad (9\text{-}3)$$

where: $PVIFA_{k,n}$ = Present Value Interest Factor for an Annuity from Table IV
$PVIF_{k,n}$ = Present Value Interest Factor for a single amount from Table II

Take Note

The determinants of nominal interest rates, or required rates of return, include the real rate of interest, the inflation premium, the default risk premium, the liquidity premium, and the maturity premium. Each person evaluating a bond will select an appropriate required rate of return, k_d, for the bond based on these determinants.

To use a calculator to solve for the value of a bond, enter the dollar value of the interest payment as [PMT], the face value payment at maturity as [FV], the number of payments as n, and the required rate of return, k_d depicted as [I/Y] on the TI BAII Plus calculator. Then compute the present value of the bond's cash flows.

Now let's apply the bond valuation model. Suppose Microsoft Corporation issues a 7 percent coupon interest rate bond with a maturity of 20 years. The face value of the bond, payable at maturity, is $1,000.

First, we calculate the dollar amount of the coupon interest payments. At a 7 percent coupon interest rate, each payment is .07 × $1,000 = $70.

Next, we need to choose a required rate of return, k_d. Remember that k_d is the required rate of return that is appropriate for the bond based on its risk, maturity, marketability, and tax treatment. Let's assume that 8 percent is the rate of return the market determines to be appropriate.

Now we have all the factors we need to solve for the value of Microsoft Corporation's bond. We know that k_d is 8 percent, n is 20, the coupon interest payment is $70 per year, and the face value payment at maturity is $1,000. Using Equation 9-2, we calculate the bond's value as follows:

$$V_B = \$70 \times \left[\frac{1 - \frac{1}{(1+.08)^{20}}}{.08}\right] + \frac{\$1,000}{(1+.08)^{20}}$$

$$= (\$70 \times 9.8181474) + \left(\frac{\$1,000}{4.660957}\right)$$

$$= \$687.270318 + \$214.548214$$

$$= \$901.82$$

Notice that the value of Microsoft Corporation's bond is the sum of the present values of the 20 annual $70 coupon interest payments plus the present value of the one-time $1,000 face value to be paid 20 years from now, given a required rate of return of 8 percent.

To find the Microsoft bond's value using present value tables, recall that the bond has a face value of $1,000, a coupon interest payment of $70, a required rate of return of 8 percent, and an n value of 20. We apply Equation 9-3 as shown:

$$V_B = (\$70 \times \text{PVIFA}_{8\%, 20 \text{ yrs}}) + (\$1,000 \times \text{PVIF}_{8\%, 20 \text{ yrs}})$$

$$= (\$70 \times 9.8181) + (\$1,000 \times .2145)$$

$$= \$687.267 + \$214.500$$

$$= \$901.77$$

We see that the sum of the present value of the coupon interest annuity, $687.267, plus the present value of the principal, $214.500, results in a bond value of $901.77. There is a five-cent rounding error in this example when the tables are used.

Here's how to find the bond's value using the TI BAII PLUS financial calculator. Enter the $70 coupon interest payment as PMT, the one-time principal payment of $1,000 as FV, the 20 years until maturity as n (N on the TI BAII PLUS), and the 8 percent required rate of return—depicted as I/Y on the TI BAII Plus. As demonstrated in chapter 8 calculator solutions, clear the time value of money TVM registers before entering the new data. Skip steps 2 and 3 if you know your calculator is set to one payment per year and is also set for end of period payment mode.

TI BAII PLUS FINANCIAL CALCULATOR SOLUTION

Step 1: Press `2nd` `CLR TVM` to clear previous values.

Step 2: Press `2nd` `P/Y` 1 `ENTER` , `2nd` `BGN` `2nd` `SET` , repeat `2nd` `SET` until END shows in the display `2nd` `QUIT` to set to the annual interest rate mode and to set the annuity payment to end of period mode.

Step 3: Input the values and compute.

1000 `FV`   8 `I/Y`   20 `N`   70 `PMT`   `CPT` `PV`   Answer: −901.82

The $901.82 is negative because it is a cash outflow—the amount an investor would pay to buy the bond today.

We have shown how to value bonds with annual coupon interest payments in this section. Next, we show how to value bonds with semiannual coupon interest payments.

Semiannual Coupon Interest Payments

In the hypothetical bond valuation examples for Microsoft Corporation, we assumed the coupon interest was paid annually. However, most bonds issued in the United States pay interest semiannually (twice per year). With semiannual interest payments, we must adjust the bond valuation model accordingly. If the Microsoft bond paid interest twice per year, the adjustments would look like this:

To see how to adjust table and calculator solutions for semiannual interest, review chapter 8, pages 195–197.

	ANNUAL BASIS	SEMIANNUAL BASIS
Coupon Interest Payments	$70	÷ 2 = $35 per six-month period
Maturity	20 yrs	× 2 = 40 six-month periods
Required Rate of Return	8%	÷ 2 = 4%

These values can now be used in Equation 9-2, Equation 9-3, or a financial calculator, in the normal manner. For example, if Microsoft's 7 percent coupon, 20-year bond paid interest semiannually, its present value per Equation 9-2 would be:

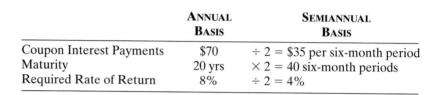

$$V_B = \$35 \times \left[\frac{1 - \frac{1}{(1 + .04)^{40}}}{.04} \right] + \frac{\$1000}{(1 + .04)^{40}}$$

$$= (\$35 \times 19.792774) + \left(\frac{\$1{,}000}{4.801021} \right)$$

$$= \$692.747090 + \$208.2890$$

$$= \$901.04$$

The value of our Microsoft bond with semiannual interest and a 4 percent per semiannual discount rate is $901.04. This compares to a value of $901.82 for the same bond if it pays annual interest and has an 8 percent annual discount rate. Note that a required rate of return of 4 percent per semiannual period is not the same as 8 percent per year. The difference in the frequency of discounting gives a slightly different answer.

The Yield to Maturity of a Bond

Most investors want to know how much return they will earn on a bond to gauge whether the bond meets their expectations. That way, investors can tell whether they should add the bond to their investment portfolio. As a result, investors often calculate a bond's yield to maturity before they buy a bond. **Yield to maturity (YTM)** represents the average rate of return on a bond if all promised interest and principal payments are made on time and if the interest payments are reinvested at the YTM rate given the price paid for the bond.

Calculating a Bond's Yield to Maturity To calculate a bond's YTM, we apply the bond valuation model. However, we apply it differently than we did when solving for a bond's present value (price) because we solve for k_d, the equivalent of YTM.

To compute a bond's YTM, we must know the values of all variables except k_d. We take the market price of the bond, P_B, as the value of a bond, V_B, examining financial sources such as *The Wall Street Journal* for current bond prices.

Once you have all variables except k_d, solving for k_d algebraically is exceedingly difficult because that term appears three times in the valuation equation. Instead, we use the trial-and-error method. In other words, we guess a value for k_d and solve for V_B using that value. When we find a k_d value that results in a bond value that matches the published bond price, P_0, we know that the k_d value is the correct YTM. The YTM is the return that bond investors require to purchase the bond.[1]

Here's an illustration of the trial-and-error method for finding YTM. Suppose that *The Wall Street Journal* reported that the Microsoft bond in our earlier example is currently selling for $1,114.70. What is the bond's YTM if purchased at this price?

Recall the annual coupon interest payments for the Microsoft bond were $70 each, the bond had a 20-year maturity, and a face value of $1,000. Applying the bond valuation model, we solve for the k_d that produces a bond value of $1,114.70.

$$\$1,114.70 = \$70 \times \left[\frac{1 - \frac{1}{(1 + k_d)^{20}}}{k_d} \right] + \frac{1,000}{(1 + k_d)^{20}}$$

Although we can try any k_d value, remember that when k was 8 percent, the bond's calculated value, V_B, was $901.82. Bond prices and yields vary inversely—the higher the YTM, the lower the bond price; and the lower the YTM, the higher the bond price. The bond's current market price of $1,114.70 is higher than $901.82, so we know the YTM must be less than 8 percent. If you pay more than $901.82 to buy the bond, your return will be less than 8 percent.

Because we know that YTM and bond prices are inversely related, let's try 7 percent in our bond valuation model, Equation 9-2. We find that a k_d value of 7 percent results in the following bond value:

$$V_B = \$70 \times \left[\frac{1 - \frac{1}{(1 + .07)^{20}}}{.07} \right] + \frac{\$1,000}{(1 + .07)^{20}}$$

$$= (\$70 \times 10.59401425) + \left(\frac{\$1,000}{3.86968446} \right)$$

$$= \$741.5809975 + \$258.4190028$$

$$= \$1,000.00$$

At a k_d of 7 percent, the bond's value is $1,000 instead of $1,114.70. We'll need to try again. Our second guess should be lower than 7 percent because at $k_d = 7\%$ the bond's calculated value is lower than the market price. Let's try 6 percent. At a k_d of 6 percent, the bond's value is as follows:

---

[1] In chapter 12 this required rate of return is called the firm's cost of debt capital, which we adjust for taxes. In this chapter, however, our main focus is finding the value of different types of securities, so k_d is referred to as the investor's required rate of return.

$$V_B = \$70 \times \left[\frac{1 - \frac{1}{(1+.06)^{20}}}{.06} \right] + \frac{\$1,000}{(1+.06)^{20}}$$

$$= (\$70 \times 11.46992122) + \left(\frac{\$1,000}{3.20713547} \right)$$

$$= \$802.8944853 + \$311.8047269$$

$$= \$1,114.70$$

With a k_d of 6 percent, the bond's value equals the current market price of $1,114.70. We conclude that the bond's YTM is 6 percent.[2]

To use the table method to find the YTM of Microsoft's 7 percent coupon rate, 20-year bond at a price of $1,114.70, use Equation 9-3 as follows:

First guess: $k_d = 7\%$:

$$V_B = (\$70 \times \text{PVIFA}_{7\%, 20 \text{ periods}}) + (\$1,000 \times \text{PVIF}_{7\%, 20 \text{ periods}})$$

$$= (\$70 \times 10.5940) + (\$1,000 \times .2584)$$

$$= \$741.58 + \$258.40$$

$$= \$999.98$$

$999.98 is too low. We must guess again. Let's try $k_d = 6\%$, as follows:

$$V_B = (\$70 \times \text{PVIFA}_{6\%, 20 \text{ periods}}) + (\$1,000 \times \text{PVIF}_{6\%, 20 \text{ periods}})$$

$$= (\$70 \times 11.4699) + (\$1,000 \times .3118)$$

$$= \$802.893 + \$311.80$$

$$= \$1,114.69$$

Close enough (to $1,114.70). The bond's YTM is about 6 percent.

Finding a bond's YTM with a financial calculator avoids the trial-and-error method. Simply plug in the values on the calculator and solve for k_d, as shown:

TI BAII PLUS FINANCIAL CALCULATOR SOLUTION

Step 1: Press **2nd** **CLR TVM** to clear previous values.

Step 2: Press **2nd** **P/Y** 1 **ENTER** , **2nd** **BGN** **2nd** **SET** , repeat **2nd** **SET** until END shows in the display **2nd** **QUIT** to set to the annual interest rate mode and to set the annuity payment to end of period mode.

Step 3: Input the values and compute.

1,114.70 **+/−** **PV**   1000 **FV**   20 **N**   70 **PMT**   **CPT** **I/Y** Ans: 6.00

Using the financial calculator, we find that the YTM of the Microsoft $1,000 face value 20-year bond with a coupon rate of 7 percent and a market price of $1,114.70 is 6 percent.

---

[2] We were lucky to find the bond's exact YTM in only two guesses. Often the trial-and-error method requires four or five guesses.

The Relationship between Bond YTM and Price

A bond's market price depends on its yield to maturity. When a bond has a YTM greater than its coupon rate, it sells at a *discount* from its face value. When the YTM is equal to the coupon rate, the market price equals the face value. When the YTM is less than the coupon rate, the bond sells at a *premium* over face value.

For instance, in our initial calculations of the Microsoft bond, we found that the present value of its future cash flows was $901.82. That price was lower than the bond's $1,000 face value. Because its market price was lower than its face value, the bond sold at a discount (from its face value). A bond will sell at a discount because buyers and sellers have agreed that the appropriate rate of return for the bond should be higher than the bond's coupon interest rate. With the Microsoft bond, investors required an 8 percent rate of return, but the fixed coupon interest rate was only 7 percent. To compensate for a coupon interest rate that is lower than the required rate, investors would be unwilling to pay the $1,000 face value. Instead, they would only be willing to pay $901.82 to buy the bond.

Now recall the trial-and-error calculations for the YTM of the Microsoft 7 percent coupon rate bond in the previous section. We found that when the YTM was 7 percent, the bond's price was $1,000. This was no coincidence. When the YTM is equal to the coupon interest rate—that is, when the bond is selling at *par*—the bond's price is equal to its face value.

We saw that when would-be buyers and sellers of Microsoft Corporation's bond agree that the appropriate yield to maturity for the bond is 6 percent instead of 7 percent the price is above $1,000. The coupon rate and the face value remain the same, at 7 percent and $1,000, respectively.

The change from a 7 percent to a 6 percent yield to maturity results in a market value of $1,114.70. That market value for the bond is higher than the $1,000 face value. Because the market price is higher than the bond's face value in our case, the bond sells at a *premium*. Why? Investors pay more to receive "extra" interest because the coupon rate paid is higher than the yield to maturity demanded.

In our example, the calculations show that investors were willing to pay $1,114.70 for a bond with a face value of $1,000 because the coupon interest was one percentage point higher than the expected rate of return.

Figure 9-1 shows the relationship between YTM and the price of a bond.

The inverse relationship between bond price and YTM is important to bond traders. Why? Because if market YTM interest rates rise, bond prices fall. Conversely, if market YTM interest rates fall, bond prices rise. The suggestion that the Fed might raise interest rates is enough to send the bond market reeling as bond traders unload their holdings.

In this section we examined bond valuation for bonds that pay annual and semiannual interest. We also investigated how to find a bond's yield to maturity, and the relationship between a bond's YTM and its price. We turn next to preferred and common stock valuation.

Preferred Stock Valuation

To value preferred stock, we adapt the discounted cash flow valuation formula, Equation 9-1, to reflect the characteristics of preferred stock. First, recall that the value of any security is the present value of its future cash payments. Second, review characteristics of preferred stock. Preferred stock has no maturity date, so it has no maturity value. Its future cash payments are dividend payments that are paid to

Figure 9-1 Bond YTM versus Bond Price

Figure 9-1 shows the inverse relationship between the price and the YTM for a $1,000 face value, 20-year, 7% coupon interest rate bond that pays annual interest.

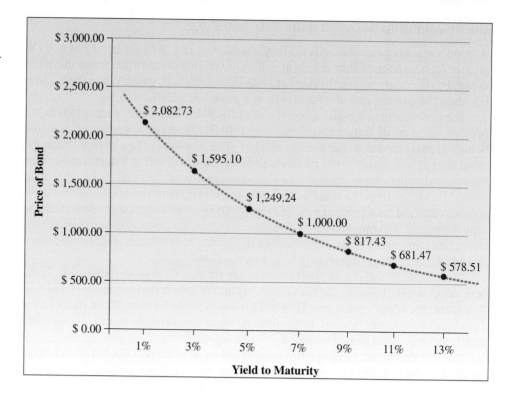

preferred stockholders at regular time intervals for as long as they (or their heirs) own the stock. Cash payments from preferred stock dividends are scheduled to continue forever. To value preferred stock, then, we must adapt the discounted cash flow model to reflect that preferred stock dividends are a perpetuity.

Finding the Present Value of Preferred Stock Dividends

To calculate the value of preferred stock, we need to find the present value of its future cash flows—which are a perpetuity. In chapter 8 we learned how to find the present value of a perpetuity. We use the formula for the present value of a perpetuity, Equation 8-5, but adapt the terms to reflect the nature of preferred stock.[3]

We discussed the present value of a perpetuity in chapter 8, page 187.

The preferred stock valuation calculations require that we find the present value (V_P) of preferred stock dividends (D_p), discounted at required rate of return, k_p. The formula for preferred stock valuation follows:

The Formula for the Present Value of Preferred Stock

$$V_P = \frac{D_p}{k_p} \qquad (9\text{-}4)$$

where: V_P = Current market value of the preferred stock
D_p = Amount of the preferred stock dividend
k_p = Required rate of return for this issue of preferred stock

---

[3]Equation 8-5 is $PV = \frac{PMT}{k}$. In Equation 9-4, V_P substitutes for PV, D_p replaces PMT, and k_p replaces k.

Let's apply Equation 9-4 to an example. Suppose investors expect an issue of preferred stock to pay an annual dividend of $2 per share. Investors in the market have evaluated the issuing company and market conditions and have concluded that 10 percent is a fair rate of return on this investment. The present value of one share of this preferred stock, assuming a 10 percent required rate of return follows:

$$V_P = \left(\frac{\$2}{.10}\right)$$
$$= \$20$$

We find that for investors whose required rate of return (k_p) is 10 percent, the value of each share of this issue of preferred stock is $20.

The Yield on Preferred Stock

The yield on preferred stock represents the annual rate of return that investors would realize if they bought the preferred stock for the current market price and then received the promised preferred dividend payments.

Like bond investors, preferred stock investors want to know the percentage yield they can expect if they buy shares of preferred stock at the current market price. That way, investors can compare the yield with the minimum they require to decide whether to invest in the preferred stock.

Fortunately, calculating the yield on preferred stock is considerably easier than calculating the YTM for a bond. To calculate the yield we rearrange Equation 9-4 so that we solve for k_p. We are not solving for the value of the preferred stock, V_P, but rather are taking the market price of the preferred stock, P_P, as a given and solving for k_p as follows:

Take Note

With bonds, an investor's annual percent return on investment is called the yield to maturity, or YTM. With preferred and common stocks, an investor's percent return on investment is simply called the yield because preferred and common stock don't have a maturity date.

Formula for the Yield on Preferred Stock

$$k_p = \frac{D_P}{P_P} \quad (9\text{-}5)$$

where: D_p = Amount of the preferred stock dividend
P_P = Current market price of the preferred stock
k_p = Yield on investment that an investor can expect if the shares are purchased at the current market price P_P and if the preferred dividend D_p is paid forever

To illustrate how to find the yield using Equation 9-5, suppose Sure-Thing Corporation's preferred stock is selling for $25 per share today and the dividend is $3 a share. Now assume you are a potential buyer of Sure-Thing's preferred stock, so you want to find the expected annual percent yield on your investment. You know that the current market value of the stock, V_p, is $25, and the stock dividend, D_p, is $3. Applying Equation 9-5, you calculate the yield as follows:

$$k_p = \frac{D_p}{P_P}$$
$$= \frac{\$3}{\$25}$$
$$= .12, \text{ or } 12\%$$

Chapter 9   Bond and Stock Valuation   217

You find that the yield for Sure-Thing's preferred stock is 12 percent. If your minimum required rate of return is less than or equal to 12 percent, you would invest in the Sure-Thing preferred stock. If your required rate of return is greater than 12 percent, you would look for another preferred stock that had a yield of over 12 percent.

Common Stock Valuation

The valuation of common stock is somewhat different than the valuation of bonds and preferred stock. Common stock valuation is complicated by the fact that common stock dividends are difficult to predict compared to the interest and principal payments on a bond, or dividends on preferred stock. Indeed, corporations may pay common stock dividends irregularly, or not pay dividends at all.

Analysts use many different approaches to common stock valuation. The most widely accepted models assume that the business is a going concern—that is, it has more value from its future earning potential than the sum of its assets less the sum of its liabilities. For this reason, financial analysts rarely use valuation methods that rely on balance sheet values.

Going concern valuation methods are typically more reliable because the cash flows associated with potential earnings of a firm are usually worth more than the firm's book or liquidation value. To illustrate, assume you were valuing a popular restaurant that is packed every night with customers who appreciate the excellent cuisine, service, and decor. The eatery's physical assets—the furniture, tableware, decorations, and kitchen equipment—are probably worth much less than the value that comes from its potential earnings and related cash flows. Thus, the going concern value of the restaurant would be a more accurate valuation measure in this instance.

Occasionally, balance sheet valuation methods are appropriate because a company is "worth more dead than alive." For example, a mining company whose mines are entirely depleted of minerals could have valuable digging, drilling, or moving equipment. The assets cannot yield any future returns for this business—there is nothing of value to mine—so their book or liquidation values are probably worth more than the firm's future earning power.

In the sections that follow, we examine the most popular going concern and balance sheet common stock valuation models.

Common Stock Going Concern Valuation Models

As with bonds and preferred stock, we value common stock by estimating the present value of the expected future cash flows from the common stock. Those future cash flows are the expected future dividends and the expected price of the stock when the stock is sold.

The discounted cash flow valuation model, Equation 9-1, adapted for common stock is shown in Equation 9-6:

The DCF Valuation Model Applied to Common Stock

$$P_0 = \frac{D_1}{(1+k_s)^1} + \frac{D_2}{(1+k_s)^2} + \frac{D_3}{(1+k_s)^3} \cdots + \cdots \frac{P_n}{(1+k_s)^n} \quad (9\text{-}6)$$

where: P_0 = Present value of the expected dividends, the current price of the common stock
$D_1, D_2, D_3,$ etc. = Common stock dividends expected to be received at the end of periods 1, 2, 3, and so on until the stock is sold
P_n = Anticipated selling price of the stock in n periods
k_s = Required rate of return on this common stock investment

In practice, however, using Equation 9-6 to value common stock is problematic because an estimate of the future selling price of a share of stock is often speculative. This severely limits the usefulness of the model.

Instead, some analysts use models that are a variation of Equation 9-6 but do not rely on an estimate of a stock's future selling price. We turn to those models next.

The Constant Growth Dividend Model Common stock dividends can grow at different rates. The two growth patterns we examine here are no growth and constant growth. In Appendix 9A, we discuss one additional common stock growth pattern: supernormal growth.

The constant dividend growth model assumes common stock dividends will be paid regularly and grow at a constant rate. The constant growth dividend model (also known as the Gordon growth model because financial economist Myron Gordon helped develop and popularize it) is shown in Equation 9-7:

The Constant Growth Dividend Valuation Model

$$P_0 = \frac{D_1}{k_s - g} \qquad (9\text{-}7)$$

where: P_0 = Current price of the common stock
D_1 = Dollar amount of the common stock dividend expected one period from now
k_s = Required rate of return per period on this common stock investment
g = Expected constant growth rate per period of the company's common stock dividends

Equation 9-7 is easy to use if the stock grows at a constant rate. For example, assume your required rate of return (k_s) for Wendy's common stock is 10 percent. Suppose your research leads you to believe that Wendy's Corporation will pay a $.25 dividend next year ($D_1$), and for every year after the dividend will grow at a constant rate (g) of 8 percent a year. Using Equation 9-7, we calculate the present value of Wendy's common stock dividends as follows:

$$P_0 = \frac{\$.25}{.10 - .08}$$
$$= \frac{\$.25}{.02}$$
$$= \$12.50$$

We find that with a common stock dividend in one year of $.25, a constant growth rate of 8 percent, and a required rate of return of 10 percent, the value of the common stock is $12.50.

In a no-growth situation, g, in the denominator of Equation 9-7 becomes zero. To value stocks that have no growth, then, we adapt the general model, as shown in the next section.

No-Growth Dividend Model When common stock dividends are not expected to grow, the cash flow patterns are similar to those of preferred stock—the stockholder receives the same dividend every year. To find the value of common stock that has no-growth dividends, we simply alter the notation of Equation 9-5, the formula for preferred stock valuation, as follows:

No-Growth Common Stock Dividend Valuation Formula

$$P_0 = \frac{D_s}{k_s} \tag{9-8}$$

where: P_0 = Current price of the common stock
D_s = Amount of the common stock dividend, a constant amount
k_s = Required rate of return per period on this common stock investment

We see that Equation 9-8 is like Equation 9-7, except that g equals zero. Because zero is subtracted from k_s, the "$-g$" is dropped in Equation 9-8.

As an example of a no-growth stock valuation, let's say that the common stock annual dividend was $2, and investors' required rate of return was 10 percent. The value (price) of the common stock, according to Equation 9-8, follows:

$$P_0 = \left(\frac{\$2}{.10}\right)$$
$$= \$20$$

We find that if market investors' required rate of return (k_s) were 10 percent, the market price of each share of this common stock would be $20.

The P/E Model Many investment analysts use the price to earnings, or P/E, ratio to value common stock. As we discussed in chapter 6, the P/E ratio, defined as follows, is the price per share of a common stock divided by the company's earnings per share:

The P/E Ratio

$$\text{P/E ratio} = \frac{\text{Price per Share}}{\text{Earnings per Share}}$$

The P/E ratio indicates how much investors are willing to pay for each dollar of a stock's earnings. So, a P/E ratio of 20 means that investors are willing to pay $20 for $1 of a stock's earnings. A high P/E ratio indicates that investors believe the stock's earnings will increase, or that the risk of the stock is low, or both.

Financial analysts often use a P/E model to estimate common stock value for businesses that are not public. First, analysts compare the P/E ratios of similar companies within an industry to determine an appropriate P/E ratio for companies in that industry. Second, analysts calculate an appropriate stock price for firms in the industry by multiplying each firm's earnings per share (EPS) by the industry average P/E ratio. The P/E model formula, Equation 9-9, follows:

The P/E Model

$$\text{Appropriate Stock Price} = \text{Industry P/E Ratio} \times \text{EPS} \tag{9-9}$$

To illustrate how to apply the P/E model, let's value the common stock of the Zumwalt Corporation. Suppose that Zumwalt Corporation has current earnings per share of $2 and, given the risk and growth prospects of the firm, the analyst has determined that the company's common stock should sell for 15 times current earnings. Applying the P/E model, we calculate the following price for Zumwalt Corporation's common stock:

$$\text{Appropriate Stock Price} = \text{Industry P/E Ratio} \times \text{EPS}$$
$$= 15 \times \$2$$
$$= \$30$$

Our P/E model calculations show that $30 per share is the appropriate price for common stock that has a $2 earnings per share and an industry P/E ratio of 15. The industry P/E ratio would be adjusted up or down according to the individual firm's growth prospects and risk relative to the industry norm.

Balance Sheet Valuation Approaches

Some valuation approaches view the firm as if it were about to cease operations immediately. These rely on the balance sheet value of the firm's assets, less liabilities, to value the firm. Such valuation approaches are *balance sheet valuation approaches*. We discuss two valuation methods that start with the balance sheet: book value and liquidation value.

Book Value One of the simplest ways to value a firm's common stock is to subtract the value of the firm's liabilities, and preferred stock, if any, as recorded on the balance sheet, from the value of its assets. The result is the **book value** or **net worth** of a company's common stock. To find the book value per share of common stock, divide the company's book value by the number of outstanding common stock shares. The solution follows:

Book Value per Share

$$\text{Book Value per Share} = \frac{(\text{Total Assets} - \text{Total Liabilities} - \text{Preferred Stock})}{\text{Number of Shares Outstanding}}$$

(9-10)

The book value approach has severe limitations. The asset values recorded on a firm's balance sheet usually reflect what the current owners originally paid for the assets, not the current value of the assets. Due to these and other limitations, the book value is rarely used.

Liquidation Value The liquidation value and book value valuation methods are similar, except that the liquidation method uses the market values of the assets and liabilities, not book values, as in Equation 9-10. The market values of the assets are the amounts the assets would earn on the open market if they were sold (or liquidated). The market values of the liabilities are the amounts of money it would take to pay off the liabilities.

The **liquidation value** is the amount each common stockholder would receive if the firm closed, sold all assets and paid all liabilities and preferred stock, and distributed the net proceeds to the common stockholders. To calculate the liquidation value per share, divide the liquidation value by the number of common shares outstanding.

Although more reliable than book value, liquidation value is a worst-case valuation assessment. A company's common stock should be worth at least the amount generated per share at liquidation. Because liquidation value does not consider the earnings and cash flows the firm will generate in the future, it may provide misleading results for companies that have significant future earning potential.

Deciding Which Stock Valuation Approach to Use

To choose between valuation approaches, determine first whether the business has a significant going concern value. The value of a company is whichever is greater, the going concern or the liquidation value. Owners will either continue operating a company or liquidate it, depending on the alternative that generates the greater value for them.

The Yield on Common Stock

We calculate the yield for common stock by rearranging the terms in Equation 9-7, to arrive at the constant growth dividend model, as shown in Equation 9-11:

Formula for the Yield on Common Stock

$$k_s = \frac{D_1}{P_0} + g \qquad (9\text{-}11)$$

where: D_1 = Amount of the common stock dividend anticipated in one period
P_0 = Current market price of the common stock
g = Expected constant growth rate of dividends per period
k_s = Expected rate of return per period on this common stock investment

Equation 9-11 is also called the formula for an investor's total rate of return from common stock. The stock's **dividend yield** is the term D_1/P_0.

To demonstrate how Equation 9-11 works, suppose you found that the price of General Motors common stock today was $52 per share. Then suppose you believe that General Motors will pay a common stock dividend next year of $4.80 a share, and the dividend will grow each year at a constant annual rate of 4 percent. If you buy one share of General Motors common stock at the listed price of $52, your expected annual percent yield on your investment will be:

$$\begin{aligned} k_s &= \frac{D_1}{P_0} + g \\ &= \frac{\$4.80}{\$52} + .04 \\ &= .0923 + .04 \\ &= .1323 \text{ or } 13.23\% \end{aligned}$$

If your minimum required rate of return for General Motors common stock, considering its risk, were less than 13.23 percent, you would proceed with the purchase. Otherwise you would look for another stock that had an expected return appropriate for its level of risk.

What's Next

In this chapter we investigated valuation methods for bonds, preferred stock, and common stock. The valuation methods applied risk and return, and time value of money techniques learned in chapters 7 and 8, respectively. In the next chapter, we will explore capital budgeting techniques.

Summary

1. Explain the importance of bond and stock valuation.

When corporations contemplate selling securities in the open market to raise long-term funds, they do not want to undervalue the securities because they want to raise the most money possible. Likewise, would-be purchasers of securities use valuation methods to avoid paying more than the securities are worth.

2. Discuss the concept of securities valuation.

To value any security, we apply risk and return and time value of money techniques. In sum, the value of a security is the present value of the security's future cash flows. Bond cash flows are the periodic interest payments and the principal at maturity. Stock cash flows come from the future earnings that the assets produce for the firm, usually leading to cash dividend payments.

To value securities, investors and financial managers use a general valuation model to calculate the present value of the future cash flows of the security. That model incorporates risk and return and time value of money concepts.

3. Compute the market value and the yield to maturity of a bond.

The market value of a bond is the sum of the present values of the coupon interest payments plus the present value of the face value to be paid at maturity, given a market's required rate of return.

The yield to maturity of a bond (YTM) is the average annual rate of return that investors realize if they buy a bond for a certain price, receive the promised interest payments and principal on time, and reinvest the interest payments at the YTM rate.

A bond's market price and its YTM vary inversely. That is, when the YTM rises, the market price falls, and vice versa. When a bond has a YTM greater than its coupon rate, it sells at a discount to its face value. When the YTM is equal to the coupon rate, the market price equals the face value. When the YTM is less than the coupon rate, the bond sells at a premium over face value.

4. Compute the market value and expected yield of preferred stock.

The market value of preferred stock is the present value of the stream of preferred stock dividends, discounted at the market's required rate of return for that investment. Because the dividend cash flow stream is a perpetuity, we adapt the present value of a perpetuity formula, Equation 8-5, to value preferred stock.

The yield on preferred stock represents the annual rate of return that investors realize if they buy the stock for the current market price and then receive the promised dividend payments on time.

5. Compute the market value and expected yield of common stock.

The market value of common stock is estimated in a number of ways, including (1) finding the present value of all the future dividends the stock is expected to pay, discounted at the market's required rate of return for that stock; (2) finding the price implied, given the level of earnings per share and the appropriate P/E ratio; (3) estimating the value based on the book value of the firm's assets as recorded on the balance sheet; and (4) estimating the value of the firm's assets if they were liquidated on the open market and all claims on the firm were paid off.

The dividend growth model and the P/E valuation approaches assume the firm will be a going concern. That is, the model values the future cash flows that a firm's assets are expected to produce. The book value and liquidation value approaches concentrate on the current value of the firm's assets and liabilities without regard to future earning potential. The book value method is less reliable than the liquidation method because the book value typically uses the original asset and liability accounting values whereas the liquidation method uses market values.

The yield on common stock is the percent return on investment investors can expect if they purchase the stock at the prevailing market price and receive the expected cash flows.

You can access your FREE PHLIP/CW On–Line Study Guide, Current Events, and Spreadsheet Problem templates, which may correspond to this chapter either through the Prentice Hall Finance Center CD–ROM or by directly going to: **http://www.prenhall.com/gallagher.**

PHLIP/CW — *The Prentice Hall Learning on the Internet Partnership–Companion Web Site*

Equations Introduced in This Chapter

Equation 9-1. The Discounted Cash Flow Valuation Model:

$$V_0 = \frac{CF_1}{(1+k)^1} + \frac{CF_2}{(1+k)^2} + \frac{CF_3}{(1+k)^3} \cdots + \frac{CF_n}{(1+k)^n}$$

where: V_0 = Present value of the anticipated cash flows from the security, its current value
$CF_{1,2,3,\text{ and } n}$ = Cash flows expected to be received one, two, three, and so on up to n periods in the future
k = Discount rate, the required rate of return

Equation 9-2. The Bond Valuation Formula (Algebraic Version):

$$V_B = INT \times \left[\frac{1 - \frac{1}{(1+k_d)^n}}{k_d} \right] + \frac{M}{(1+k_d)^n}$$

where: V_B = Current market value of the bond
INT = Dollar amount of each periodic interest payment
n = Number of times the interest payment is received (which is also the number of periods until maturity)
M = Principal payment received at maturity
k_d = Required rate of return per period on the bond debt instrument

Equation 9-3. The Bond Valuation Formula (Table Version):

$$V_B = (INT \times PVIFA_{k,n}) + (M \times PVIF_{k,n})$$

where: $PVIFA_{k,n}$ = Present Value Interest Factor for an Annuity from Table IV
$PVIF_{k,n}$ = Present Value Interest Factor for a single amount from Table II

Equation 9-4. The Formula for the Present Value of Preferred Stock:

$$V_p = \frac{D_p}{k_p}$$

where: V_p = Current market value of the preferred stock
D_p = Amount of the preferred stock dividend
k_p = Required rate of return for this issue of preferred stock

Equation 9-5. Formula for the Yield on Preferred Stock:

$$k_p = \frac{D_p}{P_P}$$

where: k_p = Yield on investment that an investor can anticipate if the shares are purchased at the current market price P_P and if the preferred dividend D_p is paid forever
D_p = Amount of the preferred stock dividend
P_P = Current market price of the preferred stock

Equation 9-6. The DCF Valuation Model Applied to Common Stock:

$$P_0 = \frac{D_1}{(1+k_s)^1} + \frac{D_2}{(1+k_s)^2} + \frac{D_3}{(1+k_s)^3} \cdots + \frac{P_n}{(1+k_s)^n}$$

where: P_0 = Present value of the expected dividends, the current price of the common stock
$D_1, D_2, D_3,$ = Common stock dividends expected to be received at the end of periods 1, 2, 3, and so on until the stock is sold
P_n = Anticipated selling price of the stock in n periods
k_s = Required rate of return on the common stock investment

Equation 9-7. The Constant Growth Version of the Dividend Valuation Model:

$$P_0 = \frac{D_1}{k_s - g}$$

where: P_0 = Current price of the common stock
D_1 = Dollar amount of the common stock dividend expected one period from now
k_s = Required rate of return per period on this common stock investment
g = Expected constant growth rate per period of the company's common stock dividends

Equation 9-8. No-Growth Common Stock Dividend Valuation Formula:

$$P_0 = \frac{D_s}{k_s}$$

where: P_0 = Current price of the common stock
D_s = Amount of the common stock dividend, a constant amount
k_s = Required rate of return on this common stock investment

Equation 9-9. The P/E Model for Valuing Common Stock:

$$\text{Appropriate Stock Price} = \text{Industry P/E Ratio} \times \text{EPS}$$

Equation 9-10. The Book Value per Share of Common Stock:

$$\text{Book Value per Share} = \frac{(\text{Total Assets} - \text{Total Liabilities} - \text{Preferred Stock})}{\text{Number of Shares Outstanding}}$$

Equation 9-11. The Yield, or Total Return, on Common Stock:

$$k_s = \frac{D_1}{P_0} + g$$

where: D_1 = Amount of the common stock dividend anticipated in one period
P_0 = Current market price of the common stock
g = Expected constant growth rate of dividends per period
k_s = Expected rate of return per period on this common stock investment

Self-Test

ST-1. The Chrysler Corporation has issued a 10.95 percent annual coupon rate bond that matures December 31, 2020. The face value is $1,000. If the required rate of return on bonds of similar risk and maturity is 9 percent, and assuming the time now is January 1, 2000, what is the current value of Chrysler's bond?

ST-2. Chrysler's 10.95 percent coupon rate, 2020 bond is currently selling for $1,115. At this price what is the yield to maturity of the bond? Assume the time is January 1, 2000.

ST-3. McDonald's is offering preferred stock that pays a dividend of $1.93 a share. The dividend is expected to continue indefinitely. If your required rate of return for McDonald's preferred stock is 8 percent, what is the value of the stock?

ST-4. Quaker Oats Corporation's next annual dividend is expected to be $1.14 a share. Dividends have been growing at a rate of 6 percent a year and you expect this rate to continue indefinitely. If your required rate of return for this stock is 9 percent, what is the maximum price you should be willing to pay for it?

ST-5. Goodyear Corporation stock is currently selling for $38. The company's next annual dividend is expected to be $1.00 a share. Dividends have been growing at a rate of 5 percent a year and you expect this rate to continue indefinitely. If you buy Goodyear at the current price, what will be your yield, or total return?

FinCoach Practice Exercises

To help improve your grade and master the mathematical skills covered in this chapter, open FinCoach on the Prentice Hall Finance Center CD-ROM and work practice problems in the following categories: 1) Bond Valuation and 2) Stock Valuation.

Review Questions

1. Describe the general pattern of cash flows from a bond with a positive coupon rate.
2. How does the market determine the fair value of a bond?
3. What is the relationship between a bond's market price and its promised yield to maturity? Explain.
4. All other things held constant, how would the market price of a bond be affected if coupon interest payments were made semiannually instead of annually?
5. What is the usual pattern of cash flows for a share of preferred stock? How does the market determine the value of a share of preferred stock, given these promised cash flows?
6. Name two patterns of cash flows for a share of common stock. How does the market determine the value of the most common cash flow pattern for common stock?
7. Define the P/E valuation method. Under what circumstances should a stock be valued using this method?
8. Compare and contrast the book value and liquidation value per share for common stock. Is one method more reliable? Explain.

Build Your Communication Skills

CS-1. Check the current price of Texas Instruments Corporation stock or a stock of your choice in the financial press. Then research financial information from last year predicting how Texas Instruments stock (or the stock you chose) would fare and analyze what valuation methods the analysts used. Next, compare the current price with the analysts' price predictions. Prepare a brief oral report that you present to the class discussing your assessment of the analysts' stock valuation.

CS-2. Choose a stock in the financial press. Check its current price. Then estimate the value of the stock using the dividend growth model, Equation 9-7. Compare the value of the stock with its current price and prepare a short memo in which you explain why the two figures might differ.

Problems

9-1. Owen Meany is considering the purchase of a $1,000 Amity Island Municipal Bond. The city is raising funds for a much needed advertising campaign to promote its East Coast resort community. The stated coupon rate is 6 percent, paid annually. The bond will mature in 10 years. The yield to maturity for similar bonds in the market is 8 percent. *Bond Valuation*

   a. How much will the annual interest payments be?
   b. What is the market price of the bond today?
   c. Is the interest received on a municipal bond generally tax free?

9-2. Assume Disney Studios is offering a corporate bond with a face value of $1,000 and an annual coupon rate of 12 percent. The maturity period is 15 years. The interest is to be paid annually. The annual yield to maturity for similar bonds in the market is currently 8 percent. *Bond Valuation*

   a. What is the amount of interest to be paid annually for each bond?
   b. What is the value of this $1,000 bond today?
   c. What would be the present value of the interest and principal paid to holders of one of their bonds if the interest payments were made semiannually instead of annually?

9-3. After a major earthquake, the San Francisco Opera Company is offering zero coupon bonds to fund the needed structural repairs to its historic building. Buster Norton is considering the purchase of several of these bonds. The bonds have a face value of $2,000 and are scheduled to mature in 10 years. Similar bonds in the market have an annual yield to maturity of 12 percent. If Mr. Norton purchases three of these bonds today, how much money will he receive 10 years from today at maturity? *Bond Valuation*

9-4. Two best friends, Thelma and Louise, are making long-range plans for a road trip vacation to Mexico. They will embark on this adventure in five years, and want to invest during the five-year period to earn money for the trip. They decide to purchase a $1,000 Grand Canyon Oil Company bond with an annual coupon rate of 10 percent with interest to be paid semiannually. The bond will mature in five years. The yield to maturity of similar bonds is 8 percent. How much should they be willing to pay for the bond if they purchase it today? *Bonds with Semiannual Interest Payments*

9-5. Clancy Submarines, Inc., is offering $1,000 par value bonds for sale. The bonds will mature 10 years from today. The annual coupon interest rate is 12 percent, with payments to be made annually. James Hobson just purchased one bond at the current market price of $1,125. *YTM*

   a. Will the yield to maturity of this bond be greater than or less than the coupon interest rate? Answer this part without doing any calculations.
   b. To the nearest whole percent, what is the yield to maturity of Mr. Hobson's bond? You'll need to crunch some numbers for this part.

Pricing
    c. What would the yield to maturity have to be to make the market price of the bond equal to the face value? No number crunching is needed to answer.

YTM

9-6. A corporate bond has a face value of $1,000 and an annual coupon interest rate of 7 percent. Interest is paid annually. Of the original 20 years to maturity, only 10 years of the life of the bond remain. The current market price of the bond is $872. To the nearest whole percent, what is the yield to maturity of the bond today?

Preferred Stock Valuation

9-7. The new Shattuck Corporation will offer its preferred stock for sale in the very near future. These shares will have a guaranteed annual dividend of $10 per share. As you research the market, you find that similar preferred stock has an expected rate of return of 12 percent. If this preferred stock could be purchased today, what price per share would you expect to pay for it?

Preferred Stock k

9-8. Lucky Jackson knows that one share of Grand Prix Enterprises preferred stock sells for $20 per share on the open market. From its annual reports, he sees that Grand Prix pays an annual dividend of $1.75 per share on this preferred stock. What is the market's required rate of return on Grand Prix's stock?

Preferred Stock Valuation

9-9. Tiny Shipping Corporation is planning to sell preferred stock that will pay an annual dividend of $8 per share. The current expected rate of return from similar preferred stock issues is 13 percent.

    a. What price per share would you expect to have to pay to purchase this stock?
    b. If the stock is actually selling for $50 per share, what is the market's required rate of return for this stock?

Common Stock Valuation

9-10. China S. Construction, Inc. is in the business of building electrical power plants in the eastern United States. Jack Godell and the rest of the board members of the firm have just announced a $4 per share dividend on the corporation's common stock to be paid in one year. Because the quality of some of its recent projects is under attack by investigative television reporters, the expected constant dividend growth rate is only estimated to be 1 percent. The required rate of return for similar stocks in this industry is 16 percent.

    a. What is the present value of the expected dividends from one share of China S. Construction's common stock?
    b. What is the stock's dividend yield (D_1/P_0)?

Challenge Problem

9-11. The current listed price per share of a certain common stock is $15. The cash dividend expected from this corporation in one year is $2 per share. All market research indicates that the expected constant growth rate in dividends will be 4 percent per year in future years. What is the rate of return on this investment that an investor can expect if shares are purchased at the current listed price?

Common Stock Valuation

9-12. Golden Manufacturing Company is expected to pay a dividend of $8 per share of common stock in one year. The dollar amount of the dividends is expected to grow at a constant 3 percent per year in future years. The required rate of return from shares of similar common stock in the present environment is 14 percent.

    a. What would you expect the current market price of a share of Golden common stock to be?
    b. Assuming the cash dividend amount and the growth rate are accurate, what is the annual rate of return on your investment in Golden common stock if you purchased shares at the stock's actual listed price of $65 per share?

Book Value

9-13. Jack and Frank Baker know their piano renditions of lounge songs have limited appeal on the night club circuit, so they work part-time as investment consultants. They are researching relatively unknown corporations, one of which is Susie Diamond Enterprises. To get a quick idea of the value of SDE's common stock, they have taken the following numbers from the most recent financial statements.

Total Assets	$675,000
Total Liabilities	$120,000
250,000 Shares of Common Stock Issued	
100,000 Shares of Common Stock Outstanding	

a. What is the book value (net worth) of Susie Diamond Enterprises?
b. What is the book value per share?

9-14. The most recent balance sheet of Free Enterprise, Inc. follows.

Book Value, Liquidation Value, and P/E Methods

Free Enterprise, Inc.
Balance Sheet
December 31, 1999
(thousands of dollars)

Assets		Liabilities + Equity	
Cash	$ 4,000	Accounts Payable	$ 4,400
Accounts Receivable	10,000	Notes Payable	4,000
Inventory	13,000	Accrued Expenses	5,000
Prepaid Expenses	400	Total Current Liabilities	13,400
Total Current Assets	27,400	Bonds Payable	6,000
Fixed Assets	11,000	Common Equity	19,000
Total Assets	$38,400	Total Liabilities + Equity	$38,400

a. What was Free Enterprise's book value (net worth) at the beginning of 2000?
b. If the company had 750,000 shares of common stock authorized and 500,000 shares outstanding, what was the book value per share of common stock at the beginning of 2000?
c. Net income of Free Enterprise, Inc. was $5,610,000 in 1999. Calculate the earnings per share of Free Enterprise's common stock.
d. The P/E Ratio for a typical company in Free Enterprise, Inc.'s industry is estimated to be 6. Using the EPS from part c) above, calculate the price of one share of common stock at the beginning of 2000 assuming that Free Enterprise commands a P/E ratio value equal to that of an average company in its industry.
e. What would you infer about the company's total assets shown on the balance sheet when comparing this calculated stock price with the company's book value per share?
f. Calculate the liquidation value of Free Enterprise's common stock assuming the market value of the total assets is $50 million and the market value of total liabilities is $20 million, as estimated by your analyst.

9-15. Lucky Jackson is trying to choose from among the best of the three investment alternatives recommended to him by his full-service investment broker. The alternatives are:

Comprehensive Problem

a. The corporate bond of Star Mining Company has a face value of $1,000 and an annual coupon interest rate of 13 percent. The bond is selling in the market at $1,147.58. Of the original 20 years to maturity, only 16 years of the life of the bond remain.
b. The preferred stock of Supernova Minerals Company has a par value of $100 per share and it offers an annual dividend of $14 per share. The market price of the stock is $140 per share.
c. The common stock of White Dwarf Ores Company sells in the market at $300 per share. The company paid a dividend of $39 per share yesterday. The company is expected to grow at 3 percent per annum in the future.

Which of the three alternatives should Lucky choose? Remember the priority of claims for bondholders, preferred stockholders, and common stockholders from chapters 1 and 4.

Answers to Self-Test

ST-1. The present value of Chrysler's 10.95 percent 2020 bond can be found using Equation 9-2:

$$V_B = INT \times \left[\frac{1 - \frac{1}{(1 + k_d)^n}}{k_d} \right] + \frac{M}{(1 + k_d)^n}$$

Face value is $1,000
The coupon interest payment is 10.95 percent of $1,000, or $109.50
n is 2020 − 1999 = 21
$k_d = 9\%$

$$V_B = \$109.50 \times \left[\frac{1 - \frac{1}{(1 + .09)^{21}}}{.09} \right] + \frac{\$1,000}{(1 + .09)^{21}}$$

$$= (\$109.50 \times 9.29224) + \left(\frac{\$1,000}{6.108801} \right)$$

$$= \$1017.50 + \$163.70$$

$$= \$1,181.20$$

So the present value of the bond is $1,181.20.

ST-2. The bond's YTM is found by trial and error. We know that the bond has a price of $1,115, face value of $1,000, coupon interest payment is $109.50 (10.95% of $1,000), and matures in 21 years (2020 − 1999 = 21). Now we can find that value of k_d that produces a P_b of $1,115. Use Equation 9-2 and solve for P_B.
First try k = 9%:

$$V_B = \$109.50 \times \left[\frac{1 - \frac{1}{(1 + .09)^{21}}}{.09} \right] + \frac{\$1,000}{(1 + .09)^{21}}$$

$$= (\$109.50 \times 9.29224) + \left(\frac{\$1,000}{6.108801} \right)$$

$$= \$1017.50 + \$163.70$$

$$= \$1,181.20 > 1,115$$

$1,181.20 is too high. Try again using a higher yield (remember, bond prices and yields vary inversely).
Second try at $k_d = 10\%$:

$$V_B = \$109.50 \times \left[\frac{1 - \frac{1}{(1 + .10)^{21}}}{.10} \right] + \frac{\$1,000}{(1 + .10)^{21}}$$

$$= (\$109.50 \times 8.64869) + \left(\frac{\$1,000}{7.40025} \right)$$

$$= \$947.03 + \$135.13$$

$$= \$1,082.60 < 1,115$$

$1,082.16 is too low. Try again using a lower yield.

Third try at $k_d = 9.65\%$:

$$V_B = \$109.50 \times \left[\frac{1 - \frac{1}{(1 + .0965)^{21}}}{.0965} \right] + \frac{\$1,000}{(1 + .0965)^{21}}$$

$$= (\$109.50 \times 8.86545) + \left(\frac{\$1,000}{6.92120} \right)$$

$$= \$970.77 + \$144.48$$

$$= \$1,115.25$$

At $k_d = 9.65\%$ the calculated value of P_B is within $.25 of the current market price. We conclude the bond's YTM is 9.65 percent.[4]

ST-3. Equation 9-5 is used to find the value of preferred stock as follows:

$$V_P = \frac{D_p}{k_p} \qquad (9\text{-}5)$$

D_p is $1.93 and k_p is 8 percent.

$$V_P = \frac{\$1.93}{.08}$$

$$= \$24.125, \text{ or } \$24\tfrac{1}{8}$$

(Stock prices are usually quoted in eighths.)

ST-4. The maximum price you are willing to pay for Quaker Oats is what it is worth to you, or its value. Because the characteristics of the stock fit the constant dividend growth model, use Equation 9-7 to compute the value.

$$P_0 = \frac{D_1}{k_s - g} \qquad (9\text{-}7)$$

D_1 is $1.14, k_e is 9 percent, and g is 6 percent.
Given these conditions, the value of Quaker Oats stock is:

$$P_0 = \frac{\$1.14}{.09 - .06}$$

$$= \frac{\$1.14}{.03}$$

$$= \$38$$

ST-5. The yield on common stock can be found using Equation 9-10:

$$k_s = \frac{D_1}{P_0} + g \qquad (9\text{-}10)$$

D_1 is $1.00, P_0 is $38, and g is 5 percent. Given these conditions, the yield on Goodyear common stock is as follows:

$$k_s = \frac{\$1.00}{\$38} + .05$$

$$= .0263 + .05$$

$$= .0763, \text{ or } 7.63\%$$

---

[4]The exact YTM, found using a financial calculator, is 9.652590645 percent.

APPENDIX 9A

Common Stock Valuation: Supernormal Growth

In addition to the constant growth and no-growth common stock dividend cash flow patterns that we discussed in the chapter, some companies have very high growth rates, known as supernormal growth of the cash flows. Valuing the common stock of such companies presents a special problem because high growth rates cannot be sustained indefinitely. A young high-technology firm, such as Netscape, for instance, may be able to grow at a 40 percent rate per year for a few years, but that growth must slow down because it is not sustainable given the population and productivity growth rates. In fact, if the firm's growth rate did not slow down, its sales would surpass the gross domestic product of the entire nation over time. Why? The company has a 40 percent growth rate that will compound annually, whereas the gross domestic product may grow at a 4 percent compounded average annual growth rate.

The constant dividend growth model for common stock, Equation 9-6, then, must be adjusted for those cases where a company's dividend grows at a supernormal rate that will not be sustained over time. We do this by dividing the projected dividend cash flow stream of the common stock into two parts: the initial supernormal growth period and the next period where normal and sustainable growth is expected. We then calculate the present value of the dividends during the fast-growth time period first. Then we solve for the present value of the dividends during the constant growth period that are a perpetuity. The sum of these two present values determines the current value of the stock.

To illustrate, suppose Supergrowth Corporation is expected to pay an annual dividend of $2 per share one year from now, and that this dividend will grow at a 30 percent annual rate during each of the following four years (taking us to the end of year 5). After this supernormal growth period, the dividend will grow at a sustainable 5 percent rate each year beyond year 5. The cash flows are shown in Figure 9A-1.

The valuation of a share of Supergrowth Corporation's common stock is described in the following three steps.

Step 1. Add the present values of the dividends during the supernormal growth period. Assume that the required rate of return, k_s, is 14 percent.

$$
\begin{aligned}
\$2.00 \times 1/1.14 &= \$\ 1.75 \\
\$2.60 \times 1/1.14^2 &= \$\ 2.00 \\
\$3.38 \times 1/1.14^3 &= \$\ 2.28 \\
\$4.39 \times 1/1.14^4 &= \$\ 2.60 \\
\$5.71 \times 1/1.14^5 &= \underline{\$\ 2.97} \\
\Sigma &= \$11.60
\end{aligned}
$$

Step 2. Calculate the sum of the present values of the dividends during the normal growth period, from t_6 through infinity in this case. To do this, pretend for a moment that t_6 is t_0. The present value of the dividend growing at the constant rate of 5 percent to perpetuity could be computed using Equation 9-7.

$$P_0 = \frac{D_1}{k_s - g}$$

Figure 9A-1 Timeline of Supergrowth Common Stock Dividend with Initial Supernormal Growth

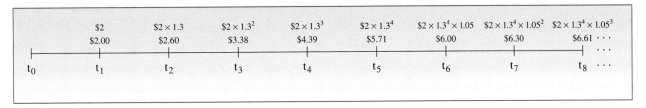

Substituting our values we would have:

$$P_0 = \frac{\$6.00}{.14 - .05} = \$66.67$$

Because the $6.00 dividend actually occurs at t_6 instead of t_1, the $66.67 figure is not a t_0 value, but rather a t_5 value.

It, therefore, needs to be discounted back five years at our required rate of return of 14 percent. This gives us $66.67 \times 1/1.14^5 = \$34.63$. The result of $34.63 is the present value of the dividends from the end of year 6 through infinity.

Step 3. Finally we add the present values of the dividends from the supernormal growth period and the normal growth period. In our example we add $11.60 + $34.63 = $46.23. The sum of $46.23 is the appropriate market price of Supergrowth Corporation's common stock, given the projected dividends and the 14 percent required rate of return on those dividends.

**PART 3:
LONG-TERM FINANCIAL
MANAGEMENT DECISIONS**

CHAPTER 10

CAPITAL BUDGETING DECISION METHODS

Everything is worth what its purchaser will pay for it.
—Publilius Syrus

TO ACCEPT OR NOT TO ACCEPT, THAT WAS THE QUESTION

In June 1998 PanAmSat Corporation celebrated the 10-year anniversary of the launch of the PAS-1 Atlantic Ocean Region satellite, the first international communications satellite ever launched by a commercial company. With PAS-1's launch on June 15, 1988, PanAmSat broke a global satellite monopoly and paved the way for today's multibillion-dollar international satellite communications industry.

"How times have changed. When we launched PAS-1, Intelsat was the global monopoly and PanAmSat had no customers, no revenue and few prospects. The pundits scoffed at the viability of private-sector international satellite service providers, particularly serving emerging markets such as Latin America," said Frederick A. Landman, PanAmSat's president and chief executive officer. "Today, the global satellite services industry is one of the fastest-growing telecommunications markets in the world. PanAmSat operates 16 satellites and has hundreds of customers worldwide. And most importantly, commercial companies and their customers, not a monopoly and its owners, are leading the industry into the 21st century."

In September 1995, when the PanAmSat Corporation announced that it planned to spend $800 million to expand its network of communications satellites,

CHAPTER OBJECTIVES

After reading this chapter, you should be able to:

- Explain the capital budgeting process.
- Calculate the payback period, net present value, and internal rate of return for a proposed capital budgeting project.
- Describe capital rationing and how firms decide which projects to select.
- Measure the risk of a capital budgeting project.
- Explain risk-adjusted discount rates.

financial analysts at the PanAmSat Corporation had to determine whether the proposed expansion project made financial sense for the firm. Because the project was a major risk—the company's future was at stake—the financial analysts could not just recommend that the firm spend $800 million willy-nilly. Analysts decided to accept the project based on objective financial decision methods, concluding that, despite the risks, the project's returns would justify the $800 million initial investment.

In this chapter you will learn some of the techniques that financial analysts like those at PanAmSat use to analyze long-term investment opportunities.

Source: PanAmSat Celebrates 10-year Anniversary of PAS-1, The World's First Private International Satellite, Press Release, PanAmSat, June 15, 1998.
⟨http://www.panamsat.com/media/p061598.htm⟩.

Chapter Overview

In this chapter we'll look at the decision methods that analysts use to approve investment projects and how they account for a project's risk. Investment projects such as the PanAmSat expansion project fuel a firm's success, so effective project selection often determines a firm's value. The decision methods for choosing investment projects, then, are some of the most important tools financial managers use.

We begin this chapter by looking at the capital budgeting process and three capital budgeting decision methods: *the payback method, the net present value method,* and the *internal rate of return method.* Then we discuss how to select projects when firms limit their budget for capital projects, a practice called *capital rationing.* Finally, we look at how firms measure and compensate for the risk of capital budgeting projects.

The Capital Budgeting Process

All businesses budget for capital projects—an airline that considers purchasing new planes, a production studio that decides to buy new film cameras, or a pharmaceutical company that invests in researching and developing a new drug. **Capital budgeting** is the process of evaluating proposed large, long-term investment projects. The projects may be the purchase of fixed assets, investments in research and development, advertising, or intellectual property. For instance, Microsoft's development of *Windows 98*™ software was a capital budgeting project.

Long-term projects may tie up cash resources, time, and additional assets. They can also affect a firm's performance for years to come. As a result, careful capital budgeting is vital to ensure that the proposed investment will add value to the firm.

Before we discuss the specific decision methods for selecting investment projects, we briefly examine capital budgeting basics: capital budgeting decision practices, types of capital budgeting projects, the cash flows associated with such projects, and the stages of the capital budgeting process.

Decision Practices

Financial managers apply two decision practices when selecting capital budgeting projects: *accept/reject* and *ranking.* The accept/reject decision focuses on the question of whether the proposed project would add value to the firm or earn a rate of return that is acceptable to the company. The ranking decision ranks competing projects in order of desirability in order to choose the best one(s).

The accept/reject decision practice determines whether a project is acceptable in light of the firm's financial objectives. That is, if a project meets the firm's basic risk and return requirements, it will be accepted. If not, it will be rejected.

Ranking compares projects to a standard measure and orders the projects based on how well they meet the measure. If, for instance, the standard is how quickly the project pays off the initial investment, then the project that pays off the investment most rapidly would be ranked first. The project that paid off most slowly would be ranked last.

Types of Projects

Firms invest in two categories of projects: independent projects and mutually exclusive projects. **Independent projects** do not compete with each other. A firm may accept none, some, or all from among a group of independent projects. Say, for example, that a firm is able to raise funds for all worthwhile projects it identifies. The firm is considering two new projects—a new telephone system and a warehouse. The telephone system and the warehouse would be independent projects. Accepting one does not preclude accepting the other.

In contrast, **mutually exclusive projects** compete against each other. The best project from among a group of acceptable mutually exclusive projects is selected. For example, if a company needed only one copier, a proposal to buy a Xerox™ copier and a second proposal to buy a Toshiba™ copier would indicate that these two projects are mutually exclusive.

Capital Budgeting Cash Flows

Although decision makers examine accounting information such as sales and expenses, capital budgeting decision makers base their decisions on relevant cash flows associated with a project. The relevant cash flows are **incremental cash flows**—cash flows that will occur if an investment is undertaken, but won't occur if it isn't.

For example, imagine that you are an animator who helps to create children's films. Your firm is considering whether to buy a computer to help you design your characters more quickly. The computer would enable you to create two characters in the time it now takes you to draw one, thereby increasing your productivity.

The incremental cash flows associated with the animation project are (1) the initial investment in the computer and (2) the *additional* cash the firm receives because you will double your animation output for the life of the computer. These cash flows will occur if the new computer is purchased, and will not occur if it is not purchased. These are the cash flows, then, that affect the capital budgeting decision.

Estimating cash flows, the subject of chapter 11, is a major component of capital budgeting decisions, as we see in the next section.

Stages in the Capital Budgeting Process

The capital budgeting process has four major stages:

1. finding projects
2. estimating the incremental cash flows associated with projects

3. evaluating and selecting projects
4. implementing and monitoring projects

This chapter focuses on stage 3—how to evaluate and choose investment projects. We assume that the firm has found projects in which to invest and has estimated the projects' cash flows effectively.

CAPITAL BUDGETING DECISION METHODS

The three formal capital budgeting decision methods we are going to examine are *payback, net present value,* and *internal rate of return.* Let's begin with the payback method.

> **Take Note**
>
> *Capital budgeting techniques are usually used only for projects with large cash outlays. Small investment decisions are usually made by the "seat of the pants." For instance, if your office supply of pencils is running low, you order more—you know that the cost of buying pencils is justified without undergoing capital budgeting analysis.*

The Payback Method

One of the simplest capital budgeting decision methods is the payback method. To use it, analysts find a project's **payback period**—the number of time periods it will take before the cash inflows of a proposed project equal the amount of the initial project investment (a cash outflow). In the PanAmSat example discussed at the beginning of this chapter, the analysts would determine how many years it would take to recoup the initial $800 million investment.

How to Calculate the Payback Period To calculate the payback period, simply add up a project's positive cash flows, one period at a time, until the sum equals the amount of the project's initial investment. That number of time periods is the payback period.

To illustrate, imagine that you work for a firm called the AddVenture Corporation. AddVenture is considering two investment proposals, Projects X and Y. The initial investment for both projects is $5,000. The finance department estimates that the projects will generate the following cash flows:

	Cash Flows				
	FOR INITIAL INVESTMENT	END OF YEAR 1	END OF YEAR 2	END OF YEAR 3	END OF YEAR 4
Project X	−$5,000	$2,000	$3,000	$500	$0
Project Y	−$5,000	$2,000	$2,000	$1,000	$2,000

By analyzing the cash flows, we see that Project X has a payback period of two years because the project's initial investment of $5,000 (a cash flow of −$5,000) will be recouped (by offsetting positive cash flows) at the end of year 2. In comparison, Project Y has a payback of three years.

Payback Method Decision Rule To apply the payback decision method, firms must first decide what payback time period is acceptable for long-term projects. In our case, if the firm sets two years as its required payback period, it will accept Project X but reject Project Y. If, however, the firm allows a three-year payback period, then both projects would be accepted if they were independent.

Source: DILBERT © 1996 United Feature Syndicate, Inc. Reprinted by permission.

Problems with the Payback Method The payback method is used in practice because of its simplicity. But it does not consider cash flows that occur after the payback period. For example, suppose Project Y's cash flows were $10 million instead of $2,000 in year 4. It wouldn't make any difference. Project Y would still be rejected under the payback method if the company's policy were to demand a payback of no more than two years. Failing to look beyond the payback period can lead to poor business decisions, as shown.

Another deficiency of the payback method is that it does not consider the time value of money. For instance, compare the cash flows of Projects A and B:

	Cash Flows		
	FOR INITIAL INVESTMENT	END OF YEAR 1	END OF YEAR 2
Project A	−$5,000	$3,000	$2,000
Project B	−$5,000	$2,000	$3,000

Assuming the required payback period is two years, we see that both projects are equally preferable—they both pay back the initial investment in two years. However, time value of money analysis indicates that if both projects are equally risky, then Project A is better than Project B because the firm will get more money sooner. Nevertheless, timing of cash flows makes no difference in the payback computation.

Because the payback method does not factor in the time value of money, nor the cash flows after the payback period, its usefulness is limited. Although it does provide a rough measure of a project's liquidity and can help to supplement other techniques, financial managers should not rely on it as a primary decision method.

The Net Present Value (NPV) Method

The project selection method that is most consistent with the goal of owner wealth maximization is the net present value method. The **net present value (NPV)** of a capital budgeting project is *the dollar amount of the change in the value of the* *firm as a result of undertaking the project.* The change in firm value may be positive, negative, or zero, depending on the NPV value.

If a project has an NPV of zero, it means that the firm's overall value will not change if the new project is adopted. Why? Because the new project is expected to generate exactly the firm's required rate of return—no more and no less. A positive NPV means that the firm's value will increase if the project is adopted because the new project's estimated return exceeds the firm's required rate of return. Conversely, a negative NPV means the firm's value will decrease if the new project is adopted because the new project's estimated return is less than what the firm requires.

Calculating NPV To calculate the net present value of a proposed project, we add the present value of a project's projected cash flows and then subtract the amount of the initial investment.[1] The result is a dollar figure that represents the change in the firm's value if the project is undertaken.

Formula for Net Present Value (NPV), Algebraic Version (10-1a)

$$NPV = \left(\frac{CF_1}{(1+k)^1}\right) + \left(\frac{CF_2}{(1+k)^2}\right) \ldots + \left(\frac{CF_n}{(1+k)^n}\right) - \text{Initial Investment}$$

where: CF = Cash flow at the indicated times
k = Discount rate, or required rate of return for the project
n = Life of the project measured in the number of time periods

To use financial tables to solve present value problems, write the NPV formula as follows:

The Formula for NPV, Table Version (10-1b)

$$NPV = CF_1(PVIF_{k,1}) + CF_2(PVIF_{k,2}) \ldots - CF_n(PVIF_{k,n}) - \text{Initial Investment}$$

where: PVIF = Present Value Interest Factor
k = Discount rate, or required rate of return for the project
n = Life of the project measured in the number of time periods

To use the TI BAII Plus financial calculator to solve for NPV, switch the calculator to the spreadsheet mode, enter the cash flow values in sequence, then the discount rate, depicted as I on the TI BAII Plus calculator, then compute the NPV.

For simplicity, we use only the algebraic equation and the financial calculator to calculate NPV. If you prefer using financial tables, simply replace Equation 10-1a with Equation 10-1b.[2]

To show how to calculate NPV, let's solve for the NPVs of our two earlier projects, Projects X and Y. First, note the following cash flows of Projects X and Y:

---

[1] We use many of the time value of money techniques learned in chapter 8 to calculate NPV.
[2] If the future cash flows are an annuity, Equations 10-1a and 10-1b can be modified to take advantage of the present value of annuity formulas discussed in chapter 8, Equations 8-4a and 8-4b.

Cash Flows

	For Initial Investment	End of Year 1	End of Year 2	End of Year 3	End of Year 4
Project X	−$5,000	$2,000	$3,000	$500	$0
Project Y	−$5,000	$2,000	$2,000	$1,000	$2,000

Assume the required rate of return for Projects X and Y is 10 percent. Now we have all the data we need to solve for the NPV of Projects X and Y.

Applying Equation 10-1a and assuming a discount rate of 10 percent, the NPV of Project X follows:

$$\begin{aligned} \text{NPV}_x &= \left(\frac{\$2,000}{(1+.10)^1}\right) + \left(\frac{\$3,000}{(1+.10)^2}\right) + \left(\frac{\$500}{(1+.10)^3}\right) - \$5,000 \\ &= \frac{\$2,000}{1.1} + \frac{\$3,000}{1.21} + \frac{\$500}{1.331} - \$5,000 \\ &= \$1,818.1818 + \$2,479.3388 + \$375.6574 - \$5,000 \\ &= -\$326.82 \end{aligned}$$

We may also find Project X's NPV with the financial calculator at a 10 percent discount rate as follows:

TI BAI I PLUS FINANCIAL CALCULATOR SOLUTIONS
Project X NPV

Keystrokes	Display
CF	CF_0 = old contents
2nd CLR Work	CF_0 = 0.00
5000 +/− ENTER	CF_0 = −5,000.00
↓ 2000 ENTER	C01 = 2,000.00
↓	F01 = 1.00
↓ 3000 ENTER	C02 = 3,000.00
↓	F02 = 1.00
↓ 500 ENTER	C03 = 500.00
↓	F03 = 1.00
NPV	I = 0.00
10 ENTER	I = 10.00
↓ CPT	NPV = −326.82

The preceding calculations show that at a 10 percent discount rate, an initial cash outlay of $5,000 and cash inflows of $2,000, $3,000, and $500 at the end of years 1, 2, and 3, the NPV for Project X is −326.82.

To find the NPV of Project Y, we apply Equation 10-1a as follows:

$$NPV_y = \left(\frac{\$2{,}000}{(1+.10)^1}\right) + \left(\frac{\$2{,}000}{(1+.10)^2}\right) + \left(\frac{\$1{,}000}{(1+.10)^3}\right) + \left(\frac{\$2{,}000}{(1+.10)^4}\right) - \$5{,}000$$

$$= \frac{\$2{,}000}{1.1} + \frac{\$2{,}000}{1.21} + \frac{\$1{,}000}{1.331} + \frac{\$2{,}000}{1.4641} - \$5{,}000$$

$$= \$1{,}818.1818 + \$1{,}652.8926 + \$751.3148 + \$1{,}366.0269 - \$5{,}000$$

$$= \$588.42$$

Using the financial calculator, we solve for Project Y's NPV at a 10 percent discount rate as follows:

TI BAII PLUS FINANCIAL CALCULATOR SOLUTIONS
Project Y NPV

KEYSTROKES	DISPLAY
CF	CF_0 = old contents
2nd CLR Work	$CF_0 = 0.00$
5000 +/− ENTER	$CF_0 = -5{,}000.00$
↓ 2000 ENTER	$C01 = 2{,}000.00$
↓ 2 ENTER	$F01 = 2.00$
↓ 1000 ENTER	$C02 = 1{,}000.00$
↓	$F02 = 1.00$
↓ 2000 ENTER	$C03 = 2{,}000$
↓	$F03 = 1.00$
NPV	$I = 0.00$
10 ENTER	$I = 10.00$
↓ CPT	$NPV = 588.42$

Our calculations show that with a required rate of return of 10 percent, an initial cash outlay of $5,000, and positive cash flows in years 1 through 4 of $2,000, $2,000, $1,000, and $2,000, respectively, the NPV for Project Y is $588.42. If we compare Project X's NPV of −326.82 to Project Y's NPV of $588.42, we see that Project Y would add value to the business and Project X would decrease the firm's value.

NPV Decision Rules NPV is used in two ways: (1) to determine if independent projects should be accepted or rejected (the accept/reject decision); and (2) to compare acceptable mutually exclusive projects (the ranking decision). The rules for these two decisions are:

- NPV Accept/Reject Decision—A firm should accept all independent projects having NPVs greater than or equal to zero. Projects with positive NPVs will add to the value of the firm if adopted. Projects with NPVs of zero will not alter the firm's value but (just) meet the firm's requirements. Projects with NPVs less than zero should be rejected because they will reduce the firm's value if adopted. Applying this decision rule, Project X would be rejected and Project Y would be accepted.

- NPV Ranking Decision—The mutually exclusive project with the highest positive NPV should be ranked first, the next highest should be ranked second, and so on. Under this decision rule, if the two projects in our previous example were mutually exclusive, Project Y would be ranked first and Project X second. (Not only is Project X second in rank here, but it is unacceptable because it has a negative NPV.)

The NPV Profile The k value is the cost of funds used for the project. It is the discount rate used in the NPV calculation because the cost of funds for a given project is that project's required rate of return. The relationship between the NPV of a project and k is inverse—the higher the k, the lower the NPV and the lower the k, the higher the NPV.[3]

We discussed the inverse relationship between k and present value (PV) in chapter 8, page 182a.

Because a project's NPV varies inversely with k, financial managers need to know how much the value of NPV will change in response to a change in k. If k is incorrectly specified, what appears to be a positive NPV could in fact be a negative NPV and vice versa—a negative NPV could turn out to be positive. Mutually exclusive project rankings could also change if an incorrect k value is used in the NPV computations.[4]

To see how sensitive a project's NPV value is to changes in k, analysts often create an NPV profile. The **NPV profile** is a graph that shows how a project's NPV changes when different discount rate values are used in the NPV computation.

Building an NPV profile is straightforward. First, the NPV of the project is calculated at a number of different discount rates. Then the results are plotted on the graph, with k values on one axis and the resulting NPV values on the other. If more than one project is included on the graph, the process is repeated for each project until all are depicted. To illustrate, we will build an NPV profile of Projects X and Y. We will plot Project X and then Project Y on the graph.

To begin, we first calculate the NPV of Project X with a number of different discount rates. The different k values may be chosen arbitrarily. For our purposes, let's use 0 percent, 5 percent, 10 percent, 15 percent, and 20 percent. The results of Project X's NPV calculations follow:

Discount Rate	Project X NPV
0%	$500.00
5%	$57.77
10%	−$326.82
15%	−$663.68
20%	−$960.65

---

[3] We are assuming here that the project is a typical one, meaning that it has an initial negative cash flow, the initial investment, followed by all positive cash flows. It is possible that if a project has negative cash flows in the future the relationship between NPV and k might not be inverse.

[4] The estimation of the cost of funds used for capital budgeting projects is covered in chapter 13.

> **Take Note**
>
> *The NPV profile line is curved, not straight. The curve is steepest at low discount rates and gets more shallow at higher rates. This shape occurs because discounting is the inverse of the exponential compounding phenomenon, as described in chapter 8.*

Now Project X's NPV values may be plotted on the NPV profile graph. Figure 10-1 shows the results.

When the data points are connected in Figure 10-1, we see how the NPV of Project X varies with the discount rate changes. The graph shows that with a k of about 5.7 percent, the value of the project's NPV is zero. At that discount rate, then, Project X would provide the firm's required rate of return, no more and no less.

Next, we add project Y to the NPV profile graph. We calculate the NPV of Project Y at a number of different discount rates, 0 percent, 5 percent, 10 percent, 15 percent, and 20 percent. The results follow:

Discount Rate	Project Y NPV
0%	$2,000.00
5%	$1,228.06
10%	$588.42
15%	$52.44
20%	−$401.23

Figure 10-2 shows these NPV values plotted on the NPV profile graph.

Notice in Figure 10-2 that Project Y's NPV profile falls off more steeply than Project X's. This indicates that Project Y is more sensitive to changes in the discount rate than is Project X. Project X's NPV becomes negative and, thus, the project is unacceptable when the discount rate rises above about 6 percent. Project Y's NPV becomes negative and, thus, the project is unacceptable when the discount rate rises above about 16 percent.

Figure 10-1 NPV Profile of Project X

The NPV profile shows how the NPV of Project X varies inversely with the discount rate, k. Project X's NPV is highest ($500) when the discount rate is zero. Its NPV is lowest (−$960.65) when the discount rate is 20 percent.

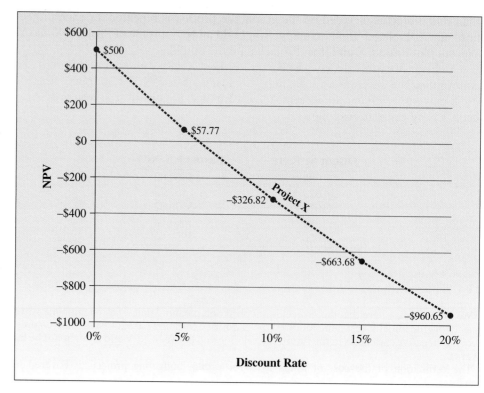

244 Part Three Long-Term Financial Management Decisions

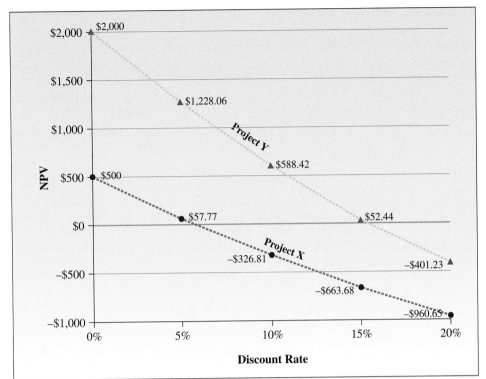

Figure 10-2 **NPV Profile of Projects X and Y**

This NPV profile shows how the NPVs of two capital budgeting projects, Projects X and Y, vary inversely with the discount rate, k.

Problems with the NPV Method Although the NPV method ensures that a firm will choose projects that add value to a firm, it suffers from two practical problems. First, it is difficult to explain NPV to people who are not formally trained in finance. Few nonfinance people understand phrases such as "the present value of future cash flows" or "the change in a firm's value given its required rate of return." As a result, many financial managers have difficulty using NPV analysis persuasively.

A second problem is that the NPV method results are in dollars, not percentages. Many owners and managers prefer to work with percentages because percentages can be easily compared to other alternatives—Project 1 has a 10 percent rate of return compared to Project 2's 12 percent return rate. The next method we discuss, the internal rate of return, provides results in percentages.

The Internal Rate of Return (IRR) Method

The **internal rate of return (IRR)** is the estimated rate of return for a proposed project, given the project's incremental cash flows. Just like the NPV method, the IRR method considers all cash flows for a project and adjusts for the time value of money. However, the IRR results are expressed as a percentage, not a dollar figure.

When capital budgeting decisions are made using the IRR method, the IRR of the proposed project is compared to the rate of return management requires for the project. The required rate of return is often referred to as the **hurdle rate.** If the project's IRR is greater than or equal to the hurdle rate (jumps over the hurdle), the project is accepted.

Chapter 10 *Capital Budgeting Decision Methods* **245**

The IRR calculation is the same trial-and-error procedure we used to find a bond's yield to maturity in chapter 9, pages 212–214.

Calculating Internal Rate of Return: Trial-and-Error Method If the present value of a project's incremental cash flows were computed using management's required rate of return as the discount rate, and the result exactly equaled the cost of the project, then the NPV of the project would be zero. When NPV equals zero, the required rate of return, or discount rate used in the NPV calculation, is the projected rate of return, IRR. To calculate IRR, then, we reorder the terms in Equation 10-1a to solve for a discount rate, k, that results in an NPV of zero.

The formula for finding IRR, Equation 10-2, follows:

Formula for IRR (10-2)

$$NPV = 0 = \left(\frac{CF_1}{(1+k)^1}\right) + \left(\frac{CF_2}{(1+k)^2}\right) \cdots + \left(\frac{CF_n}{(1+k)^n}\right) - \text{Initial Investment}$$

To find the IRR of a project using Equation 10-2, fill in the cash flows, the n values, and the initial investment figure. Then choose different values for k (all the other values are known) until the result equals zero. The IRR is the k value that causes the left-hand side of the equation, the NPV, to equal zero.

To illustrate the process, let's calculate the internal rate of return for Project X. Recall that the cash flows associated with Project X were as follows:

Cash Flows

	For Initial Investment	Year 1	Year 2	Year 3	Year 4
Project X	−$5,000	$2,000	$3,000	$500	$0

First, we insert Project X's cash flows and the times they occur into Equation 10-2.

$$0 = \left(\frac{\$2,000}{(1+k)^1}\right) + \left(\frac{\$3,000}{(1+k)^2}\right) + \left(\frac{\$500}{(1+k)^3}\right) - \$5,000$$

Next, we try various discount rates until we find the value of k that results in an NPV of zero. Let's begin with a discount rate of 5 percent.

$$0 = \left(\frac{\$2,000}{(1+.05)^1}\right) + \left(\frac{\$3,000}{(1+.05)^2}\right) + \left(\frac{\$500}{(1+.05)^3}\right) - \$5,000$$

$$= \left(\frac{\$2,000}{1.05}\right) + \left(\frac{\$3,000}{1.1025}\right) + \left(\frac{\$500}{1.157625}\right) - \$5,000$$

$$= \$1,904.76 + \$2,721.09 + \$431.92 - \$5,000$$

$$= \$57.77$$

This is close, but not quite zero. Let's try a second time, using a discount rate of 6 percent.

$$0 = \left(\frac{\$2,000}{(1+.06)^1}\right) + \left(\frac{\$3,000}{(1+.06)^2}\right) + \left(\frac{\$500}{(1+.06)^3}\right) - \$5,000$$

$$= \left(\frac{\$2,000}{1.06}\right) + \left(\frac{\$3,000}{1.1236}\right) + \left(\frac{\$500}{1.191016}\right) - \$5,000$$

$$= \$1,886.79 + \$2,669.99 + \$419.81 - \$5,000$$

$$= -\$23.41$$

This is close enough for our purposes. We conclude the IRR for Project X is slightly less than 6 percent.

Although calculating IRR by trial and error is time-consuming, the guesses do not have to be made entirely in the dark. Remember that the discount rate and NPV are inversely related. When an IRR guess is too high, the resulting NPV value will be too low. When an IRR guess is too low, the NPV will be too high.

Calculating Internal Rate of Return: Financial Calculator Finding solutions to IRR problems with a financial calculator is simple. After clearing previous values, enter the initial cash outlay and other cash flows and compute the IRR value.

The TI BAII PLUS financial calculator keystrokes for finding the IRR for Project X are shown next.

We could estimate the IRR more accurately through more trial and error. However, carrying the IRR computation to several decimal points may give a false sense of accuracy, as in the case where the cash flows estimates are in error.

TI BAII PLUS FINANCIAL CALCULATOR SOLUTION
IRR

Keystrokes	Display
CF	CF_0 = old contents
2nd CLR Work	CF_0 = 0.00
5000 +/− ENTER	−5,000.00
↓ 2000 ENTER	C01 = 2,000.00
↓	F01 = 1.00
↓ 3000 ENTER	C02 = 3,000.00
↓	F02 = 1.00
↓ 500 ENTER	C03 = 500.00
↓	F03 = 1.00
IRR	IRR = 0.00
CPT	IRR = 5.71

We see that with an initial cash outflow of $5,000, and Project X's estimated cash inflows in years 1 through 3, the IRR is 5.71 percent.

IRR and the NPV Profile Notice in Figure 10-2 that the point where Project X's NPV profile crosses the zero line is, in fact, the IRR (5.71 percent). This is no accident. When the required rate of return, or discount rate, equals the expected rate of return, IRR, then NPV equals zero. Project Y's NPV profile crosses the zero line just below 16 percent. If you use your financial calculator, you will find that the Project Y IRR is 15.54 percent. If an NPV profile graph is available, you can always find a project's IRR by locating the point where a project's NPV profile crosses the zero line.

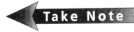

If you are using a financial calculator other than the TI BAII PLUS, the calculator procedure will be similar but the keystrokes will differ. Be sure to check your calculator's instruction manual.

IRR Decision Rule When evaluating proposed projects with the IRR method, those that have IRRs equal to or greater than the required rate of return (hurdle rate) set by management are accepted, and those projects with IRRs that are less than the required rate of return are rejected. Acceptable, mutually exclusive projects are ranked from highest to lowest IRR.

Benefits of the IRR Method The IRR method for selecting capital budgeting projects is popular among financial practitioners for three primary reasons:

1. IRR focuses on all cash flows associated with the project.
2. IRR adjusts for the time value of money.
3. IRR describes projects in terms of the *rate* of return they earn, which makes it easy to compare them to other investments and the firm's hurdle rate.

Problems with the IRR Method The IRR method has several problems, however. First, because the IRR is a percentage number, it does not show how much the value of the firm will change if the project is selected. If a project is quite small, for instance, it may have a high IRR but a small effect on the value of the firm. (Consider a project that requires a $10 investment and returns $100 a year later. The project's IRR is 900 percent, but the effect on the value of the firm if the project is adopted is negligible.)

If the primary goal for the firm is to maximize its value, then knowing the *rate* of return of a project is not the primary concern. What is most important is the *amount* by which the firm's value will change if the project is adopted, which is measured by NPV.

To drive this point home, ask yourself whether you would rather earn a 100 percent rate of return on $5 ($5) or a 50 percent rate of return on $1,000 ($500). As you can see, it is not the rate of return that is important but the dollar value. Why? Dollars, not percentages, comprise a firm's cash flows. NPV tells financial analysts how much value will be created. IRR does not.

A second problem with the IRR method is that, in rare cases, a project may have more than one IRR, or no IRR. This is shown in detail in appendix 10A.

The IRR can be a useful tool in evaluating capital budgeting projects. As with any tool, however, knowing its limitations will enhance decision making.

Conflicting Rankings between the NPV and IRR Methods

As long as proposed capital budgeting projects are independent, both the NPV and IRR methods will produce the same accept/reject indication. That is, a project that has a positive NPV will also have an IRR that is greater than the discount rate. As a result, the project will be acceptable based on both the NPV and IRR values. However, when mutually exclusive projects are considered and ranked, a conflict occasionally arises. For instance, one project may have a higher NPV than another project, but a lower IRR.

To illustrate how conflicts between NPV and IRR can occur, imagine that the AddVenture Company owns a piece of land that can be used in different ways. On the one hand, this land has minerals beneath it that could be mined, so AddVenture could invest in mining equipment and reap the benefits of retrieving and selling the minerals. On the other hand, the land has perfect soil conditions for growing grapes that could be used to make wine, so the company could use it to support a vineyard.

Clearly, these two uses are mutually exclusive. The mine cannot be dug if there is a vineyard, and the vineyard cannot be planted if the mine is dug. The acceptance of one of the projects means that the other must be rejected.

Now let's suppose that AddVenture's finance department has estimated the cash flows associated with each use of the land. The estimates are presented next:

Time	Cash Flows for the Mining Project	Cash Flows for the Vineyard Project
t_0	($736,369)	($736,369)
t_1	$500,000	$0
t_2	$300,000	$0
t_3	$100,000	$0
t_4	$20,000	$50,000
t_5	$5,000	$200,000
t_6	$5,000	$500,000
t_7	$5,000	$500,000
t_8	$5,000	$500,000

Note that although the initial outlays for the two projects are the same, the incremental cash flows associated with the projects differ in amount and timing. The Mining Project generates its greatest positive cash flows early in the life of the project, whereas the Vineyard Project generates its greatest positive cash flows later. The differences in the projects' cash flow timing has a considerable effect on the NPV and IRR for each venture.

Assume AddVenture's required rate of return for long-term projects is 10 percent. The NPV and IRR results for each project given its cash flows are summarized as follows:

	NPV	IRR
Mining Project	$65,727.39	16.05%
Vineyard Project	$194,035.65	14.00%

These figures were obtained with our TI BAII PLUS calculator, as shown on page 250.

The NPV and IRR results show that the Vineyard Project has a higher NPV than the Mining Project ($194,035.65 versus $65,727.39), but the Mining Project has a higher IRR than the Vineyard Project (16.05 percent versus 14.00 percent). AddVenture is faced with a conflict between NPV and IRR results. Because the projects are mutually exclusive, the firm can only accept one.

In cases of conflict among mutually exclusive projects, the one with the highest NPV should be chosen because NPV indicates the dollar amount of value that will be added to the firm if the project is undertaken. In our example, then, AddVenture should choose the Vineyard Project (the one with the higher NPV) if its primary financial goal is to maximize firm value.

In this section we looked at three capital budgeting decision methods: the payback method, net present value method, and the internal rate of return method. We also investigated how to resolve conflicts between NPV and IRR decision methods. Next, we turn to a discussion of capital rationing.

Capital Rationing

Our discussion so far has shown that all independent projects with NPVs greater than or equal to zero should be accepted. All such projects that are adopted will add value to the firm. To act consistently with the goal of shareholder wealth maximization, then, it seems that if a firm locates $200 billion worth of positive NPV projects, it should accept all the projects. In practice, however, many firms place dollar limits on the total amount of capital budgeting projects they will undertake. They may wish to limit spending on new projects to keep a ceiling on business size. This practice of setting dollar limits on capital budgeting projects is known as **capital rationing.**

If capital rationing is imposed, then financial managers should seek the combination of projects that maximizes the value of the firm within the capital budget limit. For example, suppose a firm called BeLimited does not want its capital budget to exceed $200,000. Seven project proposals, Proposals A to G, are available, as shown in Table 10-1.

Note that all the projects in Table 10-1 have positive net present values, so they are all acceptable. However, BeLimited cannot adopt them all as that would require the expenditure of more than $200,000, its self-imposed capital budget limit.

TI BAII PLUS FINANCIAL CALCULATOR SOLUTIONS
IRR and NPV Solutions

The Mining Project

Keystrokes	Display
CF	CF_0 = old contents
2nd CLR Work	CF_0 = 0.00
736369 +/− ENTER	−736,369.00
↓ 500000 ENTER	$C01$ = 500,000.00
↓	$F01$ = 1.00
↓ 300000 ENTER	$C02$ = 300,000.00
↓	$F02$ = 1.00
↓ 100000 ENTER	$C03$ = 100,000.00
↓	$F03$ = 1.00
↓ 20000 ENTER	$C04$ = 20,000.00
↓	$F04$ = 1.00
↓ 5000 ENTER	$C05$ = 5,000
↓ 4 ENTER	$F05$ = 4.00
NPV	I = 0.00
10 ENTER	I = 10.00
↓	NPV = 0.00
CPT	NPV = 65,727.39
IRR	IRR = 0.00
CPT	IRR = 16.05

The Vineyard Project

Keystrokes	Display
CF	CF_0 = old contents
2nd CLR Work	CF_0 = 0.00
736369 +/− ENTER	−736.369.00
↓ 0 ENTER	$C01$ = 0.00
↓ 3 ENTER	$F01$ = 3.00
↓ 50000 ENTER	$C02$ = 50,000.00
↓	$F02$ = 1.00
↓ 200000 ENTER	$C03$ = 200,000.00
↓	$F03$ = 1.00
↓ 500000 ENTER	$C04$ = 500,000.00
↓ 3 ENTER	$F04$ = 3
NPV	I = 0.00
10 ENTER	I = 10.00
↓ CPT	NPV = 194,035.65
IRR	IRR = 0.00
CPT	IRR = 14.00

250   Part Three   Long-Term Financial Management Decisions

TABLE 10-1

BeLimited Project Proposals

PROJECT	INITIAL CASH OUTLAY	NET PRESENT VALUE
A	$20,000	$8,000
B	$50,000	$7,200
C	$40,000	$6,500
D	$60,000	$5,100
E	$50,000	$4,200
F	$30,000	$3,800
G	$30,000	$2,000

Under capital rationing, BeLimited must choose the combination of acceptable Projects A through G that produces the highest total NPV without exceeding its capital budget limit of $200,000.

Under capital rationing, BeLimited's managers try different combinations of projects seeking the combination that gives the highest NPV without exceeding the $200,000 limit. For example, the combination of Projects B, C, E, F, and G costs $200,000 and yields a total NPV of $23,700. A better combination is that of Projects A, B, C, D, and F, which costs $200,000 and has a total NPV of $30,600. In fact, this combination has the highest total NPV, given the $200,000 capital budget limit (try a few other combinations to see for yourself), so that combination is the one BeLimited should choose.

In this section, we explored capital rationing. In the following section, we will examine how risk affects capital budgeting decisions.

RISK AND CAPITAL BUDGETING

Suppose you are a financial manager for AddVenture Corporation. You are considering two capital budgeting projects, Project X (discussed throughout the chapter) and Project X2. Both projects have identical future cash flow values and timing, and both have the same NPV, if discounted at the same rate, k, and the same IRR. The two projects appear equally desirable.

Now suppose you discover that Project X's incremental cash flows are absolutely certain, but Project X2's incremental cash flows, which have the same expected value, are highly uncertain. These cash flows could be higher or lower than forecast. In other words, Project X is a sure thing and Project X2 appears to be quite risky. As a risk-averse person, you would prefer Project X over Project X2.

The assumption that most investors are risk averse is explained in chapter 7, pages 144–145.

Financial managers are not indifferent to risk. Indeed, the riskier a project, the less desirable it is to the firm. To incorporate risk into the NPV and IRR evaluation techniques, we use the standard deviation and coefficient of variation (CV) measures discussed in chapter 7. We begin by examining how to measure the risk of a capital budgeting project. Then we explain how to include the risk measurement in the NPV and IRR evaluation.

Measuring Risk in Capital Budgeting

Financial managers may measure three types of risk in the capital budgeting process. The first type is *project-specific risk*, or the risk of a specific new project. The second type is *firm risk*, which is the impact of adding a new project to the existing

FINANCE AT WORK

SALES·RETAIL·SPORTS·MEDIA TECHNOLOGY·PUBLIC RELATIONS·PRODUCTION·EXPORTS

Jim Bruner, Former Maricopa County Supervisor, State of Arizona

Jim Bruner

Jim Bruner is a former Scottsdale City Council member and Maricopa County Supervisor for the state of Arizona. Currently, he is Executive Vice President for First Interstate Bank. During his tenure in public life, Jim voted for and against many long-term projects. In fact, Jim cast a key vote in favor of using a sales tax to help finance a domed stadium for the Arizona Diamondbacks baseball team.

Q. What factors are considered when making a capital budgeting decision for a private versus a public project?

In the private sector, corporations try to assess how much value—in dollars—a project will add to the firm. The goal is, of course, to maximize the wealth of the stockholders. In public ventures, measuring value becomes more complicated because value is often intangible. Members of the community are owners and stakeholders in the city, county, or state. Quantifying how much value a project will add to a community is difficult because an increase in job growth, national recognition, or public spirit can add value, but it's hard to put an exact price tag on those intangible items.

Q. How do government workers make capital budgeting decisions in a public/private venture?

A: Quite honestly, it depends on the project. Let me explain how we made a decision about building a golf course for the Tournament Players Club in Scottsdale as one example. The annual Phoenix Open, sanctioned by the Professional Golf Association tour, was looking for a larger golf course for its tournament. I was serving on the Scottsdale City Council when we decided to build a golf course for the tournament in a partnership between Scottsdale and the private sector, the PGA tour. During the other 51 weeks of the year, the course would be open to the general public for its enjoyment. Thus, the city of Scottsdale agreed to build the golf course and the PGA tour, a for-profit entity, agreed to operate the golf course, pro shop, restaurant, and so on. Scottsdale, as owner of the course, rents the golf course to the PGA and receives a percent of all receipts from all the course and retail activities. The estimates of the future receipts were our expected cash flows, and the cost of building the course was the city's initial investment. We calculated that Scottsdale would earn millions of dollars in excess of cost of building the golf course, so the two 18-hole courses and the tournament would clearly add value to the city of Scottsdale.

But we recognized that there was more to the capital budgeting equation than future cash flows from the tournament receipts. Although we couldn't estimate it with precise certainty, we knew that the tournament would increase the Scottsdale tourist industry, add new jobs, and increase the quality of life. A city is more than streets, sewers, lights, and fire and police departments. Having two 18-hole golf courses benefits a community by providing open space and entertainment.

Q. How did the Maricopa County Board of Supervisors do capital budgeting analysis before building a stadium for the Arizona Diamondbacks?

If you look at building a baseball stadium as a corporation might—what is the project's NPV for the state of Arizona—the estimates of cash flows are tough to make. The County of Maricopa instituted a tax (¼ of 1 percent for three years) to raise the initial cash outlay of $238 million for the stadium. In effect, the owners of the stadium—the public—voted to accept the project.

Had the county taken that $238 million raised to build the stadium and put it in a savings account, the county would have received more direct revenue at less risk than it will receive from the stadium. But the total economic impact through increased sales tax, property tax, employment, and so on, far exceed the direct revenues from the rent.

Q. What future cash flows can the county expect from the Diamondbacks Stadium project?

We receive an annual guaranteed revenue, even if nobody attends games. If, however, the team is successful, the taxpayers will benefit. As attendance increases, the stadium rent increases above the guaranteed revenue rate.

FINANCE AT WORK—Continued

Maricopa County will also receive a percentage of the revenues from the skyboxes, the luxury suites. In addition, the County will receive a percentage of the naming rights for the stadium. If a corporation wants its name on the stadium, the County gets one-third of the fee that corporation pays to display its name.

The Board of Supervisors estimated that its direct cash flows would be a $3.5 million annuity for the 30 years of the lease. Analyzing the amount and timing of those cash flows, and the $238 million initial investment, there's no way to justify accepting the project if direct revenue were the only means of the economic impact.

But we also looked at intangibles, such as the synergism created from having a stadium downtown, what it will do for economic development, for hotels, and for public spirit. We also knew that the stadium would generate at least 2,500 jobs during construction and believed it would generate many more jobs once the team began playing.

And because of the special tax, which stops once the stadium is complete, we will have enough money before the first pitch is thrown to ensure that the stadium is debt free and an ongoing asset to the community for years to come.

Source: Interview with Jim Bruner.

---

projects of the firm. The third is *market (or beta) risk*, which is the effect of a new project on the stockholders' nondiversifiable risk.

In this chapter we focus on measuring firm risk. To measure firm risk, we compare the coefficient of variation (CV) of the firm's portfolio before and after a project is adopted. The before-and-after difference between the two CVs serves as the project's risk measure from a firm perspective. If the CAPM approach (discussed in chapter 7) is preferred, a financial manager could estimate the risk of a project by calculating the project's beta instead of the change in firm risk.

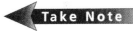

The CV is often chosen as the risk measure because the CV reflects the expected return per unit of risk. Using only the standard deviation makes it more difficult to tell whether the risk is high or low.

Computing Changes in the Coefficient of Variation Let's reconsider our earlier example, Project X. Recall that the IRR of Project X was 5.71 percent. Now suppose that after careful analysis, we determine this IRR is actually the most likely value from a range of possible values, each with some probability of occurrence. We find the expected value and standard deviation of Project X's IRR distribution using Equations 7-1 and 7-2, respectively. We find that the expected value of the IRR distribution is 5.71 percent and the standard deviation is 2.89 percent.

To see how adding Project X to the firm's existing portfolio changes the portfolio's coefficient of variation (CV), we follow a five-step procedure.

We discussed the expected value formula, Equation 7-1, and the standard deviation formula, Equation 7-2, on page 147.

1. Find the CV of the Existing Portfolio. Suppose the expected rate of return and standard deviation of possible returns of AddVenture's existing portfolio are 6 percent and 2 percent, respectively. Given this information, calculate the CV of the firm's existing portfolio using Equation 7-3:

$$\text{CV} = \frac{\text{Standard Deviation}}{\text{Mean, or expected value}} \quad (7\text{-}3)$$

$$= \frac{.02}{.06}$$

$$= .3333, \text{ or } 33.33\%$$

Chapter 10 *Capital Budgeting Decision Methods* 253

2. Find the Expected Rate of Return of the New Portfolio (the Existing Portfolio Plus the Proposed Project).
Assume that the investment in Project X represents 10 percent of the value of the portfolio. In other words, the new portfolio after adding Project X will consist of 10 percent Project X and 90 percent of the rest of the firm's assets. With these figures, solve for the expected rate of return of the new portfolio using Equation 7-4. In the calculations, Asset a represents Project X and Asset b represents the firm's existing portfolio.

$$E(R_p) = (w_a \times E(R_a)) + (w_b \times E(R_b))$$
$$= (.10 \times .0571) + (.90 \times .06)$$
$$= .00571 + .05400$$
$$= .05971, \text{ or } 5.971\%$$

3. Find the Standard Deviation of the New Portfolio (the Existing Portfolio Plus the Proposed Project).
Calculate the standard deviation of the new portfolio using Equation 7-5. Assume the degree of correlation (r) between project X (Asset a) and the firm's existing portfolio (Asset b) is zero. Put another way, there is no relationship between the returns from Project X and the returns from the firm's existing portfolio.

$$\sigma_p = \sqrt{w_a^2 \sigma_a^2 + w_b^2 \sigma_b^2 + 2w_a w_b r_{ab} \sigma_a \sigma_b} \quad (7\text{-}5)$$
$$= \sqrt{(.10^2)(.0289^2) + (.90^2)(.02^2) + (2)(.10)(.90)(0.0)(.0578)}$$
$$= \sqrt{(.01)(.000835) + (.81)(.0004) + 0}$$
$$= \sqrt{.00000835 + .000324}$$
$$= \sqrt{.000332352}$$
$$= .0182, \text{ or } 1.82\%$$

4. Find the CV of the New Portfolio (the Existing Portfolio Plus the Proposed Project).
To solve for the CV of the new portfolio with Project X included, we use Equation 7-3, as follows:

$$CV = \frac{\text{Standard Deviation}}{\text{Mean, or Expected Value}} \quad (7\text{-}3)$$
$$= \frac{.0182}{.05971}$$
$$= .3048, \text{ or } 30.48\%$$

5. Compare the CV of the Portfolio with and without the Proposed Project.
To evaluate the effect of Project X on the risk of the AddVenture portfolio, we compare the CV of the old portfolio (without Project X) to the CV of the new portfolio (with Project X). The coefficients follow:

CV OF THE PORTFOLIO WITHOUT PROJECT X	CV OF THE PORTFOLIO WITH PROJECT X	CHANGE IN CV
33.33%	30.48%	−2.85

The CV of the firm's portfolio dropped from 33.3 percent to 30.48 percent with the addition of Project X. This 2.85 percentage point decrease in CV is the measure of risk in Project X from the firm's perspective.

TABLE 10-2

Risk-Adjustment Table

If Adoption of the Proposed Capital Budgeting Project Would Change the CV of the Firm's Portfolio IRR by the Following Amount:	Then Assign the Project to the Following Risk Category:	And Adjust the Discount Rate for the Project as Follows:
More than a 1 percent increase	High Risk	+2%
Between a 1 percent decrease and a 1 percent increase	Average Risk	0
More than a 1 percent decrease	Low Risk	−2%

Adjusting for Risk

Most business owners and financial managers are risk averse, so they want the capital budgeting process to reflect risk concerns. Next, we discuss how to adjust for risk in the capital budgeting process—risk-adjusted discount rates.

Risk-Adjusted Discount Rates (RADRs)   One way to factor risk into the capital budgeting process is to adjust the rate of return demanded for high- and low-risk projects. **Risk-adjusted discount rates (RADRs),** then, are discount rates that differ according to their effect on a firm's risk. The higher the risk, the higher the RADR and the lower the risk, the lower the RADR. A project with a normal risk level would have an RADR equal to the actual discount rate.

Adjusting required rates of return to compensate for the presence of risk was discussed in chapter 7, pages 162–165.

To find the risk-adjusted discount rate for a capital budgeting project, we first prepare a risk adjustment table like the one in Table 10-2. We assume in Table 10-2 that the coefficient of variation, CV, is the risk measure to be adjusted for. The CV is based on the probability distribution of the IRR values.

Table 10-2 shows how the discount rates for capital budgeting projects might be adjusted for varying degrees of risk. The discount rate of a project that decreased the CV of the firm's portfolio of assets by 2.5 percentage points, for example, would be classified as a low-risk project. That project would be evaluated using a discount rate two percentage points lower than that used on average projects.

This has the effect of making Project X a more desirable project. That is, financial managers would calculate the project's NPV using an average discount rate that is two percentage points lower, which would increase the project's NPV. If the firm uses the IRR method, financial managers would compare the project's IRR to a hurdle rate two percentage points lower than average, which would make it more likely that the firm would adopt the project.

The data in Table 10-2 are only for illustration. In practice, the actual amount of risk adjustment will depend on the degree of risk aversion of the managers and stockholders.

RADRs are an important part of the capital budgeting process because they incorporate the risk–return relationship. All other things being equal, the more a project increases risk, the less likely it is that a firm will accept the project; the more a project decreases risk, the more likely it is that a firm will accept the project.

WHAT'S NEXT

In this chapter we explored the capital budgeting process and decision methods. We examined three capital budgeting techniques—the payback method, net present value method, and internal rate of return method. We also discussed capital rationing and how to measure and adjust for risk when making capital budgeting decisions.

In the next chapter, we'll investigate how to estimate incremental cash flows in capital budgeting.

Summary

1. Explain the capital budgeting process.

Capital budgeting is the process of evaluating proposed investment projects. Capital budgeting projects may be independent—the projects do not compete with each other—or mutually exclusive—accepting one project precludes the acceptance of the other(s) in that group.

Financial managers apply two decision practices when selecting capital budgeting projects: *accept/reject* decisions and *ranking* decisions. The accept/reject decision is the determination of which independent projects are acceptable in light of the firm's financial objectives. Ranking is a process of comparing projects to a standard measure, and ordering the mutually exclusive projects based on how well they meet the measure. The capital budgeting process is concerned only with incremental cash flows, that is, those cash flows that will occur if an investment is undertaken but won't occur if it isn't.

2. Calculate the payback period, net present value, and internal rate of return for a proposed capital budgeting project.

The payback period is defined as the number of time periods it will take before the cash inflows from a proposed capital budgeting project will equal the amount of the initial investment. To find a project's payback period, add all the project's positive cash flows, one period at a time, until the sum equals the amount of the initial cash outlay for the project. The number of time periods it takes to recoup the initial cash outlay is the payback period. A project is acceptable if it pays back the initial investment within a time frame set by firm management.

The net present value (NPV) of a proposed capital budgeting project is the dollar amount of the change in the value of the firm that will occur if the project is undertaken. To calculate NPV, total the present values of all the projected incremental cash flows associated with a project and subtract the amount of the initial cash outlay. A project with an NPV greater than or equal to zero is acceptable. An NPV profile—a graph that shows a project's NPV at many different discount rates (required rates of return)—shows how sensitive a project's NPV is to changes in the discount rate.

The internal rate of return (IRR) is the projected percent rate of return that a proposed project will earn, given its incremental cash flows and required initial cash outlay. To calculate IRR, find the discount rate that makes the project's NPV equal to zero. That rate of return is the IRR. A project is acceptable if its IRR is greater than or equal to the firm's required rate of return (the hurdle rate).

3. Describe capital rationing and how firms decide which projects to select.

The practice of placing dollar limits on the total size of the capital budget is called capital rationing. Under capital rationing the firm will select the combination of projects that yields the highest NPV without exceeding the capital budget limit.

4. Measure the risk of a capital budgeting project.

To measure the risk of a capital budgeting project, we determine how the project would affect the risk of the firm's existing asset portfolio. We compare the coefficient of variation of expected returns of the firm's asset portfolio with the proposed project and without it. The difference between the two coefficients of variation is the measure of the risk of the capital budgeting project.

5. **Explain risk-adjusted discount rates.**

To compensate for the degree of risk in capital budgeting, firms may adjust the discount rate for each project according to risk. The more risk is increased, the higher the discount rate. The more risk is decreased, the lower the discount rate. Rates adjusted for the risk of projects are called risk-adjusted discount rates (RADRs).

You can access your FREE PHLIP/CW On–Line Study Guide, Current Events, and Spreadsheet Problem templates, which may correspond to this chapter either through the Prentice Hall Finance Center CD–ROM or by directly going to: **http://www.prenhall.com/gallagher.**

PHLIP/CW — *The Prentice Hall Learning on the Internet Partnership–Companion Web Site*

Equations Introduced in This Chapter

Equation 10-1a. NPV Formula, Algebraic Version:

$$NPV = \left(\frac{CF_1}{(1+k)^1}\right) + \left(\frac{CF_2}{(1+k)^2}\right) \cdots + \left(\frac{CF_n}{(1+k)^n}\right) - \text{Initial Investment}$$

where: CF = Cash flow at the indicated times
k = Discount rate, or required rate of return for the project
n = Life of the project measured in the number of time periods

Equation 10-1b. NPV Formula, Table Version:

$$NPV = CF_1(PVIF_{k,1}) + CF_2(PVIF_{k,2}) + \ldots CF_n(PVIF_{k,n}) - \text{Initial Investment}$$

where: PVIF = Present Value Interest Factor

Equation 10-2. Formula for IRR Using Equation 10-1a:

$$NPV = 0 = \left(\frac{CF_1}{(1+k)^1}\right) + \left(\frac{CF_2}{(1+k)^2}\right) \cdots + \left(\frac{CF_n}{(1+k)^n}\right) - \text{Initial Investment}$$

where: k = the IRR value

Fill in cash flows (CFs) and periods (n). Then choose values for k by trial and error until the NPV equals zero. The value of k that causes the NPV to equal zero is the IRR.

Self-Test

ST-1. What is the NPV of the following project?

Initial investment:	$50,000
Net cash flow at the end of year 1:	$20,000
Net cash flow at the end of year 2:	$40,000
Discount rate:	11%

ST-2. What is the IRR of the project in ST-1?

ST-3. You've been assigned to evaluate a project for your firm that requires an initial investment of $200,000, is expected to last for 10 years, and is expected to produce after-tax net cash flows of $44,503 per year. If your firm's required rate of return is 14 percent, should the project be accepted?

ST-4. What is the IRR of the project in ST-3?

ST-5. Assume you have decided to reinvest your portfolio in zero-coupon bonds. You like zero-coupon bonds because they pay off a known amount, $1,000 at maturity, and involve no other cash flows other than the purchase price. Assume your required rate of return is 12 percent. If you buy some 10-year zero-coupon bonds for $400 each today, will the bonds meet your return requirements? (Hint: Compute the IRR of the investment given the cash flows involved.)

ST-6. Joe the cut-rate bond dealer has offered to sell you some zero-coupon bonds for $300. (Remember, zero-coupon bonds pay off $1,000 at maturity and involve no other cash flows other than the purchase price.) If the bonds mature in 10 years and your required rate of return for cut-rate bonds is 20 percent, what is the NPV of Joe's deal?

FinCoach Practice Exercises

To help improve your grade and master the mathematical skills covered in this chapter, open FinCoach on the Prentice Hall Finance Center CD-ROM and work practice problems in the category Project and Firm Valuation.

Review Questions

1. How do we calculate the payback period for a proposed capital budgeting project? What are the main criticisms of the payback method?
2. How does the net present value relate to the value of the firm?
3. What are the advantages and disadvantages of the internal rate of return method?
4. Provide three examples of mutually exclusive projects.
5. What is the decision rule for accepting or rejecting proposed projects when using net present value?
6. What is the decision rule for accepting or rejecting proposed projects when using internal rate of return?
7. What is capital rationing? Should a firm practice capital rationing? Why?
8. Explain how to resolve a ranking conflict between the net present value and the internal rate of return. Why should the conflict be resolved as you explained?
9. Explain how to measure the firm risk of a capital budgeting project.
10. Why is the coefficient of variation a better risk measure to use than the standard deviation when evaluating the risk of capital budgeting projects?
11. Explain why we measure a project's risk as the change in the CV.
12. Explain how using a risk-adjusted discount rate improves capital budgeting decision making compared to using a single discount rate for all projects.

Build Your Communication Skills

CS-1. Arrange to have a person who has no training in finance visit the class on the next day after the net present value (NPV) evaluation method has been covered. Before the visitor arrives, choose two teams who will brief the visitor on the results of their evaluation of a capital budgeting project. The visitor will play the role of the CEO of the firm. The CEO must choose one capital budgeting project for the firm to undertake. The CEO will choose the project based on the recommendations of Team A and Team B.

Now have each team describe the following projects:

	Team A	Team B
	Solar Energy Satellite (a plan to beam energy down from orbit)	Fresh Water Extraction Project (a plan to extract drinking water from the sea)
Initial Investment	$40 million	$40 million
NPV	$10 million	$8 million
IRR	12%	18%

After the projects have been described, have the CEO ask each team for its recommendation as to which project to adopt. The CEO may also listen to input from the class at large.

When the debate is complete, have the CEO select one of the projects (based on the class's recommendations). Then critique the class's recommendations.

CS-2. Divide the class into groups of two members each. Have each pair prepare a written description (closed book) of how to measure the risk of a capital budgeting project. When the explanations are complete, select a volunteer from each group to present their findings to the class.

Problems

10-1. Three separate projects each have an initial cash outlay of $10,000. The cash flows for Peter's Project is $4,000 per year for three years. The cash flow for Paul's Project is $2,000 in years 1 and 3, and $8,000 in year 2. Mary's Project has a cash flow of $10,000 in year 1, followed by $1,000 each year for years 2 and 3. *Payback*

   a. Use the payback method to calculate how many years it will take for each project to recoup the initial investment.
   b. Which project would you consider most liquid?

10-2. The Bedford Falls Bridge Building Company is considering the purchase of a new crane. George Bailey, the new manager, has had some past management experience while he was the chief financial officer of the local savings and loan. The cost of the crane is $17,291.42 and the expected incremental cash flows are $5,000 at the end of year 1, $8,000 at the end of year 2, and $10,000 at the end of year 3. *NPV and IRR*

   a. Calculate the net present value if the required rate of return is 12 percent.
   b. Calculate the internal rate of return.
   c. Should Mr. Bailey purchase this crane?

10-3. Lin McAdam and Lola Manners, managers of the Winchester Company, do not practice capital rationing. They have three independent projects they are evaluating for inclusion in this year's capital budget. One is for a new machine to make rifle stocks. The second is for a new forklift to use in the warehouse. The third project involves the purchase of automated packaging equipment. The Winchester Company's required rate of return is 13 percent. The initial investment (a negative cash flow) and the expected positive net cash flows for years 1 through 4 for each project follow: *NPV*

	Expected Net Cash Flow		
Year	Rifle Stock Machine	Forklift	Packaging Equipment
0	$(9,000)	$(12,000)	$(18,200)
1	2,000	5,000	0
2	5,000	4,000	5,000
3	1,000	6,000	10,000
4	4,000	2,000	12,000

a. Calculate the net present value for each project.
b. Which project(s) should be undertaken? Why?

NPV and IRR

10-4. The Trask Family Lettuce Farm is located in the fertile Salinas Valley of California. Adam Trask, the head of the family, makes all the financial decisions that affect the farm. Because of an extended drought, the family needs more water per acre than what the existing irrigation system can supply. The quantity and quality of lettuce produced are expected to increase when more water is supplied. Cal and Aron, Adam's sons, have devised two different solutions to their problem. Cal suggests improvements to the existing system. Aron is in favor of a completely new system. The negative cash flow associated with the initial investment and expected positive net cash flows for years 1 through 7 for each project follow. Adam has no other alternatives and will choose one of the two projects. The Trask Family Lettuce Farm has a required rate of return of 12 percent for these projects.

	Expected Net Cash Flow	
Year	Cal's Project	Aron's Project
0	$(100,000)	$(300,000)
1	22,611	63,655
2	22,611	63,655
3	22,611	63,655
4	22,611	63,655
5	22,611	63,655
6	22,611	63,655
7	22,611	63,655

a. Calculate the net present value for each project.
b. Calculate the internal rate of return for each project.
c. Which project should Adam choose? Why?
d. Is there a conflict between the decisions indicated by the NPVs and the IRRs?

Payback, NPV, and IRR

10-5. Dave Hirsh publishes his own manuscripts and is unsure which of two new printers he should purchase. He is a novelist living in Parkman, Illinois. Having slept through most of his Finance 300 course in college, he is unfamiliar with cash flow analysis. He enlists the help of the finance professor at the local university, Dr. Gwen French, to assist him. Together they estimate the following expected initial investment (a negative cash flow) and net positive cash flows for years 1 through 3 for each machine. Dave only needs one printer and estimates it will be worthless after three years of heavy use. Dave's required rate of return for this project is 10 percent.

	Expected Net Cash Flow	
Year	Printer 1	Printer 2
0	$(2,000)	$(2,500)
1	900	1,500
2	1,100	1,300
3	1,300	800

a. Calculate the payback period for each printer.
b. Calculate the net present value for each printer.

c. Calculate the internal rate of return for each printer.
   d. Which printer do you think Dr. French will recommend? Why?
   e. Suppose Dave's required rate of return was 16 percent. Does the decision about which printer to purchase change?

10-6. Project A has an initial investment of $11,000 and it generates positive cash flows of $4,000 each year for the next six years. Project B has an initial investment of $17,000 and it generates positive cash flows of $4,500 each year also for the next six years. Assume the discount rate, or required rate of return, is 13 percent.

Independent and Mutually Exclusive Projects

   a. Calculate the net present value of each project. If Project A and Project B are mutually exclusive, which one(s) would you select?
   b. Assume Projects A and B are independent. Based on NPV, which one(s) would you select?
   c. Calculate the internal rate of return for each project.
   d. Using IRR for your decision, which project would you select if Project A and Project B are mutually exclusive? Would your answer change if these two projects were independent instead of mutually exclusive?
   e. Project C was added to the potential capital budget list at the last minute as a mutually exclusive alternative to Project B. It has an initial cash outlay of $17,000 and it generates its only positive cash flow of $37,500 in the sixth year. Years 1 through 5 have $0 cash flow. Calculate Project C's NPV. Which alternative, Project B or C, would you choose?
   f. Now calculate the IRR of Project C and compare it to Project B's IRR. Based solely on the IRRs, which project would you select?
   g. Because Projects B and C are mutually exclusive, would you recommend that Project B or Project C be added to the capital budget for this year? Explain your choice.

10-7. Joanne Crale is an independent petroleum geologist. She is taking advantage of every opportunity to lease the mineral rights of land she thinks lies over oil reserves. She thinks there are many reservoirs with oil reserves left behind by major corporations when they plugged and abandoned fields in the 1960s. Ms. Crale knows that with today's technology, production can be sustained at much lower reservoir pressures than was possible in the 1960s. She would like to lease the mineral rights from George Hansen and Ed McNeil, the landowners. Her net cost for this current oil venture is $5 million. This includes costs for the initial leasing and the three planned development wells. The expected positive net cash flows for the project are $1.85 million each year for four years. On depletion, she anticipates a negative cash flow of $250,000 in year 5 because of reclamation and disposal costs. Because of the risks involved in petroleum exploration, Ms. Crale's required rate of return is 16 percent.

NPV and IRR

   a. Calculate the net present value for Ms. Crale's project.
   b. Calculate the internal rate of return for this project.
   c. Would you recommend the project?

10-8. The product development managers at World Series Innovations are about to recommend their final capital budget for next year. They have a self-imposed budget limit of $100,000. Five independent projects are being considered. Vernon Simpson, the chief scientist and CEO, has minimal financial analysis experience and relies on his managers to recommend the projects that will increase the value of the firm by the greatest amount. Given the following summary of the five projects, which ones should the managers recommend?

Challenge Problem

Projects	Initial Cash Outlay	Net Present Value
Chalk Line Machine	($10,000)	$4,000
Gel Padded Glove	($25,000)	3,600
Insect Repellent	($35,000)	3,250
Titanium Bat	($40,000)	2,500
Recycled Base Covers	($20,000)	2,100

Various Capital Budgeting Issues

10-9. A project you are considering has an initial investment of $5,669.62. It has positive net cash flows of $2,200 each year for three years. Your required rate of return is 12 percent.

  a. Calculate the net present value of this project.
  b. Would you undertake this project if you had the cash available?
  c. What would the discount rate have to be before this project's NPV would be positive? Construct an NPV profile with discount rates of 0 percent, 5 percent, and 10 percent to answer this question.
  d. What other method might you use to determine at what discount rate the net present value would become greater than 0?

Expected IRR

10-10. The internal rate of return (IRR) of a capital investment Project B is expected to have the following values with the associated probabilities for an economic life of five years. Calculate the expected value, standard deviation, and coefficient of variation of the IRR distribution.

IRR	Probability of Occurrence
0%	0.05
1%	0.10
3%	0.20
6%	0.30
9%	0.20
11%	0.10
12%	0.05

Capital Budgeting Risk

10-11. Four capital investment alternatives have the following expected IRRs and standard deviations.

Project	Expected IRR	Standard Deviation
A	14%	2%
B	16%	6%
C	11%	5%
D	14%	4%

If the firm's existing portfolio of assets has an expected IRR of 13 percent with a standard deviation of 3 percent, identify the lowest- and the highest-risk projects in the preceding list. You may assume the correlation coefficient of returns from each project relative to the existing portfolio is .50 and the investment in each project would constitute 20 percent of the firm's total portfolio.

Portfolio Risk

10-12. Assume that a company has an existing portfolio A, with an expected IRR of 10 percent and standard deviation of 2 percent. The company is considering adding a Project B, with expected IRR of 11 percent and standard deviation of 3 percent, to its portfolio. Also assume that the amount invested in Portfolio A is $700,000 and the amount to be invested in Project B is $200,000.

  a. If the degree of correlation between returns from Portfolio A and Project B is .90 (r = +.9), calculate:
  1. the coefficient of variation of portfolio A
  2. the expected IRR of the new combined portfolio
  3. the standard deviation of the new combined portfolio
  4. the coefficient of variation of the new combined portfolio.

Coefficient of Variation Risk in Capital Budgeting

  b. What is the change in the coefficient of variation as a result of adopting Project B?
  c. Assume the firm classifies projects as high risk if they raise the coefficient of variation of the portfolio 2 percent or more, as low risk if they lower the coefficient of variation of the portfolio 2 percent or more, and as average risk if they change the coefficient of

variation of the portfolio by less than 2 percent in either direction. In what risk classification would Project B be?

d. The required rate of return for average-risk projects for the company in this problem is 13 percent. The company policy is to adjust the required rate of return downward by 3 percent for low-risk projects and raise it 3 percent for high-risk projects (average-risk projects are evaluated at the average required rate of return). The expected cash flows from Project B are as follows:

RADRs

CF of Initial Investment	($200,000)
end of year 1	55,000
end of year 2	55,000
end of year 3	55,000
end of year 4	100,000

Calculate the NPV of Project B when its cash flows are discounted at:

1. the average required rate of return
2. the high-risk discount rate
3. the low-risk discount rate

10-13. Dorothy Gale is thinking of purchasing a Kansas amusement park, Glinda's Gulch. She already owns at least one park in each of the surrounding states and wants to expand her operations into Kansas. She estimates the IRR from the project may be one of a number of values, each with a certain probability of occurrence. Her estimated IRR values and the probability of their occurrence follow.

Risk in Capital Budgeting

Glinda's Gulch

Possible IRR Value	Probability
2%	.125
5%	.20
9%	.35
13%	.20
16%	.125

Now Dorothy wants to estimate the risk of the project. Assume she has asked you to do the following for her:

a. Calculate the mean or expected IRR of the project.
b. Calculate the standard deviation of the possible IRRs.
c. Assume the expected IRR of Dorothy's existing portfolio of parks is 8 percent, with a standard deviation of 2 percent. Calculate the coefficient of variation of Dorothy's existing portfolio.
d. Calculate the expected IRR of Dorothy's portfolio if the Glinda's Gulch project is adopted. For this calculation assume that Dorothy's new portfolio will consist of 80 percent existing parks and 20 percent new Glinda's Gulch.
e. Again assuming that Dorothy's new portfolio will consist of 80 percent existing parks and 20 percent new Glinda's Gulch, calculate the standard deviation of Dorothy's portfolio if the Glinda's Gulch project is added. Because Glinda's Gulch is identical to Dorothy's other parks, she estimates the returns from Glinda's Gulch and her existing portfolio are perfectly positively correlated (r = +1.0).
f. Calculate the coefficient of variation of Dorothy's new portfolio with Glinda's Gulch.
g. Compare the coefficient of variation of Dorothy's existing portfolio with and without the Glinda's Gulch project. How does the addition of Glinda's Gulch affect the riskiness of the portfolio?

10-14. Four capital investment alternatives have the following expected IRRs and standard deviations.

Risk in Capital Budgeting

Project	Expected IRR	Standard Deviation
A	18%	9%
B	15%	5%
C	11%	3%
D	8%	1%

The firm's existing portfolio of projects has an expected IRR of 12 percent and a standard deviation of 4 percent.

    a. Calculate the coefficient of variation of the existing portfolio.
    b. Calculate the coefficient of variation of the portfolio if the firm invests 10 percent of its assets in Project A. You may assume that there is no correlation between the returns from Project A and the returns from the existing portfolio (r = 0). (Note: In order to calculate the coefficient of variation, you must first calculate the expected IRR and standard deviation of the portfolio with Project A included.)
    c. Using the same procedure as in b, and given the same assumptions, calculate the coefficient of variation of the portfolio if it includes, in turn, projects B, C, and D.
    d. Which project has the highest risk? Which has the lowest risk?

Firm Risk

10-15. A firm has an existing portfolio of projects with a mean, or expected IRR of 15 percent. The standard deviation of this estimate is 5 percent. The existing portfolio is worth $820,000. The addition of a new project, PROJ1, is being considered. PROJ1's expected IRR is 18 percent with a standard deviation of 9 percent. The initial investment for PROJ1 is expected to be $194,000. The returns from PROJ1 and the existing portfolio are perfectly positively correlated (r = +1.0).

    a. Calculate the coefficient of variation for the existing portfolio.
    b. If PROJ1 is added to the existing portfolio, calculate the weight (proportion) of the existing portfolio in the combined portfolio.
    c. Calculate the weight (proportion) of PROJ1 in the combined portfolio.
    d. Calculate the standard deviation of the combined portfolio. Is the standard deviation of the combined portfolio higher or lower than the standard deviation of the existing portfolio?
    e. Calculate the coefficient of variation of the combined portfolio.
    f. If PROJ1 is added to the existing portfolio, will the firm's risk increase or decrease?

Risk in Capital Budgeting

10-16. VOTD Pharmaceuticals is considering the mass production of a new sleeping pill. Neely O'Hara, VOTD's financial analyst, has gathered all the available information from the finance, production, advertising, and marketing departments and has estimated that the yearly net incremental cash flows will be $298,500. She estimates the initial investment for this project will be $2 million. The pills are expected to be marketable for 10 years and Neely does not expect that any of the investment costs will be recouped at the end of the 10-year period. VOTD's required rate of return for average-risk projects is 8 percent. Two percent is added to the required rate of return for high-risk projects.

    a. Calculate the net present value of the project if VOTD management considers it to be average risk.
    b. A lot of uncertainty is associated with the potential competition of new drugs that VOTD's competitors might introduce. Because of this uncertainty, VOTD management has changed the classification of this project to high risk. Calculate the net present value at the risk-adjusted discount rate.
    c. Assuming the risk classification does not change again, should this project be adopted?
    d. What will the yearly net incremental cash flow have to be in order to have a positive NPV when using the high-risk discount rate?

Comprehensive Problem

10-17. Aluminum Building Products Company (ABPC) is considering investing in either of the two mutually exclusive projects described as follows:

Project 1. Buying a new set of roll-forming tools for its existing roll-forming line to introduce a new cladding product. After its introduction the product will need to be promoted. This means that cash inflows from additional production will start some time after and will gradually pick up in subsequent periods.

Project 2. Modifying its existing roll-forming line to increase productivity of its available range of cladding products. Cash inflows from additional production will start immediately and will reduce over time as the products move through their life cycle.

Sarah Brown, project manager of ABPC, has requested that you do the necessary financial analysis and give your opinion as to which project ABPC should select. The projects have the following net cash flow estimates:

Year	Project 1	Project 2
0	($200,000)	($200,000)
1	0	90,000
2	0	70,000
3	20,000	50,000
4	30,000	30,000
5	40,000	10,000
6	60,000	10,000
7	90,000	10,000
8	100,000	10,000

Both the foregoing projects have the same economic life of eight years and average risk characteristics. ABPC's weighted average cost of capital or hurdle rate is 7.2 percent.

  a. Which project would you recommend Ms. Brown accept to maximize value of the firm? (Hint: Calculate and compare NPVs of both the projects.)
  b. What are the IRRs of each project? Which project should be chosen using IRR as the selection criterion?
  c. Draw the NPV profiles of both the projects. What is the approximate discount rate at which both the projects would have the same NPV? What is that NPV?
  d. Does the selection remain unaffected for i) WACC > 5.4 percent; ii) WACC > 8.81 percent; and iii) WACC > 14.39 percent?
  e. Further market survey research indicates that both the projects have lower than average risk and, hence, the risk-adjusted discount rate should be 5 percent. What happens to the ranking of the projects using NPV and IRR as the selection criteria? Explain the conflict in ranking, if any.
  f. Answer questions a, d, and e, assuming the projects are independent of each other.

Answers to Self-Test

ST-1. The NPV of the project can be found using Equation 10-1a as follows:

$$\text{NPV} = \left(\frac{\$20,000}{(1+.11)^1}\right) + \left(\frac{\$40,000}{(1+.11)^2}\right) - \$50,000$$

$$= \left(\frac{\$20,000}{1.11}\right) + \left(\frac{\$40,000}{1.2321}\right) - \$50,000$$

$$= \$18,018 + \$32,465 - \$50,000$$

$$= \$483$$

ST-2. To find the IRR, we solve for the k value that results in an NPV of zero in the NPV formula (Equation 10-1a) by trial and error:
Try using k = 11%:

$$0 = \left(\frac{\$20{,}000}{(1+.11)^1}\right) + \left(\frac{\$40{,}000}{(1+.11)^2}\right) - \$50{,}000$$

$$= \left(\frac{\$20{,}000}{1.11}\right) + \left(\frac{\$40{,}000}{1.2321}\right) - \$50{,}000$$

$$= \$18{,}018 + \$32{,}465 - \$50{,}000$$

$$= \$483$$

The result does not equal zero. Try again using a slightly higher discount rate: Second, try using k = 11.65%:

$$0 = \left(\frac{\$20{,}000}{(1+.1165)^1}\right) + \left(\frac{\$40{,}000}{(1+.1165)^2}\right) - \$50{,}000$$

$$= \left(\frac{\$20{,}000}{1.1165}\right) + \left(\frac{\$40{,}000}{1.24657225}\right) - \$50{,}000$$

$$= \$17{,}913 + \$32{,}088 - \$50{,}000$$

$$= \$1$$

Close enough. The IRR of the project is almost exactly 11.65 percent.

ST-3. The NPV of the project must be calculated to see if it is greater or less than zero. Because the project in this problem is an annuity, its NPV can be found most easily using a modified version of Equation 10-1b as follows:

$$NPV = PMT(PVIFA_{k,n}) - \text{Initial Investment}$$

We include this problem (and some others that follow) to show you that Equations 10-1a and 10-1b can be modified to take advantage of the present value of an annuity formulas covered in chapter 8, Equations 8-4a and 8-4b. The problem could also be solved using the basic forms of Equations 10-1a and 10-1b but would take longer.

In this problem the initial investment is $200,000, the annuity payment is $44,503, the discount rate k is 14 percent, and the number of periods n is 10.

$$NPV = \$44{,}503(PVIFA_{14\%, 10}) - \$200{,}000$$

$$= \$44{,}503(5.2161) - \$200{,}000$$

$$= \$232{,}132 - \$200{,}000$$

$$= \$32{,}132$$

Because the NPV is positive, the project should be accepted.

ST-4. The IRR is found by setting the NPV formula to zero and solving for the IRR rate, k. In this case, there are no multiple cash flows so the equation can be solved algebraically. For convenience we use Equation 10-1b modified for an annuity:

$$0 = \$44{,}503(PVIFA_{k\%, 10}) - \$200{,}000$$

$$\$200{,}000 = \$44{,}503(PVIFA_{k\%, 10})$$

$$\frac{\$200{,}000}{\$44{,}503} = PVIFA_{k\%, 10}$$

$$4.49408 = PVIFA_{k\%, 10}$$

Finding the PVIFA in Table IV inside the book cover, we see that 4.4941 (there are only four decimal places in the table) occurs in the year 10 row in the 18 percent column. Therefore, the IRR of the project is 18 percent.

ST-5. There are only two cash flows in this problem, a $400 investment at time 0 and the $1,000 payoff at the end of year 10. To find the IRR, we set the NPV formula to zero, fill in the two cash flows and periods, and solve for the IRR rate k that makes the right-hand side of the equation equal zero. Equation 10-1b is more convenient to use in this case:

$$0 = \$1{,}000(PVIF_{k\%, 10}) - \$400$$
$$\$400 = \$1{,}000(PVIF_{k\%, 10})$$
$$\frac{\$400}{\$1{,}000} = PVIF_{k\%, 10}$$
$$.4000 = PVIF_{k\%, 10}$$

If we find the PVIF in Table II inside the book cover, we see that .4000 occurs in the year 10 row between the 9 percent and 10 percent columns. Therefore, the IRR of the project is between 9 percent and 10 percent. (The exact value of the IRR is 9.596 percent). Because your required rate of return in this problem was 12 percent, the bonds would *not* meet your requirements.

ST-6. Solve for the NPV using Equation 10-1a or b. Equation 10-1a would be the more convenient version for this problem. The cash flows are a $300 investment at time-0 and a $1,000 future value (FV) in 10 years. The discount rate is 20 percent.

$$\begin{aligned}
NPV &= \left(\frac{\$1{,}000}{(1 + .20)^{10}}\right) - \$300 \\
&= \left(\frac{\$1{,}000}{6.191736}\right) - \$300 \\
&= \$161.51 - \$300 \\
&= -\$138.49
\end{aligned}$$

Appendix 10A
Wrinkles in Capital Budgeting

In this appendix we discuss three situations that change the capital budgeting decision process. First, we examine *nonsimple projects,* projects that have a negative initial cash flow, in addition to one or more negative future cash flows. Next, we explore projects that have multiple IRRs. Finally, we discuss how to compare mutually exclusive projects with unequal project lives.

Nonsimple Projects

Most capital budgeting projects start with a negative cash flow—the initial investment—at t_0 followed by positive future cash flows. Such projects are called **simple projects. Nonsimple projects** are projects that have one or more negative future cash flows after the initial investment.

To illustrate a nonsimple project, consider Project N, the expected cash flows for a nuclear power plant project. The initial investment of $500 million is a negative cash flow at t_0, followed by positive cash flows of $25 million per year for 30 years as electric power is generated and sold. At the end of the useful life of the project, the storage of nuclear fuel and the shutdown safety procedures require cash outlays of $100 million at the end of year 31. The timeline depicted in Figure 10A-1 shows the cash flow pattern:

Figure 10A-1 Cash Flow Timeline for Project N (in millions)

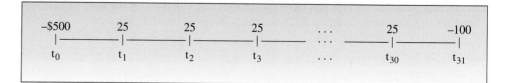

With a 20 percent discount rate, an initial investment of −$500 million, a 30-year annuity of $25 million, and a shutdown cash outlay of −$100 million in year 31, the NPV of Project N follows (in millions):

> **Take Note**
>
> The $25 million annual payment for 30 years constitutes an annuity, so we were able to adapt Equation 10-1a by using the present value of annuity formula, Equation 8-4a.

$$NPV_n = -\$500\left(\frac{1}{(1+.20)^0}\right) + \$25\left[\frac{1 - \frac{1}{(1+.20)^{30}}}{.20}\right] - \$100\left(\frac{1}{(1+.20)^{31}}\right)$$

$$= (-\$500 \times 1) + (\$25 \times 4.97894) - (\$100 \times .0035106)$$

$$= -\$500 + \$124.47341 - \$.35106$$

$$= -\$375.8776$$

We find that at a discount rate of 20 percent, Project N has a negative net present value of −375.878, so the firm considering Project N should reject it.

MULTIPLE IRRs

Some projects may have more than one internal rate of return. That is, a project may have several different discount rates that result in a net present value of zero.

Here is an example. Suppose Project Q requires an initial cash outlay of $160,000 and is expected to generate a positive cash flow of $1 million in year 1. In year 2, the project will require an additional cash *outlay* in the amount of $1 million. The cash flows for Project Q are shown on the following timeline in Figure 10A-2:

We find the IRR of Project Q by using the trial-and-error procedure, Equation 10-2. When k = 25%, the NPV is zero.

$$0 = \left(\frac{\$1,000,000}{(1+.25)^1}\right) - \left(\frac{\$1,000,000}{(1+.25)^2}\right) - \$160,000$$

$$= \$800,000 - \$640,000 - \$160,000$$

$$= \$0$$

Because 25 percent causes the NPV of Project Q to be zero, the IRR of the project must be 25 percent. But wait! If we had tried k = 400%, the IRR calculation would look like this:

$$0 = \left(\frac{\$1,000,000}{(1+4.00)^1}\right) - \left(\frac{\$1,000,000}{(1+4.00)^2}\right) - \$160,000$$

$$= \$200,000 - \$40,000 - \$160,000$$

$$= \$0$$

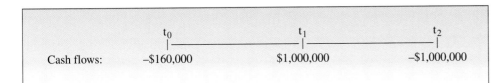

Figure 10A-2 Cash Flow Timeline for Project Q

Because 400 percent results in an NPV of zero, 400 percent must also be the IRR of the Project Q. Figure 10A-3 shows the net present value profile for Project Q. By examining this graph, we see how 25 percent and 400 percent both make the net present value equal to zero.

As the graph shows, Project Q's net present value profile crosses the horizontal axis (has a zero value) in two different places, at discount rates of 25 percent and 400 percent.

Project Q had two IRRs because the project's cash flows changed from negative to positive (at t_1) and then from positive to negative at (t_2). It turns out that a non-simple project may have (but does not have to have) as many IRRs as there are sign changes. In this case, two sign changes resulted in two internal rates of return.

Whenever we have two or more IRRs for a project, the IRR method is not a useful decision-making tool. Remember the IRR accept/reject decision rule: Firms should accept projects with IRRs higher than the discount rate and reject projects with IRRs lower than the discount rate. With more than one discount rate, decision makers will not know which IRR to use for the accept/reject decision. In projects that have multiple IRRs, then, switch to the NPV method.

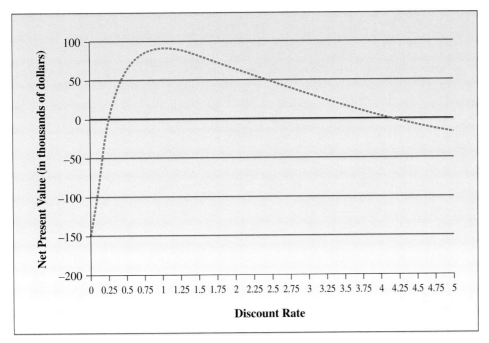

Figure 10A-3 Net Present Value Profile for Project Q

The net present value profile for Project Q crosses the zero line twice, showing that two IRRs are possible.

Chapter 10   *Capital Budgeting Decision Methods*

Mutually Exclusive Projects with Unequal Project Lives

When mutually exclusive projects have different expected useful lives, selecting among the projects requires more than comparing the projects' NPVs. To illustrate, suppose you are a business manager considering a new business telephone system. One is the Cheap Talk System, which requires an initial cash outlay of $10,000 and is expected to last three years. The other is the Rolles Voice System, which requires an initial cash outlay of $50,000 and is expected to last 12 years. The Cheap Talk System is expected to generate positive cash flows of $5,800 per year for each of its three years of life. The Rolles Voice System is expected to generate positive cash flows of $8,000 per year for each of its 12 years of life. The cash flows associated with each project are summarized in Figure 10A-4.

To decide which project to choose, we first compute and compare their NPVs. Assume the firm's required rate of return is 10 percent. We solve for the NPVs as follows:

1. **NPV of Cheap Talk.**

$$NPV_{CT} = \$5{,}800 \left[\frac{1 - \frac{1}{(1+.10)^3}}{.10} \right] - \$10{,}000$$

$$= \$5{,}800(2.48685) - \$10{,}000$$

$$= \$14{,}424 - \$10{,}000$$

$$= \$4{,}424$$

2. **NPV of Rolles Voice.**

$$NPV_{Rolles} = \$8{,}000 \left[\frac{1 - \frac{1}{(1+.10)^{12}}}{.10} \right] - \$50{,}000$$

$$= \$8{,}000(6.81369) - \$50{,}000$$

$$= \$54{,}510 - \$50{,}000$$

$$= \$4{,}510$$

We find that Project Cheap Talk has an NPV of $4,424, compared to Project Rolles' NPV of $4,510. We might conclude based on this information that the Rolles Voice System should be selected over the Cheap Talk system because it has the higher NPV. However, before making that decision, we must assess how Project Cheap Talk's NPV would change if its useful life were 12 years, not three years.

Take Note

Note that the unequal lives problem is only an issue when the projects under consideration are mutually exclusive. If the projects were independent, we would adopt all projects that had NPVs greater than or equal to zero, no matter what the lives of the projects.

Figure 10A-4 Cash Flows for the Cheap Talk and Rolles Voice Communications Systems (in thousands)

This figure shows the cash flows for two projects with unequal lives, Project Cheap Talk and Project Rolles Voice.

```
Cheap    t₀     t₁     t₂     t₃
Talk    ─$10  +$5.8  +$5.8  +$5.8

         t₀   t₁   t₂   t₃   t₄   t₅   t₆   t₇   t₈   t₉   t₁₀  t₁₁  t₁₂
Rolles  ─$50 +$8  +$8  +$8  +$8  +$8  +$8  +$8  +$8  +$8  +$8  +$8  +$8
```

Comparing Projects with Unequal Lives

Two possible methods that allow financial managers to compare projects with unequal lives are the replacement chain approach and the equivalent annual annuity (EAA) approach.

The Replacement Chain Approach The *replacement chain* approach assumes each of the mutually exclusive projects can be replicated, until a common time period has passed in which the projects can be compared. The NPVs for the series of replicated projects are then compared to the project with the longer life. An example illustrates this process. Project Cheap Talk could be repeated four times in the same time span as the 12-year Rolles Voice project. If a business replicated project Cheap Talk four times, the cash flows would look like this:

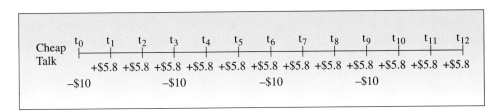

Figure 10A-5 Cash Flows for the Cheap Talk System Repeated Four Times (in thousands)

The NPV of this series of cash flows, assuming the discount rate is 10 percent, is (in thousands) $12,121. Each cash flow, be it positive or negative, is discounted back the appropriate number of years to get the NPV of the four consecutive investments in the Cheap Talk system.

The NPV of $12,121 for Cheap Talk system is the sum of the NPVs of the four repeated Cheap Talk projects, such that the project series would have a life of 12 years, the same life as the Rolles Voice System Project. We are now comparing apples to apples. Cheap Talk's replacement chain NPV is $12,121, whereas the NPV of Project Rolles Voice is $4,510 over the same 12 year period. If a firm invested in project Cheap Talk four successive times, it would create more value than investing in one project Rolles Voice.

The Equivalent Annual Annuity (EAA) The equivalent annual annuity (EAA) approach converts the NPVs of the mutually exclusive projects into their equivalent annuity values. The *equivalent annual annuity* is the amount of the annuity payment that would give the same present value as the actual future cash flows for that project. The EAA approach assumes that you could repeat the mutually exclusive projects indefinitely as each project came to the end of its life.

The equivalent annuity value (EAA) is calculated by dividing the NPV of a project by the present value interest factor for an annuity (PVIFA) that applies to the project's life span.

Formula for an Equivalent Annual Annuity (EAA) (10A-1)

$$EAA = \frac{NPV}{PVIFA_{k,n}}$$

where: k = Discount rate used to calculate the NPV
 n = Life span of the project

The NPVs of Cheap Talk ($4,424) and Rolles Voice ($4,510) were calculated earlier, assuming a required rate of return of 10 percent. With the project's NPV and the discount rate, we calculate each project's EAA, per Equation 10A-1, as follows:

1. EAA of Project Cheap Talk

$$EAA_{CT} = \frac{\$4,424}{2.48685}$$
$$= \$1778.96$$

2. EAA of Project Rolles

$$EEA_{Rolles} = \frac{\$4,510}{6.81369}$$
$$= \$661.90$$

The EAA approach decision rule calls for choosing whichever mutually exclusive project has the highest EAA. Our calculations show that Project Cheap Talk has an EAA of $1,778.96 and Project Rolles Voice System has an EAA of $661.90. Because Project Cheap Talk's EAA is higher than Project Rolles's, Project Cheap Talk should be chosen.

Both the replacement chain and the EAA approach assume that mutually exclusive projects can be replicated. If the projects can be replicated, then either the replacement chain or the equivalent annual annuity methods should be used because they lead to the same correct decision. Note in our case that the EAA method results in the same project selection (Project Cheap Talk) as the replacement chain method. If the projects cannot be replicated, then the normal NPVs should be used as the basis for the capital budgeting decision.

Equations Introduced in This Appendix

Equation 10A-1. The Formula for an Equivalent Annual Annuity:

$$EAA = \frac{NPV}{PVIFA_{k,n}}$$

where: k = Discount rate used to calculate the NPV
n = Life span of the project

CHAPTER 11

ESTIMATING INCREMENTAL CASH FLOWS

CHAPTER OBJECTIVES

After reading this chapter, you should be able to:

- Explain the difference between incremental cash flows and sunk costs.

- Identify types of incremental cash flows in a capital budgeting project.

- Explain why cash flows associated with project financing are not included in the capital budgeting analysis.

Never underestimate the value of cold cash.
—Gregory Nunn

SAVING MONEY?

Company X is a large multinational corporation with many facilities throughout the United States and the rest of the world. Employees at a variety of U.S. sites frequently are required to travel to the home office in Headquarters City. On a typical day there may be a dozen or more employees traveling to Headquarters City from the many satellite city sites.

Company X has many corporate jets at airports throughout the United States near the larger satellite cities. Senior executives routinely fly on these corporate jets when traveling to Headquarters City. Middle-level managers fly on commercial aircraft, usually located at quite a greater distance from the workplace. The reason for this is that the department of the traveling employee is "billed" $800 if the corporate jet is used. This $800 expense goes into the financial report of that department, which goes to corporate headquarters. Managers can frequently find commercial airfares under $300 for employees traveling to Headquarters City.

Because each department would rather be charged $300 instead of $800 when reporting its financial performance, only a few of the most senior executives fly the corporate jet. This means that the corporate jet typically has a dozen empty seats for its daily flights to Headquarters City.

It is clearly in the interest of each department manager to keep his or her expenses down. Is it in the interests of the stockholders to have a mostly empty plane fly each day to Headquarters City? The cost of adding an additional passenger, or 12 additional passengers, to the corporate jet is almost zero. A very small amount of additional fuel would be consumed. The stockholders would save $300 for each additional person who took an otherwise empty seat on the corporate jet instead of flying on the commercial airline.

Consider the interests of the department managers and those of the stockholders of Company X as you read chapter 11.

Chapter Overview

In chapter 10 we applied capital budgeting decision methods, taking the cash flow estimates as a given. In this chapter, we see how financial managers determine which cash flows are incremental and, therefore, relevant to a capital budgeting decision. We define incremental cash flows and distinguish incremental cash flows from sunk costs. We also examine how financial managers estimate incremental initial investment cash flows and incremental operating cash flows in the capital budgeting decision. Finally, we explore how the financing cash flows of a capital budgeting project are factored into the capital budgeting decision.

Incremental Cash Flows

We discussed the importance of cash in chapter 1, pages 6 and 7.

The capital budgeting process focuses on cash flows, not accounting profits. Recall from our discussion in chapter 1, it is cash flow that changes the value of a firm. Cash outflows reduce the value of the firm, whereas cash inflows increase the value of the firm.

In capital budgeting **incremental cash flows** are the positive and negative cash flows directly associated with a project. They occur if a firm accepts a project but do not occur if the project is rejected.

For instance, suppose that the chief financial officer of Photon Manufacturing, Mr. Sulu, is analyzing the cash flows associated with a proposed project. He finds that the CEO hired a consultant to assess the proposed project's environmental effects. The consultant will be paid $50,000 for the work. Although the $50,000 fee is related to the project, it is not an incremental cash flow because the money must be paid whether the project is accepted or rejected. Therefore, the fee should not be included as a relevant cash flow of the expansion project decision. Cash flows that have already occurred, or will occur whether a project is accepted or rejected, are **sunk costs**.

Financial managers carefully screen out irrelevant cash flows, such as sunk costs, from the capital budgeting decision process. If they include irrelevant cash flows in their capital budgeting decision, then their calculations of a project's payback period, net present value (NPV), or internal rate of return (IRR) will be distorted and inaccurate. The calculations may be so distorted that they lead to an incorrect decision about a capital budgeting project.

Types of Incremental Cash Flows

To accurately assess the value of a capital budgeting project, financial managers must identify and estimate many types of incremental cash flows. The three main

Source: DILBERT © 1996 United Feature Syndicate, Inc. Reprinted by permission.

types of incremental cash flows are *initial investment cash flows, operating cash flows,* and *shutdown cash flows.* We examine these three types of incremental cash flows in the sections that follow.

Initial Investment Cash Flows

Generally, financial managers begin their incremental cash flow estimates by assessing the costs of the initial investment. The negative cash flow associated with the initial investment occurs only if the project is accepted. Initial investment cash flows include the purchase price of the asset or materials to produce the asset, the installation and delivery costs, and the additional investment in net working capital.

Purchase Price, Installation, and Delivery Financial managers usually obtain quotes on the purchase price and installation and delivery costs from suppliers. These figures, then, can usually be estimated with a high degree of accuracy.

Changes in Net Working Capital Aside from the set-up costs and purchase price of a proposed capital budgeting project, a company may have to invest in changes in net working capital. As explained in chapter 4, net working capital is defined as current assets (working capital) minus current liabilities. If a proposed capital budgeting project will cause a positive change in net working capital (the most likely scenario), the cash outlay needed to finance this must be included in the cash flow estimates.

We defined net working capital in chapter 4, page 67.

A capital budgeting project may affect working capital. Recall that working capital consists of cash, accounts receivable, and inventory, along with other current assets, if any. Companies invest in these assets in much the same way they invest in plant and equipment. Accepting a new project often triggers an increase in cash, accounts receivable, and inventory investments. Working capital investments tie up cash the same way that investment in a new piece of equipment does.

A company's current liabilities—such as accounts payable, accrued wages, and accrued taxes—may also be affected if a firm accepts a capital budgeting project. For example, if a plant is expanded, the company may place larger orders with suppliers to accommodate the increased production. The increase in orders is likely to lead to an increase in accounts payable.

TABLE 11-1

Change in McGuffin Company Net Working Capital If New Project Accepted

Current Asset Changes	Current Liability Changes
$5,000 Increase in Cash	$8,000 Increase in Accounts Payable
$7,000 Increase in Receivables	$2,000 Increase in Accruals
$15,000 Increase in Inventory	
Total Current Asset Changes: $27,000	Total Current Liability Changes: $10,000
Increase in Net Working Capital (NWC): $27,000 − $10,000 = $17,000	
Incremental Cash Flow Due to the Increase in NWC = −$17,000	

Increases in current liabilities create cash inflows. It is unlikely that current liabilities will increase sufficiently to finance all the needed current asset buildup. This is the reason that an investment in net working capital is almost always required.

Table 11-1 shows an example of the incremental changes in net working capital that might occur with a proposed capital budgeting project for the McGuffin Company.

As Table 11-1 indicates, the McGuffin Company project has an estimated increase of $27,000 in current assets and a $10,000 increase in current liabilities, resulting in a $17,000 change in net working capital. That is, the firm will have to spend $17,000 to increase its net working capital by this amount—a negative cash flow.

Figure 11-1 depicts graphically the $17,000 change in the McGuffin Company's net working capital if the new project is accepted.

Once financial managers estimate the initial investment incremental cash flows, they analyze the operating cash flows of a capital budgeting project. We turn to those cash flows next.

Operating Cash Flows

Operating cash flows are those cash flows that the project generates after it is in operation. For example, cash flows that follow a change in sales or expenses are operating cash flows. Those operating cash flows incremental to the project under consideration are the ones relevant to our capital budgeting analysis. Incremental operating cash flows also include tax changes, including those due to changes in depreciation expense, opportunity costs, and externalities.

Taxes The change in taxes that will occur if a project is accepted is part of the incremental cash flow analysis. Tax effects are considered because a tax increase is equivalent to a negative cash flow, and a tax decrease is equivalent to a positive cash flow. In a capital budgeting decision, then, financial managers must examine whether and how much tax the firm will pay on additional income that the proposed project generates during a given period. They must also see whether and how much taxes will decrease if the project increases the firm's periodic operating expenses (such as payments for labor and materials), thereby creating additional tax deductions.

Depreciation and Taxes Financial managers estimate changes in depreciation expense as part of the incremental cash flow analysis because increases in depreciation expense may increase a firm's cash flow. How? Depreciation expense is deductible for tax purposes. The greater the depreciation expense deduction, the less tax a firm must pay, and the less cash it must give to the IRS. Financial managers estimate the

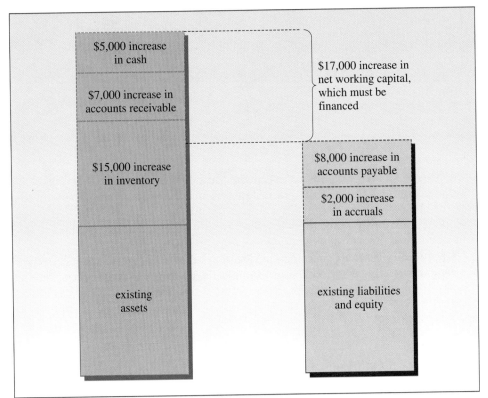

Figure 11-1 Change in McGuffin's Net Working Capital Due to New Project

Figure 11-1 shows the $17,000 increase in net working capital. The McGuffin Company must raise an additional $17,000 in funds if it accepts the new project.

amount of depreciation expense a capital budgeting project will have, therefore, to see how much the firm's taxable income and taxes owed will decrease.

Incremental depreciation expense is the change in depreciation expense that results from accepting a proposed capital budgeting project. Incremental depreciation expense affects the change in taxes attributable to a capital budgeting project.

To illustrate how incremental depreciation expense changes taxes due, recall how we converted after-tax net profits into operating cash inflows in chapter 4. We added all noncash charges (including depreciation) that were deducted as expenses on the firm's income statement to net profits after taxes. Once the tax effects of a project's depreciation expense are calculated, we add this incremental depreciation expense back to the project's net profits after taxes.

We discussed converting after-tax net profits into operating cash flows in chapter 4, page 68.

The following example demonstrates how to estimate the incremental depreciation expense of a capital budgeting project. Suppose your firm is considering a project that is expected to earn $100,000 in cash sales in year 1. Suppose that, in addition to the sales increase, the project is expected to increase cash operating expenses by $50,000 and new depreciation expense will be $10,000. Assume your firm's marginal tax rate is 40 percent. First, compute the net operating cash flows from the project for this year, as shown in Table 11-2.

Compare line 3 and line 7 in Table 11-2. Note that once we used the new project incremental depreciation expense of $10,000 to make the tax change calculations, we added the $10,000 depreciation expense back to the new project's after-tax net income to calculate the year's incremental operating cash flow for this year from the new project. The incremental depreciation expense affected cash flow only because of its effect on taxes.

Chapter 11 *Estimating Incremental Cash Flows* **277**

TABLE 11-2

Net Operating Cash Flows for New Project

1. New Project Sales (cash inflow)	$100,000
2. New Project Cash Operating Expenses (cash outflow)	− 50,000
3. New Project Depreciation Expense (noncash expense)	− 10,000
4. Net New Project Taxable Income	40,000
5. Taxes on New Project Income (40%)(cash outflow)	− 16,000
6. Net New Project After-Tax Income	24,000
7. Plus Depreciation Expense (added back)	+ 10,000
8. Net Incremental Cash Flow	$ 34,000

Opportunity Costs Sometimes accepting a capital budgeting project precludes other opportunities for the firm. For instance, if an industrial mixer, already owned by a toy company, is used to make a new product called Slime #4, then that mixer will not be available to make the Slime #2 currently produced in that mixer. The forgone benefits of the alternative not chosen are opportunity costs.

Opportunity costs are incremental cash flows that financial managers consider in a capital budgeting decision. In our example, the opportunity cost comes from the lost use of the industrial mixer for other products our firm makes. If our cash flows decrease by $30,000 due to the decrease in sales of Slime #2 that we can no longer make, then this $30,000 is the opportunity cost of choosing to produce the new product, Slime #4.

Externalities In estimating incremental operating cash flows, financial managers consider the effect a capital budgeting project might have on cash flows related to other parts of the business. **Externalities** are the positive or negative effects on existing projects if a new capital budgeting project is accepted.

For instance, suppose that a tennis ball manufacturer decides to start making tennis racquets but does not want to hire any additional managers. The current managers may become overworked because the expansion project requires manager time and oversight. This is a negative externality. Existing projects suffer due to manager inattention, but it is difficult to measure the size of those incremental costs.

On the other hand, the new racquet project may give the company more visibility than it had before and increase sales of its existing tennis ball business, thereby leading to an increase in cash flows. Because these cash flows from the increased tennis ball sales are incremental to the tennis racquet project under consideration, they should be considered in the capital budgeting analysis. This is a positive externality. Here again, however, the costs associated with the positive externalities are likely to be difficult to measure.

If the impact of externalities can be measured, they should be incorporated in the capital budgeting analysis. If the cost of externalities cannot be measured precisely—as is likely the case—most firms use a subjective analysis of externalities before making a project's final accept or reject decision. For example, if the net present value of a project is only slightly greater than zero, company officials may reject the project if they believe significant unmeasured negative externalities are present.

After estimating the operating cash flows, financial managers then estimate the shutdown cash flows of a proposed capital budgeting project. We examine these costs next.

Real Options Externalities and opportunity costs are not the only elements of the capital budgeting decision that are difficult to reduce to an incremental cash flow estimate. Many projects have options embedded in them that add to the value of the project and, therefore, of the firm. For example, a project may provide management with the option to revise a capital budgeting project at a later date. This characteristic is called a **real option.** It is a real option because it is related to a real asset such as a piece of equipment or a new plant. You may already be familiar with financial options (calls and puts) that give the holder the opportunity to buy or sell financial assets such as stocks or bonds at a later date. Real options are similar except that their value is related to the value of real assets rather than to the value of financial assets. Note that the word *option* indicates that the future alternative does not have to be taken. It will be taken only if it is seen as adding value.

The flexibility that is provided by a real option to revise a project at a later date has value. This option may be to expand a project, to abandon it, to create another project that is an offshoot of the current project, or something else. For example, a restaurant with room to expand is more valuable than one that is confined to its original fixed space, other things being equal. A project that can be shut down before its scheduled useful life if it turns out to be a failure is more valuable, other things being equal, than a similar failed project that must continue operating. An investment in a research laboratory that might develop a wonderful new drug that is completely unknown to us now is more valuable, other things being equal, than an investment in another project that has no potential future spinoffs.

Traditional NPV and IRR analysis often overlooks the value that may come from real options because this value cannot be reduced to a simple incremental cash flow estimate. Faced with this difficulty managers usually omit real options that are part of a project from capital budgeting analysis. This causes the NPV and IRR figures to be understated. As a result the value that real options add to the firm and the increase in the project rates of return they provide are not recognized.

The NPV process can be modified to reflect the value that real options add to the firm. This involves computing the traditional NPV and then adding today's value of any real options that may be present. The mechanics of this are best left to a subsequent finance course but every student of finance should be aware of the fact that flexibility, such as that connected with real options, adds value.

Cash Flows at the End of a Project's Life

Financial managers estimate the shutdown cash flows that are expected to occur at the end of the useful life of a proposed capital budgeting project. Shutdown cash flows may include those from the project's salvage value, the reduction of net working capital, and taxes tied to the sale of the used asset.

Project Salvage Value   If a project is expected to have a positive salvage value at the end of its useful life, there will be a positive incremental cash flow at that time. However, this salvage value incremental cash flow must be adjusted for tax effects.

Four possible tax scenarios may occur when the used asset is sold, depending on the asset's sale price (summarized in Table 11-3). First, the asset may be sold for more than its purchase price. In this instance, the difference between the purchase and the sale price is taxed at the capital gains tax rate. (The *capital gain* is the portion of the sale price that exceeds the purchase price.) In addition, the purchase price minus depreciation book value is taxed at the ordinary income rate.

Second, the asset may be sold for less than the purchase price but for more than its depreciation book value. The amount that the sales price exceeds the depreciation book value is ordinary income, so it is taxed at the ordinary income tax rate.

TABLE 11-3

Tax Effects of the Sale of an Asset at the End of Its Useful Life

TYPE OF SALE	TAX EFFECT
The asset is sold for more than its purchase price.	This difference is taxed at the capital gains rate. In addition, the purchase price minus the depreciation book value is ordinary income and is taxed at the ordinary income rate.
The asset is sold for less than its purchase price but for more than its depreciation book value.	The sales price minus the depreciation book value is ordinary income and is taxed at the ordinary income tax rate.
The asset is sold for its depreciation book value.	There is no tax effect.
The asset is sold for less than its depreciation book value.	The depreciation book value minus the sales price is an ordinary loss and reduces the firm's tax liability by that amount times the ordinary income tax rate.

Third, the asset may be sold for its depreciation book value. In that case the asset sale has no tax effects.

Fourth, the asset may be sold for less than its depreciation book value. The amount the depreciation book value exceeds the sales price is an ordinary loss. The firm's tax liability is reduced by the amount of the loss times the ordinary income tax rate.

Incremental Cash Flows of an Expansion Project

To practice capital budgeting cash flow estimation, let's examine a proposed expansion project. An expansion project is one in which the company adds a project and does not replace an existing one. Photon Manufacturing makes torpedoes. It is considering a project to install $3 million worth of new machine tools in its main plant. The new tools are expected to last for five years. Photon operations management and marketing experts estimate that during that five years the tools will result in a sales increase of $800,000 per year.

The Photon Manufacturing CEO has asked Mr. Sulu, the company CFO, to identify all incremental cash flows associated with the project, and to calculate the project's net present value (NPV) and internal rate of return (IRR). Based on the incremental cash flow analysis and Sulu's recommendation, the company will make an accept or reject decision about the project.

Sulu's first step is to identify the relevant (incremental) cash flows associated with the project. He begins with the initial investment in the project, then looks at the operating cash flows, and finally the shutdown cash flows.

The cash flows that will occur as soon as the project is implemented (at t_0) make up the project's initial investment. The initial investment includes the cash outflows for the purchase price, installation, delivery, and increase in net working capital.

Sulu knows that its tool supplier gave Photon a bid of $3 million to cover the cost of the new tools, including set-up and delivery. Photon inventory and accounting specialists estimate that if the tools are purchased, inventory will need to increase by $40,000, accounts receivable by $90,000, and the needed increase in cash will be $10,000. This is a $140,000 increase in current assets (working capital).

Also, Photon experts estimate that if the tools are purchased, accounts payable will increase by $20,000 as larger orders are placed with suppliers, and accruals

FINANCIAL MANAGEMENT AND YOU

The Incremental Costs of Studying Abroad

Imagine that you want to study international finance at the Universität Heidelberg in Germany during the next academic year. Your college has an exchange agreement with Heidelberg that guarantees tuition and housing at your in-state resident rate. To decide whether to spend the academic year studying abroad instead of at home, you draw up a list of costs in U.S. dollars. Here is what you find:

1. Cost of travel to and from Heidelberg $1,100.00
2. Cost of student lodging and tuition for academic year 4,985.00
3. Cost of books and other supplies 600.00
4. Cost of food ($450 per month for nine months) 4,050.00
5. Cost of summer travel in Europe (including food, supplies, rail passes, and student hostel charges) 2,000.00

Total: $12,735.00

At first glance the opportunity appears too expensive to consider. Now that you have read this chapter, however, you realize you need to screen out irrelevant costs, and consider opportunity costs and externalities. Your new list looks like this:

1. Cost of travel to and from Heidelberg $1,100.00
2. Cost of tuition and lodging in Heidelberg minus cost of lodging and tuition in the United States 0.00
3. Cost of books and other supplies ($600) minus cost of same in the United States ($500) 100.00
4. Cost of food abroad for academic year ($4,050) minus cost of same in the United States ($3,150 with cafeteria pass) 900.00
5. Cost of summer travel in Europe ($2,000) minus cost of summer room/board/travel in the United States ($1,200) 800.00
6. Opportunity cost of no summer job abroad (U.S. summer job income of $1,200 minus $0 income abroad) 1,200.00

Total: $4,100.00

You estimate that your relevant costs are $4,100 for the year. Although you can't measure them exactly, you know the study abroad has positive externalities—it will further your career plans. Can you afford to accept this opportunity?

Source: Air travel, rail passes, and lodging price information retrieved from Saber® Database.

(wages and taxes) will increase by $10,000—a $30,000 increase in current liabilities. Subtracting the increases in current liabilities from the increases in current assets ($140,000 − $30,000) results in a $110,000 increase in net working capital associated with the expansion project.

Sulu concludes after extensive research that he has found all the initial investment incremental cash flows. Those cash flows are summarized in Table 11-4.

Now Sulu examines the operating cash flows, those cash flows expected to occur from operations during the five-year period after the project is implemented (at t_1 through t_5). The Photon expansion project operating cash flows reflect changes in sales, operating expenses, and depreciation tax effects. We assume these cash flows occur at the end of each year.

Sulu learns from the vice president of sales that cash sales are expected to increase by $800,000 per year because the new tools will increase manufacturing capacity. If purchased, the tools will be used to perform maintenance on other equipment at Photon, so operating expenses (other than depreciation) are expected to decrease by $100,000 per year.

TABLE 11-4

Photon Manufacturing Expansion Project Initial Investment Incremental Cash Flows at t_0

Cost of Tools and Setup	$3,000,000
+ Investment in Additional NWC	110,000
= Total Initial Cash Outlay	$3,110,000

We listed MACRS depreciation rules in chapter 4, Table 4-1, page 73.

Depreciation is a noncash expense, but remember that Sulu must use depreciation to compute the change in income tax that Photon must pay. After taxes are computed, Sulu then will add back the change in depreciation in the operating cash flow analysis.

To calculate depreciation expense, Sulu looks at MACRS depreciation rules and finds that the new manufacturing tools are in the three-year asset class. According to the MACRS rules, 33.3 percent of the new tools' $3 million cost will be charged to depreciation expense in the tools' first year of service, 44.5 percent in the second year, 14.8 percent in the third year, and 7.4 percent in the fourth year.[1]

Now Sulu summarizes the incremental operating cash flows for the Photon capital budgeting project in Table 11-5.

Sulu is not quite through yet. He must include in his analysis additional shutdown cash flows that occur at t_5, the end of the project's life.

Photon company experts estimate that the actual economic life of the tools will be five years, after which time the tools should have a salvage value of $800,000. Under MACRS depreciation rules, assets are depreciated to zero at the end of their class life, so at t_5 the book value of the new tools is zero. Therefore, if the tools are sold at the end of year 5 for their salvage value of $800,000, Photon Manufacturing will realize a taxable gain on the sale of the tools of $800,000 ($800,000 − 0 = $800,000). The income tax on the gain at Photon's marginal tax rate of 40 percent will be $800,000 × .40 = $320,000.

The net amount of cash that Photon will receive from the sale of the tools is the salvage value minus the tax paid:

$800,000	Salvage Value
− 320,000	Taxes Paid
$480,000	Net Proceeds

The net proceeds from the tool sale in year 5, then, are $480,000.

Finally, if the new tools are sold at t_5, Sulu concludes (based on sales department information) that Photon's sales will return to the level they were before the new tools were installed. Consequently, there will be no further need for the additional investment in net working capital that was made at t_0. When the $110,000 investment in net working capital is recaptured,[2] that amount is recovered in the form of a positive cash flow.

The additional incremental cash flows from the sale of the tools and the change in net working capital are summarized in Table 11-6.

Tables 11-4, 11-5, and 11-6 contain all the incremental cash flows associated with Photon Manufacturing's proposed expansion project. Sulu's next step is to summa-

---

[1]Depreciation expenses for the tools are spread over four years instead of three because the MACRS depreciation rules apply a half-year convention—all assets are assumed to be purchased and sold halfway through the first and last years, respectively. If an asset with a three-year life is assumed to be purchased halfway through year 1, then the three years will be complete halfway through year 4.

[2]Current assets, in the amount by which they exceed current liabilities, are sold and not replaced because they are no longer needed.

TABLE 11-5 Photon Manufacturing Expansion Project Incremental Operating Cash Flows, Years 1–5

	t_1	t_2	t_3	t_4	t_5
+ Change in Sales	+ 800,000	800,000	800,000	800,000	800,000
+ Reduction in Nondepreciation Operating Expenses	+ 100,000	100,000	100,000	100,000	100,000
− Depreciation Expense	− 999,000	1,335,000	444,000	222,000	0
= Change in Operating Income	= (99,000)	(435,000)	456,000	678,000	900,000
− Tax on New Income (See Note)	− (39,600)	(174,000)	182,400	271,200	360,000
= Change in Earnings after Taxes	= (59,400)	(261,000)	273,600	406,800	540,000
+ Add Back Depreciation Expense	+ 999,000	1,335,000	444,000	222,000	0
= Net Incremental Operating Cash Flow	= 939,600	1,074,000	717,600	628,800	540,000

Note: Taxes at t_1 and t_2 are negative amounts, which means earnings after taxes on the lines above for those years is *increased* by the amount of the taxes saved. Operating losses in year 1 (−$99,000) and in year 2 (−$435,000) cause a decrease in the taxes Photon owes during years 1 and 2 of $39,600 and $174,000, respectively.

rize the total incremental net cash flows occurring at each point in time in one table. Table 11-7 shows the results.

Table 11-7 contains the bottom-line net incremental operating cash flows associated with Photon's proposed expansion project. The initial investment at t_0 is −$3,110,000. The operating cash flows from t_1 to t_5 total $4,490,000. The sum of all the incremental positive and negative cash flows for the project is $1,380,000.

In chapter 10 we discussed the NPV and IRR techniques on pages 239–249.

Now Sulu is ready to compute the NPV and IRR of the expansion project based on the procedures described in chapter 10.

Assuming that Photon's discount rate is 10 percent, Sulu computes the NPV of the project using Equation 10-1 as follows[3]:

	t_0	t_1	t_2	t_3	t_4	t_5
Net Incremental Cash Flows	(3,110,000)	939,600	1,074,000	717,600	628,800	1,130,000
PV of Cash Flows	$\dfrac{(3,110,000)}{(1+.10)^0}$ = (3,110,000)	$\dfrac{939,600}{(1+.10)^1}$ = 854,182	$\dfrac{1,074,000}{(1+.10)^2}$ = 887,603	$\dfrac{717,600}{(1+.10)^3}$ = 539,144	$\dfrac{628,800}{(1+.10)^4}$ = 429,479	$\dfrac{1,130,000}{(1+.10)^5}$ = 701,641

---

[3] The NPV Formula, Equation 10-1a, follows:

$$\text{NPV} = \left(\frac{CF_1}{(1+k)^1}\right) + \left(\frac{CF_2}{(1+k)^2}\right) \cdots + \left(\frac{CF_n}{(1+k)^n}\right) - \text{Initial Investment}$$

TABLE 11-6

Photon Manufacturing Expansion Project Additional Incremental Operating Cash Flows at t_5

Salvage Value	800,000
− Taxes on Salvage Value	− 320,000
= Net Cash Inflow from Sale of Tools	+ 480,000
+ Cash from Reduction in NWC	+ 110,000
= Total Additional Cash Flows at t_5	= 590,000

TABLE 11-7 Photon Manufacturing Expansion Project Summary of Incremental Cash Flows

	t_0	t_1	t_2	t_3	t_4	t_5
For Purchase and Setup	(3,000,000)					
For Additional NWC	(110,000)					
From Operating Cash Flows		939,600	1,074,000	717,600	628,800	540,000
From Salvage Value Less Taxes						480,000
From Reducing NWC						110,000
Net Incremental Cash Flows	(3,110,000)	939,600	1,074,000	717,600	628,800	1,130,000

$$\text{NPV} = (3{,}110{,}000) + 854{,}182 + 887{,}603 + 539{,}144 + 429{,}479 + 701{,}641$$
$$= \$302{,}049$$

Assuming a discount rate of 10 percent, the NPV of the project is $302,049.

Next, Sulu uses the trial-and-error method described in chapter 10 to find the IRR of the project.[4] He finds the IRR, the discount rate that makes the NPV equal to zero, is 13.78 percent.

$$\text{NPV}_P = 0 = \frac{-3{,}110{,}000}{(1+k)^0} + \frac{939{,}600}{(1+k)^1} + \frac{1{,}074{,}000}{(1+k)^2} + \frac{717{,}600}{(1+k)^3} + \frac{628{,}800}{(1+k)^4} + \frac{1{,}130{,}000}{(1+k)^5}$$

$$k = .1378 \therefore \text{IRR} = .1378 = 13.78\%$$

To solve for the NPV and IRR of this project with the TI BAII PLUS financial calculator, first switch to the calculator's spreadsheet mode. Then enter the cash flows, keystrokes, and compute as follows:

---

[4]The formula for finding a project's IRR, Equation 10-2, is:

$$0 = \left(\frac{CF_1}{(1+k)^1}\right) + \left(\frac{CF_2}{(1+k)^2}\right) \ldots + \left(\frac{CF_n}{(1+k)^n}\right) - \text{Initial Investment}$$

TI BAI I PLUS FINANCIAL CALCULATOR SOLUTION	
IRR and NPV	
KEYSTROKES	DISPLAY
CF	CF_0 = old contents
2nd CLR Work	CF_0 = 0.00
3110000 +/− ENTER	−3,110,000.00
↓ 939600 ENTER	C01 = 939,600.00
↓	F01 = 1.00
↓ 1074000 ENTER	C02 = 1,074,000.00
↓	F02 = 1.00
↓ 717600 ENTER	C03 = 717,600.00
↓	F03 = 1.00
↓ 628800 ENTER	C04 = 628,800.00
↓	F04 = 1.00
↓ 1130000 ENTER	C05 = 1,130,000.00
↓	F05 = 1.00
IRR	IRR = 0.00
CPT	IRR = 13.78
NPV	I = 0.00
10 ENTER	I = 10.00
↓ CPT	NPV = 302,048.58

Because the project's NPV of $302,049 is positive, and the IRR of 13.78 percent exceeds the required rate of return of 10 percent, Sulu will recommend that Photon proceed with the expansion project.

In this discussion we examined how a firm determines the incremental costs of an expansion project, and the project's NPV and IRR. We turn next to replacement projects and their incremental costs.

AN ASSET REPLACEMENT DECISION

Often a company considers replacing existing equipment with new equipment. A replacement decision is a capital budgeting decision to purchase a new asset and replace and retire an old asset, or to keep the old asset. Financial managers identify the *differences* between the company's cash flows with the old asset versus the company's cash flows if the new asset is purchased and the old asset retired. As illustrated in Figure 11-2, these differences are the incremental cash flows of the proposed new project.

As part of an asset replacement decision, financial managers must often estimate the tax changes associated with disposing of the old asset. Table 11-3 described

Figure 11-2 **Comparing Cash Flows: Replacing an Asset versus Keeping It**

Figure 11-2 illustrates how firms compare the difference between the cash flows of replacing an old asset with a new one or keeping the old asset. The scenario that generates the greater cash flow will be the one the firm chooses.

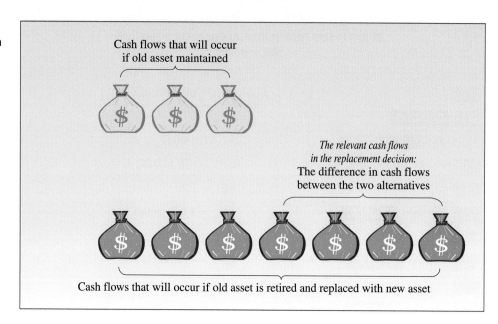

the four tax effects that can occur when an asset is sold. Table 11-8 shows a case of an old asset (that will be replaced) that sells at a price above its depreciation book value but below its purchase price. The sales price of the old asset is $80,000 and it has been depreciated over the years in the amount of $30,000. Its original depreciation basis was $90,000.

The cash received from the sale of the old asset, $72,000 as shown in Table 11-8, would be netted against any cash outflows associated with the proposed new investment. After calculating the cash received from the old asset sale, the capital budgeting analysis is the same as the expansion project discussed earlier.

FINANCING CASH FLOWS

Suppose a company planned to borrow or to sell new common stock to raise part or all of the funds needed for a proposed capital budgeting project. The company would receive a cash inflow on receipt of the loan or the sale of new common stock.

TABLE 11-8

Incremental Cash Flows at t_0 Associated with the Disposal of an Old Asset

Cash from Sale of Old Asset		$80,000
Book Value of Old Asset:		
Original Depreciation Basis	$90,000	
− Accumulated Depreciation	30,000	
= Book Value		60,000
Sales Price − Book Value = Taxable Gain (Loss)		20,000
Tax on Gain (Loss):		
$20,000 Gain (Loss) × 40% Marginal Tax Rate		8,000
Net Cash Flow from Sale of Old Asset:		
Total Cash Received − Tax on Gain (Loss)		$80,000
		− 8,000
		$72,000

Figure 11-3   The Cash Flow Estimation Process

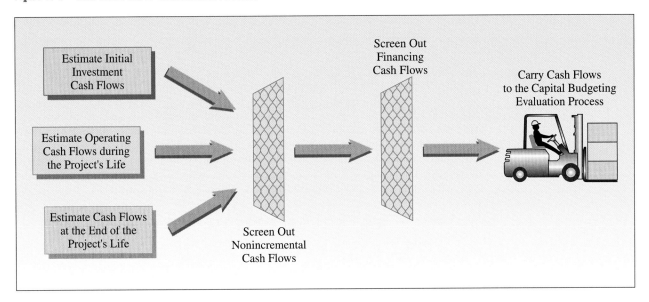

Conversely, the company must pay the interest and principal payments on the loan or may make dividend payments to stockholders. **Financing cash flows** are the cash outflows that occur as creditors are paid interest and principal and stockholders are paid dividends.

If a capital budgeting project is rejected, financing cash flows will not occur, so they are relevant cash flows in the capital budgeting decision. However, as we will see in chapter 12, financing costs are factored into the discount rate (required rate of return) in the NPV calculation. Those costs are also included in the hurdle rate of the IRR decision rule. Therefore, to avoid double counting, we do not include financing costs in our operating incremental cash flow estimates when we make capital budgeting decisions. If we did include financing costs as part of the incremental operating cash flows, then the NPV or IRR analysis would be distorted. That distortion could lead in turn to a poor capital budgeting decision.

Figure 11-3 summarizes the cash flow estimation process and its role in capital budgeting.

What's Next

In this chapter, we learned how financial managers estimate incremental cash flows as part of the capital budgeting process. We described the difference between sunk costs and incremental cash flows. We also discussed various types of incremental cash flows. In chapter 12 we examine the cost of capital.

Summary

1. *Explain the difference between incremental cash flows and sunk costs.*

Incremental cash flows are the cash flows that will occur if a capital budgeting project is accepted. They will not occur if the investment is rejected. Sunk costs are costs that will occur whether a project is accepted or rejected. Financial managers must screen out sunk costs from the capital budgeting analysis to prevent distortion in cash flow estimates. Any distortion in these estimates will, in turn, lead to inaccurate NPV or IRR values and could result in poor capital budgeting decisions.

2. Identify types of incremental cash flows in a capital budgeting project.

Financial managers must examine three main types of cash flows to estimate the incremental cash flows of a proposed capital budgeting project. First, they must assess the cost of the initial investment: the purchase price, the installation and delivery costs, and any change in net working capital. Then the financial manager must analyze incremental operating cash flows. These may include tax changes due to changes in sales and depreciation expense, opportunity costs, and externalities. Finally, a financial manager must examine the project shutdown cash flows, such as those cash flows from the project's salvage value, from the reduction of net working capital, and tax-related cash flows from the sale of the used asset.

3. Explain why cash flows associated with project financing are not included in incremental cash flow estimates.

Incremental operating cash flows are treated separately from incremental financing cash flows. The latter are captured in the discount rate used in the NPV calculation and in the hurdle rate used when applying the IRR decision rule. Financial managers do not include financing costs as incremental operating cash flows to avoid distorting the NPV or IRR calculations in the capital budgeting process. Double counting would result if financing costs were reflected in both the operating cash flows and the discount rate.

You can access your FREE PHLIP/CW On–Line Study Guide, Current Events, and Spreadsheet Problems templates, which may correspond to this chapter either through the Prentice Hall Finance Center CD–ROM or by directly going to: **http://www.prenhall.com/gallagher.**

PHLIP/CW — *The Prentice Hall Learning on the Internet Partnership–Companion Web Site*

Self-Test

ST-1. Fat Tire Corporation had $20,000 in depreciation expense last year. Assume its federal marginal tax rate is 36 percent whereas its state marginal tax rate is 4 percent. How much are the firm's taxes reduced (and cash flow increased) by the depreciation deduction on federal and state income tax returns?

ST-2. Skinny Ski Corporation had net income in 1996 of $2 million and depreciation expense of $400,000. What was the firm's operating cash flow for the year?

ST-3. Powder Hound Ski Company is considering the purchase of a new helicopter for $1.5 million. The company paid an aviation consultant $20,000 to advise them on the need for a new helicopter. The decision as to whether to buy the helicopter hasn't been made yet. If it is purchased, what is the total initial cash outlay that will be used in the NPV calculation?

ST-4. Rich Folks Ski Area is considering the replacement of one of its older ski lifts. By replacing the lift, Rich Folks expects sales revenues to increase by $500,000 per year. Maintenance expenses are expected to increase by $75,000 per year if the new lift is purchased.

Depreciation expense would be $100,000 per year. (The company uses the straight-line method.) The old lift has been fully depreciated. The firm's marginal tax rate is 38 percent. What would the firm's incremental annual operating cash flows be if the new lift is purchased?

ST-5. Snorkel Ski Company is considering the replacement of its aerial tram. Sales are expected to increase by $900,000 per year and depreciation expense is also expected to rise by $300,000 per year. The marginal tax rate is 32 percent. The purchase will be financed with a $900,000 bond issue carrying a 10 percent annual interest rate. What are the annual incremental operating cash flows if this project is accepted?

FinCoach Practice Exercises

To help improve your grade and master the mathematical skills covered in this chapter, open FinCoach on the Prentice Hall Finance Center CD-ROM and work practice problems in the category Project and Firm Valuation.

Review Questions

1. Why do we focus on cash flows instead of profits when evaluating proposed capital budgeting projects?
2. What is a sunk cost? Is it relevant when evaluating a proposed capital budgeting project? Explain.
3. How do we estimate expected incremental cash flows for a proposed capital budgeting project?
4. What role does depreciation play in estimating incremental cash flows?
5. How and why does working capital affect the incremental cash flow estimation for a proposed large capital budgeting project? Explain.
6. How do opportunity costs affect the capital budgeting decision-making process?
7. How are financing costs generally incorporated into the capital budgeting analysis process?

Build Your Communication Skills

CS-1. Pick a company and obtain a copy of its recent income statement, balance sheet, and statement of cash flows. Compare its profit and cash flow for the period covered by the income statement. Use the statement of cash flows to describe where the firm's cash came from and where it went during that period. Examine the balance sheet to assess the cash position of the firm at that point in time. Submit a written report that analyzes the firm's cash position.

CS-2. Stock market analysts often compare a firm's reported quarterly earnings and its expected quarterly earnings. Quarterly earnings that come in below expectations often mean a drop in the stock price whereas better than expected earnings usually mean a stock price increase. Is this consistent with the view that cash flows and not accounting profits are the source of firm value? Divide into small groups, and discuss these issues with the group members for 15 minutes. Select a spokesperson to present the group's key points to the class.

Problems

11-1. An asset falling under the MACRS five-year class was purchased three years ago for $200,000 (its original depreciation basis). Calculate the cash flows if the asset is sold now at

Salvage Value Cash Flows

a. $60,000
b. $80,000

Assume the applicable tax rate is 40 percent.

Operating Cash Flows

11-2. Mr. Van Orten is evaluating the purchase of new trenching equipment for Scorpio Enterprises. For now, he is only figuring the incremental operating cash flow from the proposed project for the first year. Mr. Van Orten estimates that the firm's sales of earth-moving services will increase by $10,000 in year 1. Using the new equipment will add an additional $3,000 to their operating expenses. Interest expense will increase by $100 because the machine will be partly financed by a loan from the bank. The additional depreciation expense for the new machine will be $2,000. Scorpio Enterprises' marginal tax rate is 35 percent.

a. Calculate the change in operating income (EBIT) for year 1.
b. Calculate the cash outflow for taxes associated with this new income.
c. What is the net new after-tax income (change in earnings after taxes)?
d. Calculate the net incremental operating cash flow from this project for year 1.
e. Are there any expenses listed that you did not use when estimating the net incremental cash flow? Explain.

Estimating Cash Flows

11-3. Mr. Phelps, a financial analyst at Rhodes Manufacturing Corporation, is trying to analyze the feasibility of purchasing a new piece of equipment that falls under the MACRS five-year class. The initial investment, including the cost of equipment and its start-up, would be $375,000. Over the next six years, the following earnings before depreciation and taxes (EBDT) will be generated from using this equipment:

End of Year	EBDT ($)
1	120,000
2	90,000
3	70,000
4	70,000
5	70,000
6	70,000

Rhodes's discount rate is 13 percent and the company is in the 40 percent tax bracket. There is no salvage value at the end of year 6. Should Mr. Phelps recommend acceptance of the project?

Estimating Cash Flows

11-4. Assume the same cash flows, initial investment, MACRS class, discount rate and income tax rate as given in problem 11-3. Now assume that the resale value of the equipment at the end of six years will be $50,000. Calculate the NPV and recommend whether the project should be accepted.

Initial Investment, Operating Cash Flows, and Salvage Value

11-5. George Kaplan is considering adding a new crop-dusting plane to his fleet at North Corn Corner, Inc. The new plane will cost $85,000. He anticipates spending an additional $20,000 immediately after the purchase to modify it for crop-dusting. Kaplan plans to use the plane for five years and then sell it. He estimates that the salvage value will be $20,000. With the addition of the new plane, Kaplan estimates revenue in the first year will increase by 10 percent over last year. Revenue last year was $125,000. Other first-year expenses are also expected to increase. Operating expenses will increase by $20,000 and depreciation expense will increase by $10,500. Kaplan's marginal tax rate is 40 percent.

a. For capital budgeting purposes, what is the net cost of the plane? Or, stated another way, what is the initial net cash flow?
b. Calculate the net incremental operating cash flow for year 1.
c. In which year would the salvage value affect the net cash flow calculations?

Changes in Net Working Capital

11-6. The management of the local cotton mill is evaluating the replacement of low-wage workers by automated machines. If this project is adopted, production and sales are expected to increase significantly: Norma Rae, the mill's financial analyst, expects cash will have to increase by $8,000 and the accounts receivable will increase by $10,000 in response to the increase in sales volume. Because of the higher level of production, inventory will have to increase by $12,000 with an associated $6,000 increase in accounts payable. Accrued taxes and wages, even with the decrease in the number of laborers, are estimated to increase by $2,500.

  a. Calculate the change in net working capital if the automation project is adopted.
  b. Is this change in NWC a cash inflow or outflow?
  c. Given the limited information about the duration of the project, in what year should this change affect the net incremental cash flow calculations?

11-7. Sunstone, Inc. has entrusted financial analyst Flower Belle Lee with the evaluation of a project that involves buying a new asset at a cost of $90,000. The asset falls under the MACRS three-year class and will generate the following revenue stream:

Cash Flows and Capital Budgeting

End of Year	1	2	3	4
Revenues ($)	50,000	30,000	20,000	20,000

The asset has a resale value of $10,000 at the end of the fourth year. Sunstone's discount rate is 11 percent. The company has an income tax rate of 30 percent. Should Flower recommend purchase of the asset?

11-8. Moonstone, Inc., a competitor of Sunstone, Inc. in problem 11-7, is considering purchasing similar equipment with the same revenue, initial investment, MACRS class, and resale value. Moonstone's discount rate is 10 percent and its income tax rate is 40 percent. However, Moonstone is considering the new asset to replace an existing asset with a book value of $20,000 and a resale value of $10,000. What would be the NPV of the replacement project?

Replacement Decision and Cash Flows

11-9. You have just joined Moonstone, Inc. as its new financial analyst. You have learned that accepting the project described in problem 11-8 will require an increase of $10,000 in current assets and will increase current liabilities by $5,000. The investment in net working capital will be recovered at the end of year 4. What would be the new NPV of the project?

Changes in Net Working Capital

11-10. You have been hired by Drs. Venkman, Stantz, and Spenler to help them with net present value analysis for a replacement project. These three New York City parapsychologists need to replace their existing supernatural beings detector with the new, upgraded model. They have calculated all the necessary figures but were unsure about how to account for the sale of their old machine. The original depreciation basis of the old machine was $20,000 and the accumulated depreciation was $12,000 at the date of the sale. They can sell the old machine for $18,000 cash. Assume the tax rate for their company is 30 percent.

Challenge Problem

  a. What is the book value of the old machine?
  b. What is the taxable gain (loss) on the sale of the old equipment?
  c. Calculate the tax on the gain (loss).
  d. What is the net cash flow from the sale of the old equipment? Is this a cash inflow or outflow?
  e. Assume the new equipment costs $40,000 and they do not expect a change in net working capital. Calculate the incremental cash flow for t_0.
  f. Assume they could only sell the old equipment for $6,000 cash. Recalculate parts b through e.

11-11. Mitch and Lydia Brenner own a small factory located in Bodega Bay, California. They manufacture rubber snakes used to scare birds away from houses, gardens, and playgrounds. The recent and unexplained increase in the bird population in northern California has significantly increased the demand for the Brenners' products. To take advantage of this marketing opportunity, they plan to add a new molding machine that will double the output of their existing facility. The cost of the new machine is $20,000. The machine set-up fee is

Cash Flows and Capital Budgeting

Chapter 11 *Estimating Incremental Cash Flows* **291**

$2,000. With this purchase, current assets must increase by $5,000 and current liabilities will increase by $3,000. The economic life of the new machine is four years and it falls under the MACRS three-year depreciation schedule. The machine is expected to be obsolete at the end of the fourth year and have no salvage value.

The Brenners anticipate recouping 100 percent of the additional investment in net working capital at the end of year 4. Sales are expected to increase by $20,000 each year in years 1 and 2. By year 3, the Brenners expect sales to be mostly from repeat customers purchasing replacements instead of sales to new customers. Therefore, the increase in sales for years 3 and 4 is estimated to only be $10,000 in each year. The increase in operating expenses is estimated to be 20 percent of the annual change in sales. Assume the marginal tax rate is 40 percent.

a. Calculate the initial net incremental cash flow.
b. Calculate the net incremental operating cash flows for years 1 through 4. Round all calculations to the nearest whole dollar. Use Table 5-2 to calculate the depreciation expense.
c. Assume the Brenners' discount rate is 14 percent. Calculate the net present value of this project. Would you recommend the Brenners add this new machine to their factory?

Cash Flows and Capital Budgeting

11-12. The RHPS Corporation specializes in the custom design, cutting, and polishing of stone raw materials to make ornate building facings. These stone facings are commonly used in the restoration of older mansions and estates. Janet Weiss and Brad Majors, managers of the firm, are evaluating the addition of a new stone-cutting machine to their plant. The machine's cost to RHPS is $150,000. Installation and calibration costs will be $7,500. They do not anticipate an increase in sales but the reduction in the operating expenses is estimated to be $50,000 annually. The machine falls under the MACRS three-year depreciation schedule. The machine is expected to be obsolete after five years. At the end of five years, Weiss and Majors expect the cash received (less applicable capital gains taxes) from the sale of the obsolete machine to offset the shutdown and dismantling costs. The RHPS cost of capital is 10 percent and the marginal tax rate is 35 percent.

a. Calculate the net present value for the addition of this new machine. Round all calculations to the nearest whole dollar.
b. Would you recommend that Weiss and Majors go forward with this project?

Comprehensive Problem

11-13. The Chemical Company of Baytown purchased new processing equipment for $40,000 on December 31, 1997. The equipment had an expected life of four years and was classified in the MACRS three-year class. Due to changes in environmental regulations, the operating cost of this equipment has increased. The company is considering replacing this equipment with a more efficient process line at the end of 1999. The salvage value of the old equipment is estimated to be $4,000. The marginal tax rate is 40 percent.

a. Calculate the cash flow from the sale of this equipment. Assume the sale occurred at the end of 1999. Use Table 4-1 (page 73) to calculate the depreciation.
b. The new process line has a higher capacity than the old one and is expected to boost sales. As a result, the cash requirement will increase by $1,000, accounts receivable by $5,000, and inventory by $10,000. It will also increase accounts payable by $6,000 and accrued expenses by $3,000. Calculate the incremental cash flow due to the change in the net working capital.
c. The new equipment will cost $180,000, including installation and start-up costs. Calculate the net cash outflow in 1999 if the new process line is installed and is ready to operate by the end of 1999.
d. Beginning in January 2000, this new equipment is expected to generate additional sales of $60,000 each year for the next four years. It will have an economic life of four years and will fall under the MACRS three-year classification. Being more efficient, the new

equipment will reduce yearly operating expenses by $6,000. Calculate the net incremental operating cash flows for the years 2000 through 2003. Assume the marginal tax rate will remain at 40 percent. Round calculations to the nearest whole dollar.

e. At the end of its economic life, the new process line is expected to be sold for $20,000. The cost of capital for the company is 6 percent. Calculate the net present value and the internal rate of return (only if you have a financial calculator) for this project. Round calculations to the nearest whole dollar. Recommend whether the replacement project should be adopted or rejected. (Hint: Preparation of a summary of incremental cash flows similar to Table 11-5 may be helpful.)

f. Draw an NPV profile for the project.

Answers to Self-Test

ST-1. $20,000 \times (.36 + .04)$
= $20,000 \times .40 = $8,000 tax savings

ST-2. $2,000,000 + $400,000 = $2,400,000 operating cash flow

ST-3. $1,500,000 total initial cash outlay (The $20,000 for the consultant is a sunk cost.)

ST-4. ($500,000 − $75,000 − $100,000) × (1 − .38)
= $325,000 × .62 = $201,500 incremental net income
$201,500 + $100,000 depreciation expense
= $301,500 incremental operating cash flow

ST-5. ($900,000 − $300,000) × (1 − .32)
= $600,000 × .68 = $408,000 incremental net income
$408,000 + $300,000 depreciation expense
= $708,000 incremental operating cash flow
(The finance costs are not part of operating cash flows. They will be reflected in the required rate of return.)

CHAPTER 12

THE COST OF CAPITAL

There's no such thing as a free lunch.
— Milton Friedman

AMCOR TAKES STEPS TO REDUCE ITS COST OF CAPITAL

Amcor Ltd., a leading paper and packaging firm in Australia, announced in September 1998 that it was selling its softwood sawmilling business and using the proceeds to implement a share buy-back program. In a company news release Amcor's managing director, Russell Jones, said, "A share buy-back is an attractive investment opportunity at this time and as part of Amcor's active capital management program will reduce the company's cost of capital while interest rates are low."

The sale of the sawmilling business was expected to raise $35 million. At roughly the same time Amcor also announced the divestment of its Australian Graphics operation, its Australian Serv-Pack thermoforming operation, and its New Zealand–headquartered Kiwiplan software business. In total, these three transactions were expected to yield proceeds of about $12 million. Thus, when the transactions were complete, about $45 million would be available for the stock buy-back program.

Why would Amcor be interested in buying back its stock, and why would the program reduce the company's cost of capital? The answer is that buying back stock changes the relative amount of debt and equity financing used to fund the

CHAPTER OBJECTIVES

After reading this chapter, you should be able to:

- Describe the sources of capital and how firms raise capital.

- Estimate the cost of capital for each financing source.

- Estimate the weighted average cost of capital.

- Use the marginal cost of capital (MCC) schedule to make capital budgeting decisions.

- Explain the importance of the marginal cost of capital (MCC) schedule in financial decision making.

company's assets. As you will see from your study of the chapter ahead, debt financing is cheaper than equity financing, so using relatively more debt financing and relatively less equity financing tends to lower the overall cost of the company's capital. Naturally, therefore, Amcor's managers would be interested in buying back the company's stock as long as the increase in the percent of debt capital used was not seen as increasing the firm's risk too much.

Sources: ⟨http://www.amcor.com.au/docs/today.htm⟩;
⟨http://www.amcor.com.au/docs/news17.htm⟩. Amcor press release, September 14, 1998.

Chapter Overview

In capital budgeting decisions, financial managers must analyze many factors to determine whether a project will add value to a firm, including estimated incremental cash flows, the timing of those cash flows, risk, and the project's required rate of return.

One of the key components of the capital budgeting decision is the cost of capital. **Capital** is the term for funds that firms use. Businesses, such as Amcor Ltd., raise capital from creditors and owners. All capital raised has a cost because the suppliers of capital demand compensation for the funds provided.

In this chapter we examine the cost of different types of capital. We see how to estimate the cost of capital from a particular source, as well as the overall cost of capital for a firm. We also see how estimating a firm's cost of capital affects a firm's financing and investment decisions.

The Cost of Capital

To properly evaluate potential investments, firms must know how much their capital costs. Without a measure of the cost of capital, for example, a firm might invest in a new project with an expected return of 10 percent, even though the capital used for the investment costs 15 percent. If a firm's capital costs 15 percent, then the firm must seek investments that return at least that much. It is vital, then, that managers know how much their firm's capital costs *before* committing to investments.

Suppliers and users of capital use cost estimates before making short- or long-term financial decisions. As we saw in chapter 9, investors determine their required rate of return, k, to value either a bond or stock before they invest. That required rate of return, k, for each type of security issued is the cost of capital for that source. Overall, the cost of capital is the compensation investors demand of firms that use their funds, adjusted for taxes and transaction costs in certain cases, as we will explain later in this chapter.

In chapter 10 we saw that firms determine their discount rate, k, to solve for a project's NPV or use k as the IRR hurdle rate before deciding whether to accept a capital budgeting project. The discount rate or hurdle rate, k, is the firm's cost of capital for that project. Investors supply capital, so they require a return, and firms use capital, so they must pay suppliers of capital for the use of those funds.

Sources of Capital

A firm's capital is supplied by its creditors and owners. Firms raise capital by borrowing it (issuing bonds to investors or promissory notes to banks), or by issuing preferred or common stock. The overall cost of a firm's capital depends on the return demanded by each of these suppliers of capital.

To determine a firm's overall cost of capital, the first step is to determine the cost of capital from each supplier. The cost of capital from a particular source, such as bondholders or common stockholders, is known as the **component cost of capital.**

In the following sections, we estimate the cost of debt capital, k_d; the cost of capital raised through a preferred stock issue, k_p; and the cost of equity capital supplied by common stockholders, k_s for internal equity and k_n for external equity.

The Cost of Debt

When a firm borrows money at a stated rate of interest, determining the cost of debt, k_d, is relatively straightforward. As shown in Figure 12-1, the lender's cost of capital is the required rate of return on either a company's new bonds or a promissory note. The firm's **cost of debt** when it borrows money by issuing bonds is the interest rate demanded by the bond investors. When borrowing money from an individual or financial institution, the interest rate on the loan is the firm's cost of debt.

The After-Tax Cost of Debt (AT k_d) The **after-tax cost of debt, AT k_d,** is the cost to the company of obtaining debt funds. Because the interest paid on bonds or bank loans is a tax-deductible expense for a business, a firm's AT k_d is less than the required rate of return of the suppliers of debt capital. For example, suppose Ellis Industries borrowed $100,000 for one year at 10 percent interest paid annually. The interest rate on the loan is 10 percent, so Ellis must pay the lender $10,000 in interest each year the loan is outstanding (10 percent of $100,000). However, look at what happens when Ellis takes its taxes for the year into account:

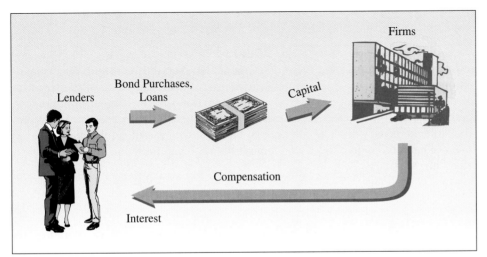

Figure 12-1 **The Flow of Debt Capital from Investors to Firms**

Figure 12-1 shows how debt investors supply capital to and receive interest from the firm.

	BEFORE BORROWING	AFTER BORROWING
Operating Income	$50,000	$50,000
Interest Expense	0	10,000
Before-Tax Income	50,000	40,000
Income Tax (40% rate)	20,000	16,000
Net After-Tax Income	30,000	24,000

The $10,000 interest charge caused a $6,000 decrease in Ellis's net after-tax income ($30,000 − $24,000 = $6,000). Therefore, assuming a tax rate of 40 percent, the true cost of the loan is only $6,000 (6 percent), not $10,000 (10 percent).

The following formula converts k_d into AT k_d, the true after-tax cost of borrowing:

Formula for the After-Tax Cost of Debt
$$AT\ k_d = k_d\,(1 - T) \tag{12-1}$$

where: k_d = The before-tax cost of debt
T = The firm's marginal tax rate[1]

To solve for AT k_d in our Ellis Industries example, recall that Ellis's before-tax cost of debt is 10 percent and its marginal tax rate is 40 percent. The after-tax cost of debt, according to Equation 12-1, follows:

$$\begin{aligned} AT\ k_d &= .10\,(1 - .40) \\ &= .10 \times .60 \\ &= .06,\ \text{or}\ 6\% \end{aligned}$$

Our calculations show that for Ellis Industries the after-tax cost of debt on a $100,000 loan at a 10 percent rate of interest is 6 percent.

We have seen that the cost of using borrowed money is the interest rate charged by the lender. In addition, we discussed how the tax deductibility of interest lowers the firm's true cost of debt. Next, we turn to the cost of preferred stock and common stock equity.

The Cost of Preferred and Common Stock Funds

When corporations raise capital by issuing preferred or common stock, these investors expect a return on their investments. If that return is not realized, investors will sell their stock, driving the stock price down. Although the claim of preferred and common stockholders may not be contractual, as it is for bondholders, there is a cost nonetheless. To calculate the cost of using preferred and common stockholders' money, then, the firm must estimate the rate of return that these investors demand. Figure 12-2 shows how firms raise capital from, and compensate, equity investors.

---

[1]The tax rate used here should reflect the firm's total combined federal, state, and/or local income tax rate.

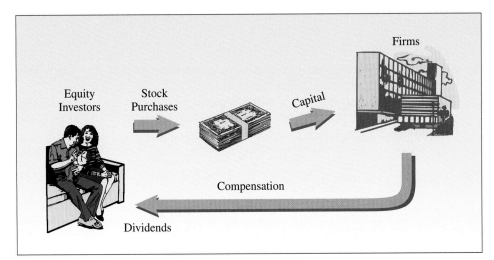

Figure 12-2 The Flow of Capital from Equity Investors to Firms

Figure 12-2 shows how equity investors supply capital to the firm and receive dividends from it.

The Cost of Preferred Stock (k_p) The **cost of preferred stock (k_p)** is the rate of return investors require on a company's new preferred stock, plus the cost of issuing the stock. Therefore, to calculate k_p, a firm's managers must estimate the rate of return that preferred stockholders would demand, and add in the cost of the stock issue. Because preferred stock investors normally buy preferred stock to obtain the stream of constant preferred stock dividends associated with the preferred stock issue, their return on investment can normally be measured by dividing the amount of the firm's expected preferred stock dividend by the price of the shares. The cost of issuing the new securities, known as **flotation cost,** includes investment bankers' fees and commissions, and attorneys' fees. These costs must be deducted from the preferred stock price paid by investors to obtain the net price paid to the firm. Equation 12-2 shows how to estimate the cost of preferred stock:

Formula for the Cost of Preferred Stock, k_p

$$k_p = \frac{D_p}{(P_p - F)} \qquad (12\text{-}2)$$

where: k_p = The cost of the preferred stock issue; the expected return
D_p = The amount of the expected preferred stock dividend
P_p = The current price of the preferred stock
F = The flotation cost per share

The cost of using the company's preferred stock is k_p. The value of k_p, the expected return from the preferred stock issue, is the minimum return the firm's managers must earn when they use the money supplied by preferred stockholders. If they cannot earn this return, the preferred stockholders will sell their shares, causing the preferred stock's price to fall. This means the firm must issue more shares of the stock than before the stock price fell to raise the same amount of funds.

Suppose Ellis Industries issued preferred stock that has been paying annual dividends of $2.50 and is expected to continue to do so indefinitely. The current price of Ellis's preferred stock is $22 a share and the flotation cost is $2 per share. According to Equation 12-2, the cost of Ellis's preferred stock is as follows:

$$k_p = \frac{\$2.50}{(\$22 - \$2)}$$
$$= .125, \text{ or } 12.5\%$$

We see that Ellis Industries' cost of new preferred stock, assuming that stock pays dividends of $2.50 per year and has a market price of $22, is 12.5 percent.

The cost of preferred stock, k_p, is higher than the before-tax cost of debt, k_d, because a company's bondholders and bankers have a prior claim on the earnings of the firm and on its assets in the event of a liquidation. Preferred stockholders, as a result, take a greater risk than bondholders or bankers and demand a correspondingly greater rate of return.

> **Take Note**
>
> *There is no tax adjustment in the cost of preferred stock calculation. Unlike interest payments on debt, firms may not deduct preferred stock dividends on their tax returns. The dividends are paid out of after-tax profits.*

The Cost of Internal Common Equity (k_s) The **cost of internal common equity (k_s)** is the required rate of return on funds supplied by existing common stockholders. The cost of equity depends on the rate of return the common stockholders demand for holding the company's common stock. Calculating k_s is tougher than calculating k_p because common stockholders do not receive a constant stream of dividends. Instead, investors own the firm, including the corporate earnings left over after other claimants on the firm have been paid. Neither creditors nor preferred stockholders have a claim on these residual earnings. Corporations may either retain the residual earnings, or return them in the form of common stock dividends. Retained earnings have a cost, however. The cost of retained earnings, another name for the cost of internal equity, is the rate of return that the company must earn to justify retaining the earnings instead of paying them as dividends. These are internally generated (within the firm) equity funds.

As noted, common stock dividend payments change from year to year or may not be paid at all. Ultimately, however, dividends are the only payments a corporation makes to its common stockholders. The corporation may pay regular dividends, or it may pay a liquidation dividend some time in the future. For companies that pay regular dividends that grow at a constant rate, the constant dividend growth model, introduced in chapter 9, may be used to estimate the cost of equity. For companies that do not pay regular dividends, or when the market approach to risk is more appropriate, the CAPM may be used to estimate the cost of equity.

We discussed the constant dividend growth model in chapter 9, page 219.

Using the Dividend Growth Model to Estimate k_s. Recall the dividend growth model introduced in chapter 9:

$$P_0 = \frac{D_1}{k_s - g}$$

where: P_0 = The current price of the common stock
 D_1 = The dollar amount of the common stock dividend expected one period from now
 k_s = Required rate of return per period on this common stock investment
 g = Expected constant growth rate per period of the company's common stock dividends[2]

---

[2] This is the *constant growth version* of the dividend growth model. It assumes that the company's dividends grow at the same rate indefinitely.

Rearranging the terms in the dividend growth model to solve for k_s, we rewrite the formula as follows:

Formula for the Cost of Common Stock Equity

$$k_s = \frac{D_1}{P_0} + g \qquad (12\text{-}3)$$

By making use of Equation 12-3, we can solve for k_s, assuming we know the values of the terms P_0, D_1, and g. The term D_1/P_0 in Equation 12-3 represents the stock's *dividend yield,* and the g term represents dividend growth rate from year to year.

To apply Equation 12-3, suppose that Ellis Industries' common stock is selling for $40 a share. Next year's common stock dividend is expected to be $4.20, and the dividend is expected to grow at a rate of 5 percent per year indefinitely. Given these conditions, Equation 12-3 tells us that the expected rate of return on Ellis's common stock is as follows:

$$\begin{aligned}k_s &= \frac{\$4.20}{\$40} + .05 \\ &= .105 + .05 \\ &= .155, \text{ or } 15.5\%\end{aligned}$$

At a stock share price of $40, a dividend of $4.20, and an expected dividend growth rate of 5 percent, the expected return from Ellis's common stock is 15.5 percent. The expected return of 15.5 percent is the minimum return the firm's managers must earn when they use money supplied by common stockholders. If they cannot achieve this return, common stockholders will sell their shares, causing the stock's price to fall. This will make it necessary to sell more shares to raise the desired amount of funds. The cost of using money supplied by the company's common stockholders, then, is 15.5 percent. Because dividends paid are not tax deductible to the corporation, there is no tax adjustment to the k_s calculation.

The CAPM Approach to Estimating k_s. A firm may pay dividends that grow at a changing rate, pay no dividends at all, or the managers of the firm may believe that market risk is the relevant risk. In such cases the firm may choose to use the capital asset pricing model (CAPM) to calculate the rate of return that investors require for holding common stock. The CAPM solves for the rate of return that investors demand for holding a company's common stock according to the degree of nondiversifiable risk[3] present in the stock. The CAPM formula, Equation 12-4, follows:

CAPM Formula for the Cost of Common Stock Equity

$$k_s = k_{RF} + (k_M - k_{RF}) \times \beta \qquad (12\text{-}4)$$

where: k_s = The required rate of return from the company's common stock equity
k_{RF} = The risk-free rate of return
k_M = The expected rate of return on the overall stock market
β = The beta of the company's common stock, a measure of the amount of nondiversifiable risk

We discussed the CAPM in chapter 7, pages 163–165.

---

[3]According to the CAPM, common stockholders hold well-diversified portfolios, so the only relevant risk measure is nondiversifiable (market) risk.

Suppose Ellis Industries has a beta of 1.39, the risk-free rate as measured by the rate on short-term U.S. Treasury bills is 3 percent, and the expected rate of return on the overall stock market is 12 percent. Given those market conditions, according to Equation 12-4, the required rate of return for Ellis's common stock is as follows:

$$k_s = .03 + (.12 - .03) \times 1.39$$
$$= .03 + (.09 \times 1.39)$$
$$= .03 + .1251$$
$$= .1551, \text{ or about } 15.5\%$$

According to the CAPM, we see that the cost of using money supplied by Ellis's common stockholders is about 15.5 percent, given a company beta of 1.39, a risk-free rate of 3 percent, and an expected market rate of return of 12 percent.

Deciding How to Estimate k_s. Should you use the dividend growth model, Equation 12-3, or the CAPM, Equation 12-4, to estimate a firm's cost of common equity? The choice depends on the firm's dividend policy, available data, and management's view of risk. As a financial manager, if you were confident that your firm's dividends would grow at a fairly constant rate in the future, you could apply the dividend growth model to calculate k_s. If your firm's growth rate were erratic or difficult to determine, you might use the CAPM instead, assuming that you agreed with the CAPM's underlying hypothesis that common stockholders hold well-diversified portfolios and that nondiversifiable risk is what is priced in the market. Where possible, practitioners apply both models and use their business judgment to reconcile differences between the two outcomes.

The Cost of Equity from New Common Stock (k_n) The cost incurred by a company when new common stock is sold is the **cost of equity from new common stock (k_n)**. In the preceding section, we discussed the cost of using funds supplied by the firm's existing stockholders. Capital from existing stockholders is internal equity capital. That is, the firm already has these funds. In contrast, capital from issuing new stock is external equity capital. The firm is trying to raise new funds from outside sources.

New stock is sometimes issued to finance a capital budgeting project. The cost of this capital includes not only stockholders' expected return on their investment, but also the **flotation** costs incurred to issue new securities. Flotation costs make the cost of using funds supplied by new stockholders slightly higher than using retained earnings supplied by the existing stockholders.

To estimate the cost of using funds supplied by new stockholders, we use a variation of the dividend growth model that includes flotation costs:

Formula for the Cost of New Common Stock

$$k_n = \frac{D_1}{P_0 - F} + g \qquad (12\text{-}5)$$

where: k_n = The cost of new common stock equity
P_0 = The price of one share of the common stock
D_1 = The amount of the common stock dividend expected to be paid in one year
F = The flotation cost per share
g = The expected constant growth rate of the company's common stock dividends

Equation 12-5 shows mathematically how the cost of new common stock, k_n, is greater than the cost of existing common stock equity, k_s. By subtracting flotation costs, F, from the common stock price in the denominator, the k_n term becomes larger.

Let's look at the cost of new common stock for Ellis Industries. Suppose again that Ellis Industries' anticipated dividend next year is $4.20 a share, its growth rate is 5 percent a year, and its existing common stock is selling for $40 a share. New shares of stock can be sold to the public for the same price. But to do so Ellis must pay its investment bankers 5 percent of the stock's selling price, or $2 per share. Given these conditions, we use Equation 12-5 to calculate the cost of Ellis Industries' new common equity as follows:

$$k_n = \frac{\$4.20}{\$40 - \$2} + .05$$

$$= .1105 + .05$$

$$= .1605, \text{ or } 16.05\%$$

Because of $2 flotation costs, Ellis Industries only keeps $38 of the $40 per share paid by investors. As a result, the cost of new common stock is higher than the cost of existing equity—16.05 percent compared to 15.5 percent.

If the cost of new common equity is higher than the cost of internal common equity, the cost of preferred stock, and the cost of debt, why use it? Sometimes corporations have no choice. Take Duramed Pharmaceuticals. Duramed issued new stock to raise $25 million to invest in a promising project to develop a new generic version of a drug used to fight osteoporosis and the effects of menopause. In addition, Duramed needed new funds to pay $7 million it owed to Provident Bancorp. If Duramed did not repay Provident within a month, the interest rate would leap from 10.75 percent to 36.5 percent. With further delays, that rate could increase to 53.6 percent. Facing skyrocketing interest rates, Duramed had to raise capital through a new stock issue.[4]

Also, if the amount of debt a firm has incurred continues to increase, and internal equity funds have run out, it may be necessary to issue new common stock to bring the weight of debt and equity on the balance sheet into line.[5]

We have examined the sources of capital and the cost of each capital source. Next, we investigate how to measure the firm's overall cost of capital.

The Weighted Average Cost of Capital (WACC)

To estimate a firm's overall cost of capital, the firm must first estimate the cost for each component source of capital. The component sources include the after-tax cost of debt, AT k_d; the cost of preferred stock, k_p; and the cost of common stock equity, k_s, and the cost of new common stock equity, k_n. In the following section, we first discuss all component sources except k_n, which we discuss separately.

---

[4]Floyd Norris, "Duramed Pharmaceuticals Faces Some Harsh Terms to Raise Cash," *The New York Times* (November 16, 1995): D10.
[5]This issue will be discussed in detail in chapter 13.

To illustrate the first step in estimating a firm's overall cost of capital, let's review Ellis Industries' component costs of capital. From our previous calculations, we know the following costs of capital:

$$AT\ k_d = 6\%$$
$$k_p = 12.5\%$$
$$k_s = 15.5\%$$

The next step in finding a firm's overall cost of capital is assessing the firm's *capital structure*. In practice, the assets of most firms are financed with a mixture of debt, preferred stock, and common equity. The mixture of capital used to finance a firm's assets is called the **capital structure** of the firm. To analyze the capital structure of a business, we must find the percentage of each type of capital source.

To illustrate how to assess a firm's capital structure, assume that Ellis Industries finances its assets through a mixture of capital sources, as shown on its balance sheet:[6]

Total Assets:	$1,000,000
Long- and Short-Term Debt:	$400,000
Preferred Stock:	100,000
Common Equity:	500,000
Total Liabilities and Equity:	$1,000,000

In percentage terms, then, the mixture of capital used to finance Ellis's $1 million worth of assets is as follows:

Debt:	400,000 / 1,000,000 = .40, or 40%
Preferred Stock:	100,000 / 1,000,000 = .10, or 10%
Common Equity:	500,000 / 1,000,000 = .50, or 50%

Our calculations show that Ellis Industries' capital structure consists of 40 percent debt, 10 percent preferred stock, and 50 percent common equity. If Ellis Industries thinks that this mixture is optimal and wants to maintain it, then it will finance new capital budgeting projects with a mixture of 40 percent debt, 10 percent preferred stock, and 50 percent common equity. This mixture might not be used for each and every project. But in the long run, the firm is likely to seek this capital structure if it is believed to be optimal.

The final step in estimating a firm's overall cost of capital is to find the weighted average of the costs of each individual financing source. The **weighted average cost of capital (k_a or WACC)** is the mean of all component costs of capital, weighted according to the percentage of each component in the firm's optimal capital structure. We find the WACC by multiplying the individual source's cost of capital times its

---

[6]Ideally, the percentage of each component in the capital structure would be measured on the basis of market values instead of accounting values. For the sake of simplicity, we use accounting values here, as do many real-world companies.

percentage of the firm's capital structure, then adding these results. For Ellis Industries, the weighted average of the financing sources follows:

$$(.40 \times AT\ k_d) + (.10 \times k_p) + (.50 \times k_s)$$
$$= (.40 \times .06) + (.10 \times .125) + (.50 \times .155)$$
$$= .024 + .0125 + .0775$$
$$= .114, \text{ or } 11.4\%$$

Ellis Industries' weighted average cost of capital is 11.4 percent. The general formula for any firm's WACC is shown in Equation 12-6:

Formula for the Weighted Average Cost of Capital (WACC)
$$k_a = (WT_d \times AT\ k_d) - (WT_p \times k_p) - (WT_s \times k_s) \qquad (12\text{-}6)$$

where: k_a = The weighted average cost of capital (WACC)
        WT_d = The weight, or proportion, of debt used to finance the firm's assets
        $AT\ k_d$ = The after-tax cost of debt
        WT_p = The weight, or proportion, of preferred stock being used to finance the firm's assets
        k_p = The cost of preferred stock
        WT_s = The weight, or proportion, of common equity being used to finance the firm's assets
        k_s = The cost of common equity

A firm must earn a return equal to the weighted average cost of capital (WACC) to pay suppliers of capital the return they expect. In the case of Ellis Industries, for instance, its average-risk capital budgeting projects must earn a return of 11.4 percent to pay its capital suppliers the return they expect.

To illustrate how earning the WACC ensures that all capital suppliers will be paid their required cost of capital, let's return to our example. Suppose Ellis Industries undertakes a plant expansion program that costs $1 million and earns an annual return of 11.4 percent, equal to Ellis's WACC. Capital for the project is supplied as follows:

- 40 percent of the $1 million, or $400,000, is supplied by lenders expecting a return equal to before-tax k_d, 10 percent.
- 10 percent of the $1 million, or $100,000, is supplied by preferred investors at a cost equal to k_p, 12.5 percent.[7]
- 50 percent of the $1 million, or $500,000, is supplied by common stockholders expecting a return equal to k_s, 15.5 percent.

If the project does in fact produce the expected 11.4 percent return, will all these suppliers of capital receive the return they expect? The computations that follow show how Ellis will pay its capital suppliers:

---

[7]In this example the firm sold 5,000 shares of preferred stock at $22 a share for a total of $110,000; $2 a share, or $10,000, went to the investment bankers and attorneys in the form of flotation costs, leaving $100,000 for the capital budget.

First-year return from the project:	$1,000,000 \times .114 = \$114,000$
Interest at 10 percent paid to the bondholders:	$\$400,000 \times .10 = \$40,000$
Less tax savings on interest expense:	$\$40,000 \times$ firm's tax rate of 40% = $\underline{\$16,000}$
Net interest cost to the firm:	$\$40,000 - \$16,000 = \underline{\$24,000}$
Amount remaining to repay other sources of capital:	$\$114,000 - \$24,000 = \$90,000$
Preferred dividend paid to preferred stockholders:	5,000 shares $\times \$2.50 = \underline{\$12,500}$
Amount remaining for common stockholders:	$\$90,000 - \$12,500 = \$77,500$
Summary:	
Return realized by lenders	$\$40,000/\$400,000 = .10$ or 10%
Return realized by preferred stock investors	$\$12,500/\$100,000 = .125$ or 12.5%[8]
Return realized by common stockholders	$\$77,500/\$500,000 = .155$ or 15.5%

We see that Ellis was able to pay all its capital suppliers by earning an overall return of 11.4 percent, its WACC.

In the long run, companies generally try to maintain an optimal mixture of capital from different sources. In the short run, however, one project may be financed entirely from one source. Even if a particular project is financed entirely from one source, the WACC should still be used as the required rate of return for an average-risk project. Say, for instance, such a project is entirely financed with debt, a relatively cheap source of capital. The cost of debt should not be used as that project's cost of capital. Why? Because the firm's risk would increase with the increase in debt, and the costs of all sources of capital would increase.

The Marginal Cost of Capital (MCC)

A firm's weighted average cost of capital will change if one component cost of capital changes. Often, a change in WACC occurs when a firm raises a large amount of capital. For example, lenders may increase the interest rate they charge, k_d, if they think the firm's debt load will be too heavy. Or a firm's cost of equity may increase when new stock is issued after new retained earnings run out. This is because of the flotation costs incurred when new stock is issued.

Firms, then, must consider how increasing component costs of capital affect the WACC. The weighted average cost of the next dollar of capital to be raised is the **marginal cost of capital (MCC).** To find the MCC, financial managers must: (1) assess at what point a firm's cost of debt or equity will change the firm's WACC; (2) estimate how much the change will be; and (3) calculate the cost of capital up to and after the points of change.

---

[8]To be precise, 12.5 percent is the return to the preferred stockholders plus the investment bankers' and attorneys' fees for issuing the stock ($2). The net return realized by the preferred stock investors is the preferred dividend they receive ($2.50) divided by the price they paid for the stock ($22), which is 11.36 percent.

FINANCE AT WORK

SALES-RETAIL-SPORTS-MEDIA TECHNOLOGY-PUBLIC RELATIONS-PRODUCTION-EXPORTS

Interview with Fred Higgins: Minit Mart Foods, Inc., CEO

Fred Higgins

Fred Higgins was a law student when he launched his first Minit Mart. He owned six stores by the time he was out of law school. Now his company owns over a hundred Minit Marts spread throughout Kentucky. Fred realized the importance of raising capital and assessing its cost as he built his Minit Mart empire.

Q. Tell us about your first experience in trying to raise capital.

When I was in law school trying to build my first Minit Mart, I had convinced the people who owned the land to build, and then lease, the building back to me. I was convinced the site had a great location because of its heavy traffic count and apartments nearby. And it was a great location; it's still making money today.

However, I needed capital to buy inventory, so I went to a bank. I explained the whole scenario to the banker. Then the banker asked what assets I had. I said, "Me." The banker threw me out of his office. I was lucky, though. I was finally able to obtain financing with the help of my father, who knew another banker.

If someone is starting in business, he or she needs credibility. The first time I talked with a banker, I had no credibility, no track record. The banker didn't know me from a hole in the ground. Today when I walk into the bank, the bankers can accurately judge the risk of loaning money to me because they can look at my past experience.

Q. How do you raise debt capital now?

Aside from bank loans, I sometimes use insurance companies for long-term debt financing, particularly when I want to buy some real estate. But as a small business, it's tough to have insurance companies work with you. It costs them the same to loan millions of dollars as it does $100,000. They'd rather make a big loan and be done with it. Again, I was lucky because I knew people who could provide introductions to insurance company lenders.

Q. Why is it important to measure your capital costs?

At Minit Mart, we want our projects to generate a positive return. We need to be able to estimate how much the cost of our capital is to be sure that our investments earn a certain amount of dollars, after all the costs, including capital, have been deducted. There are a lot of projects out there, and if we can't get our desired rate of return with one, we will bypass it and move onto the next one.

Q. What outside factors affect your cost of capital?

So many outside factors can increase your cost of debt capital. A few years ago, Circle K went bankrupt, and Seven-Eleven and several other convenience store chains were not performing as well as they had been. The financial markets became very leery of our industry. And that raised the cost of debt capital.

Another time our cost of capital increased due to the Environmental Protection Agency. The EPA started closely scrutinizing businesses that sold gas for potential environmental violations. Minit Marts sell gas, so because of the EPA's new scrutiny, banks either wouldn't loan to us or demanded a higher interest rate.

Source: Interview with Fred Higgins.

The Firm's MCC Schedule

The marginal cost of the *first* dollar of capital a firm raises is the same as the firm's basic WACC. However, as the firm raises more capital, a point is reached at which the marginal cost of capital changes. Why? Because one of the component sources of capital changes. This point is the *break point* in the firm's MCC schedule. Capital above the break point can only be raised at a higher cost.

Finding the Break Points in the MCC Schedule To find break points in the MCC schedule, financial managers determine what limits, if any, there are on the firm's ability to raise funds from a given source at a given cost. Suppose that Ellis Industries' financial managers, after consulting with bankers, determined that the firm can borrow up to $300,000 at an interest rate of 10 percent, but any money borrowed above that amount will cost 12 percent. To calculate Ellis Industries' after-tax cost of debt, assume the firm's tax rate is 40 percent. We apply Equation 12-1 as follows:

$$\text{AT } k_d = k_d (1 - T) \quad (12\text{-}1)$$
$$\text{AT } k_d = .10 \times (1 - .40)$$
$$= .06, \text{ or } 6.0\%$$

We see that at an interest rate of 10 percent, and a tax rate of 40 percent, that Ellis's after-tax cost of debt is 6 percent. However, if Ellis Industries borrows more than $300,000, then its interest rate increases to 12 percent. At a tax rate of 40 percent and interest rate of 12 percent, the firm's after-tax k_d for amounts borrowed over $300,000 is:

$$\text{AT } k_d \text{ (over \$300,000 borrowed)} = .12 \times (1 - .40)$$
$$= .072, \text{ or } 7.2\%$$

Because Ellis's AT k_d increases when it borrows more than $300,000, its MCC will also increase when it borrows more than $300,000. (We'll see how much it increases in the next section.) The financial managers at Ellis Industries want to know how much total capital they can raise before the debt portion reaches $300,000, causing an increase in the AT k_d and the MCC.

Ellis Industries' marginal cost of capital break point is not $300,000 because the firm's capital structure is 40 percent debt, 10 percent preferred stock, and 50 percent common stock. At $300,000, then, only 40 percent of that capital is debt. Instead, the financial managers of Ellis Industries must figure out at what point the firm will use $300,000 of debt capital.

To find a firm's marginal cost of capital break point, we use Equation 12-7:

Formula for the MCC Break Point

$$BP = \frac{\text{Limit}}{\text{Proportion of Total}} \quad (12\text{-}7)$$

where: BP = The capital budget size at which the MCC changes (break point)
Limit = The point at which the cost of the source of capital changes
Proportion of Total = The percentage of this source of capital in the firm's capital structure

In our Ellis Industries example, we know that the firm has a $300,000 debt limit before its after-tax cost of debt will increase, and that debt is 40 percent of the firm's capital structure. Applying Equation 12-7 to our Ellis example, we see that its debt break point is the following:

$$BP_d = \frac{\$300,000}{.40}$$
$$= \$750,000$$

We find that Ellis Industries' break point is $750,000. By applying Equation 12-7, Ellis's financial managers know that they may raise up to $750,000 in capital before their borrowing costs rise from 6 percent to 7.2 percent. Any capital raised over $750,000 will reflect the higher cost of borrowing, as shown in Figure 12-3.

Notice we used subscript $_d$ with BP in Equation 12-7. That was to identify the break point as a *debt break point*. There could be other debt break points for Ellis Industries. If, for instance, the company's lenders set additional limits on the company's borrowing, the debt break points would be denoted as BP_{d1}, BP_{d2}, and so on. For our example, let's assume that Ellis's bankers will lend the firm an unlimited amount of money over $300,000 at 7.2 percent so there are no further debt break points.

The Equity Break Point. The equity break point is the point at which the marginal cost of capital changes because the cost of equity changes. Equity costs may change because firms exhaust the supply of funds from the firm's existing common stockholders—that is, they exhaust additions to retained earnings. After firms exhaust their supply of internal equity that has a capital cost of k_s, they will have to raise additional equity funds by issuing new stock that has a higher cost, k_n. This additional equity is external equity capital. The marginal cost of capital increases accordingly.

Let's illustrate how the marginal cost of capital increases because of changes in the cost of equity. We'll assume that Ellis Industries expects to realize $600,000 in income this year after it pays preferred stockholders their dividends. The $600,000 in earnings belong to the common stockholders. The firm may either pay dividends or retain the earnings. Let's assume Ellis retains the $600,000. The finite supply of capital from the existing common stockholders is the $600,000 addition to retained earnings. To find the equity break point, then, Ellis's managers must know at what point the firm will exhaust the common equity capital of $600,000, assuming existing common stock equity is 50 percent of the firm's capital budget. Figure 12-4 graphically depicts the Ellis equity break point analysis.

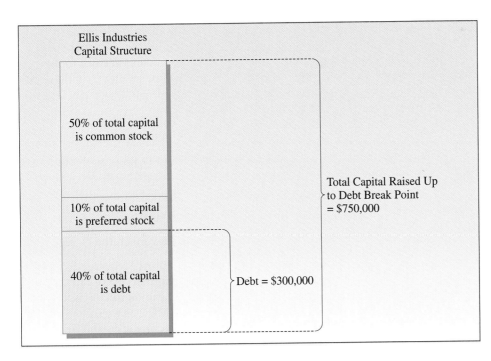

Figure 12-3 Ellis Industries Debt Break Point

Figure 12-3 shows how Ellis Industries can raise up to $750,000 of total capital before the $300,000 of lower-cost debt is exhausted.

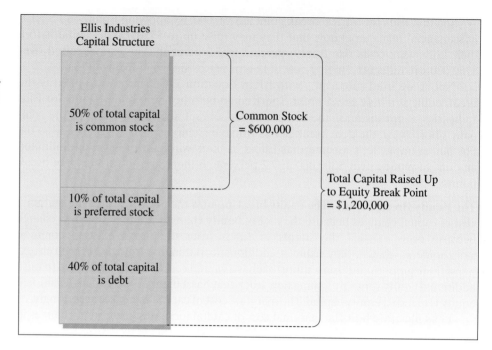

Figure 12-4 Ellis Industries Equity Break Point

Figure 12-4 shows how Ellis Industries can raise up to $1,200,000 of total capital before the $600,000 of lower-cost internal equity is exhausted.

To find the equity break point, we apply Equation 12-7, the marginal cost of capital break point formula. We know that the existing common stock capital limit is $600,000, and that common equity finances 50 percent of the total capital budget. Using Equation 12-7, we solve for the equity break point, BP_e, as follows:

$$BP_e = \frac{\$600,000}{.50}$$

$$= \$1,200,000$$

Our calculations show that the Ellis equity break point is $1,200,000. If the capital budget exceeds $1,200,000, the portion financed with common equity will exceed $600,000. At that point, Ellis will need to issue new common stock to raise the additional capital. The new common stock's cost, k_n, will be greater than the cost of internal common equity, k_s, so the marginal cost of capital, MCC, will rise when the capital budget exceeds $1,200,000 as shown in Figure 12-4.

Calculating the Amount the MCC Changes To calculate MCC changes, we must first identify the break points. In our Ellis Industries example, we identified two break points at which the firm's MCC will change:

Debt Break Point, BP_d = $ 750,000
Equity Break Point, BP_e = $1,200,000

The next step in the MCC analysis is to estimate how much the change in the MCC will be for varying amounts of funds raised.

The MCC Up to the First Break Point. Because the marginal cost of capital, MCC, is simply the weighted average cost of the next dollar of capital to be raised, we can use the WACC formula, Equation 12-6, to calculate the MCC as well.

We assume that Ellis Industries wants to maintain its current capital structure of 40 percent debt, 10 percent preferred stock, and 50 percent common equity. We assume further that its after-tax cost of debt, AT k_d, is 6 percent, its cost of preferred stock, k_p, is 12.5 percent, and its cost of internal common equity, k_s, is 15.5 percent. With these values, we use Equation 12-6 to find the Ellis MCC for capital raised up to the first break point, BP_d, as follows:

$$\begin{aligned} \text{MCC up to BP}_d\ (\$750{,}000) &= (.40 \times \text{AT } k_d) + (.10 \times k_p) + (.50 \times k_s) \\ &= (.40 \times .06) + (.10 \times .125) + (.50 + .155) \\ &= .024 + .0125 + .0775 \\ &= .114, \text{ or } 11.4\% \end{aligned}$$

We see from our calculations that up to the first break point, the Ellis MCC is 11.4 percent—the WACC we calculated earlier. We know, however, that the Ellis lenders will raise the interest rate to 10 percent if the firm raises more than $750,000, at which point the AT k_d increases from 6 percent to 7.2 percent. So between the first break point, BP_d, and the second break point, BP_e, the MCC is:

$$\begin{aligned} \text{MCC between BP}_d\ (\$750{,}000) \\ \text{and BP}_e\ (\$1{,}200{,}000) \end{aligned} = \begin{aligned} &(.40 \times \text{AT } k_d) + (.10 \times k_p) + (.50 \times k_s) \\ &= (.40 \times .072) + (.10 \times .125) + (.50 + .155) \\ &= .0288 + .0125 + .0775 \\ &= .1188, \text{ or } 11.88\% \end{aligned}$$

Our calculations show that the MCC between the first and second break points, $750,000 and $1,200,000, is 11.88 percent.

At the second break point, BP_e, we know from our earlier Ellis discussion that k_s of 15.5 percent changes to k_n that has a value of 16.05 percent. Applying Equation 12-6, the MCC for amounts raised over $1,200,000 follows:

$$\begin{aligned} \text{MCC over BP}_e\ (\$1{,}200{,}000) &= (.40 \times \text{AT } k_d) + (.10 \times k_p) + (.50 \times k_n) \\ &= (.40 \times .072) + (.10 \times .125) + (.50 + .1605) \\ &= .0288 + .0125 + .08025 \\ &= .1216, \text{ or } 12.16\% \end{aligned}$$

We find that the Ellis MCC with a capital budget exceeding $1,200,000, is 12.16 percent. A graph of Ellis Industries' marginal cost of capital is shown in Figure 12-5. Now that we have learned to estimate the MCC for a firm, we examine how MCC estimates affect capital budgeting decisions.

The MCC Schedule and Capital Budgeting Decisions

Firms use the MCC schedule to identify which new capital budgeting projects should be selected for further consideration and which should be rejected. For example, Ellis Industries has identified the following projects for possible adoption:

Project	Initial Investment Required	Project's IRR
A	$500,000	18.00%
B	$300,000	14.00%
C	$200,000	12.05%
D	$300,000	11.50%
E	$200,000	9.00%

Figure 12-5 Ellis Industries Marginal Cost of Capital Schedule

Figure 12-5 shows the marginal cost of capital (MCC) schedule that reflects the cost of debt and cost of equity break points.

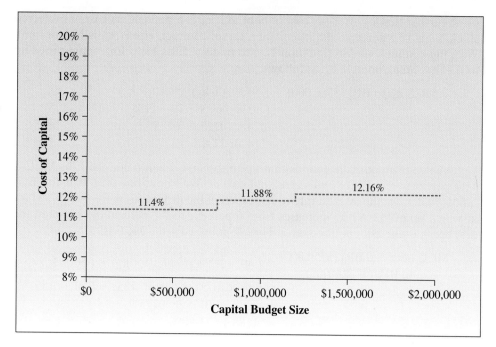

Take Note

Remember that in chapter 10 we learned that capital budgeting projects whose IRRs exceed the cost of capital are acceptable.

The projects are ranked from highest to lowest IRR. The list of proposed capital budgeting projects ranked by IRR is the firm's *investment opportunity schedule (IOS)*. To determine which proposed projects will be accepted, the Ellis financial managers determine which projects have IRRs that exceed their respective costs of capital. To compare the projects' IRRs to the firm's cost of capital, the financial managers plot the investment opportunity schedule (IOS) on the same graph as the MCC. Figure 12-6 shows this technique.

Figure 12-6 Ellis Industries MCC and IOS Schedules

Figure 12-6 shows the MCC and IOS schedules. Those projects on the IOS schedule above the MCC schedule are accepted.

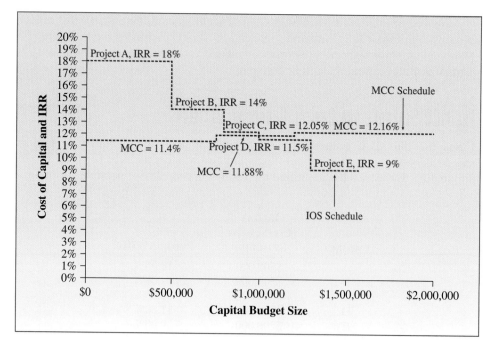

312   Part Three   Long-Term Financial Management Decisions

The projects with the highest IRRs are plotted first. The Ellis financial managers should start with Project A, which has an IRR of 18 percent. That project requires a capital investment of $500,000. Next, they should add Project B, a project with an IRR of 14 percent and an investment of $300,000. The total capital budget with Projects A and B is $800,000. Then Project C, with an IRR of 12.05 percent, should be added. Project C's investment requirement of $200,000 increases the capital budget to $1,000,000.

The addition of Project D, a project with an IRR of 11.5 percent and investment of $300,000, results in a capital budget to $1,300,000. Notice, however, that Project D's IRR is less than the marginal cost of capital at the point where it is added. If Project D were adopted, Figure 12-6 shows that it would have to be financed with capital that costs 11.88 percent, even though the Project's IRR is only 11.5 percent. The Ellis financial managers, then, should reject Project D. Project E, a project with an even lower IRR, is also rejected. Combining the IOS and MCC schedules is an effective tool to see whether a firm should accept or reject a project.

The Optimal Capital Budget

When we integrate the investment opportunity schedule and the marginal cost of capital schedule, as shown in Figure 12-6, we see that Ellis Industries' optimal capital budget is $1 million, consisting of Projects A, B, and C. The **optimal capital budget** is the list of all accepted projects and the total amount of initial cash outlays for these projects. All projects on the IOS schedule that are above the MCC schedule are accepted; the rest are rejected.

Table 12-1 summarizes the seven steps to calculate the optimal capital budget.

TABLE 12-1

Determining the Optimal Capital Budget

STEPS	ACTIONS
Step 1	Calculate the costs of the firm's sources of capital, AT k_d, k_p, k_s, and k_n. Record any borrowing limits and the resulting changes in AT k_d with those limits.
Step 2	Calculate the break points in the capital budget size at which the MCC will change. There will always be an equity break point, BP_e, and there may be one or more debt break points, BP_{d1}, BP_{d2}, and so on.
Step 3	Calculate the MCC up to, between, and above all the break points. The MCC increases at each break point. Record the MCC values before and after each break point.
Step 4	Plot the MCC values on a graph with the capital budget size on the X axis and cost of capital/IRR on the Y axis.
Step 5	Identify the firm's potential investment projects. Record each investment project's initial investment requirement and IRR. Make an investment opportunity schedule (IOS) by lining up the projects from highest IRR to lowest.
Step 6	Plot the IOS on the same graph with the MCC.
Step 7	Note the point where the IOS and MCC schedules cross. Projects on the IOS line above the MCC line should be accepted, and those below the MCC line rejected.

The Importance of MCC to Capital Budgeting Decisions

Analyzing the combined IOS and MCC schedule allows financial managers to examine many projects at once, rather than each project in isolation. This way they choose the best projects.

To demonstrate how important the MCC schedule is to capital budgeting decisions, look at Figure 12-6 again. Notice that Project D is rejected because it is below the MCC line. The graph shows us that if Project D is accepted, it will have to be financed in part with capital costing 11.88 percent and in part with capital costing 12.16 percent. Because Project D's IRR is only 11.5 percent, it's a poor investment. However, this statement is only true because Projects A, B, and C were already considered *before* Project D. Together Projects A, B, and C require $1 million of capital investment. Given that $1 million has already been spent for Projects A, B, and C, the $300,000 required for Project D can only be raised at a cost of 11.88 percent (the first $200,000) and 12.16 percent (the last $100,000).

If Project D were considered by itself, as if it were Ellis Industries' only capital budgeting project, then the $300,000 investment the project requires could have been raised at the company's initial WACC of 11.4 percent. Because the Project's IRR of 11.5 percent exceeds that WACC rate by .1 percent, the project looks like a good investment. But because its IRR was lower than Project's A, B, and C, Project D is not a good investment, given the changes in Ellis's cost of capital due to the MCC break points.

This example illustrates the importance of using a firm's marginal cost of capital (MCC) and not the firm's initial weighted average cost of capital (WACC) to evaluate investments. If all investment projects are treated in isolation, and evaluated using the firm's initial WACC, then some of them may be overvalued. The discrepancy will become apparent when the firm tries to raise the entire amount of capital to support the complete capital budget and finds that the cost of the last dollar raised exceeds the IRR of the last project adopted.

What's Next

In this chapter we learned how to calculate a firm's individual component costs of capital and how to calculate its weighted average cost of capital and its marginal cost of capital. In chapter 13 we discuss how a firm should structure its sources of capital.

Summary

1. Describe the sources of capital and how firms raise capital.

Firms raise debt capital from lenders or bondholders. They also raise funds from preferred stockholders, from current stockholders, and from investors who buy newly issued shares of common stock. All suppliers of capital expect a rate of return proportionate to the risk they take. To ensure a supply of capital to meet their capital budgeting needs, firms must pay that return to capital suppliers. To compensate creditors, firms must pay the interest and principal on loans. For bondholders, firms must pay the market required interest rate. For preferred stockholders, the dividend payments serve as compensation for investors. To compensate common stock investors, firms pay dividends or reinvest the stockholders' earnings.

2. Estimate the cost of capital for each financing source.

To find a firm's overall cost of capital, a firm must first estimate how much each source of capital costs. The after-tax cost of debt, AT k_d, is the market's required rate of return on the firm's debt, adjusted for the tax savings realized when interest payments are deducted from taxable income. The before-tax cost of debt, k_d, is multiplied by one minus the tax rate $(1 - T)$ to arrive at the firm's after-tax cost of debt.

The cost of preferred stock, k_p, is the investor's required rate of return on that security. The cost of common stock equity, k_s, is the opportunity cost of new retained earnings, the required rate of return on the firm's common stock. The cost of new common stock, k_n, (external equity) is the required rate of return on the firm's common stock, adjusted for the flotation costs incurred when new common stock is sold in the market.

3. Estimate the weighted average cost of capital.

The weighted average cost of capital, k_a or WACC, is the overall average cost of funds considering each of the component capital costs and the weight of each of those components in the firm's capital structure. To estimate WACC, we multiply the individual source's cost of capital times its percentage of the firm's capital structure, and then add the results.

4. Use the marginal cost of capital schedule (MCC) to make capital budgeting decisions.

A firm's weighted average cost of capital changes as the cost of debt or equity increases as more capital is raised. Financial managers calculate the break points in the capital budget size at which the MCC will change. There will always be an equity break point, BP_e, and there may be one or more debt break points, BP_{d1}, BP_{d2}, and so on. Financial managers then calculate the MCC up to, between, and above all the break points, and plot the MCC values on a graph showing how the cost of capital changes as capital budget size changes.

Financial managers create an investment opportunity schedule (IOS) by ranking all potential capital budgeting projects from the highest internal rate of return to the lowest, and then plotting the IOS on the same graph with the MCC. To increase the value of the firm, projects on the IOS line above the MCC line should be accepted and those below the MCC line rejected.

5. Explain the importance of the marginal cost of capital (MCC) schedule in financial decision making.

The marginal cost of capital schedule forces financial managers to match the project's rate of return with the cost of funds for that specific project. This marginal analysis prevents financial managers from estimating a project's value incorrectly because of faulty cost of capital estimates that fail to consider how a larger capital budget increases capital costs.

You can access your FREE PHLIP/CW On–Line Study Guide, Current Events, and Spreadsheet Problems templates, which may correspond to this chapter either through the Prentice Hall Finance Center CD–ROM or by directly going to: **http://www.prenhall.com/gallagher.**

PHLIP/CW — *The Prentice Hall Learning on the Internet Partnership–Companion Web Site*

Equations Introduced in this Chapter

Equation 12-1. Formula for the After-Tax Cost of Debt (AT k_d):

$$AT\ k_d = k_d (1 - T)$$

where: k_d = The before-tax cost of debt
T = The firm's marginal tax rate

Equation 12-2. Formula for the Cost of Preferred Stock (k_p):

$$k_p = \frac{D_p}{(P_p - F)}$$

where: k_p = The cost of the preferred stock issue; the expected return
D_p = The amount of the expected preferred stock dividend
P_p = The current price of the preferred stock
F = Flotation cost per share

Equation 12-3. Formula for the Cost of Common Stock Equity (k_s):

$$k_s = \frac{D_1}{P_0} + g$$

where: P_0 = The current price of the common stock
D_1 = The amount of the common stock dividend expected one period from now
g = The expected constant growth rate of the company's common stock dividends

Equation 12-4. CAPM Formula for the Cost of Common Equity (k_s):

$$k_s = k_{RF} + (k_M - k_{RF}) \times \beta$$

where: k_s = The required rate of return from the company's common stock equity
k_{RF} = The risk-free rate of return
k_M = The expected rate of return on the overall stock market
β = The beta of the company's common stock, a measure of the amount of nondiversifiable risk

Equation 12-5. Formula for the Cost of New Common Equity (k_n):

$$k_n = \frac{D_1}{P_0 - F} + g$$

where: k_n = The cost of new common stock equity
P_0 = The price of one share of common stock
D_1 = The amount of the common stock dividend expected to be paid in one year
F = The flotation cost per share
g = The expected constant growth rate of the company's common stock dividends

Equation 12-6. Formula for the Weighted Average Cost of Capital, (WACC):

$$k_a = (WT_d \times AT\ k_d) + (WT_p \times k_p) + (WT_s \times k_s)$$

where: k_a = The weighted average cost of capital (WACC)
WT_d = The weight, or proportion, of debt used to finance the firm's assets
$AT\ k_d$ = The after-tax cost of debt
WT_p = The weight, or proportion, of preferred stock being used to finance the firm's assets
k_p = The cost of preferred stock
WT_s = The weight, or proportion, of common equity being used to finance the firm's assets
k_s = The cost of common equity

Equation 12-7. Formula for the Marginal Cost of Capital (MCC) Break Point:

$$BP = \frac{Limit}{Proportion\ of\ Total}$$

where: BP = The capital budget size at which the MCC changes (break point)
limit = The point at which the cost of the source of capital changes
Proportion of Total = The percentage of this source of capital in the firm's capital structure

Self-Test

ST-1. Jules' Security Company can issue new bonds with a market interest rate of 14 percent. Jules' marginal tax rate is 32 percent. Compute the after-tax cost of debt, AT k_d, for this company.

ST-2. Mr. White's company, The Problem Solvers, wants to issue new preferred stock. The preferred dividend is $3.00 per share, the stock can be sold for $30, and the flotation costs are $1. What is the cost of preferred stock, k_p?

ST-3. Vincent's Dance Studio Incorporated has a beta of 1.9. The risk-free rate of interest is 4 percent. The market portfolio has an expected rate of return of 10 percent. What is the cost of internal equity, k_s, for this company using the CAPM approach?

ST-4. Marsalis's Entertainment Corporation has an after-tax cost of debt of 8 percent, a cost of preferred stock of 12 percent, and a cost of equity of 16 percent. What is the weighted average cost of capital, k_a or WACC, for this company? The capital structure of Marsalis's company contains 20 percent debt, 10 percent preferred stock, and 70 percent equity.

ST-5. Quinten's Movie Company has been told by its investment banking firm that it could issue up to $8 million in bonds at an after-tax cost of debt of 10 percent. But after that, additional bonds would have a 12 percent after-tax cost. Quinten's company uses 40 percent debt and 60 percent equity for major projects. How much money could this company raise, maintaining its preferred capital structure, before the after-tax cost of debt would jump to 12 percent?

FinCoach Practice Exercises

To help improve your grade and master the mathematical skills covered in this chapter, open FinCoach on the Prentice Hall Finance Center CD-ROM and work practice problems in the category Cost of Capital.

Review Questions

1. Which is lower for a given company: the cost of debt or the cost of equity? Explain. Ignore taxes in your answer.
2. When a company issues new securities, how do flotation costs affect the cost of raising that capital?
3. What does the "weight" refer to in the weighted average cost of capital?
4. How do tax considerations affect the cost of debt and the cost of equity?
5. If dividends paid to common stockholders are not legal obligations of a corporation, is the cost of equity zero? Explain your answer.
6. What is the investment opportunity schedule (IOS)? How does it help financial managers make business decisions?
7. What is a marginal cost of capital schedule (MCC)? Is the schedule always a horizontal line? Explain.
8. For a given IOS and MCC, how do financial managers decide which proposed capital budgeting projects to accept, and which to reject?

Build Your Communication Skills

CS-1. Prepare a brief paper in which you explain to the CEO of a fictional company why the cost of equity is greater than either the cost of debt or the cost of preferred stock. Be sure to explain why equity funds have a cost even though those costs are not reflected on the income statement. Finally, discuss why it is important for a firm to know its cost of capital.

CS-2. Write a short report that analyzes the role of the marginal cost of capital (MCC) schedule in making capital budgeting decisions. In your report, explain how the investment opportunity schedule (IOS) affects this decision-making process. Be sure to explain your rationale for choosing some proposed projects while rejecting others.

Problems

Cost of Debt

12-1.
   a. What would be the after-tax cost of debt for a company with the following yields to maturity for its new bonds, if the applicable tax rate were 40 percent?
   - (i) YTM = 7%
   - (ii) YTM = 11%
   - (iii) YTM = 13%

   b. How would the cost of debt change if the applicable tax rate were 34 percent?

Cost of Debt

12-2. Calculate the after-tax cost of debt for loans with the following effective annual interest rates and corporate tax rates.

   a. Interest rate, 10%; tax rate 0%.
   b. Interest rate, 10%; tax rate 22%.
   c. Interest rate, 10%; tax rate 34%.

Cost of Debt

12-3. What would be the cost of debt for the following companies given their yields to maturity (YTM) for new bonds and the applicable corporate tax rates?

Company	YTM	Tax Rate
A	8%	34%
B	11%	40%
C	14%	30%

Cost of Debt

12-4. Mary Lynn Eatenton is the Chief Financial Officer of Magnolia Steel, Inc. She has asked Trudy Jones, one of the financial analysts, to calculate the after-tax cost of debt based on different bond yield to maturity rates. Magnolia Steel's current tax rate is 34 percent, but increasing sales and profits will put them in the 40 percent tax bracket by the end of the year. Calculate the after-tax cost of debt figures that will be shown in Ms. Jones's report at each tax rate for the following YTM rates.

   a. Yield to maturity, 8%.
   b. Yield to maturity, 14%.
   c. Yield to maturity, 16%.

Cost of Debt

12-5. A firm is issuing new bonds that pay 8 percent annual interest. The market required annual rate of return on these bonds is 13 percent. The firm has a tax rate of 40 percent.

   a. What is the before-tax cost of debt?
   b. What is the after-tax cost of debt?

Cost of Preferred Stock

12-6. A company can sell preferred stock for $26 per share and each share of stock is expected to pay a dividend of $2. If the flotation cost per share of stock is $0.75, what would be the estimate of the cost of capital from this source?

12-7. Leo Bloom, the treasurer of a manufacturing company, thinks that debt (YTM = 11%, tax rate = 40%) will be a cheaper option for acquiring funds compared to issuing new preferred stock. The company can sell preferred stock at $61 per share and pay a yearly preferred dividend of $8 per share. The cost of issuing preferred stock is $1 per share. Is Leo correct?

Cost of Debt and of Preferred Stock

12-8. One-Eyed Jacks Corporation needs money to fund a new production line of playing cards. Rio Longworth, manager of the finance department, suggests they sell preferred stock for $50 per share. They expect to pay $6 per share annual dividends. What is the estimate of the cost of preferred stock if the flotation cost is $2.25 per share?

Cost of Preferred Stock

12-9. El Norte Industries will issue $100 par, 12 percent preferred stock. The market price for the stock is expected to be $89 per share. El Norte must pay flotation costs of 5 percent of the market price. What is El Norte's cost of preferred stock?

Cost of Preferred Stock

12-10. Twister Corporation is expected to pay a dividend of $7 per share one year from now on its common stock, which has a current market price of $143. Twister's dividends are expected to grow at 13 percent.

Cost of Equity

    a. Calculate the cost of the company's retained earnings.
    b. If the flotation cost per share of new common stock is $4, calculate the cost of issuing new common stock.

12-11. Amy Jolly is the treasurer of her company. She expects the company will grow at 4 percent in the future, and debt securities (YTM = 14%, tax rate = 30%) will always be a cheaper option to finance the growth. The current market price per share of its common stock is $39 and the expected dividend in one year is $1.50 per share. Calculate the cost of the company's retained earnings and check if Amy's assumption is correct.

Cost of Retained Earnings

12-12. Pedro Muzquiz and Tita de la Garza are the CEOs of a large bakery chain, Chocolates, Inc. The common stock sells on the NASDAQ with a current market price of $65 per share. A $7 dividend is planned for one year from now. Business has been good and they expect the dividend growth rate of 10 percent to continue.

Challenge Problem

    a. Calculate the cost of the corporation's retained earnings.
    b. At the beginning of the year, 1 million shares were authorized with 500,000 issued and outstanding. They plan to issue another 200,000 shares. Calculate the cost of capital of the new common stock if the flotation cost per share is $3. Do you expect the cost of new common equity (external) to be higher than the cost of the internal equity? Why?

12-13. Margo Channing, the financial analyst for Eve's Broadway Production Company, has been asked by management to estimate a cost of equity for use in the analysis of a project under consideration. In the past, dividends declared and paid have been very sporadic. Because of this, Ms. Channing elects to use the CAPM approach to estimate the cost of equity. The rate on the short-term U.S. Treasury bills is 3 percent, and the expected rate of return on the overall stock market is 11 percent. Eve's Broadway Production Company has a beta of 1.6. What will Ms. Channing report as the cost of equity?

Cost of Equity, CAPM Approach

12-14. African Queen River Tours, Inc., has capitalized on the renewed interest in riverboat travel. Charlie Allnut, the lone financial analyst, estimates the firm's earnings, dividends, and stock price will continue to grow at the historical 5 percent rate. AQRT's common stock is currently selling for $30 per share. The dividend just paid was $2. They pay dividends every year. The rate of return expected on the overall stock market is 12 percent.

Cost of Equity, CAPM Approach

    a. What is AQRT's cost of equity?
    b. If they issue new common stock today and pay flotation costs of $2 per share, what is the cost of new common equity?
    c. If AQRT has a risk-free rate of 3 percent and a beta of 1.4, what will be AQRT's cost of equity using the CAPM approach?

Weighted Average Cost of Capital

12-15. Alvin C. York, the founder of York Corporation, thinks that the optimal capital structure of his company is 30 percent debt, 15 percent preferred stock, and the rest common equity. If the company is in the 40 percent tax bracket, compute its weighted average cost of capital given that:

YTM of its debt is 10 percent.
New preferred stock will have a market value of $31, a dividend of $2 per share, and flotation costs of $1 per share.
Price of common stock is currently $100 per share and new common stock can be issued at the same price with flotation costs of $4 per share. The expected dividend in one year is $4 per share and the growth rate is 6 percent.

Assume the addition to retained earnings for the current period is zero.

Weighted Average Cost of Capital

12-16. A company has an optimal capital structure as follows:

Total Assets	$600,000
Debt	$300,000
Preferred Stock	$100,000
Common Equity	$200,000

What would be the minimum expected return from a new capital investment project to satisfy the suppliers of the capital? Assume the applicable tax rate to be 40 percent, YTM of its debt to be 11 percent, flotation cost per share of preferred stock to be $0.75, and flotation cost per share of common stock $4. The preferred stock and common stock are selling in the market for $26 and $143, respectively, and are expected to pay a dividend of $2 and $7, respectively, in one year. The company's dividends are expected to grow at 13 percent per year. The firm would like to maintain the foregoing optimal capital structure to finance the new project.

Weighted Average Cost of Capital

12-17. Great Expectations, a wedding and maternity clothing manufacturer, has a cost of equity of 16 percent, and a cost of preferred stock of 14 percent. Its before-tax cost of debt is 12 percent and its marginal tax rate is 40 percent. Assume that the most recent balance sheet shown here reflects the optimal capital structure. Calculate Great Expectations' after-tax weighted average cost of capital.

Great Expectations
Balance Sheet
December 31, 1999

Assets		Liabilities and Equity	
Cash	$ 50,000		
Accounts Receivable	90,000	Long-Term Debt	$ 600,000
Inventories	300,000	Preferred Stock	250,000
Plant and Equipment, net	810,000	Common Stock	400,000
Total Assets	$1,250,000	Total Liabilities and Equity	$1,250,000

Weighted Average Cost of Capital

12-18. Puppet Masters is considering a new capital investment project. The company has an optimal capital structure and plans to maintain it. The yield to maturity on Puppet Masters' debt is 10 percent and its tax rate is 35 percent. The market price of the new issue of preferred stock is $25 per share with an expected per share dividend of $2 at the end of this year. Flotation costs are set at $1 per share. The new issue of common stock has a current market price of $140 per share with an expected dividend in one year of $5. Flotation costs for issuing new common stock are $4 per share. Puppet Masters' dividends are growing at 10 percent per year, and this growth is expected to continue for the foreseeable future. Selected figures from last year's balance sheet follow:

Total Assets	$1,000,000
Long-Term Debt	300,000
Preferred Stock	100,000
Common Stock	600,000

Calculate the minimum expected return from the new capital investment project needed to satisfy the suppliers of the capital.

12-19. Fans By Fay Company has a capital structure of 60 percent debt and 40 percent common equity. The company expects to realize $200,000 in net income this year and will pay no dividends. The effective annual interest rate on its new borrowings increases by 3 percent for amounts over $500,000.

Marginal Cost of Capital Schedule

  a. At what capital budget size will Fans By Fay's cost of equity increase? In other words, what is its equity break point?
  b. At what capital budget size will its cost of debt increase (debt break point)?

12-20. Babe's Dog Obedience School, Inc. wants to maintain its current capital structure of 50 percent common equity, 10 percent preferred stock, and 40 percent debt. Its cost of common equity is 13 percent and the cost of preferred stock is 12 percent. The bank's effective annual interest rate is 11 percent for amounts borrowed that are less than or equal to $1 million, and 13 percent for amounts between $1 million and $2 million. If more than $2 million is borrowed, the effective annual interest rate charged is 15 percent. Babe's tax rate is 40 percent. The firm expects to realize $2,750,000 in net income this year after preferred dividends have been paid.

Marginal Cost of Capital Schedule

  a. Calculate the marginal cost of capital if $900,000 is needed for an upcoming project.
  b. Calculate the marginal cost of capital if $1,500,000 is needed for the project instead.
  c. If a different project is adopted, and $2,005,000 is needed for it, what is the marginal cost of capital?

12-21. Stone Wood Products has a capital structure of 35 percent debt and 65 percent common equity. The managers consider this mix to be optimal and want to maintain it in the future. Net income for the coming year is expected to be $1.2 million dollars. Duke Mantee, the loan officer at the local bank, has set up the following schedule for Stone Wood Products' borrowings. There are 40,000 shares of common stock outstanding. The firm's tax rate is 40 percent.

Comprehensive Problem

Loan Amount	Interest Rate
$0 to $750,000	10%
> $750,000	12%

The market price per share of Stone Wood Product's common stock is $50 per share. They have declared a $5 dividend to be paid in one year. The company's expected growth rate is 9 percent. The flotation costs for new common stock issued are set at 8 percent of the market price.

The managers are considering several investment opportunities for the upcoming year. They have asked the senior financial analyst, Gabrielle Maple, to recommend which of the following projects the firm should undertake. Because you are the newest member of her team and need the experience, she has passed this management request on to you.

Investment Opportunities

Project	Initial Investment (in millions)	Rate of Return
A	$0.5	16%
B	1.6	12%
C	0.6	15%
D	1.5	18%

a. Calculate all of Stone Wood Products' component costs of capital (after-tax cost of debt, cost of equity, and cost of new equity).
b. Calculate all of the MCC break points.
c. Calculate all of the marginal cost of capital figures.
d. Make an investment opportunity schedule (IOS) by listing the projects from the highest to the lowest internal rates of return.
e. Plot the MCC values and the IOS values on the same graph.
f. Which projects will you recommend management adopt for the next year?

Answers to Self-Test

ST-1. $.14 \times (1 - .32) = .0952 = 9.52\%$ AT k_d

ST-2. Using Equation 12-7, $k_p = D_p/(P_p - F)$ $\$3/(\$30 - \$1) = .1034$, or 10.34%

ST-3. $.04 + (.10 - .04) \times 1.9 = .154 = 15.4\%$ k_s

ST-4. $(.08 \times .2) + (.12 \times .10) + (.16 \times .70) = .14 = 14\%$ k_a

ST-5. The break point in the MCC schedule caused by the increase in the cost of debt, BP_d, after $\$8,000,000$ is borrowed, equals $\$8,000,000 \div .40 = \$20,000,000$.

CHAPTER 13

CAPITAL STRUCTURE BASICS

The Lord forbid that I should be out of debt, as if, indeed, I could not be trusted.
—Francois Rabelais

A POPCORN VENTURE

Jason is a college student who wants to start his own business. Jason's business idea is to sell popcorn from a cart, just as he has seen in the downtown area of the city in which he lives. The downtown vendor sells about 500 bags of popcorn a day, and Jason thinks he might be able to do as well with a similar popcorn stand near the college in his town.

However, the wagon contains both a popcorn-making machine and a storage room for supplies, so it isn't cheap. Also, if Jason went into the business, he would need an expensive business operator's license from the city. The downtown vendor charges only $1 for a bag of popcorn, so Jason would have to sell a lot of popcorn to recoup the high price of the wagon and the license.

Is this a viable business idea or not? What are the risks and the potential returns of this business? Is this a better path than taking a McJob, as many of Jason's friends have done? In this chapter we'll look at some of these issues.

Source: Jason's popcorn venture is based on actual events. The entrepreneur's name has been changed and approximate numbers have been used, because data about this private company are confidential.

CHAPTER OBJECTIVES

After reading this chapter, you should be able to:

- Find the breakeven level of sales for a firm.

- Explain operating, financial, and combined leverage effects and the resulting risks.

- Describe the risks and returns of a leveraged buyout.

- Explain how changes in capital structure affect a firm's value.

Chapter Overview

In this chapter we investigate how fixed costs affect the volatility of a firm's operating and net income. We see how fixed operating costs create *operating leverage,* which magnifies the effect of sales changes on operating income. We also examine how fixed financial costs create *financial leverage,* which magnifies the effect of changes in operating income on net income. Then we analyze the risk and return of leveraged buy outs (LBOs). Finally, we see how changes in a firm's capital structure affect the firm's overall value.

Breakeven Analysis and Leverage

We introduced the topic of leverage in chapter 7, page 153.

Investments in projects may change a firm's fixed operating and financing costs, thereby affecting firm value. Fixed costs may affect firm value because of *leverage effects* and the resulting risk from those leverage effects.

To understand a firm's potential for risk and return, then, financial managers must understand two types of leverage effects: operating leverage and financial leverage.

Breakeven analysis is a key to understanding *operating leverage.* In breakeven analysis we examine fixed and variable operating costs. **Fixed costs** are those costs that do not vary with the company's level of production. **Variable costs** are those costs that change as the company's production levels change.

We discussed operating income in chapter 4, page 63.

In breakeven analysis the **sales breakeven point** is the level of sales that a firm must reach to cover its operating costs. Put another way, it is the point where the operating income (earnings before interest and taxes) equals zero.

A company with high fixed operating costs must generate high sales revenue to reach the sales breakeven point. A company with low fixed operating costs requires relatively low sales revenue to reach its sales breakeven point.

We usually observe a high/low trade-off in breakeven analysis. Firms with high fixed operating costs tend to have low variable costs, and vice versa. A company that automates a factory, for instance, commits to significant fixed costs—the expensive equipment. But the company's variable labor costs are likely to be low at a highly automated plant that operates with relatively few employees. In contrast, a company that produces handmade pottery with little overhead, and hires hourly workers as needed, is likely to have low fixed costs but high variable costs.[1]

To demonstrate the high/low trade-off, we gather data for a sales breakeven chart for two firms. The first firm has high fixed and low variable costs. The second firm has low fixed and high variable costs.

Constructing a Sales Breakeven Chart

A breakeven chart shows graphically how fixed costs, variable costs, and sales revenue interact. Analysts construct the chart by plotting sales revenue and costs at various unit sales levels on a graph. To illustrate, let's construct the breakeven chart for Jason's Popcorn wagon, featured in the opening of the chapter.

---

[1]Labor costs can be either fixed or variable. If workers are guaranteed pay for a certain minimum number of hours per week, as might be called for in a union contract, the labor costs associated with this minimum guaranteed pay would be fixed costs. The costs associated with hourly worker pay with no guaranteed minimum are variable.

TABLE 13-1

Jason's Relevant Figures for Breakeven Analysis

Fixed Costs:	
Wagon (annual rental)	$8,000
City License (annual fee)	$4,000
Total	$12,000
Variable Costs per Unit:	
One Paper Bag	$0.020
Oil	$0.005
Salt	$0.003
Popcorn	$0.012
Total	$0.040
Sales Price per Unit:	$1.00

The first step in constructing the breakeven chart is to find the breakeven point for the business. Let's look at some of the numbers for Jason's business and calculate the level of sales Jason must achieve to break even. Recall that at the breakeven point, operating income equals zero. If sales are below the breakeven point, Jason suffers an operating loss. If sales are above the breakeven point, Jason enjoys an operating profit. (Interest and taxes, subtracted after finding operating income, will be discussed in the last section of the chapter.)

Jason wants to know that his business venture has the potential for a positive operating profit, so he is keenly interested in finding his breakeven point. To find this point, we need to know how many bags of popcorn he must sell before the sales revenue contributed by each bag sold just covers his fixed and variable operating costs. The relevant sales breakeven figures for Jason's proposed business are shown in Table 13-1.

The numbers in Table 13-1 show that Jason's fixed costs are high compared to his sales price of $1 per bag of popcorn. The fixed costs include the $8,000 annual rental fee for the wagon and the $4,000 annual license fee. Jason must pay these costs no matter how much popcorn he produces and sells.

In contrast to the high fixed operating costs, Jason's variable operating costs per unit are a tiny fraction of his sales price of $1 per unit. The bag, oil, salt, and popcorn that help produce one bag of popcorn cost a total of $0.04. Each bag of popcorn that is sold, then, contributes $0.96 to cover the fixed costs, variable costs, and ultimately profit of the business ($1.00 − $0.04 = $0.96). The sales price per unit minus the variable cost per unit, $.96 in this case, is the **contribution margin.**

From the numbers presented in Table 13-1, we can calculate the breakeven level of sales for Jason's business. We find the level of sales needed to reach the operating income breakeven point by applying the following formula:

The Breakeven Point in Unit Sales

$$Q_{b.e.} = \frac{FC}{p - vc} \quad (13\text{-}1)$$

where: $Q_{b.e.}$ = Quantity unit sales breakeven level
FC = Total fixed costs
p = Sales price per unit
vc = Variable cost per unit

For Jason's business, the total fixed costs are $12,000, the price per unit is $1, and the variable cost per unit is $.04. According to Equation 13-1, Jason's popcorn business has the following sales breakeven point:

$$Q_{b.e.} = \frac{\$12{,}000}{\$1.00 - \$.04}$$

$$= \frac{\$12{,}000}{\$.96}$$

$$= 12{,}500$$

We find that Jason's sales breakeven point with $12,000 in fixed costs, $.04 per unit in variable costs, and a $1 per bag sales price, is 12,500 units. At $1 per bag, this is $12,500 in sales to reach the breakeven point.

Now that we know Jason's sales breakeven point, we need revenue and cost information to construct the breakeven chart.

Revenue Data At any given level of unit sales, Jason's total sales revenue can be found using Equation 13-2:

Total Revenue, TR

$$TR = p \times Q \qquad (13\text{-}2)$$

where: p = Sales price per unit
Q = Unit sales (Quantity sold)

Table 13-2 shows how to calculate Jason's sales revenues at different sales levels. For instance, we see that if Jason sells 5,000 bags of popcorn at the price of $1 per bag, his total revenue will be 5,000 × $1.00 = $5,000. If Jason sells 10,000 bags, his total revenue will be $10,000.

Cost Data By definition, Jason's fixed costs will remain $12,000, regardless of the level of unit sales. His variable costs, however, increase by $0.04 for each unit sold. Jason's total costs for any given level of unit sales can be found using Equation 13-3 as follows:

Total Costs, TC

$$TC = FC + (vc \times Q) \qquad (13\text{-}3)$$

where: FC = Fixed costs
vc = Variable costs per unit
Q = Unit sales (Quantity sold)

Table 13-3 demonstrates how we use Equation 13-3 to calculate Jason's total costs for different sales levels. For instance, we see that if Jason sells 5,000 bags of popcorn at a variable cost of $0.04 per bag and fixed costs of $12,000, his total cost will be $12,200. At 10,000 bags, his total cost will be $12,400.

TABLE 13-2

Sales Revenues at Different Unit Sales Levels

Unit Sales (Q)	×	Price (p)	=	Total Revenue (TR)
0	×	$1	=	$0
5,000	×	$1	=	$5,000
10,000	×	$1	=	$10,000
15,000	×	$1	=	$15,000
20,000	×	$1	=	$20,000
25,000	×	$1	=	$25,000
30,000	×	$1	=	$30,000

Fixed Costs (FC)	+	(Variable Cost/Unit (VC)	×	Unit Sales (Q))	=	Total Cost (TC)
$12,000	+	($.04	×	0)	=	$12,000
$12,000	+	($.04	×	5,000)	=	$12,200
$12,000	+	($.04	×	10,000)	=	$12,400
$12,000	+	($.04	×	15,000)	=	$12,600
$12,000	+	($.04	×	20,000)	=	$12,800
$12,000	+	($.04	×	25,000)	=	$13,000
$12,000	+	($.04	×	30,000)	=	$13,200

TABLE 13-3

Jason's Total Costs for Different Sales Levels

Plotting Data on the Breakeven Chart   Jason's breakeven chart is shown in Figure 13-1. The chart is constructed with unit sales (Q) on the horizontal axis and cost and revenue dollars on the vertical axis. Total revenues from Table 13-2 are shown on the TR line, and total costs from Table 13-3 are shown on the TC line.

We see from the chart that to break even, Jason has to sell $12,500 worth of popcorn at $1 per bag—a quantity of 12,500 bags.

Applying Breakeven Analysis

Although 12,500 bags of popcorn may seem like a lot of sales just to break even, Jason has watched another vendor downtown sell on average 500 bags of popcorn a day. Jason plans to sell for three months during the summer, four weeks a month, five days a week. He estimates that he could sell 30,000 bags of popcorn (500 bags × 3 months × 4 weeks × 5 days) during the summer. At this sales level, Jason expects $30,000 in gross sales revenue at $1 per bag and $16,800 in operating income [$30,000 total revenue − $12,000 fixed costs − ($30,000 × .04 variable costs) = $16,800 operating income]. Not a bad summer job income.

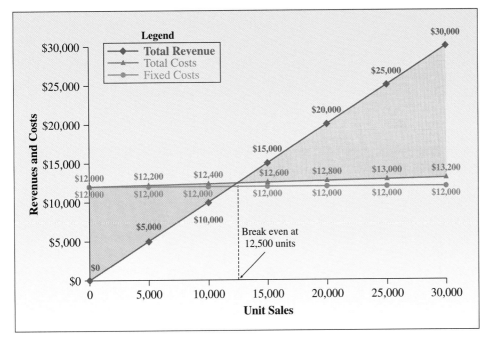

Figure 13-1 Breakeven Chart for Jason's Popcorn Wagon

Figure 13-1 shows the fixed costs, total costs, and total revenues of Jason's Popcorn Wagon at different levels of unit sales. The breakeven sales point, the intersection of the total costs and total revenues lines, is 12,500 units.

How is it possible to make so much money selling popcorn? Note how once Jason passes the breakeven point in sales, each additional $1 bag of popcorn he sells generates $0.96 of operating profit. The $0.04 in variable costs incurred in the production of that bag of popcorn represents a small part of the $1 in revenue generated. Operating profit rises rapidly as sales climb above the breakeven point of 12,500 units, as shown by Figure 13-1. Were Jason's sales potential not so promising, however, his risk of loss (negative operating income) would be high. The blue area of the graph in Figure 13-1 shows Jason's loss potential.

The breakeven chart allows Jason to see the different sales scenarios to understand his profit and loss potential. Because the total revenue line in Figure 13-1 is much steeper than the total cost line (because the sales price per unit is much greater than the variable cost per unit) the profit potential is great. Because of the high fixed costs, however, the loss potential is great, too. What happens depends on how much popcorn Jason can sell.

To illustrate what happens with a low breakeven business, let's construct a breakeven chart for Carey, another college student, who wants to sell hotplate mini-cookbooks (only five pages long) to college students.[2]

Because Carey plans to operate from her apartment and use her own recipes for the mini-cookbook, her only fixed cost would be a $1,000 printer's design fee. Her variable costs consist of her paper printing costs at $0.60 per unit. Carey plans to sell her cookbook for $1 per unit.

This is a low-risk business. The design fee is modest and there are no other fixed costs. The contribution margin is $0.40 ($1.00 sales price − $0.60 variable cost per unit). We can find Carey's breakeven point using Equation 13-1:

$$Q_{b.e.} = \frac{FC}{p - vc}$$

$$= \frac{\$1,000}{\$1.00 - \$.60}$$

$$= \frac{\$1,000}{\$.40}$$

$$= 2,500$$

We find that Carey's breakeven point is 2,500 units. Carey figures she can sell to friends in the dorm. Beyond that, however, the sales potential is uncertain. She may or may not reach the breakeven point.

To find Carey's breakeven point, we find her total revenue and total costs at different sales levels and plot them on a breakeven chart.

Note how small the loss potential is for Carey's business, as shown in the blue area in Figure 13-2, compared to Jason's loss potential, shown in the blue area in Figure 13-1. Carey's loss potential is small because her breakeven level of sales ($0 operating income) is 2,500 units, compared to Jason's 12,500 unit breakeven point. Even if she sold nothing, Carey would lose only the $1,000 in fixed costs that she had to pay (compared to Jason's $12,000). Table 13-4 shows the profit and loss potential for Jason and Carey.

---

[2]Believe it or not, Carey's business is also inspired by a true story. Oliver Stone would be proud.

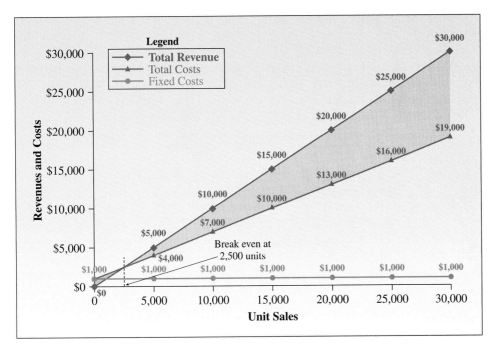

Figure 13-2 Breakeven Chart for Carey's Mini-Cookbooks

Figure 13-2 shows the fixed costs, total costs, and total revenues of Carey's Mini-Cookbooks at different levels of unit sales. The breakeven sales level of 2,500 units is the point where the total costs line crosses the total revenues line.

The risk of Jason's business is also evident when we look at sales of 15,000 units for each business. Jason has a profit of only $2,400, whereas Carey would earn a profit of $5,000; at 30,000 units sold, however, Jason earns a profit of $16,800 and Carey earns only $11,000, as shown in Table 13-4. Jason's profits are much more dependent on selling a large number of units than are Carey's.

Now compare the profit potential for the two proposed businesses. Jason has the potential to make much more profit (operating income) than Carey. At a sales level of 30,000, Table 13-4 shows Jason makes $16,800 whereas Carey would make only $11,000. Even though Jason's business has more risk—he stands to lose much more if sales don't go well—he has the potential for greater returns.

TABLE 13-4

Jason's and Carey's Profit and Loss Potential

UNITS PRODUCED AND SOLD	JASON			CAREY		
	$Total Costs	$Total Revenue	$Operating Income	$Total Costs	$Total Revenue	$Operating Income
0	12,000	0	−12,000	1,000	0	−1,000
5,000	12,200	5,000	−7,200	4,000	5,000	1,000
10,000	12,400	10,000	−2,400	7,000	10,000	3,000
15,000	12,600	15,000	2,400	10,000	15,000	5,000
20,000	12,800	20,000	7,200	13,000	20,000	7,000
25,000	13,000	25,000	12,000	16,000	25,000	9,000
30,000	13,200	30,000	16,800	19,000	30,000	11,000
35,000	13,400	35,000	21,600	22,000	35,000	13,000
40,000	13,600	40,000	26,400	25,000	40,000	15,000

> **Take Note**
>
> *Long-distance telephone and cable companies are examples of firms with high fixed costs and low variable costs per unit. A consulting firm would be an example of a firm with low fixed costs and high variable costs.*

Whether the high fixed cost and low variable cost per unit business (like Jason's) is better than the low fixed cost and high variable cost per unit business (like Carey's) depends on two factors: how many units you think you can sell, and how much tolerance you have for risk. High fixed costs and low variable costs per unit mean high profit potential and high loss potential, as in the case of Jason's proposed business. Conversely, low fixed costs and high variable costs per unit mean low profit potential and low loss potential, as in the case of Carey's proposed business.

Leverage

In physics the term *leverage* describes how a small force can be magnified to create a larger force. For example, a farmer wants to move a large boulder in a field. He wedges a long board (a lever) between the large boulder and a small rock (a fulcrum), which gives him enough leverage to push down on the end of the long board and easily move the boulder.

The power of leverage can also be harnessed in a financial setting. Its magnifying power can help or hurt a business. A firm that has leverage will earn or lose more than it would without leverage. In the sections that follow, we investigate operating and financial leverage and the risks associated with each type.

Operating Leverage

Operating leverage refers to the phenomenon whereby a small change in sales triggers a relatively large change in operating income (or earnings before interest and taxes, also known as EBIT). Operating leverage occurs because of fixed costs in the operations of the firm. A firm with fixed costs in the production process will see its EBIT rise by a larger percentage than sales when unit sales are increasing. If unit sales drop, however, the firm's EBIT will decrease by a greater percentage than does its sales.

Table 13-5 illustrates the operating leverage effect for a firm in which all production costs, a total of $5,000, are fixed. Observe how the presence of the fixed costs causes a 10 percent change in sales to produce a 20 percent change in operating income.

Calculating the Degree of Operating Leverage The degree of operating leverage, or DOL, measures the magnitude of the operating leverage effect. The **degree of operating leverage** is the percentage change in earnings before interest and taxes (%ΔEBIT) divided by the percentage change in sales (%ΔSales):

Degree of Operating Leverage (DOL)

$$\text{DOL} = \frac{\%\Delta \text{EBIT}}{\%\Delta \text{Sales}} \qquad (13\text{-}4)$$

TABLE 13-5

The Operating Leverage Effect—Fixed Costs Only

	Period 1	Period 2	Percent Change
Sales	$10,000	$11,000	10%
Fixed Costs	−5,000	−5,000	
Operating Income	$ 5,000	$ 6,000	20%

where: %△EBIT = Percentage change in earnings before interest and taxes
%△Sales = Percentage change in sales

According to Equation 13-4, the DOL for the firm in Table 13-5 is:

$$DOL = \frac{20\%}{10\%}$$
$$= 2.0$$

We see that, for a firm with a 10 percent change in sales and a 20 percent change in EBIT, the DOL is 2.0. A DOL greater than 1 shows that the firm has operating leverage. That is, when sales change by some percentage, EBIT will change by a greater percentage.

The Effect of Fixed Costs on DOL   Table 13-6 shows the projected base-year and second-year income statement for Jason's Popcorn Wagon. The income statement allows us to analyze Jason's operating leverage. (Note that Table 13-6 divides the operating expenses into two categories, fixed and variable.) We see that sales and operating expenses are likely to change in the second and subsequent years. We also see the predicted impact on EBIT, given the sales forecast.

We see from Table 13-6 that Jason's percentage change in sales is 10 percent, and his percentage change in EBIT (or operating income) is 17.1 percent. We use Equation 13-4 to find Jason's DOL, as follows:

$$DOL = \frac{\%\triangle EBIT}{\%\triangle Sales}$$
$$= \frac{(19{,}680 - 16{,}800)/16{,}800}{(33{,}000 - 30{,}000)/30{,}000}$$
$$= \frac{.171}{.10}$$
$$= 1.71$$

Our DOL calculations indicate that if Jason's Popcorn Wagon business sales increase by 10 percent from the base year to the next year, EBIT will increase by 17.1 percent. This larger percentage increase in EBIT is caused by the company's fixed operating costs. No matter how much popcorn Jason produces and sells, his wagon and license costs stay the same. The fixed costs cause the EBIT to increase faster than sales. If sales were to decrease, the fixed costs must still be paid. As a result, the fixed costs cause EBIT to drop by a greater percentage than sales.

The Alternate Method of Calculating DOL   Instead of using Equation 13-4, we may also find the degree of operating leverage (DOL) by using only numbers found

TABLE 13-6

Jason's Popcorn Wagon Projected Income Statement (Fixed Costs are $12,000 and Variable Costs $0.04 per Unit)

	BASE YEAR	YEAR 2		
Sales	$30,000	$33,000	%△ = $\frac{33{,}000 - 30{,}000}{30{,}000} = 10\%$
− VC	− 1,200	− 1,320		
− FC	− 12,000	− 12,000		
= EBIT	= $16,800	= $19,680	%△ = $\frac{19{,}680 - 16{,}800}{16{,}800} = 17.1\%$

in the base-year income statement. Subtract total variable costs from sales, divide that number by sales minus total variable costs minus fixed costs, and solve for DOL. The formula for the alternate method of finding DOL, Equation 13-5, follows:

Degree of Operating Leverage, DOL (Alternate Formula)

$$DOL = \frac{Sales - VC}{Sales - VC - FC} \qquad (13\text{-}5)$$

where: VC = Total variable costs
FC = Total fixed costs

From Table 13-6 we know that in the base year, Jason's Popcorn Wagon is projected to have sales of $30,000, variable costs of $1,200, and fixed costs of $12,000. Using the alternate formula, we find that Jason has the following DOL:

$$DOL = \frac{Sales - VC}{Sales - VC - FC}$$

$$= \frac{30{,}000 - 1{,}200}{30{,}000 - 1{,}200 - 12{,}000}$$

$$= \frac{28{,}800}{16{,}800}$$

$$= 1.71$$

We find a DOL of 1.71, just as we did with Equation 13-4. How did this happen? The alternate formula, Equation 13-5, uses only numbers from the base year income statement, whereas Equation 13-4 requires information from the base year and year 2.[3] Why use two different ways to calculate DOL when they both give the same answer? Because each method reveals different information about operating leverage.

The percentage change version of the DOL formula, Equation 13-4, shows the effect of the leverage—sales change by a certain percentage, triggering a greater percentage change in operating income if the DOL is greater than 1. The percentage change in operating income, then, is the product of the percentage change in sales and this degree of operating leverage.

The alternate DOL formula, Equation 13-5, shows that fixed costs cause the leveraging effect. Whenever fixed costs are greater than 0, DOL is greater than 1, indicating a leverage effect (the percentage change in EBIT is greater than the percentage change in sales). The larger the amount of fixed costs, then, the greater the leveraging effect.

Taken together, then, the two formulas demonstrate that leverage has the effect of triggering a greater percentage change in operating income when a percentage change in sales occurs and that fixed costs cause operating leverage. Equation 13-6 shows how changes in sales and DOL combine to determine the change in EBIT.

Percentage Change in EBIT

$$\% \Delta EBIT = \% \Delta Sales \times DOL \qquad (13\text{-}6)$$

where: $\%\Delta Sales$ = Percentage change in sales
DOL = Degree of operating leverage

---

[3]Equations 13-4 and 13-5 give the same numeric result when sales price per unit, fixed costs, and variable costs per unit are constant.

The Risk of Operating Leverage As we know from chapter 7, the risk associated with operating leverage is business risk. Recall that business risk refers to the volatility of operating income. The more uncertainty about what a company's operating income will be, the higher its business risk. Volatility of sales triggers business risk. The presence of fixed costs, shown by the amount of DOL, magnifies business risk. The total degree of business risk that a company faces is a function of both sales volatility and the degree of operating leverage.

We discussed business risk and how to measure it in chapter 7, pages 151–153.

Financial Leverage

Fluctuations in sales and the degree of operating leverage determine the fluctuations in operating income (also known as EBIT). Now let's turn our attention to financial leverage. **Financial leverage** is the additional volatility of net income caused by the presence of fixed-cost funds (such as fixed-rate debt) in the firm's capital structure. Interest on fixed-rate debt is a fixed cost because a firm must pay the same amount of interest, no matter what the firm's operating income.

Calculating the Degree of Financial Leverage (DFL) The **degree of financial leverage (DFL)** is the percentage change in net income ($\%\triangle$NI) divided by the percentage change in sales ($\%\triangle$EBIT). The formula for DFL follows:

$$\text{Degree of Financial Leverage, DFL}$$

$$\text{DFL} = \frac{\%\triangle\text{NI}}{\%\triangle\text{EBIT}} \tag{13-7}$$

where: $\%\triangle$NI = Percentage change in net income
$\%\triangle$EBIT = Percentage change in earnings before interest and taxes

If net income changes by a greater percentage than does EBIT, then the DFL will have a value greater than 1, and this indicates a financial leverage effect.

Table 13-7 shows the entire base-year income statement for Jason's Popcorn Wagon and the projections for year 2. Notice that the lower portion of the income statements contains fixed interest expense, so we would expect the presence of financial leverage.

TABLE 13-7

Jason's Income Statement—Base Year and Projected Year 2

	BASE YEAR	YEAR 2	
Sales	$30,000	$33,000	$\%\triangle = \dfrac{33,000 - 30,000}{30,000} = 10\%$
− VC	− 1,200	− 1,320	
− FC	− 12,000	− 12,000	
= EBIT	= 16,800	= 19,680	$\%\triangle = \dfrac{19,680 - 16,800}{16,800} = 17.1\%$
− I	− 800	− 800	
= EBT	= 16,000	= 18,880	
− Tax (15%)	− 2,400	− 2,832	
= NI	= 13,600	= 16,048	$\%\triangle = \dfrac{16,048 - 13,600}{13,600} = 18\%$

As shown in Table 13-7, the percentage change in EBIT from the base year to year 2 is 17.1 percent, and the percentage change in net income from the base year to year 2 is 18 percent. Jason's degree of financial leverage according to Equation 13-7 follows:

$$DFL = \frac{\%\Delta NI}{\%\Delta EBIT}$$

$$= \frac{.18}{.1714}$$

$$= 1.05$$

Our calculations show that Jason's Popcorn Wagon business has a degree of financial leverage of 1.05.

Another Method of Calculating Financial Leverage   Just as with DOL, there are two ways to compute DFL. Instead of using Equation 13-7, the percentage change in NI divided by the percentage change in EBIT, we could instead calculate the DFL using only numbers found in the base-year income statement. By dividing EBIT by EBIT minus interest expense (I), we can find DFL. The equation looks like this:

Degree of Financial Leverage, DFL (Alternate Formula)

$$DFL = \frac{EBIT}{EBIT - I} \tag{13-8}$$

where: EBIT = Earnings before interest and taxes
I = Interest expense

The base-year income statement numbers in Table 13-7 show that Jason's EBIT is $16,800 and his interest expense is $800. To find the degree of financial leverage, we apply Equation 13-8 as follows:

$$DFL = \frac{EBIT}{EBIT - I}$$

$$= \frac{16,800}{16,800 - 800}$$

$$= \frac{16,800}{16,000}$$

$$= 1.05$$

Equation 13-8 yields the same DFL for Jason's business as Equation 13-7.[4] Both formulas are important because they give us different but equally important insights about financial leverage. Equation 13-7 shows the effect of financial leverage—net income (NI) will vary by a larger percentage than operating income (EBIT). Equation 13-8 pinpoints the source of financial leverage—fixed interest expense. The degree of financial leverage, DFL, will be greater than 1 if interest expense (I) is greater than 0. In sum, interest expense magnifies the volatility of NI as operating income changes.

How Interest Expense Affects Financial Leverage   To illustrate the financial leverage effect, suppose that to help start his business, Jason borrowed $10,000 from

---

[4]Equations 13-7 and 13-8 give the same DFL value only if the fixed financial costs (interest expense) and the tax rate are constant.

a bank at an annual interest rate of 8 percent. This 8 percent annual interest rate means that Jason will have to pay $800 ($10,000 × .08) in interest each year on the loan. The interest payments must be made, no matter how much operating income Jason's business generates. In addition to Jason's fixed operating costs, he also has fixed financial costs (the interest payments on the loan) of $800.

The fixed financial costs magnify the effect of a change in operating income on net income. For instance, even if Jason's business does well, the bank interest payments do not increase, even though he could afford to pay more. If Jason's business does poorly, however, he cannot force the bank to accept less interest simply because he cannot afford the payments.

The Risk of Financial Leverage The presence of debt in a company's capital structure and the accompanying interest cost create extra risk for a firm. As we know from chapter 7, the extra volatility in NI caused by fixed interest expense is financial risk. The financial risk of the firm compounds the effect of business risk and magnifies the volatility of net income. Just as fixed operating expenses increase the volatility of operating income and business risk, so too fixed financial expenses increase the volatility of NI and magnify financial risk. This is shown in Equation 13-9.

Percentage Change in Net Income

$$\%\triangle NI = \%\triangle EBIT \times DFL \quad (13\text{-}9)$$

where: $\%\triangle EBIT$ = Percentage change in earnings before interest and taxes
DFL = Degree of financial leverage

We discussed financial risk and how to measure it in chapter 7, pages 153–155.

We explore the combined effect of operating and financial leverage next.

Combined Leverage

The combined effect of operating leverage and financial leverage is known as **combined leverage.** Combined leverage occurs when net income changes by a larger percentage than sales, which occurs if there are any fixed operating or financial costs. The following combined leverage formula solves for the net income change due to sales changes that occur when fixed operating and financial costs are present.

The **degree of combined leverage (DCL)** is the percentage change in net income (NI) divided by the percentage change in sales, as shown in Equation 13-10:

Degree of Combined Leverage Formula, DCL

$$DCL = \frac{\%\triangle NI}{\%\triangle Sales} \quad (13\text{-}10)$$

where: $\%\triangle NI$ = Percentage change in net income
$\%\triangle Sales$ = Percentage change in sales

The alternate DCL formula follows:

Degree of Combined Leverage, DCL

$$DCL = \frac{Sales - VC}{Sales - VC - FC - I} \quad (13\text{-}11)$$

where: VC = Total variable costs
FC = Total fixed costs
I = Interest expense

We can also calculate the degree of combined leverage (DCL) a third way: multiplying the degree of operating leverage (DOL) by the degree of financial leverage (DFL). The third DCL formula is shown in Equation 13-12.

Degree of Combined Leverage, DCL

$$DCL = DOL \times DFL \qquad (13\text{-}12)$$

where: DOL = Degree of operating leverage
DFL = Degree of financial leverage

Equation 13-13 shows the combined effect of DOL and DFL on net income (NI).

Percentage Change in Net Income

$$\%\triangle NI = \%\triangle Sales \times DOL \times DFL \qquad (13\text{-}13)$$

where: %△Sales = Percentage change in sales
DOL = Degree of operating leverage
DFL = Degree of financial leverage

Equation 13-13 shows how the change in net income is determined by the change in sales and the compounding effects of operating and financial leverage.

Fixed Costs and Combined Leverage Fixed operating costs create operating leverage, fixed financial costs create financial leverage, and these two types of leverage together form combined leverage. If fixed operating costs (FC) and fixed interest costs (I) were both zero, there would be no leverage effect. The percentage change in net income (NI) would be the same as the percentage change in sales. If either, or both, fixed operating costs and fixed financial costs exceed zero, a leverage effect will occur (DCL > 1).

Firms that have high operating leverage need to be careful about how much debt they pile onto their balance sheets, and the accompanying interest costs they incur, because of combined leverage effects. Remember that for Jason's Popcorn Wagon, the degree of operating leverage (DOL) was 1.71 and the degree of financial leverage was 1.05. The degree of combined leverage for Jason's business according to Equation 13-11 is 1.80 (1.71 × 1.05 = 1.80 rounded to two decimal places). Jason is quite confident that his sales will be high enough so that this high leverage will not be a problem. If the sales outlook were questionable, though, the combined leverage effect could magnify poor sales results.

Leverage is helpful when sales increase (positive percentage changes). Magnifying this positive change benefits the firm. However, leverage is harmful when sales decrease because it magnifies the negative change. Because future sales for most companies are uncertain, most companies view leverage with mixed feelings.

LBOs

Many publicly owned corporations have been bought out by a small group of investors, including top management of the firm, using a large amount of borrowed money. Such a purchase is called a *leveraged buyout,* or LBO. The leverage referred to is financial leverage.

In an LBO, investment banking firms work to identify attractive target companies. These investment banking firms solicit investors to acquire the target. To take over the target, the purchasing group raises cash, mostly borrowed, to purchase the common stock shares from the general public. The stock purchase converts the publicly owned corporation to a privately owned one. The investment banking firm would collect fees for its advice and for underwriting the bond issue that helped raise the additional debt capital.

Because of the dramatic increase in financial leverage, some LBOs have worked out well for investors and others have been disasters. For instance, Kohlberg, Kravis, & Roberts (KKR) made a 50 percent annual rate of return on its $1.34 billion investment after the Beatrice Company LBO. In contrast, the 1986 $1.4 billion LBO of Revco Drug Stores didn't fare as well. Two years later the company filed for bankruptcy when it was unable to generate enough cash flow to pay the interest and principal due on its bonds. Other companies purchased through LBOs include Borg-Warner, Montgomery Ward, Safeway, and Southland.

Ethical Connections

Let's Rip Off the Bondholders?

The board of directors of a corporation is elected by the common stockholders. The bondholders generally have no say about who serves on the board. If the board is considering a proposal that will create considerable value for the stockholders, while at the same time raking the bondholders over the coals, what do you think the board will do? Is there anything wrong with this? After all, the board members have a fiduciary responsibility to the stockholders, not to the bondholders.

The RJR Nabisco bondholders at Metropolitan Life Insurance Company, in the example on page 338, should not have been shocked that the board of RJR Nabisco voted to approve the LBO offered to the company. In almost any takeover a significant premium over the current market value is offered to stockholders. You may be able to buy a few hundred shares at the current market price but a significant premium, perhaps 30 percent to 50 percent or more, will usually have to be paid if you want to buy the whole company. You are buying real control when you buy the whole company whereas you are buying an insignificant amount of control when you buy a few hundred shares.

The directors of RJR Nabisco clearly created extra value for the stockholders. Because the buyout was achieved with a huge amount of borrowed money, the bondholders found themselves suddenly with a debt claim against a very risky company, while the day before the LBO was announced the firm had had a much lower level of risk. High-risk companies have higher interest rates demanded by the market for their bonds. If the market's required rate of return for your bond goes up, the price of your bond goes down.

The bondholders clearly thought that the LBO was unfair so they sued. It was certainly unpleasant even if it was not unfair.

Questions to Consider

▶ Did the RJR Nabisco board members have any responsibility to the bondholders of that company?
▶ Did RJR Nabisco make any promises that it would not take on more debt after the earlier bonds were sold?
▶ When you obtain a mortgage loan or car loan, do you promise not to borrow more money in the future?
▶ If the bondholders wanted protection from future bond issues, shouldn't this have been brought up when those bonds were initially sold rather than years later?
▶ Is it not true that the bondholders are being taken advantage of by the common stockholders through the latter's representatives on the board of directors?

TABLE 13-8

Effect of a Leveraged Buyout on RJR Nabisco

	BEFORE THE BUYOUT (1988)	AFTER THE BUYOUT (1989)
Long-Term Debt	$4,975 million	$21,948 million
Total Equity	$5,819 million	$1,237 million
Debt to Equity Ratio	85.5%	1,774%

Source: Value-Line Investment Survey

When a company with a normal debt load goes through an LBO, investors holding the company's bonds issued before the LBO are often hurt. The surge in the company's debt results in more financial risk. With higher risk, the market requires a higher rate of return, so the bonds issued before the LBO will see their market interest rates rise—and their market prices fall—after the company announces an LBO.

To illustrate the effects on bondholders, consider the 1989 LBO of RJR Nabisco. Table 13-8 shows how the holders of the RJR Nabisco bonds suddenly had a claim on a much riskier company.

The bondholders, led by Metropolitan Life Insurance Company, sued when the value of the bonds they held dropped precipitously after the firm announced its LBO. The lawsuit was settled out of court in 1991.

The risk of an LBO is large because of financial leverage effects. As the Beatrice and Revlon examples indicate, potential returns from an LBO may be large positive or negative values because of financial leverage effects. Bondholders may suddenly see the value of their bonds drop precipitously after an LBO announcement. In chapter 14 we discuss how bondholders can protect themselves against this risk.

Now that we have analyzed how fixed operating and financial costs can create leverage effects and risk, we will consider the optimal capital structure for a firm.

Capital Structure Theory

A central question in finance is how much debt and equity a firm should have in its capital structure. **Capital structure** is the mixture of sources of funds a firm uses (debt, preferred stock, common stock). The amount of operating leverage a company has is largely determined by the original equity investors who decide what kind of business to operate. The amount of financial leverage a company has, however, is much more controllable.

Every time a company borrows, it increases its financial leverage and financial risk. New equity financing decreases financial leverage and risk. Changes in financial leverage, we have seen, bring the potential for good and bad results. How then do financial managers decide on the right balance of debt and equity? Financial managers analyze many factors, including the tax effects of interest payments, and how the comparative costs of debt and equity affect firm value.

Tax Deductibility of Interest

Debt in a firm's capital structure can be beneficial. First, debt creates the potential for leveraged increases in net income (NI) when operating income (EBIT) is rising. Second, debt gives the company a tax deduction for the interest that is paid on the debt. In contrast to debt, an issue of common stock to raise equity funds results in no tax break. In short, interest paid on business debt is tax deductible, but dividends paid to common stockholders are not. The tax laws, therefore, give companies an incentive to use debt in their capital structures.

Although the tax deductibility of interest payments on debt is a benefit, debt has costs, too. We know that the financial risk of the firm increases as debt increases. As financial risk increases, including an increasing risk of bankruptcy, a company will incur costs to deal with this risk. For example, suppliers may refuse to extend trade credit to the company; and lawyers' fees may drain funds that could go to either bondholders or common stock investors.

Modigliani and Miller

How does a company balance the costs and benefits of debt? In 1958 Franco Modigliani and Merton Miller wrote a seminal paper that has influenced capital structure discussion ever since. Modigliani and Miller (known in economics and finance circles as M&M) concluded that when interest payments are tax deductible to a firm, a capital structure of all debt is optimal.

In reaching this conclusion, M&M assumed the following:

1. There were no transaction costs.
2. Purchasers of a company's bonds or common stock paid no income tax.
3. Corporations and investors can borrow at the same rate of interest.
4. Investors and management have the same information about the firm.
5. Debt the firm issues is riskless.
6. Operating income is not affected by the use of debt.

In such an environment, M&M showed that the tax benefits to the firm from issuing debt were so beneficial that the benefits allowed the company to increase its value by issuing more and more debt. Given the assumptions, a 100 percent debt capital structure is optimal.

The assumptions do not, of course, exist in the real world. Companies don't seek a 100 percent debt capital structure, suggesting that capital structure is not optimal. In the real world, capital structures vary widely.

Toward an Optimal Capital Structure

Firms seek to balance the costs and benefits of debt to reach an optimal mix that maximizes the value of the firm. Figure 13-3 shows the component costs and weighted cost of capital according to the view of most financial managers. Given the way suppliers of capital react in the real world, many financial managers believe this view is more realistic than the M&M model.

Figure 13-3 illustrates what many believe happens to the cost of debt, equity, and the weighted average cost of capital (WACC) as the capital structure of the firm changes. First, the graph shows that debt is cheaper than equity capital. Second, it shows that the weighted average cost of capital equals the cost of equity when the firm has no debt. Third, it shows that at point Z firms minimize the weighted average cost of capital, so at that point the capital structure maximizes the value of the firm. The cost advantage that debt has over equity dominates the increasing risk up to point Z. At this point the greater risk begins to dominate and causes the weighted average cost of capital to begin to turn upward.

To review how to estimate the costs of capital and the WACC, see chapter 12, pages 297–306.

Figure 13-3 **Cost of Capital and Capital Structure**

Figure 13-3 shows the cost of debt, cost of equity, and weighted average cost of capital (WACC) for different capital structures. The WACC is minimized at point Z.

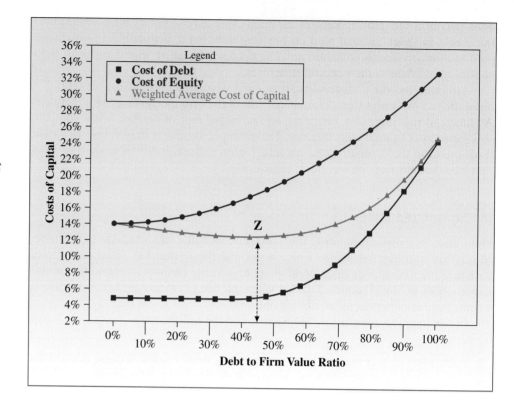

We learned in chapter 12 how to estimate the costs of debt and equity and weighted average figures. Here we study how capital structure changes may affect the firm's cost of capital and its value.

The Lower Cost of Debt Figure 13-3 shows that debt capital has a lower cost than equity capital. Debt is cheaper than equity for two reasons. As mentioned earlier, interest payments made by a firm are tax deductible and dividend payments made to common stockholders are not. Even without the tax break, debt funds are cheaper than equity funds. The required rate of return on a bond is lower than the required rate of return on common stock for a given company because debt is less risky than equity to investors. Debt is less risky because bondholders have a claim superior to that of common stockholders on the earnings and assets of the firm.

How Capital Costs Change as Debt Is Added If we examine the WACC line in Figure 13-3, we see that the weighted average cost of capital equals the cost of equity when the firm has no debt. Then, as debt is added, the cost advantage (to the issuing company) of debt over equity makes the weighted average cost of capital decrease, up to a point, as more of the cheaper debt funds and less of the more expensive equity funds are used. The effect of adding debt to capital structure is shown in Figure 13-3 as we move along the horizontal axis from the origin.

The Effect of Risk What causes the WACC to increase, as shown in Figure 13-3, beyond point Z? As the firm moves to a capital structure with higher debt (moves to the right along the horizontal axis of Figure 13-3), the risk of the firm increases. As financial risk rises with additional debt, the required return of both debt and equity investors increases. Notice that the cost of equity curve starts to climb sooner than does the cost of debt curve. This is because common stockholders get paid *after* bondholders.

As both the cost of debt and the cost of equity curves turn upward, the curve depicting the weighted average of the cost of debt and the cost of equity eventually turns upward, too. According to the capital structure view depicted in Figure 13-3, if a firm has less debt than the amount at Z, the WACC is higher than it needs to be. Likewise, if a firm has more debt than the amount at Z, the WACC is higher than it needs to be. Only at the capital structure at point Z do firms minimize the weighted average cost of capital. This is the capital structure, then, that maximizes the value of the firm.

Establishing the Optimal Capital Structure in Practice In the real world, it is unlikely that financial managers can determine an exact point for Z where the WACC is minimized. Many financial managers try instead to estimate Z and set a capital structure close to it. Unfortunately, no formula can help estimate point Z. The optimal capital structure for a firm depends on the future prospects of that firm.

For example, say a company has a product in great demand that is protected by a patent with many years to expiration. The company will find that bond and common stock investors are comfortable with a large amount of debt. This firm's Z value will be high. But a firm in a competitive industry, with some quality control problems and soft demand for its product, is in a different position. It will find that bond and common stock investors get nervous (and demand higher returns) when the debt to total value[5] ratio is above even a moderate level. This firm's Z value will be much lower than that of the first firm.

So the answer to the question, "What is the optimal capital structure for a firm?" is, "It depends." With no formula to use to estimate the firm's Z value, management examines the capital structure of similar companies and the future prospects of the firm. Financial managers must balance the costs and benefits of debt and use expertise and experience to develop the capital structure they deem optimal.

What's Next

In this chapter, we examined breakeven analysis, leverage effects, leveraged buyouts, and the effects of changes in a firm's capital structure. In chapter 14 we look at corporate bonds, preferred stock, and leasing.

Summary

1. **Find the breakeven level of sales for a firm.**

The costs of operating a business can be categorized as fixed or variable. Operating costs that do not vary with the level of production are fixed; operating costs that do vary with the level of production are variable. High fixed costs are usually tied to low variable costs per unit, and low fixed costs are usually tied to high variable costs per unit.

The breakeven point is the level of sales that results in an operating income of zero. At sales levels above the breakeven point, a firm begins to make a profit. A company with high

---

[5]Total value here refers to the total market value of the firm's outstanding debt and equity.

fixed operating costs must generate high sales revenue to cover its fixed costs (and its variable costs) before reaching the sales breakeven point. Conversely, a firm with low fixed operating costs will break even with a relatively low level of sales revenue.

2. *Explain operating, financial, and combined leverage effects and the resulting business and financial risks.*

Firms with high fixed costs have high operating leverage—that is, a small change in sales triggers a relatively large change in operating income. Firms with low fixed costs have less operating leverage. The effect of low operating leverage is that small changes in sales do not cause large changes in operating income.

Business risk refers to the volatility of a company's operating income. Business risk is triggered by sales volatility and magnified by fixed operating costs.

If a company uses fixed-cost funds (such as fixed interest rate bonds) to raise capital, financial leverage results. With financial leverage, fixed interest costs cause net income to change by a greater percentage than a concurrent change in EBIT.

The presence of financial leverage creates financial risk for a firm—the risk that the firm will not be able to make its interest payments if operating income drops. Financial risk compounds the business risk already present.

The total effect of operating leverage and financial leverage is called combined leverage. The value of the degree of financial leverage is multiplied by the value of the degree of operating leverage to give the degree of combined leverage (DCL). The DCL gives the percentage change in net income for a given percentage change in sales.

3. *Describe the risks and returns of a leveraged buyout.*

LBOs, or leveraged buyouts, occur when publicly owned corporations are bought out by a small group of investors using mostly borrowed funds. The purchase is leveraged because the investors finance it with a large amount of borrowed money. Consequently, when a firm is purchased in an LBO, it is saddled with a large amount of debt in its capital structure, and a large amount of financial leverage and financial risk.

4. *Explain how changes in capital structure affect a firm's value.*

Capital structure theory deals with the mixture of debt, preferred stock, and equity a firm utilizes. Because interest on business loans is a tax-deductible expense, and because lenders demand a lower rate of return than do stockholders for a given company (because lending money is not as risky as owning shares), debt capital is cheaper than equity capital. However, the more a company borrows, the more it increases its financial leverage and financial risk. The additional risk causes lenders and stockholders to demand a higher rate of return. Financial managers use capital structure theory to help determine the mix of debt and equity at which the weighted average cost of capital is lowest.

You can access your FREE PHLIP/CW On–Line Study Guide, Current Events, and Spreadsheet Problems templates, which may correspond to this chapter either through the Prentice Hall Finance Center CD–ROM or by directly going to: **http://www.prenhall.com/gallagher.**

PHLIP/CW — *The Prentice Hall Learning on the Internet Partnership–Companion Web Site*

Equations Introduced in This Chapter

Equation 13-1. The Breakeven Point in Unit Sales, $Q_{b.e.}$:

$$Q_{b.e.} = \frac{FC}{p - vc}$$

where: $Q_{b.e.}$ = Quantity unit sales breakeven level
FC = Total fixed costs
p = Sales price per unit
vc = Variable costs per unit

Equation 13-2. Total Revenue, TR:

$$TR = p \times Q$$

where: p = Sales price per unit
Q = Unit sales (Quantity sold)

Equation 13-3. Total Costs, TC:

$$TC = FC + (vc \times Q)$$

where: FC = Fixed costs
vc = Variable costs per unit
Q = Unit sales (Quantity sold)

Equation 13-4. Degree of Operating Leverage (DOL):

$$DOL = \frac{\%\triangle EBIT}{\%\triangle Sales}$$

where: $\%\triangle EBIT$ = Percentage change in earnings before interest and taxes
$\%\triangle Sales$ = Percentage change in sales

Equation 13-5. Degree of Operating Leverage (DOL) (alternate):

$$DOL = \frac{Sales - VC}{Sales - VC - FC}$$

where: VC = Total variable costs
FC = Total fixed costs

Equation 13-6. Percentage Change in EBIT:

$$\%\triangle EBIT = \%\triangle Sales \times DOL$$

where: $\%\triangle Sales$ = Percentage change in sales
DOL = Degree of operating leverage

Equation 13-7. Degree of Financial Leverage (DFL):

$$DFL = \frac{\%\triangle NI}{\%\triangle EBIT}$$

where: $\%\triangle NI$ = Percentage change in net income
$\%\triangle EBIT$ = Percentage change in earnings before interest and taxes

Equation 13-8. Degree of Financial Leverage (DFL) (alternate)

$$DFL = \frac{EBIT}{EBIT - I}$$

where: EBIT = Earnings before interest and taxes
I = Interest expense

Equation 13-9. Percentage Change in Net Income

$$\%\Delta NI = \%\Delta EBIT \times DFL$$

where: %ΔEBIT = Percentage change in EBIT
DFL = Degree of financial leverage

Equation 13-10. Degree of Combined Leverage (DCL)

$$DCL: = \frac{\%\Delta NI}{\%\Delta Sales}$$

where: %ΔNI = Percentage change in net income
%ΔSales = Percentage change in sales

Equation 13-11. Degree of Combined Leverage (DCL) (alternate 1):

$$DCL = \frac{Sales - VC}{Sales - VC - FC - I}$$

where: VC = Total variable costs
FC = Total fixed costs
I = Interest expense

Equation 13-12. Degree of Combined Leverage (DCL) (alternate 2):

$$DCL = DOL \times DFL$$

where: DOL = Degree of operating leverage
DFL = Degree of financial leverage

Equation 13-13. Percentage Change in Net Income (NI):

$$\%\Delta NI = \%\Delta Sales \times DOL \times DFL$$

where: %ΔSales = Percentage change in sales
DOL = Degree of operating leverage
DFL = Degree of financial leverage

Self-Test

ST-1. Mr. Marsalis's firm has fixed costs of $40,000, variable costs per unit of $4, and a selling price per unit of $9. What is Mr. Marsalis's breakeven level of sales (in units)?

ST-2. HAL's computer store has sales of $225,000, fixed costs of $40,000, and variable costs of $100,000. Calculate the degree of operating leverage (DOL) for this firm.

ST-3. HAL's computer store has operating income (EBIT) of $85,000 and interest expense of $10,000. Calculate the firm's degree of financial leverage (DFL).

ST-4. Kane Newspapers, Inc. has an after-tax cost of debt of 6 percent. The cost of equity is 14 percent. The firm believes that its optimal capital structure is 30 percent debt and 70 percent equity, and maintains its capital structure according to these weights. What is the weighted average cost of capital?

ST-5. Johnny Ringo's Western Shoppe expects its sales to increase by 20 percent next year. If this year's sales are $500,000, and the degree of operating leverage (DOL) is 1.4, what is the expected level of operating income (EBIT) for next year if this year's EBIT is $100,000?

ST-6. Marion Pardoo's Bookstore has a degree of operating leverage (DOL) of 1.6 and a degree of financial leverage (DFL) of 1.8. What is the company's degree of combined leverage (DCL)?

Review Questions

1. What is the operating leverage effect and what causes it? What are the potential benefits and negative consequences of high operating leverage?
2. Does high operating leverage always mean high business risk? Explain.
3. What is the financial leverage effect and what causes it? What are the potential benefits and negative consequences of high financial leverage?
4. Give two examples of types of companies likely to have high operating leverage. Find examples other than those cited in the chapter.
5. Give two examples of types of companies that would be best able to handle high debt levels.
6. What is an LBO? What are the risks for the equity investors and what are the potential rewards?
7. If an optimal capital structure exists, what are the reasons that too little debt is as undesirable as is too much debt?

Build Your Communication Skills

CS-1. Obtain balance sheet data from the library, the Internet, or some other source, and identify four companies having very high debt ratios and four having very low debt ratios. Write a one- to two-page report describing the characteristics of the companies with high debt ratios and those with low debt ratios. Can you identify characteristics that seem to be common to the four high-debt firms? What are characteristics common to the four low-debt firms?

CS-2. Interview the owner of a small business in your community. Ask that person to describe the fixed operating costs and the variable costs of the business. Write a report or give an oral presentation to your class on the nature of the business risk of this firm.

Problems

13-1. Howard Beal Co. manufactures molds for casting aluminum alloy test samples. Fixed costs amount to $20,000 per year. Variable costs for each unit manufactured is $16. Sales price per unit is $28.   *Breakeven Analysis*

   a. What is the contribution margin of the product?
   b. Calculate the breakeven point in unit sales and dollars.
   c. What is the operating profit (loss) if the company manufactures and sells
      (i) 1,500 units per year?
      (ii) 3,000 units per year?
   d. Plot a breakeven chart using the foregoing figures.

13-2. UBC Company, a competitor of Howard Beal Co. in problem 13-1, has a comparatively labor-intensive process with old equipment. Fixed costs are $10,000 per year and variable costs are $20 per unit. Sales price is the same, $28 per unit.   *Breakeven Analysis*

a. What is the contribution margin of the product?
   b. Calculate the breakeven point in unit sales and dollars.
   c. What is the operating profit (loss) if the company manufactures and sells
      (i) 1,500 units per year?
      (ii) 3,000 units per year?
   d. Plot a breakeven chart using the foregoing figures.
   e. Comment on the profit and loss potential of UBC Company compared to Howard Beal Co.

Operating Leverage

13-3. Use the same data given in problem 13-1 (fixed cost = $20,000 per year, variable cost = $16 per unit, and sales price = $28 per unit) for Howard Beal Co. The company sold 3,000 units in 1999 and expects to sell 3,300 units in 2000. Fixed costs, variable costs per unit, and sales price per unit are assumed to remain same in 1999 and 2000.

   a. Calculate the percentage change in operating income and compare it with the percentage change in sales.
   b. Comment on the operating leverage effect.
   c. Calculate the degree of operating leverage using
      (i) data for the years 1999 and 2000
      (ii) data for the year 1999 only
   d. Explain what the results obtained in (c) tell us.

Financial Leverage

13-4. Use the same data given in problems 13-1 and 13-3 (fixed cost = $20,000 per year, variable cost per unit = $16, sales price per unit = $28, 1999 sales = 3,000 units, and expected 2000 sales = 3,300 units) for Howard Beal Co. Fixed costs, variable costs per unit, and sales price per unit are assumed to remain the same in 1999 and 2000. The company has an interest expense of $2,000 per year. Applicable income tax rate is 30 percent.

   a. Calculate the percentage change in net income and compare it with the percentage change in operating income (EBIT).
   b. Comment on the financial leverage effect.
   c. Calculate the degree of financial leverage using
      (i) data for the years 1999 and 2000
      (ii) data for the year 1999 only
   d. Explain what the results obtained in (c) tell us about financial leverage.

Breakeven Analysis

13-5. Tony Manero owns a small company that refinishes and maintains the wood flooring of many dance clubs in Brooklyn. Because of heavy use, his services are required at least quarterly by most of the clubs. Tony's annual fixed costs consist of depreciation expense for his van and polishing equipment, and other tools. These expenses were $9,000 this year. His variable costs include wood staining products, wax, and other miscellaneous supplies. Tony has been in this business since 1977 and accurately estimates his variable costs at $1.50 per square yard of dance floor. Tony charges a rate of $15 per square yard.

   a. How many square yards of dance floor will he need to work on this year to cover all of his expenses but leave him with zero operating income?
   b. What is this number called?
   c. Calculate the breakeven point in dollar sales.
   d. Tony has little competent competition in the Brooklyn area. What happens to the breakeven point in sales dollars if Tony increases his rate to $18 per square yard?
   e. At the $18 per square yard rate, what is Tony's operating income and net income if he completes work on 14,000 square yards this year? Assume his tax rate is 40 percent and he has a $25,000 loan outstanding on which he pays 12 percent interest annually.

Operating Leverage and Breakeven Analysis

13-6. Otis Day's company manufactures and sells men's suits. His trademark gray flannel suits are popular on Wall Street and boardrooms throughout the East. Each suit sells for $800. Fixed costs are $200,000 and variable costs are $250 per suit.

a. What is the firm's operating income on sales of 600 suits? On sales of 3,000 suits?
b. What is Mr. Day's degree of operating leverage (DOL) at a sales level of 600 suits? At a sales level of 3,000 suits?
c. Calculate Mr. Day's breakeven point in sales units and sales dollars.
d. If the cost of the gray flannel material increases so that Mr. Day's variable costs are now $350 per suit, what will be his new breakeven point in sales units and sales dollars?
e. Considering the increase in variable costs, by how much will he need to increase the selling price per suit to reach the original operating income for sales of 3,000 suits calculated in part a?

13-7. Company A, Company B, and Company C all manufacture and sell identical products. They each sell 12,000 units annually at a price of $10 per unit. Assume Company A has $0 fixed costs and variable costs of $5 per unit. Company B has $10,000 in fixed costs and $4 in variable costs per unit. Company C has $40,000 fixed costs and $1 per unit variable costs. *Operating Leverage*

a. Calculate the operating income (EBIT) for each of the three companies.
b. Before making any further calculations, rank the companies from highest to lowest by their relative degrees of operating leverage. Remember what you read about how fixed costs affect operating leverage.

13-8. Faber Corporation, a basketball hoop manufacturing firm in Hickory, Indiana, plans to branch out and begin producing basketballs in addition to basketball hoops. It has a choice of two different production methods for the basketballs. Method 1 will have variable costs of $6 per ball and fixed costs of $700,000 for the high-tech machinery, which requires little human supervision. Method 2 employs many people to hand-sew the basketballs. It has variable costs of $16.50 per ball, but fixed costs are estimated to only be $100,000. Regardless of which method CEO Norman Dale chooses, the basketballs will sell for $30 each. Marketing research indicates sales in the first year will be 50,000 balls. Sales volume is expected to increase to 60,000 in year 2. *Operating Leverage*

a. Calculate the sales revenue expected in years 1 and 2.
b. Calculate the percentage change in sales revenue.
c. Calculate the earnings before interest and taxes for each year for both production methods.
d. Calculate the percentage change in EBIT for each method.
e. Calculate the year 1 degree of operating leverage for each method, using your answers from parts b and d above.
f. Calculate the degree of operating leverage again. This time use only revenue, fixed costs and variable costs from year 1 (your base year) for each production method.
g. Under which production method would EBIT be more adversely affected if the sales volume did not reach the expected levels?
h. What would drive this adverse effect on EBIT?
i. Recalculate the year 1 base year EBIT and the degree of operating leverage for both production methods if year 2 sales are expected to be only 53,000 units.

13-9. Three companies manufacture and sell identical products. They each have earnings before interest and taxes of $100,000. Assume Company A is an all-equity company and, therefore, has zero debt. Company B's capital structure is 10 percent debt and 90 percent equity. It makes annual interest payments of $2,000. Company C's capital structure is just the opposite of B. It has 90 percent debt and 10 percent equity. Company C has annual interest expense of $40,000. The tax rate for each of the three companies is 40 percent. *Financial Leverage*

a. Before making any calculations, rank the companies from highest to lowest by their relative degrees of financial leverage (DFL). Remember what you read about how debt and the interest expense which comes with it affects financial leverage.
b. Calculate the degree of financial leverage for each company. Was your answer to a correct?
c. Calculate the net income for each company.

Financial Leverage

13-10. Michael Dorsey and Dorothy Michaels each own their own companies. They design and supply custom-made costumes for Broadway plays. The income statement from each company shows they each have earnings before interest and taxes of $50,000 this year. Mr. Dorsey has an outstanding loan for $70,000, on which he pays 13 percent interest annually. When she started her business, Ms. Michaels only needed to borrow $10,000. She is still paying 9 percent annual interest on the loan. Each company expects EBIT for next year to be $60,000. The tax rate for each is 40 percent and is not expected to change for next year.

a. Calculate the net income for each company for this year and next year.
b. Calculate the percentage change in net income for each company.
c. Calculate the percentage change in EBIT for each company.
d. Calculate this year's degree of financial leverage for each company using your answers from parts b and c above.
e. Calculate the degree of financial leverage for each company again. This time use only EBIT and interest expense for this year.
f. If earnings before interest and taxes do not reach the expected levels, in which company would net income be more adversely affected?
g. What would drive this adverse effect on net income?
h. Recalculate the degree of financial leverage and the net income expected for next year for both companies if EBIT only increases to $53,000.

Challenge Problem

13-11. Fanny Brice, owner of Funny Girl Comics, has sales revenue of $200,000, earnings before interest and taxes of $95,000, and net income of $30,000 this year. She is expecting sales to increase to $225,000 next year. The degree of operating leverage is 1.35 and the degree of financial leverage is relatively low at 1.09.

a. Calculate the percentage change in EBIT Ms. Brice can expect between this year and next year.
b. How much will EBIT be next year in dollars?
c. Calculate the percentage change in net income Ms. Brice can expect between this year and next year.
d. How much net income should Ms. Brice expect next year?
e. Calculate this year's degree of combined leverage (DCL).
f. Ms. Brice is considering a price increase. This would mean the percentage change in sales revenue between this year and next year would be 20 percent. If this is true, what net income (in dollars) can she expect for next year?

Degree of Combined Leverage

13-12. Clint Reno owns Real Cowboy, a western wear store that has current annual sales of $2,800,000 per year. The degree of operating leverage (DOL) is 1.4. EBIT is $600,000. Real Cowboy has $2 million in debt, on which it pays 10 percent annual interest. Calculate the degree of combined leverage for Real Cowboy.

DOL, DFL, and DCL Interactions

13-13. Chad Gates owns Strings Attached, a store that sells guitars. The company has $5 million in current annual sales, fixed operating costs of $300,000 and $700,000 in variable operating costs, for a total EBT of $2.5 million. The firm has debt of $16,666,666.67, on which it pays 9 percent annual interest. The degree of combined leverage (DCL) for Strings Attached is 1.72.

a. Calculate the degree of operating leverage (DOL).
b. What is the degree of financial leverage (DFL) for Strings Attached? Calculate your answer using the EBIT and interest expense figures and your knowledge of how DOL and DFL jointly determine DCL.
c. If sales next year increase by 20 percent, what will be the percent change in net income?

Comprehensive Problem

13-14. Soccer International, Inc. produces and sells soccer balls. Partial information from the income statement for the years 1999 and 2000 follows.

Soccer International, Inc.
Income Statement for the Year Ending December 31

	1999	2000
Sales Revenue	$560,000	616,000
Variable Costs	240,000	264,000
Fixed Costs	160,000	160,000
EBIT	—	—
Interest Expense	40,000	40,000
EBT	—	—
Income Taxes (30%)	—	—
Net Income	—	—

Soccer International sells each soccer ball for $16.

    a. Fill in the missing values in the income statements of 1999 and 2000.
    b. Calculate Soccer International's breakeven point in sales units 1999 and 2000.
    c. Calculate the breakeven point in dollar sales for 1999 and for 2000.
    d. How many soccer balls need to be sold to have an operating income of $200,000 in 1999?
    e. What is the operating profit (loss) if the company sells (i) 18,000 and (ii) 24,000 balls in 1999?
    f. Calculate the degree of operating leverage for the year 1999 and for 2000.
    g. If sales revenue is expected to increase by 10 percent in 2000, calculate the percentage increase in EBIT and the dollar amount of EBIT for 2000.
    h. Calculate the degree of financial leverage for the year 1999 and for 2000.
    i. Calculate the percentage change in net income and the dollar amount of net income expected in 2000.
    j. Calculate the degree of combined leverage for the year 1999 and for 2000.
    k. Assume Soccer International raises its selling price and that sales revenue increases to $650,000 in 2000. How much net income can be expected in 2000?

Answers to Self-Test

ST-1. Sales breakeven point (in units)
= $40,000 ÷ ($9.00 − $4.00) = 8,000 units

ST-2. Degree of operating leverage (DOL)
= ($225,000 − $100,000) ÷ ($225,000 − $100,000 − $40,000) = 1.47

ST-3. Degree of financial leverage (DFL)
= $85,000 ÷ ($85,000 − $10,000) = 1.13

ST-4. Weighted average cost of capital, k_a
= (.3 × 6%) + (.7 × 14%) = 11.6%

ST-5. Next year's change in EBIT equals (this year's × 20% × 1.4) + this year's EBIT
= ($100,000 × 20% × 1.4) + $100,000 = $128,000

ST-6. Degree of combined leverage (DCL) = DOL × DFL
= 1.6 × 1.8 = 2.88

CHAPTER 14
CORPORATE BONDS, PREFERRED STOCK, AND LEASING

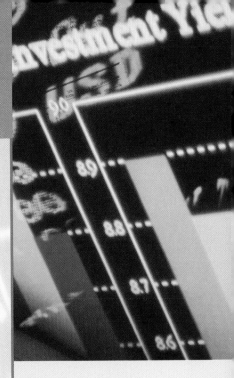

The borrower is servant to the lender.
—*Proverbs 22:7*

CONGRATS! HERE'S YOUR DEGREE AND $250,000. NOW MAKE US MONEY.

When Yael Alkalay, 30, graduated from business school at Columbia University, she decided to launch her own candle and toiletries business. But rather than rely on the typical entrepreneur's assets—maxed-out gold cards and loved ones' savings—Alkalay looked toward her own alma mater: She snagged $250,000 from Columbia's Eugene Lang Entrepreneurial Fund—a remarkable show of confidence on the school's part. Sure, any B-school can claim that its graduates are savvier than the rest, but how many are willing to back up their *U.S. News* rankings with cash?

Quite a few, actually. In addition to Columbia, Northwestern University's Kellogg School and the University of Michigan also have venture capital funds, and New York University's Stern School of Business and the Anderson School at UCLA are about to begin funding their alumni.

These funds aren't mere "learning experiences": Their administrators expect their investments to pay off and in some cases require that recipients first receive funding from outside sources. Steven Rogers, one of a three-member team investing for Kellogg's $150,000 Nice/Goldman Fund, says that he's searching for companies that will become profitable in five to seven years. "We're looking to . . . get a

CHAPTER OBJECTIVES

After reading this chapter, you should be able to:

- Describe the contract terms of a bond issue.

- Distinguish the various types of bonds and describe their major characteristics.

- Describe the key features of preferred stock.

- Compare and contrast a genuine lease and a disguised purchase contract.

- Explain why some leases must be shown on the balance sheet.

good return to keep the fund running," he says. Rogers expects 1998's pick, Outlier, which makes nylon backpacks for in-line skates, to net $10 million by 2001.

Similarly, Michigan's Wolverine Fund seeks companies that could reach $50 million in revenues within five years. Wolverine made its first investment in July, planting $50,000 in Intralase, a firm developing laser technologies for eye surgery. Karen Bantel, a professor of corporate strategy and entrepreneurship at Michigan, is quick to dismiss any notion that the university is backing frivolous enterprises: "We're not funding restaurants or pet shops," she says.

But campus venture capital funds do allow a certain degree of flexibility—the equivalent of getting to retake a test. Take Alkalay and her candles. Though she made good on test sales in museums, Alkalay learned she needed broader distribution. So she revised her business plan—a luxury few entrepreneurs get. "I think a typical venture capital fund wouldn't give you the opportunity to learn on someone else's dollar," she says.

Source: Jane Hodges, *Fortune*, October 26, 1998, p. 70. Reprinted by special permission; copyright 1998, Time, Inc.

Chapter Overview

In chapter 2 we examined the basic characteristics and terminology of corporate bonds. In chapter 9 we learned how to estimate the value of bonds. In this chapter we investigate how corporate bonds and preferred stock play a role in the financing decisions of a corporation. We also explore how leasing decisions affect a firm's finances.

To review bond information, see chapter 2, pages 27–30, and chapter 9, pages 210–215.

Bond Basics

A **corporate bond** is a security that represents a promise by the issuing corporation to make certain payments to the owner of the bond, according to a certain schedule. The corporation that issues a bond is the debtor, and the investor who buys the bond is the creditor.

The **indenture** is the contract between the issuing corporation and the bond's purchaser. It spells out the various provisions of the bond issue, including the face value, coupon interest rate, interest payment dates, maturity date, and other details. The yield to maturity is not in the indenture because it is market determined and changes with market conditions. The major features of bond indentures are described in the next section.

The investment bankers who underwrite the new bond issue help the firm set the terms of that issue. This usually means obtaining a rating for the bonds from one or more of the major rating companies, such as Moody's and Standard & Poor's. Bond ratings shown in Table 14-1 reflect the likelihood that the issuer will make the promised interest and principal payments on time. Many institutional investors, the main purchasers of bonds, are prohibited, either by law or by client demands, from purchasing unrated bonds.

Bonds rated Baa3 or above by Moody's and BBB− or above by Standard & Poor's are called **investment-grade bonds.** Bonds with lower than investment-grade

TABLE 14-1

Moody's and Standard & Poor's Bond Rating Categories

Moody's	Standard & Poor's	Remarks
Aaa	AAA	Best Quality
Aa1	AA+	
Aa2	AA	High Quality
Aa3	AA−	
A1	A+	
A2	A	Upper Medium Grade
A3	A−	
Baa1	BBB+	
Baa2	BBB	Medium Grade
Baa3	BBB−	
Ba1	BB+	
Ba2	BB	Speculative
Ba3	BB−	
B1	B+	
B2	B	Very Speculative
B3	B−	
Caa	CCC	
Ca	CC	Very, Very Speculative
C	C	
	D	In Default

ratings (Ba1 or below by Moody's and BB+ or below by Standard & Poor's) are called **junk bonds.** We'll have more to say about junk bonds later in the chapter.

Features of Bond Indentures

In addition to the basic characteristics of the bond (interest, principal, maturity, and specific payment dates), the bond indenture specifies other features of the bond issue. These features include:

- any security to be turned over to the bond's owner in the event the issuing corporation defaults
- the plan for paying off the bonds at maturity
- any provisions for paying off the bonds ahead of time
- any restrictions the issuing company places on itself to provide an extra measure of safety to the bondholder
- the name of an independent trustee to oversee the bond issue

Thus, every feature of the bond issue is spelled out in the bond indenture.

Security

A person who buys a newly issued bond is, in effect, lending money to the issuing corporation for a specified period of time. Like creditors, bondholders are concerned about getting their money back. A provision in the loan agreement (the indenture) that provides security[1] to the lender in case of default will increase the bond's value, compared to a loan agreement without a security provision. The value is higher because the investor has an extra measure of protection. A bond that has a security provision in the indenture is a **secured bond.** A bond that does not pledge any specific asset(s) as security is a **debenture.** Debentures are backed only by the company's ability and willingness to pay.

Plans for Paying off the Bond Issue

Bonds are paid off, or retired, by a variety of means. Some of the more popular methods include *staggered maturities, sinking funds,* and *call provisions.*

Staggered Maturities   Some bond issues are packaged as a series of different bonds with different or staggered maturities. Every few years a portion of the bond issue matures and is paid off. Staggering maturities in this fashion allows the issuing company to retire the debt in an orderly fashion without facing a large one-time need for cash, as would be the case if the entire issue were to mature at once. **Serial payments** pay off bonds according to a staggered maturity schedule.

Sinking Funds   Although *sinking* is not an appealing word, sending your debt to the bottom of the ocean has its appeal. When a **sinking fund** provision is included in the bond's indenture, the issuing company makes regular contributions[2] to a fund that is used to buy back outstanding bonds. Putting aside a little money at a time in this fashion ensures that the amount needed to pay off the bonds will be available.

Call Provisions

Many corporate bonds have a provision in their indentures that allows the issuing corporation to pay off the bonds before the stated maturity date, at some stated price. This is known as a **call provision.** The price at which the bonds can be purchased before the scheduled maturity date is the *call price.*

Call provisions allow issuing corporations to refinance their debt if interest rates fall, just as homeowners refinance their mortgage loans when interest rates fall. For example, a company that issued bonds in 1988 with a 9 percent coupon interest rate would be making annual interest payments of 9 percent of $1,000, or $90 on each bond. Suppose that in 1999 the market rate of the company's bonds was 7 percent. If the bond indenture contained a call provision,[3] the company could

---

[1]Chapter 2 used a different definition of security (a financial claim such as a stock or bond). Security has another definition, as used here. This definition is any asset (such as a piece of equipment, real estate, or a claim on future profits) that is promised to the investor in the event of a default.
[2]These are called "contributions" in the same sense that the government refers to the money you pay into the Social Security system as contributions. You have no choice; you must pay. If a company fails to make its required contributions to a sinking fund as described in the indenture, the bond issue can be declared to be in default.
[3]If a bond issue is callable, there is usually a certain amount of time that must pass before the bonds can be called. This is known as a *deferred call provision.* The indenture specifies the *call date.* The issuer can call bonds in from investors on, or after, this call date.

issue 7 percent bonds in 1999 and use the proceeds from the issue to "call in" or pay off the 9 percent bonds. The company's new interest payments would be only 7 percent of $1,000, or $70, thus saving the company $20 on each bond each year.

When bonds are called, convention is that the call price the issuer must pay is generally more than the face value. This excess of the call price over the face value is known as the **call premium.** The call premium may be expressed as a dollar amount or as a percentage of par.

Figure 14-1 shows a notice in which the Washington Suburban Sanitary District of Maryland called for all owners of the listed bonds to send in their bonds for redemption at 102 percent of par. Note also that only the bonds with the listed serial numbers are being called in, not the entire issue. Calling in part of a bond issue is one of many procedures that may be used in calling bonds. These bonds, with original maturity dates ranging from 1999 to 2011, were redeemed on December 1, 1998.

Issuing new bonds to replace old bonds is known as a **refunding** operation. Remember, the option to call a bond is held by the issuing corporation. The owners of the bonds have no choice in the matter. If investors don't turn in their bonds when the bonds are called, they receive no further interest payments.

A company approaches a bond refunding the same way it does any capital budgeting decision. The primary incremental cash inflows come from the interest savings realized when old high-interest debt is replaced with new low-interest debt. The primary incremental cash outflows are the call premium, if any, and the flotation costs associated with the new bond issue. All these variables must be adjusted for taxes and then evaluated by the firm. If the net present value of the incremental cash flows associated with the refunding is greater than or equal to zero, the refunding is done. If the NPV is negative, the company allows the old bonds to remain outstanding. This is the same thing you do when deciding whether to refinance a mortgage loan on a home.

Occasionally, a corporation will refund a bond issue even though there are no significant interest savings to be had. If the outstanding bonds have indenture provisions that the issuing company now finds oppressive or too limiting, the old bonds could be called and new bonds issued without the offending features. Appendix 14A covers refunding in detail, and solves a sample refunding problem.

Restrictive Covenants

A company that seeks to raise debt capital by issuing new bonds often makes certain promises to would-be investors to convince them to buy the bonds being offered, or to make it possible to issue the bonds at a lower interest rate. These promises made by the issuer to the investor, to the benefit of the investor, are **restrictive covenants.** They represent something like a courtship. If the suitor does not give certain assurances as to the way the other party will be treated, there is little chance the relationship will blossom.

In a bond issuer–bond investor relationship, these assurances may include limitations on future borrowings, restrictions on dividends, and minimum levels of working capital maintained.

Limitations on Future Borrowings Investors who lend money to a corporation by buying its bonds expect that the corporation will not borrow excessively in the future. A company in too much debt may be unable to pay bond principal and interest

Figure 14-1 Washington Suburban Sanitary District Notice of Bond Redemption

Figure 14-1 is an announcement by the Washington Suburban Sanitary District calling part of a bond issue for 102 percent of the face value of the bonds.

Source: *Washington Suburban Sanitary District,* The Wall Street Journal *(October 29, 1998): C18.*

Notice of Redemption
To the Holders of

Washington Suburban Sanitary District
Maryland

General Construction Bonds of 1988

NOTICE IS HEREBY GIVEN that bonds of the Washington Suburban Sanitary District designated "General Construction Bonds of 1988" dated November 15, 1988 and stated to mature in annual installments on December 1, in the years 1999 through 2011, inclusive, have all been called for redemption on December 1, 1998, at a redemption price (expressed as a percentage of the principal amount) of 102%, plus accrued and unpaid interest thereon to said redemption date, as follows:

Date of Maturity December 1,	CUSIP No.	Principal Amount	Rate of Interest	Principal Amount Payable Stated as a Percentage of Principal
1999	940155RA4	$2,485,000	7.00%	102
2000	940155RE6	2,655,000	7.00	102
2001	940155RJ5	2,840,000	7.00	102
2002	940155RN6	3,035,000	7.10	102
2003	940155RS5	3,255,000	7.10	102
2004	940155RW6	3,485,000	7.20	102
2005	940155SA3	3,735,000	7.20	102
2006	940155SE5	4,005,000	7.25	102
2007	940155SJ4	4,305,000	7.25	102
2008	940155SN5	4,625,000	7.25	102
2009	940155SR6	4,970,000	7.25	102
2010	940155SS4	5,340,000	7.25	102
2011	940155ST2	5,740,000	7.25	102

On said redemption date, said refunded bonds will be due and payable at the Chase Manhattan Bank, Paying Agent.

By Registered Mail:
The Chase Manhattan Bank
c/o Chase Bank of Texas NA
Corporate Trust Services
P.O. Box 219052
Dallas, Texas 75221-9052

In Person:
The Chase Manhattan Bank
Room 234—North Building
Corporate Trust
  Securities Window
55 Water Street
New York, New York 10041

By Courier or Brinks:
The Chase Manhattan Bank
c/o Chase Bank of Texas NA
Corporate Trust Services
1201 Main Street, 18th Floor
Dallas, Texas 75202

OR

At the offices of First National Bank of Maryland, Co-Paying Agent, upon presentation and surrender of the Bonds.

First National Bank of Maryland
Dept. Number 109754
110 South Paca Street
Baltimore, Maryland 21201

From and after said Redemption Date, interest on said bonds will cease to accrue.

WASHINGTON SUBURBAN SANITARY COMMISSION
Mary J. Kirby, Treasurer

By:
The Chase Manhattan Bank,
formerly Chemical Bank,
as Escrow Agent

Dated: October 29, 1998

payments on time. Bond investors would be worried if, after buying the bonds of a firm with a 20 percent debt to total assets ratio, the company then issued $100 million of additional bonds, increasing that ratio to over 90 percent. The new debt would make the earlier issued bonds instantly more risky and lower their price in the market.

A restrictive covenant in which the corporation promises not to issue a large amount of future debt would protect the company's current bondholders from falling bond ratings and plunging market prices. A bond issue with this restriction in the indenture will have more value than a bond issue without this guarantee. As a result, the bonds could be issued at a lower coupon interest rate than bonds without the restriction in the indenture.

Restrictions on Dividends An indenture may also include restrictions on the payment of common stock dividends if a firm's times interest earned ratio drops below a specified level. This restriction protects the bondholders against the risk of the common stockholders withdrawing value (cash for dividends that may be needed to make future interest payments) from the firm during difficult times. The bondholders are supposed to have priority over common stockholders. A bond issue with this sort of protection for investors can be issued at a lower interest rate than a bond issue without it.

We explained the debt to total assets ratio in chapter 5, page 91.

Minimum Levels of Working Capital Current assets can generally be quickly and easily converted to cash to pay bills. Having a good liquidity position protects all creditors, including bondholders. Minimum working capital guarantees in an indenture provide an additional margin of protection for bondholders and, therefore, reduce the interest rate required on such bonds.

The Independent Trustee of the Bond Issue

Violations of any of the provisions included in the indenture could constitute a default. Therefore, an independent **trustee** is named in the indenture to oversee the bond issue and to make sure all the provisions spelled out in the indenture are adhered to. The trustee is usually a commercial bank.

Most people think a default is a failure to make a scheduled interest or principal payment on time. Actually, this is only one possible type of default because the promise to pay interest and principal on their due dates is only part of the promise made by the bond issuer in the indenture. Failure to keep any of the substantive promises mentioned in the indenture constitutes a default.

TYPES OF BONDS

Some of the more innovative new financial instruments have been developed in the bond market. Let's now look more closely at the many kinds of bonds, both traditional and new.

Secured Bonds

A secured bond is backed by specific assets pledged by the issuing corporation. In the event of a default, the investors in these secured bonds would have a claim on these assets.

Mortgage Bonds A bond backed by real assets (not financial assets) is known as a **mortgage bond.** When you buy a house and finance the purchase with a mortgage loan, you are pledging your house (a real asset) as collateral for that loan. You are issuing a mortgage bond to the lender. That is what corporations do when they pledge real assets, such as airplanes and railroad cars, as collateral for the bonds issued to purchase those assets.

Different mortgage bonds can be issued that pledge the same real assets as collateral. Different classes of mortgage bonds signal the priority each investor has on the asset. An investor in a **first-mortgage** bond has first claim on the proceeds from the sale of the pledged assets if there is a default. A later lender may be an investor in a **second-mortgage** bond. In the event of default, the holder of the second mortgage receives proceeds from the sale of the pledged assets only after the first-mortgage bond investors have received all payments due to them. Similarly, third-mortgage bonds, fourth-mortgage bonds, and so forth can be issued with correspondingly lower priorities.

Unsecured Bonds (Debentures)

A bond that is not backed by any collateral is called a debenture. A debenture is backed only by the ability and willingness of the issuing corporation to make the promised interest and principal payments as scheduled. If a debenture were to go into default, the bondholders would be unsecured creditors. They would only have a general claim on the issuing company, not a right to the firm's specific assets.

There may be different classes of debentures. Certain issues may have a higher priority for payment than do others. If bond issue A has priority for payment over bond issue B, according to their respective indentures, then bond issue A is said to be a **senior debenture** and bond issue B is said to be a **subordinated debenture.** A senior debenture has a prior claim to the earnings and liquidation proceeds from the general assets of the firm (those assets not specifically pledged as security for other bonds) relative to the claim of subordinated debenture investors.

Subordinated debentures have a lower-priority claim on the firm's earnings and assets. Because subordinated debentures are riskier than senior debentures, investors demand and issuers pay higher interest rates on them. This higher interest rate is consistent with the risk–return relationship—the greater the risk of a security, the greater the required rate of return. Holders of first-mortgage bonds assume less risk than holders of second-mortgage bonds. Debenture holders have more risk than do secured bondholders, and subordinated debenture holders have more risk than senior debenture holders. Preferred stock investors take more risk than a bond investor, and common stock investors take more risk still for a given company. This risk hierarchy, reflecting the relative priority of claims, is shown in Figure 14-2.

Convertible Bonds

One of the special types of bonds available is called a **convertible bond.** A convertible bond is a bond that may be converted, at the option of the bond's owner, into a certain amount of another security issued by the same company. In the vast majority

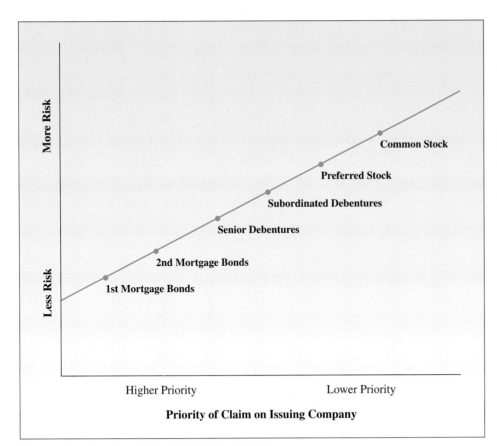

Figure 14-2 **The Risk Hierarchy**

Figure 14-2 shows the different priorities of claims that creditors and investors have on a company in default. Mortgage bondholders are paid first, whereas common stockholders are paid last.

of the cases, the other security is common stock.[4] This means that the investor who bought the convertible bond may send it back to the issuing company and "convert" it into a certain number of shares of that company's common stock.

Features of Convertible Bonds Convertible bonds have a face value, coupon rate, interest payment frequency, and maturity spelled out in their indenture, as do regular nonconvertible bonds. The indenture also spells out the terms of conversion, if the investor chooses to exercise that option. If the bond's owner elects not to convert the bond, the owner continues to receive interest and principal payments as with any other bond.

The Conversion Ratio. Each convertible bond has a conversion ratio. The **conversion ratio** is the number of shares of common stock that an investor would get if the convertible bond were converted. For example, California Microwave, Inc. issued a convertible $1,000 bond that matures in 2003, with a conversion ratio of 35.1648. That means the bond's owner can trade in the bond for 35.1648 shares of California Microwave common stock at any time.

---

[4]Some convertible bonds can be converted into a certain amount of preferred stock or some other security issued by the company.

Chapter 14 *Corporate Bonds, Preferred Stock, and Leasing* 359

TABLE 14-2

California Microwave, Inc. Convertible Bond Characteristics

Maturity Date:	December 15, 2003
Face Value:	$1,000
Type of Interest:	Semiannual, paid on June 15 and December 15
Coupon Interest Rate:	5¼%
Conversion Ratio:	35.1648

Source: Prospectus, $50,000,000 California Microwave, Inc., 5¼% Convertible Subordinated Notes Due 2003, Bear, Stearns & Co. Inc., and Oppenheimer & Co., Inc., Underwriters

The Conversion Value. To find the conversion value of a bond, multiply the conversion ratio by the market price per share of the company's common stock, as shown in Equation 14-1:

Formula for the Conversion Value of a Convertible Bond

$$\text{Conversion Value} = \text{Conversion Ratio} \times \text{Stock Price} \quad (14\text{-}1)$$

The **conversion value** is the amount of money bond owners receive if they convert the bond to common stock and then sell the common stock. For example, if the California Microwave, Inc. stock were selling for $20, then the conversion value of the convertible bond described in Table 14-2 would be as follows:

$$\begin{aligned}\text{Conversion Value} &= \text{Conversion Ratio} \times \text{Stock Price} \\ &= 35.1648 \times \$20.00 \\ &= \$703.30\end{aligned}$$

Equation 14-1 shows us that at a rate of $20 per common stock share, the conversion value of the convertible bond is $703.30.

The Straight Bond Value. If a convertible bond is not converted into stock, then it is worth at least the sum of the present values of its interest and principal payments.[5] The value coming from the interest and principal is called the convertible bond's **straight bond value.** The discount rate used to compute this straight bond value is the required rate of return for a nonconvertible bond having characteristics (risk, maturity, tax treatment, and liquidity) similar to the convertible bond.

As shown in Table 14-2, California Microwave's convertible bond has a coupon interest rate of 5¼ percent and a maturity date of December 15, 2003. The bond was issued in 1996. With a face value of $1,000, the annual interest payments will be $52.50 ($1,000 × .0525). Because interest is paid semiannually, actual interest payments of $26.25 ($52.50 ÷ 2) are made twice a year. If the required rate of return on similar nonconvertible bonds is 8 percent annual interest (4 percent semiannually),[6] then according to Equation 9-2, the bond's straight bond value as of December 15, 1999, follows:[7]

---

[5]Equation 9-2 in chapter 9 gives the present value of a bond's interest and principal payments.
[6]If 8 percent were the annual required market rate of return, then the corresponding semiannual required rate of return would be 3.923 percent ($1.03923^2 - 1 = .08$ or 8 percent) and not 4 percent. We will round this to 4 percent to simplify the calculation.
[7]December 15, 2003 is four years (eight semiannual periods) in the future from December 15, 1999.

$$V_{Bond} = \$26.25 \times \left[\frac{1 - \frac{1}{(1.04)^8}}{.04}\right] + \$1,000 \times \frac{1}{(1.04)^8}$$

$$= \$26.25 \times 6.732745 + \$1,000 \times .73069$$

$$= \$176.73 + \$730.69$$

$$= \$907.42$$

We find that the present value of a four-year, semiannual convertible bond with a face value of $1,000, and an 8 percent annual coupon interest rate is $907.42.

A rational investor may convert a bond if advantageous, but will not convert it if disadvantageous. A convertible bond, then, is always worth the conversion value or the straight bond value, whichever is greater.

Take Note

A convertible bond does not have to be converted to reap the benefits of a high conversion value. The mere fact that the bond could be converted into common stock having a certain value makes the convertible bond worth at least that conversion value.

Variable-Rate Bonds

Although most corporate bonds pay a fixed rate of interest (the coupon interest rate is constant), some pay a variable rate. With a **variable-rate bond**, the initial coupon rate is adjusted according to an established timetable and a market rate index. Figure 14-3 shows an ad for a variable-rate bond issued by the Student Loan Marketing Association (SallieMae).

The interest rate on the SallieMae bond in Figure 14-3 is tied to 91-day Treasury bills. However, variable bond rates could be tied to any rate, such as a Treasury bond rate or the London Interbank Offer Rate (LIBOR). Bond issuers check this market rate on every adjustment date specified in the indenture and reset the coupon rate accordingly.

The variable rate protects investors from much of the interest rate risk inherent in fixed-rate bonds. Rising inflation hurts investors in fixed-rate bonds because the price of fixed-rate bonds falls as rising inflation increases an investor's required rate of return. In times of rising inflation, the price of a variable-rate bond does not fall as much because the investor knows that a coupon rate adjustment will occur to adjust to new, higher interest rates. However, investors who buy bonds with fixed coupon interest rates will be better off when market interest rates are falling. A variable-rate bond would have its coupon rate drop as market interest rates fell.

An issuing corporation can benefit from issuing variable-rate bonds if market rates are historically high and a drop in rates is expected. Of course, high rates can rise even further, in which case the issuing company could lose money.

Putable Bonds

A **putable bond** is a bond that can be cashed in before maturity at the option of the bond's owner. This is like the callable bond described earlier in the chapter except that the positions of the issuer and the bond's owner with respect to the option have been reversed. Investors may exercise the option to redeem their bonds early when it is in their best interest to do so. Investors usually redeem fixed-rate bonds if interest rates have risen. The existing, lower-interest-rate bond can be redeemed and the proceeds used to buy a new higher-interest-rate bond.

Figure 14-3
Announcement of SallieMae Bonds with Variable Rates

Figure 14-3 is an announcement of a variable rate bond issue offered by the Student Loan Marketing Association. The interest rates will vary according to the rates for U.S. 91-day Treasury bills.

Source: Student Loan Marketing Association, May 9, 1995.

New Issue   May 9, 1995

SallieMae

Short Term Floating Rate Notes
$600,000,000

Dated May 11, 1995   Due November 9, 1995
Price 100%   Cusip #86387T EC8   Series 11-95

The interest rate of the Notes will be subject to weekly adjustment on the calendar day following each auction of 91-Day Treasury bills, and will be equal to 15 basis points above the "91-Day Treasury Bill Rate" (expressed on a bond equivalent basis). Interest on the Notes is paid at maturity and accrues from May 11, 1995. The Notes will be issued only in book-entry form.

These Notes are the obligations of the Student Loan Marketing Association, a federally chartered, stockholder-owned corporation, and are not obligations of or guaranteed by the United States. The Notes are legal for investment by savings banks, insurance companies, trustees and other fiduciaries under the law of many states.

This offering is made by the Student Loan Marketing Association with the assistance of a designated Selling Group of securities dealers.

Denise McGlone
Executive Vice President
and Chief Financial Officer

For more details, contact the Corporate Finance Department at 1-800-321-7179.

Student Loan Marketing Association
1050 Thomas Jefferson Street, N.W., Washington, D.C. 20007
This announcement appears as a matter of record only.

Junk Bonds

Another type of bond that has become popular (and controversial) is the junk bond. Junk bonds, also known as *high-yield bonds*, have a bond rating below investment grade. As shown earlier in this chapter, according to Moody's ratings, a junk bond would have a rating of Ba1 or below; according to Standard & Poor's ratings, it would have a rating of BB+ or below. The name *junk* is perhaps unfairly applied because these bonds are usually not trash; they are simply riskier than bonds having an investment grade. For instance, many bonds used to finance corporate takeovers have below investment-grade ratings.

Some junk bonds start out with investment-grade ratings but then suffer a downgrade—the issuing company may have fallen on hard financial times, or may have gone through a major financial restructuring that increased the risk of the outstanding bonds. Such junk bonds are known as *fallen angels*. One example of a fallen angel is the January 1996 downgrade of K-Mart bonds to junk bond status because the company's financial health had deteriorated considerably.

ETHICAL CONNECTIONS

SEC Charges 50 Municipalities in Bid to Stop Fraud in Muni-Bond Market

The Securities and Exchange Commission [recently] brought enforcement actions against 50 municipalities . . . as part of its crackdown on fraud and abuse in the municipal-bond market.

The cases are unusual, SEC officials say. Prior to 1996, the SEC did not charge a single "general purpose" municipal-bond issuer—a city, county, state—for securities fraud. Instead, when it brought cases in the muni-bond market, the SEC focused attention on bond dealers and, in some isolated cases, on special districts, or agencies that issue tax-exempt debt.

But in recent years, as the SEC has made cleaning up the $1.4 trillion municipal-bond market a priority, municipalities have found themselves in regulators' cross hairs. The SEC's Office of Municipal Securities says that it has brought cases of securities fraud against 50 municipal-bond issuers in the past two years, and SEC officials say more are likely in the months ahead.

The first case in the SEC crackdown came in a January 1996 action against Orange County, California, for failing to disclose to investors problems that contributed to its financial debacle two years earlier. Other municipalities that have been charged by the SEC and reached settlements include Maricopa County, Arizona; Syracuse, New York; and Nevada County, California. Recently, about three dozen counties and towns in Mississippi have been accused by the SEC of failing to disclose potential tax problems in their bond offerings.

Questions to Consider

▶ If a city or county is fined for municipal bond abuses, the cost is ultimately borne by the taxpayers in that municipality. Is this fair?

▶ Taking the preceding question into account, do you think the overall outcome of the SEC's policy will be positive or negative?

Source: Charles Gasparino and John Connor, staff reporters of *The Wall Street Journal*, Wall Street Journal Interactive Edition ⟨http://interactive.wsj.com⟩ accessed on October 15, 1998.

RJR Nabisco bonds were also downgraded to junk bond status after a leveraged buyout (LBO). The bondholders sued RJR Nabisco alleging that the company did not treat bondholders fairly or in good faith. It can be argued, however, that bondholders can reduce the risk of fallen angels by insisting on a protective provision in the indenture. Indeed, because of the RJR Nabisco LBO and similar restructurings, bond investors have been reading their indentures more closely.

International Bonds

An **international bond** is a bond sold in countries other than where the corporate headquarters of the issuing company is located. The bonds may be denominated in the currency of the issuing company's country, or in the currency of the country in which the bonds are sold. Foreign corporations issue bonds in the United States, sometimes denominated in their home currencies and sometimes in U.S. dollars. In turn, U.S. corporations frequently issue bonds outside the United States. These bonds may be denominated in U.S. dollars or in some other currency.

Eurobonds are bonds denominated in the currency of the issuing company's home country, and sold in another country. For example, if General Motors issued a dollar-denominated bond in Italy it would be called a Eurobond. Similarly, if Ferrari, an Italian company, issued a lira-denominated bond in the United States, the

> **Take Note**
>
> *Be careful not to confuse Eurobonds with Eurodollars. Eurodollars are dollar-denominated deposits in banks outside the United States.*

bond would be called a Eurobond. If the Ferrari bond were denominated in dollars instead of lira, it would be referred to as a *Yankee bond*.

Super Long-Term Bonds

The Disney Corporation, Coca-Cola, and recently Nacional Electricidad SA, the largest power company in Chile, have issued bonds with a maturity of 100 years, which is a much longer maturity than is typical among corporate bond issuers. Investors who purchase these bonds must have confidence in the future cash flow of these companies.

In this section we have described types of bonds. Next, we examine preferred stock, its characteristics, and those who purchase it.

PREFERRED STOCK

Preferred stock is so called because owners of preferred stock have a priority claim over common stockholders to the earnings and assets of a corporation. That is, preferred stockholders receive their dividends before common stockholders do. Preferred stock is not issued by many corporations except in certain industries, such as public utilities.

The preferred stock dividend is usually permanently fixed, so the potential return on investment for a preferred stockholder is not as high as it is for a common stockholder. Common stockholders are entitled to all residual income of the firm (which could be considerable).

Preferred stock is known as a hybrid security. It is a hybrid because it has both debt and equity characteristics. Preferred stock is like debt primarily because preferred stockholders do not have an ownership claim, nor do they have any claim on the residual income of the firm. It is also like equity because it has an infinite maturity and a lower-priority claim against the firm than the bondholders have.

Preferred Stock Dividends

Issuers of preferred stock generally promise to pay a fixed dollar amount of dividends to the investor. This promise, however, does not result in bankruptcy if it is broken. Unlike failure to make a scheduled interest or principal payment to bondholders, failure to pay a scheduled dividend to preferred stockholders is not grounds for bankruptcy of the company that issued the preferred stock.

Occasionally, *participating preferred stock* is issued. This type of preferred stock offers the chance for investors to share the benefits of rising earnings with the common stockholders. This is quite rare, however.

Preferred stock can be either *cumulative* or *noncumulative* with respect to its dividends. With cumulative preferred stock, if a dividend is missed, it must be paid at a later date before dividends may resume to common stockholders. Seldom is any interest paid, however, to compensate preferred stockholders for the fact that when dividends are resumed, they are received later than when promised. Noncumulative preferred stock does not make up missed dividends. If the dividends are skipped, they are lost forever to the investors.

Preferred Stock Investors

Corporations can generally exclude 70 percent of the dividend income received on preferred stock issued by another corporation from their taxable income. As a result, corporations are the major investors in preferred stock. The tax exclusion is higher if the investor corporation owns more than 20 percent of the common stock of the other corporation.

Because of the favorable tax treatment corporations receive on this dividend income, they bid up the price on preferred stock, thus lowering the expected rate of return. The lower expected rate of return is the price they pay for receiving the preferential tax treatment. Individuals cannot exclude any dividend income on their personal tax returns and must pay taxes on all of it, so preferred stock is not often recommended by financial planners as a good investment for individuals.

LEASING

Debt is often incurred to acquire an asset. An alternative to borrowing and buying an asset is to lease the asset. A **lease** is an arrangement in which one party that owns an asset contracts with another party to use that asset for a specified period of time, without conveying legal ownership of that asset. The party who owns the asset is known as the **lessor.** The party who uses the asset is the **lessee.** The lessee makes lease payments to the lessor for the right to use the asset for the specified time period.

A lease contract that is long term and noncancelable is very similar to a debt obligation from the perspective of the lessee. Some contracts that look like genuine leases are not, according to federal tax laws. There are different types of lease contracts. These different types have different accounting treatments, which we turn to next.

Genuine Leases versus Fakes

When a business leases an asset, the entire amount of the lease payments made by the lessee to the lessor is tax deductible to the lessee. When bonds are issued, or a bank loan obtained, only the interest portion of the loan payment is tax deductible. This sometimes leads a company to enter into a contract that looks like a lease, to obtain the large tax deductions, but which is not in fact a lease. The IRS is ever vigilant in ferreting out these fake lease contracts and denying the associated deductions.

To illustrate, suppose you needed a new truck for your business. The purchase price of the truck you want is $40,000. If you buy the truck and depreciate it over five years (ignore the half-year convention), you would have tax-deductible depreciation expense of $8,000 per year for five years. What if instead of buying the truck for $40,000, you leased it from the truck dealer for $40,000 in up-front cash, followed by additional lease payments of $1 per year for four years and then an option to buy the truck at the end of five years for $10? The extra $14 paid, with the exercising of the purchase option, would be a drop in the bucket compared to the tax savings you would realize in year 1 from the $40,000 tax deduction for the "lease payments." Because money has time value, a $40,000 deduction in year 1 is much preferred to deductions of $8,000 per year for five years.

The foregoing lease is a sham, a fake. What we have here is an installment purchase, disguised as a lease. In an audit, the IRS would deny the $40,000 year 1 "lease payment" and reclassify the deduction as a (much lower) depreciation expense. Your business would probably also be hit with interest charges and penalties.

The IRS standards for a genuine lease are as follows:

1. The remaining life of the asset at the end of the lease term must be the greater of 20 percent of the original useful life or one year.
2. The term of the lease must be 30 years or less.
3. The lease payments must provide the lessor with a reasonable rate of return.
4. Renewal options must contain terms consistent with the market value of the asset.
5. Purchase options must be for an amount close to the asset's fair market value at the time the option is exercised.
6. The property must not be limited use (custom made for only one firm's use) property.

You can see immediately that in our truck example, the purchase option specifies a price ($10) that is much lower than the fair market value of the truck in five years. The IRS would consider this "lease" to be a sham.

Operating and Financial (Capital) Leases

Once a lease passes the tests for being classified as a genuine lease, it must be further classified for accounting purposes as an *operating* or *financial (capital) lease*. An **op-**

FINANCE AT WORK

SALES·RETAIL·SPORTS·MEDIA TECHNOLOGY·PUBLIC RELATIONS·PRODUCTION·EXPORTS

Hybrid Financing—Mezzanine Capital

Mezzanine is a term used to describe debt that is junior in repayment obligation to traditional bank financing or senior debt. The need for this type of financing occurs when a gap exists between the amount the senior lender will provide and the amount of equity capital that is available.

Corporate borrowers often use more than one type of debt. A typical capital structure might include a secured term loan, a secured revolving line of credit from a bank, equipment leases, mezzanine debt, and conventional equity (common and preferred stock). Mezzanine debt (sometimes called second-tier debt) is unsecured subordinated debt in which the lender also receives some rights to acquire equity. It provides an additional layer of financing between the senior debt and the company's equity.

Mezzanine financing may be an attractive way to borrow funds beyond the amount that secured lenders will lend, although at somewhat higher interest rates. Borrowers should be aware, however, that there are both costs and benefits that are less visible and less susceptible to quantification than the financial considerations such as interest rates, amortization, security, and equity costs.

Mezzanine investors are subordinate to collateralized senior lenders but are senior to the equity investors. Mezzanine capital, therefore, looks like equity to the debtor's commercial bank lender and looks like debt to the company. It combines cash flow and risk characteristics of both senior debt and common stock. Mezzanine capital is typically supplied by venture capital limited partnerships and other nonbank financial institutions.

Sources: Douglas L. Batey, *Hidden Costs and Benefits of Mezzanine Level Financing* ⟨http://www.businesscity.com/doc/nvart5.htm⟩ and ⟨http://www.norwest.com/business-structfinance/doc/mezz_capital.htm⟩.

erating lease** has a term substantially shorter than the useful life of the asset and is usually cancelable by the lessee. A **financial (capital) lease** is long term and noncancelable. The lessee uses up most of the economic value of the asset by the end of the lease's term with a financial lease.

If you went on a business trip and leased a car for the week to make your business calls, this would be an operating lease. This same car will be leased again to many other customers and in one week you will use up a small fraction of the car's economic value. Your company, which is paying your travel expenses, would deduct these lease payments as business expenses on the income statement.

If your company signed a 10-year, noncancelable lease on a $20 million supercomputer, this is likely to be a financial lease (also known as a capital lease). After 10 years the supercomputer is likely to be obsolete. Your company would have used up most, if not all, its economic value by the end of the 10-year lease period. The lessor surely would demand lease payments high enough to recognize this fact and also to compensate for the time value of money that is paid over a 10-year period. The fact that the payments are spread out over time means that the lessor must be compensated for the cost of the asset and also for the delay in the receipt of the lease payments.

Accounting Treatment of Leases Both operating and financial (capital) lease payments show up on the income statement. Assuming that the lease is genuine, payments made by the lessee to the lessor are shown on the income statement as tax-deductible business expenses for both types of leases. These are costs of doing business for the lessee.

Financial leases have another accounting impact, however, that operating leases do not. A financial lease also shows up on the company's balance sheet because it is functionally equivalent to buying the asset and financing the purchase with borrowed money. If the asset had been purchased and financed with debt, the asset and the liability associated with the debt would both show up on the balance sheet. Because a financial lease is functionally equivalent to a purchase financed with debt, the Financial Accounting Standards Board (FASB) has ruled that the accounting treatment should be similar.

Failure to make a bond payment can lead to bankruptcy, as can failure to make a contractual lease payment on a noncancelable lease. The leased asset is shown in the asset section of the lessee's balance sheet, with a corresponding liability entry in the amount of the present value of all the lease payments owed to the lessor.

A lease is classified as a financial (capital) lease if it meets *any one* of the following four criteria:

1. Ownership of the asset is transferred to the lessee at the end of the lease's term.
2. There is an option for the lessee to buy the asset at a bargain purchase price at the end of the lease period.
3. The lease period is greater than or equal to 75 percent of the estimated useful life of the asset.
4. The present value of the lease payments equals 90 percent or more of the fair market value of the asset at the time the lease is originated, using the lower of the lessee's cost of debt or the lessor's rate of return on the lease as the discount rate.

Only if none of these four criteria applies is the lease considered an operating lease, with no balance sheet entry.

Lease or Buy?

Leasing is growing in popularity. Whether an asset should be leased or bought depends on the relative costs of the two alternatives. Leasing is most nearly comparable to a buy–borrow alternative. Because signing a debt contract is similar to signing a lease contract, comparisons are usually made between the lease option and the buy with borrowed funds option.

The alternative that has the lower present value of after-tax costs is usually chosen. The tax factor considered for the lease alternative would be the tax deductibility of the lease payments that would be made (assuming the lease is genuine and passes IRS muster). The tax factors for the buy with borrowed funds alternative would come primarily from two sources. One is the tax deduction that comes with the payment of interest on the borrowed funds. The other is the tax deduction that comes with the depreciation expense on the purchased asset.

Companies that are likely to see a clear advantage to leasing instead of buying are those that are losing money. Such companies, because they have negative taxable income, pay no taxes. The deductions for interest and depreciation expenses could not be used (subject to carry back and carry forward provisions of losses to earlier or later years). If the asset is leased instead from a profitable lessor, the lessor can take advantage of the interest and depreciation expense tax deductions and the lessee can negotiate a lease payment that is lower than it otherwise would be, so that these tax benefits are shared by the lessor and the lessee.

Many airlines are losing money. The next time you take a plane trip, look for a small sign just inside the passenger entrance of the plane. It may say that the airplane you are about to travel on is owned by a leasing company (the lessor) and leased by the airline (the lessee).

What's Next

In this chapter we learned about the basic characteristics of bonds and the different types of bonds available. We also examined preferred stock and leasing. In chapter 15 we will discuss common stock, how it is issued, and the nature of the equity claim of the common stockholders.

Summary

1. Describe the contract terms of a bond issue.

The indenture is the contract that spells out the terms of the bond issue. A call provision gives the issuer the option to buy back the bonds before the scheduled maturity date. A conversion provision gives the bondholder the option to exchange the bond for a given number of shares of stock. Restrictive covenants may include limits on future borrowings by the issuer, minimum working capital levels that must be maintained, and restrictions on dividends paid to common stockholders.

2. Distinguish the various types of bonds and describe their major characteristics.

All bonds are debt instruments that give the holder a liability claim on the issuer. A mortgage is a bond secured by real property. A debenture is an unsecured bond. A convertible bond is convertible, at the option of the bondholder, into a certain number of shares of common stock (sometimes preferred stock or another security).

A variable-rate bond has a coupon interest rate that is not fixed but is tied to a market interest rate indicator. A putable bond can be cashed in by the bondholder before maturity. Bonds that are below investment grade are junk bonds. An international bond, a bond sold in a country other than the country of the corporate headquarters of the issuing company, differs from a Eurobond. A Eurobond is a bond denominated in the currency of the issuing company's home country and sold in another country. A super long-term bond is one that matures in 100 years.

3. Describe the key features of preferred stock.

Preferred stock is a hybrid security that has debt and equity characteristics. Preferred stockholders have a superior claim relative to the common stockholders to a firm's earnings and assets, and their dividend payments are usually fixed. Those traits resemble debt. In addition, preferred stock has an infinite maturity and lower-priority claim to assets and earnings than bondholders. Because of the corporate tax exclusion of some preferred stock dividends from taxable income, corporations are much more likely to invest in stock than individuals.

4. Compare and contrast a genuine lease and a disguised purchase contract.

Because lease payments are entirely tax deductible, many attempt to disguise purchase contracts as genuine leases. A lease is an arrangement in which the owner of an asset contracts to allow another party the use of the asset over time. In order for the lease to be genuine, the lessee (the party to whom the asset is leased) may not have an effective ownership of the asset. The IRS has six standards that a lease must meet to qualify for the lease tax deductions. Failure to comply with IRS rules will result in the less favorable tax treatment of a purchase contract.

5. Explain why some leases must be shown on the balance sheet.

Operating leases are usually short term and cancelable. Financial (capital) leases are long term and noncancelable. Both operating and financial leases appear on the income statement of the lessee because they are tax-deductible business expenses. Because financial leases are functionally equivalent to a purchase financed with debt, FASB rules require that businesses treat them similarly for accounting purposes. Financial leases, therefore, appear on the balance sheet.

You can access your FREE PHLIP/CW On–Line Study Guide, Current Events, and Spreadsheet Problems templates, which may correspond to this chapter either through the Prentice Hall Finance Center CD–ROM or by directly going to: **http://www.prenhall.com/gallagher.**

PHLIP/CW — *The Prentice Hall Learning on the Internet Partnership–Companion Web Site*

Equations Introduced in This Chapter

Equation 14-1. Conversion Value of a Convertible Bond:

$$\text{Conversion Value} = \text{Conversion Ratio} \times \text{Stock Price}$$

Self-Test

ST-1. Explain the features of a bond indenture.

ST-2. What is a callable bond?

ST-3. What is the straight bond value of a convertible bond?

ST-4. What is cumulative preferred stock?

ST-5. Which financial statement(s) would a financial lease affect? Why?

ST-6. What is the conversion value of a convertible bond having a current stock price of $15 and a conversion ratio of 20?

Review Questions

1. How does a mortgage bond compare to a debenture?
2. How does a sinking fund function in the retirement of an outstanding bond issue?
3. What are some examples of restrictive covenants that might be specified in a bond's indenture?
4. Define the following terms that relate to a convertible bond: *conversion ratio, conversion value,* and *straight bond value.*
5. If a convertible bond has a conversion ratio of 20, a face value of $1,000, a coupon rate of 8 percent, and the market price for the company's stock is $15 per share, what is the convertible bond's conversion value?
6. What is a callable bond? What is a putable bond? How do each of these features affect their respective market interest rates?

Build Your Communication Skills

CS-1. Select a corporate bond and research its indenture provisions. What must the issuer do and what must the issuer not do? Write a brief report of one to two pages on your findings.

CS-2. Does a bond issuer owe potential investors full disclosure of plans that might affect the value of the bonds, or is the issuer's duty to investors only what is explicitly stated in the indenture contract? Divide into small groups to debate this issue.

Problems

Straight Bond Value

14-1. Sean Thornton has invested in a convertible bond issued by Cohan Enterprises. The conversion ratio is 20. The market price of Cohan common stock is $60 per share. The face value is $1,000. The coupon rate is 8 percent and the annual interest is paid until the maturity date 10 years from now. Similar nonconvertible bonds are yielding 12 percent (YTM) in the marketplace. Calculate the straight bond value of this bond.

Straight Bond Value

14-2. Use the same data given in problem 14-1. Now assume that interest is paid semiannually ($40 every six months). Calculate the straight bond value.

Conversion Value

14-3. Using the data in problem 14-1, calculate the conversion value of the Cohan Enterprises convertible bond.

Sinking Fund

14-4. Two years ago a company issued $10 million in bonds with a face value of $1,000 and a maturity of 10 years. The company is supposed to put aside $1 million in a sinking fund each year to pay off the bonds. Dolly Frisco, the finance manager of the company, has found out that the bonds are selling at $800 apiece in the open market now when a deposit to the sinking fund is due. How much would Dolly save (before transaction costs) by purchasing 1,000 of these bonds in the open market instead of calling them in at $1,000 each?

14-5. A company where Michael Kirby works as the vice president of finance issued 20,000 bonds 10 years ago. The bonds had a face value of $1,000, annual coupon rate of 10 percent, and a maturity of 20 years. This year the market yield on the company's bond is 8 percent. The bonds are callable after five years at par. If Mr. Kirby decides in favor of exercising the call option, financing it through a refunding operation, what would be the annual savings in interest payments for the company? (Interest is paid annually.) *Call Provision*

14-6. Use the same information given in problem 14-5. Now assume that the call premium is 5 percent and the bonds were called back today. J. B. Brooks purchased 10 bonds when they were originally issued at $950 per bond. Calculate the realized rate of return for Brooks. *Call Premium*

14-7. J. B. Brooks of problem 14-6, after getting her bonds called back by the original issuing company, can now invest in a $1,000 par, 8 percent annual coupon rate, 10-year maturity bond of equivalent risk selling at $950. (Interest is paid annually.) *Total Return on Investment*

   a. What is the overall return for Brooks over the 20 years assuming the bond is held until maturity?
   b. Compare the overall return with the return on the bond in problem 14-4 if they had not been called. Did Brooks welcome the recall?

14-8. Captain Nathan Brittles invested in a $1,000 par, 20-year maturity, 9 percent annual coupon rate convertible bond with a conversion ratio of 20 issued by a company six years ago. What is the conversion value of Captain Brittles's investment if the current market price for the company's common stock is $70 per share? (Interest is paid annually.) *Conversion Value*

14-9. Use the same information given in problem 14-8. If the current required rate of return on a similar nonconvertible bond is 7 percent, what is the straight bond value for the bond? Should Captain Brittles convert the bond into common stock now? *Straight Value*

14-10. Tom Dunston invested in a $1,000 par, 10-year maturity, 11 percent coupon rate convertible bond with a conversion ratio of 30 issued by a company five years ago. The current market price for the company's common stock is $30 per share. The current required rate of return on similar but nonconvertible bonds is 13 percent. Should Mr. Dunston consider converting the bond into common stock now? *Bond Conversion*

14-11. Six years ago Ruby Carter invested $1,000 in a $1,000 par, 20-year maturity, 9 percent annual coupon rate putable bond, which can be redeemed at $900 after five years. If the current required rate of return on similar bonds is 13 percent, should Ruby redeem the bond? What is the realized rate of return after redeeming? (Interest is paid annually.) *Putable Bond*

14-12. Five years ago Diana Troy invested $1,000 in a $1,000 par, 10-year maturity, 9 percent annual coupon rate putable bond, which can be redeemed at $900 after five years. If the current required rate of return on similar bonds is 14 percent, should Diana redeem the bond? What is the realized rate of return after redeeming? If Diana reinvests the sum in a $1,000 par, five years to maturity, 13 percent annual coupon rate bond selling at $900 and holds it until maturity, what is her realized rate of return over the next five years? What is her realized rate of return over the entire 10 years? *Reinvesting Putable Bond*

14-13. Hot Box Insulators, a public company, initially issued investment-grade, 20-year maturity, 8 percent annual coupon rate bonds 10 years ago at $1,000 par. A group of investors bought all of Hot Box's common stock through a leveraged buyout, which turned the bonds overnight into junk bonds. Similar junk bonds are currently yielding 25 percent in the market. Calculate the current price of the original bonds. *Challenge Problem*

14-14. Profit Unlimited Company is in bankruptcy. The company has the following liability and equity claims: *Priority of Claim*

First-Mortgage Bonds	$5 million
Second-Mortgage Bonds	5 million
Senior Debentures	10 million
Subordinated Debentures	4 million
Common Stock	10 million (par value)

Mortgaged assets have been sold for $7 million and other assets for $13 million. According to priority of claims, determine the distribution of $20 million obtained from the sale proceeds.

Answers to Self-Test

ST-1. A bond indenture is the contract that spells out the provisions of a bond issue. It always contains the face value, coupon rate, interest payment dates, and maturity date. It may also include terms of security in the case of default, if any; the plan for paying off the bonds at maturity; provisions for paying off the bonds ahead of time; restrictive covenants to protect bondholders; and the trustee's name.

ST-2. A callable bond is a bond that can be paid off early by the issuer at the issuer's option.

ST-3. The straight bond value of a convertible bond is the value a convertible bond would have without its conversion feature. It is the present value of the interest and principal using the required rate of return on a similar nonconvertible bond as the discount rate.

ST-4. Cumulative preferred stock is preferred stock for which missed dividends must be made up (paid) by the issuing company before common stock dividends may be resumed.

ST-5. A financial (capital) lease would show up on both the income statement and the balance sheet. Lease payments are business expenses that belong on the income statement and FASB rules call for financial leases to be reflected on the balance sheet also.

ST-6. $15 market price of the stock x 20 conversion ratio = $300 conversion value of the convertible bond.

Appendix 14A
Bond Refunding

When a corporation issues a callable bond, it needs to decide when it is advantageous to exercise that call option. This generally occurs when interest rates drop significantly and the issuing company seeks to replace the old high-interest rate bonds and to issue new bonds at the lower interest rate. This process is known as bond refunding.

The decision whether to refund or not is a capital budgeting decision. There are cash outflows involved in the form of the call premium that are usually required when bonds are called, flotation costs incurred when new bonds are sold, and tax deductions lost when lower-interest-rate bonds replace higher-interest-rate bonds. The inflows come mainly from the interest savings (lower payouts of interest represent cash inflows) that come when high-interest-rate debt is retired and replaced with lower-interest-rate debt.

A Sample Bond Refunding Problem

Suppose the Mega-Chip Corporation has $50 million worth of bonds outstanding with an annual coupon interest rate of 10 percent. However, market interest rates have fallen to 8 percent since the bonds were issued five years ago. Accordingly, the Mega-Chip Corporation would like to replace the old 10 percent bonds with a new issue of 8 percent bonds. In so doing, the firm could save 2 percent × $50 million = $1 million a year in interest payments. The original maturity of the 10 percent bonds was 20 years. The relevant financial data are summarized in Table 14A-1.

The Mega-Chip Corporation will be issuing new bonds having the same maturity as the number of years remaining to maturity on the old bonds (15 years). The *call premium* on the old bonds is the amount specified in the original bond indenture that the company must pay the bond owners if the bonds are called. The call premium is expressed as a percentage of the bond's face value. Thus, in this case, Mega-Chip will have to pay the old bondholders 5 percent of $50 million, or $2.5 million, in addition to the face value of the bonds, if it calls the bonds in.

The amount of the underwriting costs and the interest savings realized as a result of the refund are numbers that are essentially certain, so a very low discount rate is called for in this capital budgeting problem. The usual custom is to use the after-tax cost of debt for the discount rate. In this case the number is 4.8 percent.[1]

The calculations for this refunding capital budgeting problem are shown in Table 14A-2.

The cash outflows in Table 14A-2 are fairly straightforward. The 5 percent call premium on the old bonds and the 3 percent underwriting costs on the new bonds add up to a total outflow of $4 million.

TABLE 14A-1

Mega-Chip Bond Refunding Problem

Old Bond Issue:	$50,000,000; 10% annual interest rate; interest paid semiannually; 20 years original maturity, 15 years remaining to maturity
Call Premium on Old Bond:	5%
Underwriting Costs on Old Bonds When Issued 5 Years Ago:	2% of amount issued
New Bond Issue:	$50,000,000; 8% annual interest rate; interest paid semiannually; 15 years to maturity
Underwriting Costs on New Bonds:	3% of amount issued
Marginal Tax Rate:	40%
After-Tax Cost of Debt:	AT k_d = 4.8%

---

[1] This number was found using Equation 12-1. Mega-Chip's current before-tax cost of debt is 8 percent and its marginal tax rate is 40 percent. Per Equation 12-1, its after-tax cost of debt is:

$$\text{After-Tax } k_d = \text{Before-Tax } k_d \times (1 - \text{Tax Rate})$$
$$= .08 \times (1 - .40)$$
$$= .048, \text{ or } 4.8\%$$

TABLE 14A-2 Mega-Chip Bond Refunding Calculations

Cash Outflows	Calculations	Incremental Cash Flows
Call Premium Paid	$50,000,000 × .05 =	$2,500,000
New Bond Underwriting Costs	$50,000,000 × .03 =	$1,500,000
Total Outflows		$4,000,000

Cash Inflows		
Interest Savings	Interest on old bonds: $50,000,000 × .10 = $5,000,000 Interest on new bonds: $50,000,000 × .08 = $4,000,000 $1,000,000 difference each year for 15 years Less taxes on the additional income at 40%: $1,000,000 × .40 = ($400,000) Net Savings = $600,000 per year Present value of the net savings for 15 years at 4.8%: $600,000 \times \left[\dfrac{1 - \dfrac{1}{1.048^{15}}}{.048} \right] = \$600,000 \times 10.5214 =$	$6,312,840
Tax Savings on Call Premium Paid (the call premium is a tax-deductible expense amortized over the life of the bond issue)	$50,000,000 × .05 × .40 = $1,000,000 Amortized over 15 years = $1,000,000/15 years = $66,666.67 per year Present value of the savings for 15 years at 4.8%: $66,666.67 × 10.5214 =	$701,426.70
Tax Savings from Writing Off Balance of Old Bond Underwriting Costs	Unamortized Amount = $50,000,000 × .02 × (15/20)[a] = $750,000 Current Deduction PV of unamortized amount if bond is not called:[a] ($750,000 / 15) × 10.5214 = $526,070 Net Tax Savings = $750,000 − 526,070 =	$223,930
Tax Savings from New Bond Underwriting Costs	($50,000,000 × .03) / 15 = $100,000 per Year Writeoff Tax Savings = $100,000 × .4 = $40,000 PV of Tax Savings = $40,000 × 10.5214 =	$420,856
Total Inflows		$7,659,052.70
Net Present Value = $7,659,052.70 − $4,000,000 =		$3,659,052.70

Note: There are 5 of the original 20 years' worth of underwriting costs on the old bonds that have been written off. This leaves 15 of the 20 years, which is all written off immediately if the bonds are refunded.

[a] This is the PV of the tax savings you would have received anyhow without the refunding. The difference is the incremental cash flow from the refunding associated with the underwriting costs on the old bonds that have not yet been written off.

The cash inflows are more complicated. There is an annual interest saving of $1 million per year for 15 years. This amounts to $600,000 after taxes. The present value of these after tax annual savings is $6,312,840 using the after-tax cost of debt of 4.8 percent as the discount rate. The tax savings on the call premium paid on the old bonds is $701,426.70. The tax deductions from the underwriting costs on the old bonds were amortized over the original 20-year scheduled life of the bond. Thus, if the bonds are called now, the entire balance of the underwriting costs not yet claimed as a tax deduction will become immediately deductible.

The difference between the immediate tax savings from this deduction and the present value of the tax savings that would have been realized over the next 15 years is the incremental cash inflow relating to these underwriting costs. That figure is shown to be $223,930 in this case. We see from our calculations, then, that the present value of the tax savings from the amortization of the underwriting costs on the new bonds is $420,856.

Netting out all the incremental cash outflows and inflows gives a net present value figure of $3,659,052.70. Because this NPV figure is greater than zero, Mega-Chip will accept the project and proceed with the bond refunding.[2]

---

[2]This assumes that the management of Mega-Chip Corporation does not expect interest rates to fall further in the months to come. If managers are confident in a forecast for even lower interest rates to come, they may wait, expecting an even greater NPV in the near future.

CHAPTER 15

COMMON STOCK

Where your treasure is, there will your heart be also.
—*Matthew 6:21*

AN IPO IN A JITTERY MARKET

Some initial public offerings (IPOs) involve very large companies. On October 21, 1998, Conoco, Inc. went public. It was priced at $23 per share. Morgan Stanley was the main underwriter for the new stock issue.

Financial markets were very turbulent in September and October of 1998. Several planned IPOs were postponed at that time because of fears that the new issue would not be received well at that time. Goldman Sachs is a prime example of a company that had an IPO planned but withdrew it until a more favorable market environment developed.

At the time, Conoco was the eighth largest energy company in the United States. Conoco was a wholly owned subsidiary of DuPont until $4.4 billion was raised in the stock sale to the public. DuPont still indirectly owned about 75% of Conoco after the IPO. DuPont used the cash raised to reduce its debt load. Conoco common stock trades on the New York Stock Exchange and its ticker symbol is COC.

Sources: "Conoco IPO Prices at $23, Top End of Range," Reuters, http://yahoo.com/. David Chance, "FOCUS-Conoco IPO to Proceed, at $20–24/shr, http://pathfinder.com/money/latest/rbus/RB/1998Sep28/473.html.

CHAPTER OBJECTIVES

After reading this chapter, you should be able to:

- Describe the characteristics of common stock.
- Explain the disadvantages and advantages of equity financing.
- Explain the process of issuing new common stock.
- Describe the features of rights and warrants.

Chapter Overview

In this chapter we explore the characteristics and types of common stock, types of common stock owners, and the pros and cons of issuing stock to raise capital. Then we investigate how firms issue common stock. Finally, we examine *rights* and *warrants,* and their risk and return features.

The Characteristics of Common Stock

We discussed common stock in chapter 2, page 30.

As we learned in chapter 2, **common stock** is a security that represents an equity claim on a firm. Having an equity claim means that the one holding the security (the common stockholder) is an owner of the firm, has voting rights, and has a claim on the *residual income* of the firm. **Residual income** is income left over after other claimants of the firm have been paid. Residual income can be paid out in the form of a cash dividend to common stockholders, or it can be reinvested in the firm. Reinvesting this residual income increases the market value of the common stock due to the new assets acquired or liabilities reduced.

Corporations sometimes have different classes of stockholders. For example, a corporation's charter may provide for a certain class of stockholders to have greater voting rights than other classes. Or one class of stock may receive its dividends based on the performance of only a certain part of the company.

The Adolph Coors Brewing Company, for example, has two different classes of stock. Class A shares, owned by a trust for the benefit of Coors family members, have voting rights. Class B shares, available to the general public and traded on the NASDAQ stock market, have no voting rights. Separate voting and nonvoting classes of stock are often the case for public corporations, such as Coors, that started out as private family businesses and in which family members continue to actively manage the company. (There is no consistent standard, however, as to which designation, A or B, has the greater voting rights and which designation the lesser.)

Another special class of common stock is issued by General Motors. Its Class H common stock pays dividends tied to the performance of its subsidiary, Hughes Aircraft. General Motors bought Hughes Aircraft and created this special class of common stock as part of the financing arrangement of the purchase.

A relatively new special class of stock is sometimes created when a company that has long been in one line of business expands into a new, often riskier line. The company will then issue a new class of common stock that represents a claim only on the new business. This stock is called **target stock** because its value is targeted toward specific (nontraditional) assets. In 1995, for example, U.S. West Communications Corporation (one of the Baby Bells spun off in the AT&T divestiture) issued target stock to finance its venture in the cellular, cable, and other nontelephone businesses. The idea is that a different kind of stockholder is likely to be attracted to the newer, nontraditional businesses than the stockholder interested in "plain old telephone service" (POTS).

To review stock exchanges and over-the-counter exchanges, see chapter 2, pages 23–24.

All classes of stock have a value that is determined when that stock is traded from one investor to another at the various stock exchanges and in the over-the-counter market, as was described in chapter 2. The market takes into account the characteristics of a given class and values each class accordingly. Figure 15-1 shows a certificate of ownership for 288 shares of the common stock of Central Jersey Bancorp.

378 Part Three Long-Term Financial Management Decisions

Figure 15-1 Stock Certificate for 288 Shares of Central Jersey Bankcorp Common Stock

Common stockholders are paid dividends determined by the ability and willingness of the firm to pay. This dividend decision is made by the board of directors of the corporation. Residual income not paid out to the common stockholders in the form of dividends is reinvested in the firm. It benefits the common stockholders there as well (because they are the owners of the corporation).

All corporations issue stock reflecting the owners' claim. But some corporations are privately owned, whereas others are owned by members of the general public. The rules for private and public corporations differ, as we see next.

Stock Issued by Private Corporations

Private corporations (also known as closely held corporations) are so called because their common stock is not traded openly in the marketplace. Private corporations do not report financial information to the government through the Securities and Exchange Commission. (Tax returns, of course, are filed with the IRS, but this information is confidential.) Privately held corporations are usually small, and the stockholders are often actively involved in the management of the firm. The corporate form of organization is attractive to many small firms because the owners face only limited liability.

We discussed limited liability for corporate owners in chapter 1, pages 13–14.

Stock Issued by Publicly Traded Corporations

AT&T. McDonald's. Motorola. These are just some examples of well-known publicly traded corporations. A **publicly traded corporation** is a corporation whose common stock can be bought by any interested party, and that must release audited financial statements to the public. It is typically run by a professional management team, which likely owns only a tiny fraction of the outstanding shares of common stock.

The professional management team that handles the operations of the firm reports to a group called the **board of directors.** The board of directors, in turn, is elected by the common stockholders to represent their interests. The board is an especially important body for large public corporations, in which management typically owns such a tiny percentage of the firm. The agency problem discussed in chapter 1 described the conflict of interest that can occur when those who run a firm own very little of it. The common stockholders elect the board members, and the board members oversee the management of the company.

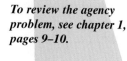

To review the agency problem, see chapter 1, pages 9–10.

Members of the board of directors have a fiduciary responsibility to the common stockholders who elected them. **Fiduciary responsibility** is the legal duty to act in the best interests of the person that entrusted you with property or power. When stockholders elect board members to represent them, they entrust the board members with the management of their company. Those board members owe it to common stockholders to act in the common stockholders' interest. Stockholders may vote directly on some major issues, such as a proposal to merge or liquidate the company.

Institutional Ownership of Common Stock

Much of the common stock of publicly traded corporations is owned by **institutional investors**—financial institutions that invest in the securities of other companies. Money management firms handling pension fund money, bank trust departments, insurance companies, mutual funds, and the like are major common stockholders. The link between ownership and control is likely to be a loose one in such cases because the individual shareholder is several layers away from the corporation. For instance, a worker may have a claim on a pension fund that is managed by a money management firm that has invested funds in another company's common stock.

Source: DILBERT © 1996 United Feature Syndicate, Inc. Reprinted by permission.

FINANCE AT WORK

SALES・RETAIL・SPORTS・MEDIA TECHNOLOGY・PUBLIC RELATIONS・PRODUCTION・EXPORTS

Chris Heller, Corporate Communications Consultant

Chris Heller

Chris Heller, a philosophy major in college, is a consultant specializing in corporate communications. Chris is a former assistant vice president at Aetna Life and Casualty and was responsible for all financial communications to investors and customers. These communications included annual reports, quarterly reports, and reports for financial analysts.

Q. *How important is it for a professional communicator to understand financial information?*

If you're going to be a practitioner in corporate public relations, especially in this day and age, you are inevitably going to have to communicate earning surprises, stock price changes, takeovers, divestitures, accounting charges, and bond rating declines—there are any number of financial transactions at the corporate level that you must be able to explain to stockholders.

You must understand financial concepts completely so you can interpret them in the company's official materials, be it a press release or an annual report. You must also be able to explain abstract financial concepts clearly to a reporter so he or she can describe them to the reader or listener in understandable and favorable terms.

Q. *Can you share one of the toughest financial situations you had to explain?*

Aetna was going through a tough time when it lost $2 billion in the commercial real estate market. That affected its ratings, which meant it went from a AAA− rated company to a AA− rated company. A rating decrease of this magnitude is a major problem for a financial services company because people lose trust in you. If customers start pulling their money out of the business—sell shares, cash in their bonds, and so forth—that could lead to serious liquidity problems.

My job was to explain to investors and customers what was happening, what wasn't happening, and what it meant, so they had a good grasp of the situation and how it affected their investments.

The challenge was to send information in a form that a variety of customers could understand. Aetna's customers ranged from a sophisticated 401K retirement plan sponsor representative to a school teacher or firefighter who was in a municipal plan for tax-deferred annuities. Aetna's debt and equity securities are also widely held and, therefore, most financial institutions watch the securities closely.

Aetna's real estate problems were covered in the press on a regular basis, in publications ranging from *US News Today* to *The Wall Street Journal*. We not only had to try to influence the content of these stories as they were being written, but we also had to give our customer service people the ability to explain the company's situation in a credible way to customers.

Q. *Are the stockholders with whom you communicated changing?*

Yes, there has been a major shift in the type of stockholders. With rare exceptions, the times when an individual stockholder has a majority of shares are over. Many individual stockholders now own their stock through a 401K retirement plan. Now, the majority of shares are owned by financial institutions, such as mutual funds.

Source: Interview with Chris Heller.

In recent years many institutional investors have begun to take a more active role in overseeing the companies in which they own common stock. Institutional investors usually have substantial amounts of funds, so they can buy a large number of shares of stock and become major shareholders. As a result, they can exercise more control than widely dispersed individual investors. Fidelity Investments, for example, a large mutual fund company, has been seeking seats on the board of directors of companies in which Fidelity is a major shareholder.

In this section, we investigated common stock characteristics, including classes of stock for stock issued by private and public corporations. We also looked at institutional ownership of common stock. Next, we examine the voting rights of common stockholders.

Voting Rights of Common Stockholders

Common stockholders have power to vote according to the number of shares they own. The general rule is "one vote per share." A stockholder or stockholder group holding more than 50 percent of the voting shares has a *majority interest* in the firm. The stockholder or group of stockholders that owns enough voting shares to control the board and operations of the firm has a *controlling interest* in the firm. The stockholder or stockholder group gains control when it elects a majority of its supporters to the board of directors.

In practice, a group can gain control with much less than 50 percent of the voting shares. This can happen if the remaining voting shares are widely distributed among many thousands of stockholders (each of whom owns a tiny percentage of the outstanding voting shares) who do not act in concert with each other. Many firms are controlled by groups of common stockholders owning as little as 5 percent or 10 percent of the voting shares, sometimes less.

Proxies

In large publicly traded corporations, the typical shareholder is likely to be uninterested in the details of the company's operations. It is not worth going to the stockholders' meeting in another part of the country if you hold only a few hundred shares. Such stockholders will typically allow others to vote their shares for them by *proxy*. This means that another group—usually the management of the company, but sometimes a group opposing management—will vote the shares for the stockholder who has given his or her proxy.

To give permission to another to vote your shares, you sign a card sent out by the group seeking this permission. In contested votes, where several competing groups may solicit shareholder proxies, each group may send out a card of a different color.

Board of Directors Elections

Corporate elections typically use one of two different sets of voting rules to fill seats on the board of directors. These are *majority voting* and *cumulative voting* rules. Under majority voting rules, a given number of seats are to be filled in a given election. The number of voting shares held, plus proxy votes held, represents the number of votes a person may cast for a given candidate for each separate seat. If multiple seats are contested, the candidates for a given seat will compete against each other for that seat. The person receiving the greatest number of votes wins that seat. With majority voting, whoever controls the majority of the votes will get their candidates elected to every seat to be filled.

Under cumulative voting rules, all the candidates run against each other but do not run for a particular seat. If there are five seats to be filled in the election, the top five vote getters among all the candidates win those seats. Votes are cast—one vote per share times the number of seats being contested—for as many or as few candidates as they wish. This means that stockholders with shares and proxies for less than a majority of the number of voting shares can "accumulate" their votes by casting them all for only a few candidates (even casting all votes for one candidate).

Cumulative voting makes it more likely that those shareholders with less than a majority of the voting shares will get some representation on the board. With majority voting rules, in which candidates run for specified seats, minority stockholders would get outvoted by the majority stockholders in each of these separate elections.

Suppose Burgerworld Corporation has a ten-member board, and terms for three of the ten members are expiring. The firm uses cumulative voting rules. Seven candidates are competing for the right to fill these three seats. There are 100,000 voting shares of common stock outstanding for Burgerworld. This means that 300,000 total votes will be cast (100,000 shares × 3 contested seats = 300,000 total votes).

The stockholders are divided into two camps of differing corporate management philosophy. The majority group controls 60 percent of the voting shares, whereas the minority group of common stockholders controls the other 40 percent. One of the seven candidates was nominated by the minority group. The minority stockholder group knows that with only 40 percent of the votes, they have no hope of winning two or three of the three seats contested. Does the minority group of stockholders have enough voting power to get their one candidate on the board?

The majority stockholders would like to get three of their people elected to the three seats available. If they want to succeed, they will have to spread their 180,000 votes (60 percent of 300,000) among their three favorite candidates. Spreading the votes evenly among their preferred candidates, each candidate supported by the majority group would receive 60,000 votes (180,000 ÷ 3 = 60,000). If the minority stockholders cast all their votes for their candidate, that person will receive 120,000 (300,000 total − 180,000 majority votes = 120,000 minority votes) votes and win a seat on the board.

The formula for determining the number of directors that a stockholder group could elect, given the number of voting shares they control, is shown next in Equation 15-1.

The Number of Directors That Can Be
Elected under Cumulative Voting Rules (15-1)

$$\text{NUM DIR} = \frac{(\text{SHARES CONTROLLED} - 1) \times (\text{TOT NUM DIR T.B.E.} + 1)}{\text{TOT NUM VOTING SHARES}}$$

where: NUM DIR = Number of directors that can be elected by a given group
SHARES CONTROLLED = The number of voting shares controlled by a given group
TOT NUM DIR T.B.E. = Total number of directors to be elected
TOT NUM VOTING SHARES = Total number of voting shares in the election

Using the number of shares owned by the minority stockholders described in our Burgerworld example (40,000 of 100,000 shares outstanding), we can calculate the number of directors that this minority group could elect. Recall that the number of directors to be elected is three. The calculations follow:

$$\text{NUM DIR} = \frac{(40{,}000 - 1) \times (3 + 1)}{100{,}000}$$

$$= 1.60$$

This group can elect one of their people to the board out of the three to be elected. Note that we rounded down to get the answer. Because people cannot be divided, the minority group can't elect .6 (60 percent) of a person to the board.

The formula for determining the number of shares needed by a given group to elect a given number of directors is shown next in Equation 15-2.

The Number of Shares Needed to Elect a
Given Number of Directors under Cumulative Voting Rules (15-2)

$$\text{NUM VOTING SHARES NEEDED} = \frac{\text{NUM DIR DESIRED} \times \text{TOT NUM VOTING SHARES}}{\text{TOT NUM DIR T.B.E.} + 1} + 1$$

where: NUM DIR DESIRED = Number of directors a given group of stockholders desires to elect
TOT NUM VOTING SHARES = Total number of voting shares in the election
TOT NUM DIR T.B.E. = Total number of directors to be elected in the election

For example, to calculate the number of voting shares needed to elect two of the three directors in the election described earlier, we could plug in the appropriate numbers into Equation 15-2. The calculation is shown next:

$$\text{NUM VOTING SHARES NEEDED} = \frac{2 \times 100{,}000}{3 + 1} + 1$$

$$= 50{,}001$$

We find that a group would need control of 50,001 voting shares to guarantee the election of two of the three directors in this election. This number is equivalent to 150,003 votes spread evenly between two director candidates. This would be 75,001.5 votes per candidate (50,001 voting shares × 3 total directors to be elected ÷ 2 directors sought to be elected). The other shareholders, holding 49,999 voting shares, would have the remaining 149,997 votes (49,999 voting shares × 3 total directors to be elected). If these 149,997 votes were divided between two candidates, that would be only 74,998.5 votes per candidate.

In this section, we reviewed the voting rights of common stockholders. We examine the advantages and disadvantages of equity financing next.

The Pros and Cons of Equity Financing

Selling new common stock has advantages and disadvantages for a corporation. Some disadvantages include a dilution of power and earnings per share of existing stockholders, flotation costs, and possible unfavorable market perceptions about the firm's financial prospects. The advantages for a new stock issue include additional capital for the firm, lower risk, and the potential to borrow more in the future.

Disadvantages of Equity Financing

Selling new common stock is like taking in new partners (although we are referring to a corporation rather than to a partnership). When you sell new common stock, you must share the profits and the power with the new stockholders. When new common stock is issued, the ownership position of the existing common stockholders is diluted because the number of shares outstanding increases.

The dilution may result in a lower earnings per share for a profitable company. Losses, of course, would be shared, too, resulting in a less negative earnings per share figure for a money-losing company. The voting power of the existing common stockholders would also be diluted. Firms concerned with losing control through diluted voting power, then, often avoid raising funds through a stock issue.

Also, when new common stock is sold, flotation costs are incurred. As we discussed in chapter 12, **flotation costs** are fees paid to investment bankers, lawyers, and others when new securities are issued. The flotation costs associated with new common stock issues are normally much higher than those associated with debt.

We discussed flotation costs in chapter 12, page 299.

Another reason that common stock issues are often a last resort for many corporations is because of signaling effects. **Signaling** is a message a firm sends, or investors infer, about a financial decision.

It is reasonable to suggest that the internal corporate managers have better insight about a firm's future business prospects than does the average outside investor. If we accept this proposition as true, then we would not expect a company to sell additional shares of common stock to the general public unless its managers know that the future prospects for the company are worse than is generally believed. How do we know this? Equity financing is expensive and often used as a last resort. The inference drawn by investors who agree with this view is that a corporation issuing new stock wants more "partners" with whom to share future bad times. A company that expected good times would attempt to preserve the benefits for the current owners alone. Instead of issuing new stock, then, the firm would issue more debt securities.

Management may issue new common stock even though the future financial prospects for the firm are bright. However, if the market believes otherwise, the price of the common stock will drop when the new common stock is issued.

Advantages of Equity Financing

Why would any corporation issue new stock? One big reason is that corporations do not pay interest (and are not legally obligated to pay any dividends) to common stockholders. Unlike interest payments on debt, dividends can be skipped without incurring legal penalties. Interest payments on debt reduce a firm's earnings, whereas dividend payments to stockholders do not.

Some firms choose equity financing because they do not like borrowing. Some business people view being "in debt" as undesirable. They avoid it if possible and pay off unavoidable debts as soon as possible. Companies whose managers and owners hold this view will tend to favor equity financing.

A final reason that firms choose equity instead of debt financing is that the firm may have so much debt that borrowing more may be difficult or too expensive. Suppose, for example, that the "normal" ratio of debt to assets in your firm's industry is 20 percent, and your firm's debt to assets ratio is 40 percent. If this is the case, lenders may be reluctant to lend your firm any more money at an affordable interest rate; your firm might be forced to issue stock to raise funds. In this situation a

We discussed debt ratios in chapter 5, pages 91–92. We discussed debt levels in chapter 13, pages 338–341.

new stock issue could bring the firm's debt ratios down to more normal industry levels. This would make it easier for the firm to borrow in the future.

In this section we described the pros and cons of a new stock issue. We turn to the process of issuing common stock next.

Issuing Common Stock

When a firm wishes to raise new equity capital, it must first decide whether to try to raise the capital from the firm's existing stockholders or whether to seek new investors. Private companies usually raise additional equity capital by selling new shares to existing common stockholders. This generally satisfies these stockholders because they continue to exercise complete control over the firm. However, when a large amount of equity capital must be raised, the existing stockholders may find that their only recourse is to sell shares of the firm's stock to the general public. A firm that sells its private shares of stock to the general public "goes public." The issuance of common stock to the public for the first time is known as an **initial public offering (IPO).** Figure 15-2 describes the Yahoo IPO.

Netscape, the company that makes the popular World Wide Web browser program for surfing the Internet, went public in late 1995. It was widely popular with investors, who bid its price up considerably from its issue price. The company's earnings at the time were quite limited, so investors were looking toward future profits they expected (and hoped) to see.[1]

Institutional investors are major buyers of new equity issues. Investment bankers who try to sell initial shares typically prefer to sell large blocks of shares to institutional investors as opposed to selling many small blocks of shares to individual investors. The institutional investors do well because a new issue is generally sold for 10 percent to 15 percent below value to ensure a favorable reception.[2]

Figure 15-2   IPO of Yahoo

YAHOO GOES PUBLIC

In April 1996 Yahoo Corporation went public by offering 2.6 million shares of stock at $13 per share, $3.25 adjusted for the two subsequent 2–1 stock splits. In December 1998 Yahoo was selling for $250 per share. That's a return of almost 7,600 percent over two and one-half years for those lucky enough to buy the stock at its offering price. Its total market value, price per share times number of shares outstanding, was over $10 billion.

This isn't bad for a company that had lost money until 1998 when it managed a few quarters of negligible profits. The market is clearly betting that this Internet portal company has a very bright future, its meager beginnings not withstanding. Yahoo is trying to branch out so that it can provide a wide range of Internet services. It doesn't want to be known simply as a "portal."

Yahoo is traded on the NASDAQ market and has the symbol YHOO. Goldman Sachs was the managing underwriter for the initial public offering (IPO).

---

[1]Christopher Farrell, "The Boom in IPO Capitalism," *Business Week* (December 18, 1995): 68–72.
[2]Ibid.

The Function of Investment Bankers

When a corporation does decide to sell stock to the public, its first step is to contact an investment bank to handle the issue. Some of the names of investment banking firms that a corporation might contact would include Merrill Lynch, Goldman Sachs, Morgan Stanley, and First Boston, to mention only a few.

Investment bankers handle all the details associated with pricing the stock and marketing it to the public. A potential investor in a new security must be given a **prospectus.** A prospectus is a disclosure document that describes the security and the issuing company. Investment bankers typically announce a new issue and the availability of the prospectus in a large boxed-in ad called a *tombstone ad.* It is so named because the large box with large print identifying the new issue looks like a tombstone. Figure 15-3 shows a tombstone ad for 4,750,000 shares of Geocities Corporation common stock.

Underwriting versus Best Efforts Investment bankers take on the job of marketing a firm's stock to the public on one of two bases: **underwriting** or **best efforts.** When an investment banker *underwrites* a stock issue, it means the investment banker agrees to buy a certain number of shares from the issuing company at a certain price. Usually, a group of investment bankers will form a temporary alliance called a **syndicate** when underwriting a new security issue. The head of the investment banking syndicate is known as the **manager.** The manager has the primary responsibility for advising the security issuer. Extra fees are collected for this advice. It is up to the syndicate to sell the shares to the public at whatever price it can get. An underwriting poses the least risk to the firm whose stock is being issued. This is because the firm gets the stock issue proceeds from the investment bankers all at once, up front. However, because the investment bankers bear the risk that they might not be able to sell the firm's shares at the price they expect, they charge a rather substantial fee for underwriting.

A cheaper alternative to underwriting is called a *best efforts offering.* In this arrangement the investment banker agrees to use its "best efforts" to sell the issuing company's shares at the desired price, but it makes no firm promises to do so. If the shares can only be sold at a lower price than was expected, then the issuing firm must either issue more shares to make up the difference or be satisfied with lower proceeds from the stock issue. Not surprisingly, the fees investment bankers charge for marketing stock on a best efforts basis are considerably less than those they charge for underwriting.

Pricing New Issues of Stock

When new shares of stock in a company are to be sold to the public, someone must decide at what price to offer them for sale. This is not a significant problem when the company's shares are already publicly traded. The new shares are simply sold at the same price as the old shares, or perhaps a little lower.[3] However, if the company is going public, there is no previous market activity to establish what the shares are worth. In this situation the investment banker, in conjunction with the issuing

---

[3]There is usually a little dilution (downward movement) in the price of a company's common stock when new shares are sold.

Figure 15-3 Tombstone Ad for Geocities Corporation Common Stock

company's managers, must put a price on the shares and hope that the market will agree that the price represents fair value. This is a daunting task indeed, and frequently investment bankers and firm managers miss the mark. Often an IPO stock will fluctuate wildly in price after it is issued. Table 15-1 shows four successful initial public offerings that involved stock price fluctuations after the IPO.

Netscape and Pixar were two of the hottest IPOs in late 1995 and Lucent Technologies and Yahoo Inc were hot IPOs in early 1996. Netscape and Yahoo attracted investors looking for Internet plays. Pixar made animation breakthroughs and Lucent Technologies was spun off from AT&T.

The figures in Table 15-1 show the differing fortunes of these five new public offering stocks.

TABLE 15-1

Recent Prices of Five Initial Public Offerings (IPOs)

IPO Date	Stock	Initial Price	Price on 12/23/98
8/95	Netscape	$14*	$59 9/16
11/95	Pixar Animation	$22	$36 5/16
4/96	Lucent Technologies	$13 1/2*	$109 3/4
4/96	Yahoo Inc.	$3.25*	$250

*Adjusted for stock splits.

Valuing the Stock of a Company That Is Not Publicly Traded Before investment bankers offer a company's stock for sale to the public, they must have some idea of how the public will value the stock to predict the new issue market price. The trouble is, when a company's stock has not been sold to anyone before, it is perilously difficult to say how much it is worth.

Suppose you have been creating oil paintings for a few years and have become pretty good at it. One day the art club you belong to has a show, and one of your paintings is included. When you deliver the painting to the gallery, you are asked what sale price you wish posted on the painting. Now you face the same question firms and investment bankers face. How much is the painting worth? How much can you get for it?

Naturally, you want to sell the painting for as high a price as possible but the potential buyers want to pay as low a price as possible. If you post too high an asking price, no one will buy your painting and you will leave empty handed. If you post too low a price, however, someone will snatch it up and may well resell it to someone else for a substantial profit. You can see that if you had previously sold a number of similar paintings in the past, you would have an idea of what to ask for this one. The first painting you try to sell is the one that presents the pricing problem.

In chapter 9 we presented some of the methods that companies and investment bankers use to estimate the market value of a company's stock. These methods include calculating the present value of expected cash flows, multiplying earnings per share by the appropriate P/E ratio, the book value approach, and the liquidation value approach.

To review stock valuation approaches, see chapter 9, pages 218–222.

Rights and Warrants

Rights and warrants are securities issued by a corporation that allow investors to buy new common stock. They originate in different ways and have somewhat different characteristics, as explained in the following sections.

Preemptive Rights

When some companies plan to issue new stock, they establish procedures to protect the ownership interest of the original stockholders. The existing stockholders are given securities that allow them to preempt other investors in the purchase of new shares. This security is called a preemptive right. A **preemptive right,** sometimes referred to simply as a *right,* gives the holder the option to buy additional shares of common stock at a specified price (the *subscription price*) until a given expiration date. Current stockholders who do not wish to exercise their rights can sell them in the open market.

The Number of Rights Required to Buy a New Share Suppose that Right Stuff Corporation has 100,000 shares of common stock currently outstanding. An additional 20,000 shares of common stock are to be sold to existing shareholders by means of a rights offering. Because one right is sent out to existing shareholders for each share held, 100,000 rights must be sent out. There are five shares of common stock outstanding for each new share to be sold (100,000/20,000 = 5). Five rights are therefore needed, along with the payment of the subscription price, to purchase a new share of common stock through the rights offering.

The Value of a Right We know that five rights are required to buy a new share of Right Stuff Corporation's common stock through the rights offering. To determine the value of each right, we must also know the subscription price and the market price of Right Stuff's common stock.

This information, along with our knowledge of the number of rights required to buy a new share, will allow us to estimate the value of one of these rights.[4]

Suppose that the current market price of Right Stuff Corporation's common stock is $65 and that the subscription price is set at $50. This means that you are saving $15 ($65 − $50 = $15) when you send in your five rights to receive one of the new shares (known as "exercising your rights"). This means that each right would be worth $3 ($15 ÷ 5 = $3) before dilution effects are considered.

The approximate value of a right can be determined by two formulas. The formula used depends on the status of the stock as it trades in the marketplace, relative to the timing of the issuance of the rights. Timing is the key to determining which approximation formula to use.

Rights are generally sent out several weeks after the announcement of the rights offering is made. This initial period is called the **rights-on** period, and the stock is said to *trade rights-on* during this time. This means that if the common stock is purchased during the rights-on period, the investor will receive the forthcoming rights.

At the opening of trading on the day following the rights-on period, the stock is said to be trading **ex-rights.** This means that the purchaser buys the stock without (*ex* is Latin for "without") receiving the entitlement to the preemptive rights if the purchase is on or after the ex-rights date.

Trading Rights-On. If the stock is trading rights-on, then we calculate the approximate market value of the right as depicted in Equation 15-3:

Approximate Value of a Right, Stock Trading Rights-On

$$R = \frac{M_0 - S}{N + 1} \qquad (15\text{-}3)$$

where: R = Approximate market value of a right
M_0 = Market price of the common stock, selling rights-on
S = Subscription price
N = Number of rights needed to purchase one of the new shares of common stock

---

[4]The actual pricing of a right is somewhat more complicated than what we are presenting here. A right is an option to buy the new stock at the specified subscription price. Option pricing is discussed in the following section on warrants. The rights valuation presented here should be considered an approximation only.

We call R the approximate market value of the right because rights are securities that can be traded just like stock and bonds. Once the rights are sent to the existing common stockholders (those who bought the stock before it "went ex-rights"), the rights can be traded in the marketplace at the option of the owner. The actual market price of the right may be slightly different than shown in the formulas presented here because of the option characteristics of the rights, discussed later.

Table 15-2 shows the calculation of the approximate market value of a Right Stuff Corporation right. This is the value of the right as determined by the rights-on formula, Formula 15-3.

We see from the calculations in Table 15-2 that the market value of the Right Stuff right is $2.50, given a market price of common stock of $65, a subscription price of $50, and five rights required to purchase one new share.

Selling Ex-Rights. To find the approximate value of the right when the stock is selling ex-rights, the ex-rights formula must be used. This formula is presented as follows in Equation 15-4:

Approximate Value of a Right, Stock Trading Ex-Rights

$$R = \frac{M_x - S}{N} \qquad (15\text{-}4)$$

where: R = Approximate market value of a right
M_x = Market price of the common stock, selling ex-rights
S = Subscription price
N = Number of rights needed to purchase one of the new shares of common stock

When the common stock begins trading ex-rights, the entitlement to the forthcoming rights is lost. Thus, the price of the common stock in the marketplace will drop by the value of the right now lost on the ex-rights date (other factors held constant).

Suppose Right Stuff Corporation common stock begins selling ex-rights today. When the opening bell rings on the exchange, the price of Right Stuff common stock will drop by the amount of the value of the right that has been lost. Holding other factors constant (no news overnight to otherwise affect the value of the common stock), the price of the common stock will drop by $2.50 from $65 to $62.50.

Table 15-3 shows the calculation of the approximate market value of the right when the common stock is selling ex-rights, using Formula 15-4.

When the common stock is selling ex-rights, Formula 15-4 gives the approximate market value of a right. The formula reflects the loss of the entitlement of the rights, $2.50 in our example.

TABLE 15-2

Right Valuation with Stock Selling Rights-On

M_0, Market Price of the Common Stock, Rights-On	$65
S, Subscription Price	$50
N, Number of Rights Required to Purchase One New Share	5
R, Approximate Market Price of One Right	$R = \dfrac{\$65 - 50}{5 + 1}$
	$= \dfrac{\$15}{6}$
	$= \$2.50$

TABLE 15-3

Right Valuation with Stock Selling Ex-Rights

M_x, Market Price of the Common Stock, Ex-Rights	$62.50
S, Subscription Price	$50
N, Number of Rights Required to Purchase One New Share	5
R, Approximate Market Price of One Right	$R = \dfrac{\$62.50 - \$50}{5}$ $= \dfrac{\$12.50}{5}$ $= \$2.50$

Warrants

A **warrant** is a security that gives its owner the option to buy a certain number of shares of common stock from the issuing company, at a certain exercise price, until a specified expiration date. The corporation benefits from issuing warrants because the issue raises funds. It also creates the possibility of a future increase in the company's number of common stock shares. The investor values warrants because of the option to buy the company's stock.

Warrants are similar to rights except they are sold to investors instead of given away to existing shareholders. They typically have longer maturities than rights and are often issued with bonds as part of a security package.

Warrant Valuation Warrants have value only until the expiration date, at which time they become worthless. Before a warrant expires, the value of a warrant depends on how the price of the common stock compares to the warrant's *exercise price*—the price the firm sets for exercising the right to buy common stock shares—and on other factors, described next.

To value warrants, investors must be able to find the exercise value. The exercise value is the amount saved by purchasing the common stock by exercising the warrant rather than buying the common stock directly in the open market. If there is no saving, the exercise value is zero.

The formula for calculating the exercise value of a warrant is described in Equation 15-5 as follows:

The Exercise Value of a Warrant

$$XV = (M - XP) \times \# \tag{15-5}$$

where: XV = Exercise value of a warrant
M = Market price of the stock
XP = Exercise price of a warrant
\# = Number of shares that may be purchased if the warrant is exercised

Suppose that the McGuffin Corporation warrant entitles the investor to purchase four shares of common stock, at an exercise price of $50/share, during the next three years. If the current common stock price is $60, the exercise value according to Equation 15-5 follows:

$$XV = (M - XP) \times \#$$
$$= (\$60 - \$50) \times 4$$
$$= \$40$$

Our calculations show that the exercise value is $40.

If the market price of the common stock price were $50 or less, the exercise value of the warrant would be zero. This is because you would have an option to buy the common stock at a price that is no better than the regular market price of the stock. You would have no reason to exercise the warrant, and a rational investor would not do so.

Note that the time remaining until the expiration of the warrant does not affect the exercise value. For the McGuffin Corporation warrants, the investor saves $10/share on four shares of common stock, creating an exercise value of $40.

As long as there is still time remaining until expiration, the actual market price of a warrant will be greater than the exercise value. The difference between the market price and the exercise value is called the warrant's time value. Warrants have time value because if the price of the common stock goes up, the exercise value increases with leverage and without limit (because of the fixed exercise price). If, on the other hand, the price of the common stock falls, the exercise value cannot dip below zero. Once the common stock price is at or below the exercise price, no further damage can be done to the exercise value.

The exercise value is zero if the common stock price is at or below the exercise value, but it can never be negative. Table 15-4 shows the exercise value for a McGuffin Corporation warrant for different possible stock values.

No matter how much below $50 the market price of the common stock goes, the exercise value stays at zero. As the stock value goes above $50, however, the exercise value increases at a much faster rate than the corresponding increase in the stock price. The difference between the potential benefit if the stock price increases (unlimited and leveraged) compared to the potential loss (limited) is what gives a warrant its time value.

Because of this time value, warrants are seldom exercised until they near maturity, even when the exercise value is high. This is because if a warrant is exercised, only the exercise value is realized. If the warrant is sold to another investor, the seller realizes the exercise value plus the time value. The time value approaches zero as the expiration date nears. A warrant approaching its expiration date, having a positive exercise value, would be exercised by the investor before the value goes to zero on the expiration date.

The greater the volatility of the stock price, and the greater the time to expiration, the greater the market value of the warrant. If the stock price is volatile, the stock price could easily increase. This would give the warrant owner the benefit of an even greater increase due to leverage. If the common stock price decreases, no

TABLE 15-4

McGuffin Corporation Warrant Exercise Value

MARKET PRICE OF COMMON STOCK	EXERCISE PRICE	NUMBER OF SHARES	EXERCISE VALUE
$100	$50	4	$200
$ 90	$50	4	$160
$ 80	$50	4	$120
$ 70	$50	4	$ 80
$ 60	$50	4	$ 40
$ 50	$50	4	$ 0
$ 40	$50	4	$ 0
$ 30	$50	4	$ 0
$ 20	$50	4	$ 0

more than the price paid can be lost. The more time left before expiration, the better the chance for a major stock price change, up or down. Again, the warrant value upside is substantial if the stock price moves up, and the downside potential limited if the stock price decreases. The asymmetry of the warrant's upside and downside potential gives the warrant greater value.

Option pricing has many applications in finance. We can apply the principles described here for warrant pricing to certain types of capital budgeting and even common stock valuation. What if a proposed capital budgeting project gives us an option to undertake future projects that are tied to the first? Common stock has unlimited upside price potential, coupled with limited downside risk, just like a warrant. These are issues you may explore further in other finance courses.

What's Next

In this chapter we examined types and traits of common stock, and the advantages and disadvantages of issuing common stock. We also explored rights and warrants. In chapter 16 we will look at how a corporation determines the amount of cash dividends to pay, and the timing of those dividend payments.

Summary

1. Describe the characteristics of common stock.

Common stock is a security that represents an ownership claim on a corporation. The shareholders are entitled to the residual income of the firm, resulting in a high-risk position relative to other claimants and a relatively high-return potential. Common stock may come in different classes with different voting rights or dividend payments.

The professional management team that handles the operations of the firm reports to the board of directors. The board of directors, in turn, is elected by the common stockholders to represent their interests. Because the stockholders have entrusted the board to represent their interests, board members have a fiduciary duty to act on stockholders' behalf. Stockholders usually vote according to the number of shares held. The two main types of voting rules are the majority voting rules, under which candidates run for specific seats and the cumulative voting rules, under which all the candidates run against each other but do not run for a particular seat.

2. Explain the disadvantages and advantages of equity financing.

Disadvantages of equity financing include the dilution of existing shareholder power and control, flotation costs incurred when new common stock is sold, and the negative signal investors often perceive (rightly or wrongly) when new common stock is sold. Equity financing has several advantages. It reduces the risk of a firm because common stockholders, as opposed to debtors, have no contractual entitlement to dividends. Equity financing can also increase the ability to borrow in the future.

3. Explain the process of issuing new common stock.

Once a firm decides that the benefits outweigh the cost of issuing stock, the firm almost always seeks the help of an investment banking firm. The investment banker usually underwrites the new issue, which means that it purchases the entire issue for resale to investors. Sometimes the investment banker will try to find investors for the new common stock without a guarantee to the issuing company that the stock will be sold. This arrangement is known as a best efforts offering.

4. Describe the features of rights and warrants.

Rights are securities given to existing common stockholders that allow them to purchase additional shares of stock at a price below market value. Corporations issue rights to safeguard the power and control of existing shareholders in the event of a new stock issue. Warrants are securities that give the holder the option to buy a certain number of shares of common stock of the issuing company at a certain price for a specified period of time. Warrants have high-return potential because if the stock price increases, the value of the warrant increases at a much higher rate due to leverage. The downside risk of a warrant is limited because the maximum loss potential is the price of the warrant. As a result of the high-return and low-risk potential, warrants have time value that is greatest when the stock price is volatile and the time to maturity is great.

You can access your FREE PHLIP/CW On–Line Study Guide, Current Events, and Spreadsheet Problems templates, which may correspond to this chapter either through the Prentice Hall Finance Center CD–ROM or by directly going to: **http://www.prenhall.com/gallagher.**

PHLIP/CW — *The Prentice Hall Learning on the Internet Partnership–Companion Web Site*

Equations Introduced in This Chapter

Equation 15-1. The Number of Directors That Can Be Elected under Cumulative Voting Rules:

$$\text{NUM DIR} = \frac{(\text{SHARES CONTROLLED} - 1) \times (\text{TOT NUM DIR T. B. E.} + 1)}{\text{TOT NUM VOTING SHARES}}$$

where: NUM DIR = Number of directors that can be elected by a given group
SHARES CONTROLLED = The number of voting shares controlled by a given group
TOT NUM DIR T.B.E. = Total number of directors to be elected
TOT NUM VOTING SHARES = Total number of voting shares in the election

Equation 15-2. The Number of Shares Needed to Elect a Given Number of Directors under Cumulative Voting Rules:

$$\text{NUM VOTING SHARES NEEDED} = \frac{\text{NUM DIR DESIRED} \times \text{TOT NUM VOTING SHARES}}{\text{TOT NUM DIR T. B. E.} + 1} + 1$$

where: NUM DIR DESIRED = Number of directors a given group of stockholders desires to elect
TOT NUM VOTING SHARES = Total number of voting shares in the election
TOT NUM DIR T.B.E. = Total number of directors to be elected in the election

Equation 15-3. Approximate Value of a Right, Stock Trading Rights-On:

$$R = \frac{M_0 - S}{N + 1}$$

where: R = Approximate market value of a right
M₀ = Market price of the common stock, selling rights-on
S = Subscription price
N = Number of rights needed to purchase one of the new shares of common stock

Equation 15-4. Approximate Value of a Right, Stock Trading Ex-Rights:

$$R = \frac{M_x - S}{N}$$

where: R = Approximate market value of a right
M_x = Market price of the common stock, selling ex-rights
S = Subscription price
N = Number of rights needed to purchase one of the new shares of common stock

Equation 15-5. The Exercise Value of a Warrant:

$$XV = (M - XP) \times \#$$

where: XV = Exercise value of a warrant
M = Market price of the stock
XP = Exercise price of a warrant
\# = Number of shares that may be purchased if the warrant is exercised

Self-Test

ST-1. What is residual income and who has a claim on it?

ST-2. Is a new common stock issue usually perceived as a good or bad signal by the market? Explain.

ST-3. What does it mean when a company's common stock is said to be trading ex-rights?

ST-4. If a company's common stock is selling at $80 per share and the exercise price is $60 per share, what would be the exercise value of a warrant that gives its holder the right to buy 10 shares at the exercise price?

Review Questions

1. What are some of the government requirements imposed on a public corporation that are not imposed on a private, closely held corporation?
2. How are the members of the board of directors of a corporation chosen and to whom do these board members owe their primary allegiance?
3. What are the advantages and the disadvantages of a new stock issue?
4. What does an investment banker do when underwriting a new security issue for a corporation?
5. How does a preemptive right protect the interests of existing stockholders?
6. Explain why warrants are rarely exercised unless the time to maturity is small.
7. Under what circumstances is a warrant's value high? Explain.

Build Your Communication Skills

CS-1. Review a financial publication, such as *The Wall Street Journal,* for a tombstone advertisement. Contact one of the investment banking firms you see listed in a tombstone ad announcing a new issue. Request a prospectus for that new issue. Once you receive the prospectus, write a report describing its key elements and what those elements reveal about the new security and its issuer.

CS-2. Research a company that is having a contested stockholder vote. You will find notice of such a vote in business publications, such as *The Wall Street Journal*. Different groups will typically run their own advertisements soliciting proxies so that group can vote the shares of other stockholders. Analyze the positions of the opposing sides, break into small groups, and debate the direction the company should take on the issue in contention.

Problems

15-1. Sonny owns 20,000 shares of common stock in QuickFix Company. The company has 1 million shares of common stock outstanding at a market value of $50 per share. What percentage of the firm is owned by Sonny?

If the company issues 500,000 new shares at $50 per share to new stockholders, how does Sonny's ownership change?

Ownership Claim

15-2. Terence Mann is considering buying some shares of common stock of an initial public offering by NewAge Communications Corporation. The privately held company is going public by issuing 2 million new shares at $20 per share. Terence gathered the following information about NewAge:

Valuation of IPO

Total assets:	$200 million (historical cost)
Total liabilities:	150 million (market value)
Number of shares retained by pre-IPO owners	3 million (5 million shares outstanding after IPO)
Estimated liquidation value:	250 million
Estimated replacement value of assets:	400 million
Expected dividend in one year:	$2 per share
Expected dividend growth rate:	8%

The required rate of return for Terence from a share of common stock for this type of company is 13 percent.

Compare the selling price of the stock with its value as obtained from different valuation methods. Would you recommend that Terence buy the stock?

15-3. Iowa Corn Corporation has nine board members. Three of these seats are up for election every three years. What is the length of the term served by each board member?

Term of Board Members

15-4. The stockholders of Blue Sky, Inc. are divided into two camps of different corporate management philosophy. The majority group controls 65 percent and the minority group controls 35 percent of the voting shares. The total number of shares of common stock outstanding is 1 million. The total number of directors to be elected in the near future is four. What is the maximum number of directors the minority group can possibly elect, assuming that the company follows cumulative voting procedure?

Cumulative Voting

15-5. Ms. O'Niel owns 26,000 shares of Tri Star Corporation out of 200,000 shares of common stock outstanding. The board has seven members, and all seven seats are up for election now. Ms. O'Niel has long wanted to serve as a member of the board. Assuming that the company follows cumulative voting procedures, can Ms. O'Niel get herself elected to the board on the strength of her own votes?

Cumulative Voting

15-6. The Rainbow Corporation had traditionally been a constant dollar dividend paying company, with the board enjoying the support of retired investors holding 65 percent of the

Dissident Group and Cumulative Voting

voting shares. A dissident group of high-salaried young investors holding 30 percent of the voting shares prefers reinvestment of earnings to save personal taxes and, hence, wants to elect board members supportive of its cause. The company has 600,000 shares of common stock outstanding and the board has 13 members—all to be reelected shortly.

    a. How many directors can the young stockholders elect under
        (i) cumulative voting rules?
        (ii) majority voting rules?
    b. What percentage of voting shares and/or proxies must the dissident group have to be able to elect seven out of the 13 board members?

Rights Offering

15-7. Fargo Corporation has 500,000 shares of common stock currently outstanding. The company plans to sell 50,000 more shares of common stock to the existing shareholders through a rights offering. How many rights will it take to buy one share?

Value of Rights

15-8. A company with 2 million shares of common stock currently outstanding is planning to sell 500,000 new shares to its existing shareholders through a rights issue. Current market price of a share is $65, and the subscription price is $55. If the stock is selling rights-on, calculate the value of a right.

Value of Rights

15-9. Use the same information given in problem 15-8. Now calculate the value of a right if the stock were selling ex-rights.

Value of Rights

15-10. Fillsulate Products, a manufacturer of refractory powders, is about to declare a rights issue. The subscription price is $65. Seven rights in addition to the subscription price are required to buy one new share of stock. Rights-on market price of the stock is $77. Calculate the value of one right. Also calculate the new stock price once it goes ex-rights.

Rights Offering

15-11. Johnny Rocco owns 700 shares of stock of East-West Tobacco Company, which is offering a rights issue to its existing shareholders. To buy one new share of stock, Johnny will need four rights plus $60. Rights-on market price of the stock is $72.

    a. Calculate the value of a right.
    b. What is the maximum number of new shares Johnny can buy?
    c. How much would he have to spend if he decides to buy all the new stock he can?
    d. If he decides not to buy the new stock, how much would he able to sell his rights for?

Challenge Problem

15-12. Armand Goldman owns 60 shares of East Asia Shipping Company stock and has $750 in cash for investment. The company has offered a rights issue in which purchasing a new stock would require four rights plus $50 in cash. Current market value of the stock is $62.

    a. Calculate the value of a right, if the stock is selling rights-on.
    b. Should Armand participate in the rights offering by buying as many shares as he can, or sell his rights and keep the shares he already owns at a diluted value?

Warrants

15-13. The current market price of a share of common stock of SkyHigh, Inc. is $100. The company had issued warrants earlier to its new bond investors that gave the investors an option to buy five shares of common stock at an exercise price of $85. Calculate the exercise value of a warrant. What happens to the exercise value of the warrant if the stock price changes to:

    a. $110?
    b. $80?

Comprehensive Problem

15-14. The current market price of Digicomm's common stock is $40 per share. The company has 600,000 common shares outstanding. To finance its growing business, the company needs to raise $2 million. Due to its already high debt ratio, the only way to raise the funds is to sell new common stock. Alvin C. York, the vice president of finance of Digicomm, has decided to go ahead with a rights issue but he is not sure at what price the existing shareholders would be willing to buy a share of new stock. Digicomm's investment banker has suggested that an analysis based on a wide range of possible prices be carried out and the subscription prices agreed upon were $36, $33, $29, and $26 per share of new stock. Digicomm's net income for the year is $1 million.

Based on the preceding information, Mr. York has asked you to carry out the following analysis:

a. For each of the possible subscription prices, calculate the number of shares that would have to be issued and the number of rights required to buy one share of new stock.
b. For each of the possible subscription prices, calculate the earnings per share immediately before and immediately after the rights offering.
c. Guy Hamilton owns 10,000 shares of stock of Digicomm. For each of the possible subscription prices, calculate the maximum number of new shares Guy would be able to buy. Under each of these cases, calculate Guy's total claim to earnings before and after the offering.

Answers to Self-Test

ST-1. Residual income is income that is left over after all claimants, except for common stockholders, have been paid. This leftover income belongs to the common stockholders, who receive this income in either the form of a dividend or by having it reinvested in the corporation that they own.

ST-2. The market usually infers bad news when new common stock is issued. Investors ask themselves why the current owners would want to share their profits with new owners if management expected good news ahead. The market often infers (rightly or wrongly) that there must be bad news coming that the management of the firm wants to "share" with new stockholders. New stock issued is, therefore, usually perceived as a negative signal.

ST-3. A stock is selling ex-rights when the purchase of that stock no longer carries with it entitlement to the rights that are soon to be sent out to stockholders.

ST-4. ($80 stock price − $60 exercise price) × 10 shares purchased per warrant = $200 exercise value of the warrant

CHAPTER 16

DIVIDEND POLICY

Finance is the art of passing currency from hand to hand until it finally disappears.
—Robert W. Sarnoff

DIVIDEND RESTRUCTURING NEEDED TO HELP FUND AGGRESSIVE GROWTH STRATEGIES

In August 1998 Washington Water Power Company announced details of a dividend restructuring plan that was designed to "strengthen the company's financial position, provide needed capital to help fund growth initiatives and new investment opportunities, and allow the company to maintain its record of service excellence to its current customers."

The restructuring calls for a 61 percent reduction in the company's annual common stock dividend, from $1.24 per share to 48 cents per share. The reduction was to be effective with the payment of the common stock cash dividend expected on December 15, 1998.

"The decision to change our dividend policy was not an easy one, but it was a decision our board deemed necessary to allow our company to grow and perform at the highest level of competition," said Tom Matthews, Washington Water Power's board chairman and chief executive officer. "This change in our dividend policy immediately improves our cash flows, enhances our ability to acquire needed capital in a cost-effective manner, and establishes a solid foundation for our continued growth and superior financial performance."

CHAPTER OBJECTIVES

After reading this chapter, you should be able to:

■ Explain the importance of and identify factors that influence the dividend decision.

■ Compare the major dividend theories.

■ Describe how a firm pays dividends.

■ Identify alternatives to paying cash dividends.

Based on a recent closing stock price of $20.875 per share, Washington Water Power's dividend yield would be 2.3 percent, placing it more in line with growth-oriented utility companies and still above the dividend yield of the average Standard & Poor's 500 company. "By having a dividend level more in line with growth-oriented utilities, our financial position is strengthened and we broaden our ability to make substantial investments in all our businesses—whether it's our traditional core energy business or our new ventures," Matthews said.

Recognizing the impact the dividend reduction could have on shareholders—particularly those with an income orientation—the company's board also approved development of an exchange offer to be open to holders of the company's common stock. Subject to regulatory approvals, which the company hopes to obtain by mid-October, shareholders will be provided the opportunity to exchange their common shares for an equal number of mandatorily convertible preferred shares, each of which will pay an annual dividend of $1.24 per share for a period of about three years.

After three years, the new-issue shares will automatically convert back to common stock on a one-for-one basis. The company has the option of converting some or all of the new-issue shares to common stock prior to the end of the three-year period. Shareholders who choose not to participate in the exchange plan will retain their ownership in Washington Water Power common stock. The exchange plan, Matthews said, is intended to allow income-oriented shareholders the opportunity to adapt to the Washington Water Power's more aggressive, growth-focused profile.

In order to grow its core energy business, Matthews said the company needs access to physical assets, specifically power generation assets and electric and gas transmission and distribution assets. Initial growth will come at local and regional levels, with national growth to follow.

Source: ⟨http://www.prnewswire.com⟩ (October 15, 1998), Washington Water Power, ⟨http://www.wwpco.com⟩.

Chapter Overview

In this chapter we explore the importance of dividends, the factors that determine a firm's dividend policy, and leading dividend policy theories. We then examine how a firm makes dividend payments to shareholders. We finish by identifying alternatives to paying cash dividends.

Dividends

Dividends are the cash payments that corporations make to their common stockholders. Dividends provide the return common stockholders receive from the firm for the equity capital they have supplied.[1] Even companies that do not currently pay

---

[1] You may wonder about this statement because common stock investors can always receive a return by selling their stock. Remember, however, that when investors sell their stock they are paid by other investors, not by the corporation. Except when a corporation buys back its own stock (which is a form of dividend payment), the only cash corporations pay to investors is a dividend payment.

dividends reinvest in the firm the earnings they generate. In this way they increase the ability of the firm to pay dividends in the future.

The board of directors decides what dividend policy best serves the common stockholders of the firm. Should a dividend be paid now, or should the earnings generated be reinvested for the future benefit of the common stockholders? If dividends are paid now, how much should be paid? These are some of the questions addressed in the following sections.

Although only corporations officially pay dividends to owners, sole proprietorships and partnerships also distribute profits to owners. Many of the same considerations examined in this chapter for corporate dividend policy can also be used to help make proper profit distribution decisions for these other forms of business organization.

Why a Dividend Policy Is Necessary

Why does a company need a strategic policy relating to dividend payments? Why not just "wing it" each year (or quarter, or other span of time) and pay the dividend that "feels right" at that time? Because market participants (current and potential stockholders) generally do not like surprises. An erratic dividend policy means that those stockholders who liked the last dividend cannot be sure that the next one will be to their liking. This uncertainty can result in a drop in the company's stock price. When stockholders do not get what they expect, they often show their displeasure by selling their stock. A well-planned policy, appropriate for the firm and its business strategy, can prevent unpleasant surprises for market participants and protect the stock price.

Factors Affecting Dividend Policy

Dividend policy is based on the company's need for funds, the firm's cash position, its future financial prospects, stockholder expectations, and contractual restrictions with which the firm may have to comply.

Need for Funds

Dividends paid to stockholders use funds that the firm could otherwise invest. Therefore, a company with ample capital investment opportunities may decide to pay little or no dividends. Alternatively, there may be an abundance of cash and a dearth of good capital budgeting projects available. This could lead to very large dividend payments.

Management Expectations and Dividend Policy

If a firm's managers perceive the future as relatively bright, on the one hand, they may begin paying large dividends in anticipation of being able to keep them up during the good times ahead. On the other hand, if managers believe that bad times are coming, they may decide to build up the firm's cash reserves for safety instead of paying dividends.

Stockholders' Preferences

Reinvesting earnings internally, instead of paying dividends, would lead to higher stock prices and a greater percentage of the total return coming from capital gains. *Capital gains* are profits earned when the price of a capital asset, such as common stock, increases.

Common stockholders in high tax brackets may prefer to receive their return from the company in the form of capital gains instead of dividends. Some stockholders prefer capital gains because the federal income tax rate on capital gains is limited to 20 percent.[2] Returns in the form of dividends may be taxed at a stated rate as high as 39.6 percent. The effective tax rate on dividends, due to the phasing out of some deductions and exemptions for high-income taxpayers, may be even higher. If the stock is not sold, capital gains taxes can be postponed indefinitely. If, however, the stockholders are mainly retired people looking for current income from their investments, then they may prefer a high dividend payment policy. The board of directors should consider such stockholder preferences when establishing the firm's dividend policy.

Restrictions on Dividend Payments

A firm may have dividend payment restrictions in its existing bond indentures or loan agreements. For example, a company's loan contract with a bank may specify that the company's current ratio cannot drop below 2.0 during the life of the loan. Because payment of a cash dividend draws down the company's cash account, the current ratio may fall below the minimum level required.[3] In such a case, the size of a dividend may have to be cut or omitted. In addition, many states prohibit dividend payments if they would create negative retained earnings on the balance sheet. This restriction is a prohibition against "raiding the initial capital." Figure 16-1 summarizes the factors that influence the dividend decision.

Cash versus Earnings

Dividends are often discussed in relation to a firm's earnings. The dividend payout ratio is often cited as an indicator of the generosity (or lack thereof) of the firm's dividend policy. The dividend payout ratio is calculated by dividing the total dollar amount of dividends paid by net income, as seen in Equation 16-1.

Formula for Dividend Payout Ratio

$$\text{Dividend Payout Ratio} = \frac{\text{Dividends Paid}}{\text{Net Income}} \quad (16\text{-}1)$$

---

[2] This was the maximum rate applied to capital gains for most individuals in the Taxpayer Relief Act of 1997.
[3] Recall that the current ratio is found by dividing current assets (of which cash is a part) by current liabilities. Thus, decreasing cash to pay a dividend will lower the ratio.

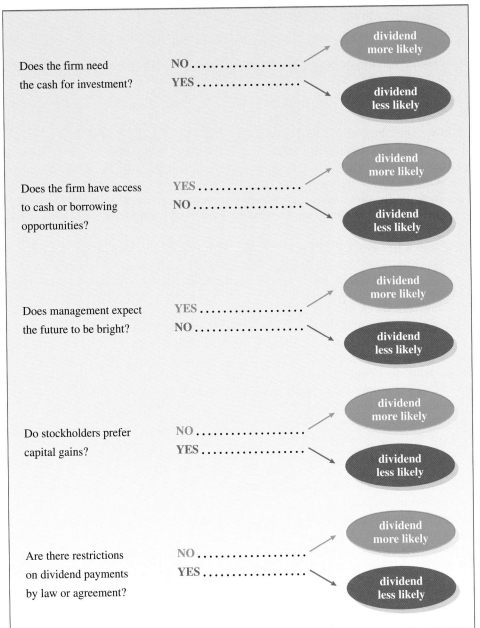

Figure 16-1 **Dividend Decision Factors**

This figure identifies key elements that make a dividend payment more or less likely.

If Calvin Corporation, for example, earns a net income of $10,000 and pays $3,000 in dividends, then its dividend payout ratio would be as follows:

$$\text{Dividend Payout Ratio} = \frac{\$3,000}{\$10,000}$$
$$= .30, \text{ or } 30\%$$

We see from our calculations that Calvin Corporation has a dividend payment ratio of 30 percent.

TABLE 16-1

Easy Credit Corporation Selected Financial Data for Current Year

Sales (all on credit payments due in next year)	$1,000,000
Total expenses	400,000
Net income	600,000
Cash received this year	$0

A caution, however: By focusing on reported earnings and the dividend payout ratio, we ignore the key to paying dividends. That key is *cash*. When a company generates earnings, this usually results in cash flowing into the firm. The earnings and the cash flows do not necessarily occur at the same time, however. Table 16-1 illustrates these timing differences.

Table 16-1 shows us that Easy Credit Corporation reported $600,000 of earnings this year but did not receive any cash. Unless cash was acquired from previous earnings, the firm could not pay dividends.

Even if the company reports negative earnings, it could pay dividends if it had, or could raise, enough cash to do so. If a company believed that a certain dividend payment were critical to the preservation of the firm's value, it might even choose to borrow so as to obtain the cash needed for the dividend payment. Corporate borrowing to obtain cash for a dividend payment happens occasionally, when the dividend payment expected by the common stockholders is believed to be crucial.

For instance, failure to make an expected dividend payment, in spite of a cash shortage, cost Consolidated Edison Company of New York (Con Ed) a third of its stock value. In 1974 Con Ed omitted its quarterly dividend for the first time since 1885. Con Ed's common stockholders included many retired people who counted on receiving their Con Ed dividends as a source of income. When the board of directors announced that the company would not pay its second-quarter dividend, the price of the common stock dropped by one-third from $18 to $12. This incident demonstrates what can happen when stockholders receive unpleasant dividend surprises.

Some would argue that in the long run, the Con Ed decision was sound. After all, regulators later allowed Con Ed to raise its rates, perhaps in response to the crisis created by its failure to pay dividends. The rate increase ultimately helped the company and its stockholders because profits increased due to the rate hikes. However, at least in the short run, the dividend omission was a disaster for stockholders—especially those who sold their holdings at greatly depressed prices after the dividend omission was announced.

LEADING DIVIDEND THEORIES

We've investigated how corporations consider many different factors when they decide what their dividend policies should be. Financial experts attempt to consolidate these factors into theories about how dividend policy affects the value of the firm. We turn to some of these theories next.

The Residual Theory of Dividends

The residual dividend theory is widely known. The theory hypothesizes that the amount of dividends should not be the focus of the company. Instead, the primary issue should be to determine the amount of earnings *retained* within the firm for investment. The amount of earnings retained, according to this view, depends on the

		TABLE 16-2
Investment Needed for New Projects	$10,000,000	**Residual Corporation—Applying the Residual Dividend Theory**
Optimal Capital Structure	30% Debt − 70% Equity	
Equity Funds Needed	70% × $10,000,000 = $7,000,000	
Earnings Available	$12,000,000	
Residual	$12,000,000 − $7,000,000 = $5,000,000	
Amount of Dividends to Be Paid	$5,000,000	

number and size of acceptable capital budgeting projects and the amount of earnings available to finance the equity portion of the funds needed to pay for these projects. Any earnings left after these projects have been funded are paid out in dividends. Because dividends arise from residual, or leftover earnings, the theory is called the *residual theory*. Table 16-2 shows how to apply this theory.

We see in Table 16-2 that Residual Corporation needs $10 million to finance its acceptable capital budgeting projects. It has earnings of $12 million. It needs equity funds in the amount of 70 percent of the $10 million needed or $7 million. This leaves residual earnings of $5 million for dividends.

If the earnings available had been $20 million instead of $12 million, then the dividend payment would have been $13 million ($20 million − $7 million). However, if earnings available had been $6 million instead of $12 million, then no dividends would have been paid at all. In fact, $1 million in additional equity funding would need to be raised by issuing new common stock. The amount of dividends to be paid is an afterthought, according to this theory. The important decision is to determine the amount of earnings to retain.

The residual dividend theory focuses on the optimal use of earnings generated from the perspective of the firm itself. This may appeal to some, but it ignores stockholders' preferences about the regularity of and the amount of dividend payments. If a firm followed the residual theory, when earnings are large and the acceptable capital budgeting projects small and few, dividends would be large. Conversely, when earnings are small and many large, acceptable projects are waiting to be financed, there may be no dividends if the residual theory is applied. The dividend payments would be erratic and the amounts unpredictable.

The Clientele Dividend Theory

The **clientele dividend theory** is based on the view that investors are attracted to a particular company in part because of its dividend policy. For example, young investors just starting out may want their portfolios to grow in value from capital gains rather than from dividends, so they seek out companies that retain earnings instead of paying dividends. Stock prices tend to increase as earnings are retained, and the resulting capital gain is not taxed until the stock is sold. If it is sold, it is taxed at a maximum rate of 20 percent,[4] which is lower than the maximum tax rate that applies to dividend income received.

---

[4]There are some exceptions to the 20 percent rate depending on the nature of the assets sold and the holding period.

Elderly investors, in contrast, may want to live off the income their portfolios provide. They would tend to seek out companies that pay high dividends rather than reinvesting for growth. According to the clientele dividend theory, each company, therefore, has its own clientele of investors who hold the stock in part because of its dividend policy.

If the clientele theory is valid, then it doesn't much matter what a company's dividend policy is so long as it has one and sticks to it. If the policy is changed, the clientele that liked the old policy will probably sell their stock. A new clientele will buy the stock based on the firm's new policy. When a dividend policy change is contemplated, managers must ask whether the effect of the new clientele's buying will outweigh the effect of the old clientele's selling. The new clientele cannot be sure that the most recent dividend policy implemented will be repeated in the future.

The Signaling Dividend Theory

The **signaling dividend theory** is based on the premise that the management of a company knows more about the future financial prospects of the firm than do the stockholders. According to this theory, if a company declares a dividend larger than that anticipated by the market, this will be interpreted as a signal that the future financial prospects of the firm are brighter than expected. Investors presume that management would not have raised the dividend if it did not think that this higher dividend could be maintained. As a result of this signal of good times ahead, investors buy more stock, causing a jump in the stock price.

Conversely, if a company cuts its dividend, the market takes this as a signal that management expects poor earnings and does not believe that the current dividend can be maintained. In other words, a dividend cut signals bad times ahead for the business. The market price of the stock drops when the firm announces a lower dividend because investors sell their stock in anticipation of future financial trouble for the firm.

If a firm's managers believe in the signaling theory, they will always be wary of the message their dividend decision may send to investors. Even if the firm has some attractive investment opportunities that should be financed with retained earnings, management may turn them down if adopting them would prevent paying the expected dividend and send an unfavorable signal to the market.

The Bird-in-the-Hand Theory

The **bird-in-the-hand theory** claims that stockholders prefer to receive dividends instead of having earnings reinvested in the firm on their behalf. Although stockholders should expect to receive benefits in the form of higher future stock prices when earnings are retained and reinvested in their company, there is uncertainty as to whether the benefits will actually be realized. However, if the stockholders were to receive the earnings now, in the form of dividends, they could invest them now in whatever they desired. In other words, "a bird in the hand is worth two in the bush."

If the bird-in-the-hand theory is correct, then stocks of companies that pay relatively high dividends will be more popular—and, therefore, will have relatively higher stock prices—than are stocks of companies that reinvest their earnings.

Modigliani and Miller's Dividend Theory

Franco Modigliani and Merton Miller (commonly referred to as M&M) theorized in 1961 that dividend policy is irrelevant.[5] Given some simplifying assumptions, M&M showed how the value of a company is determined by the income produced from its assets, not by its dividend policy. According to the **M&M dividend theory,** the way a firm's income is distributed (in the form of future capital gains or current dividends) doesn't affect the overall value of the firm. Stockholders are indifferent as to whether they receive their return on their investment in the firm's stock from capital gains or dividends—so dividends don't matter.

M&M's arguments have been critiqued for decades. Most often, financial theorists who disagree with M&M maintain that M&M's assumptions are unrealistic. The validity of a theory, however, lies with its ability to stand up to tests of its predictions. The results of these tests are mixed, and modern financial theorists continue to argue as to what dividend policy a company should pursue.

The Mechanics of Paying Dividends

We've seen how the board of directors decides whether the firm will pay a dividend. Next, let's consider what happens when companies pay dividends and the timing of those payments.

The board's decision to pay a dividend is called *declaring* a dividend. This occurs on the **declaration date.** At that date a liability, called **dividends payable,** is created on the firm's balance sheet.

Because the common stock of public corporations typically is traded every day in the marketplace, the board of directors must select a cutoff date, or **date of record,** to determine who will receive the dividend. At the end of business on this date, the company stockholder records are checked. All owners of the common stock at that time receive the forthcoming dividend.

When stock is traded on an exchange or in the over-the-counter market, it takes several days to process the paperwork necessary to record the change of ownership that occurs when the stock changes hands. On the date of record, then, the company's transfer agent will not yet know of stock trading that occurred in the days immediately preceding the date of record.

The **transfer agent** is the party, usually a commercial bank, that keeps the records of stockholder ownership for a corporation. The transfer agent pays dividends to the appropriate stockholders of record after the company has deposited the required money with the transfer agent.

Because it takes time for news of a stock trade to reach the transfer agent, the rules of trading dictate that two days before the date of record, common stock that has an upcoming dividend payment will begin to trade **ex-dividend.** (The prefix *ex* is a Latin word meaning "without.") Investors who buy the stock on or after the ex-dividend date will be buying it "without" entitlement to the forthcoming dividend. The two-day period gives exchange officials enough time to notify the transfer agent of the last batch of stock trades that occurred before the ex-dividend period. The extra time ensures that the stockholder records will be correct on the date of record.

---

[5]Merton Miller and Franco Modigliani, "Dividend Policy, Growth, and the Valuation of Shares," *Journal of Business* (October 1961): 411–33.

Figure 16-2 **The Dividend Payment Time Line**

This figure shows the sequence of events for a dividend payment.

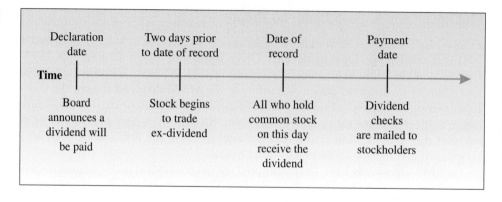

TABLE 16-3

Meditrust Company Dividend Payment Sequence

Dividend announced	July 9, 1998
Stock begins trading ex-dividend	July 29, 1998
Date of record	July 31, 1998
Payment date	August 8, 1998

Source: Meditrust News Releases July 9 and July 31, 1998.

A few weeks after the date of record, the checks are mailed out to the common stockholders. The date of mailing is called the **payment date.**

Figure 16-2 shows a time line for the dividend payment sequence.

Table 16-3 summarizes the dividend payment sequence for the Meditrust Company's second-quarter 1998 dividend. The table shows the sequence of events that led to the payment of a 1998 second-quarter dividend of $61.625 cents per share.

Dividend Reinvestment Plans

Many corporations offer a **dividend reinvestment plan (DRIP),** under which stockholders reinvest their dividends rather than receive them in cash. DRIPs are popular because they allow stockholders to purchase additional shares of stock without incurring the commission costs that accompany regular stock purchases made through a stockbroker. The new shares, including fractional shares, are purchased at the price prevailing in the market at that time. The amount of the dividend paid and reinvested is still taxable income to the stockholder.

ALTERNATIVES TO CASH DIVIDENDS

Sometimes corporations want to give something to the stockholders even though there is insufficient cash available to pay a cash dividend. At other times corporations don't want to pay a cash dividend because they want to build up their cash position. Let's look at some of the options available for giving stockholders something without using precious cash.

Stock Dividends and Stock Splits

Instead of sending out checks to stockholders, the firm could issue additional shares. Many stockholders view the receipt of these extra shares as a positive event, similar

Financial Management and You

Dividend Reinvestment Records Can Avoid Tax Headaches

If you sign up for a dividend reinvestment plan, be sure to keep good tax records. The IRS treats the dividends reinvested in new shares of stock as though you had received the cash and used that cash to buy the shares. Here are some record-keeping tips to help you avoid tax headaches.

1. **Know how often your dividends are reinvested.** Most companies that pay dividends do so quarterly. This means that under a reinvestment plan, you will be buying new shares every quarter, including fractional shares.
2. **Know how each share is taxed.** When you sell your shares, you will have a capital gain in the amount of the sales price minus the basis for each share, including fractional shares. Over 10 years you could have shares with 40 different tax bases (10 years times four quarters per year), not including the basis of the original shares you bought. Keep track of each basis.
3. **Keep your records.** You should never throw away the records establishing the basis for your dividend reinvestment shares until you sell them (and not for several years after that in case you get audited). These records may be in your files for decades.

---

to the receipt of a check. In the sections that follow we will question the validity of this widely held view.

New shares of common stock can be distributed to existing shareholders with no cash payment in two different ways: stock dividends and stock splits. Both increase the number of shares of common stock outstanding, and neither raises additional equity capital. There is an accounting difference between them, however. Let's examine these two alternatives to cash dividends next.

Stock Dividends When a **stock dividend** is declared, the existing common stockholders receive new shares, proportionate to the number of shares currently held. This is usually expressed in terms of a percentage of the existing holdings, such as 5 percent, 10 percent, or 20 percent, but usually less than 25 percent. For example, if a 20 percent stock dividend were declared, one new share would be sent out to existing stockholders for every five currently owned. The following example illustrates how the process works and how the transaction would be accounted for on the firm's balance sheet.

Suppose that Bob and Bill own a chain of bed and breakfast lodges called BB Corporation. Now suppose the BB Corporation declares a 20 percent stock dividend so that each stockholder receives 20 new shares for every 100 shares they hold. This will increase the total number of shares outstanding from 100,000 to 120,000. Assume that the market price of BB's stock at the time of this stock dividend is $16. The equity section of BB's balance sheet, before and after the 20 percent stock dividend, is shown in Table 16-4.

First, note in Table 16-4 that after the 20 percent stock dividend the common stock account changed from $100,000 to $120,000, an increase of 20 percent or $20,000. This change occurred because as of January 1, 2000, the firm had 20,000 more shares outstanding and each new share had the same $1 par value as the old shares (20,000 shares × $1 = $20,000). Next, note that the capital in excess of par account changed from $900,000 to $1,200,000, an increase of $300,000. This increase happened because the new shares were issued when the market price of the stock

TABLE 16-4

BB Corporation Capital Account as of December 31, 1999

Common Stock (100,000 shares, $1 par)	$ 100,000
Capital in Excess of Par	$ 900,000
Retained Earnings	$5,000,000
Total Common Stock Equity	$6,000,000

BB Corporation Capital Account as of January 1, 2000 (after a 20% stock dividend)

Common Stock (120,000 shares, $1 par)	$ 120,000
Capital in Excess of Par	$1,200,000
Retained Earnings	$4,680,000
Total Common Stock Equity	$6,000,000

was $16, which is $15 in excess of the $1 par value. The $15 "in excess" figure is multiplied by the 20,000 additional shares to get the $300,000 increase in the capital in excess of par account ($15 × 20,000 shares = $300,000). The $20,000 increase in the common stock account and the $300,000 increase in the capital in excess of par account total $320,000.

Finally, note that the retained earnings account changed from $5 million to $4,680,000, a decrease of $320,000. This change reflects the transfer of $20,000 to the common stock account and $300,000 to the capital in excess of par account ($20,000 + $300,000 = $320,000). Retained earnings must decrease because the 20 percent stock dividend did not alter the firm's total equity capital.

Thus, we see that a stock dividend is just an accounting transfer from retained earnings to the other capital accounts. The number of shares of common stock in-

Source: *Reprinted with special permission of King Features Syndicate.*

creased by 20,000 in this case, but the overall economic effect was zero. Neither profits nor cash flows changed, nor did the degree of risk in the firm. The firm's ownership "pie" was cut into more pieces, but the pie itself is the same size.

On receiving their new shares of stock in the mail, some stockholders may think they have received something of economic value because they have new stock certificates for which they did not pay. But who issued these new stock certificates? The corporation did. Who owns the corporation? The stockholders do. The stockholders gave themselves new shares of stock, but continue to hold the same percentage of the firm as they held before the stock dividend.

The price per share of common stock will drop because of the increase in the number of shares outstanding. The price decrease happens or the total market value of the common stock would increase (price per share × number of shares outstanding) while no economically significant event has occurred.

Adjustment of a Stock's Market Price after a Stock Dividend. Investors in the stock market generally recognize that a stock dividend simply increases the number of shares of a firm, but it does not otherwise affect the total value of the firm. After a firm declares a stock dividend, then, the market price of the firm's stock will adjust accordingly. Table 16-5 illustrates the expected stock price adjustment for BB Corporation.

With the information from Table 16-5, we can find the new price of BBs' stock is as follows:

$$\text{New number of shares: } 120,000$$
$$\text{Total value of firm's stock: } \$1,600,000$$

Now let X = the new stock price. We solve for the new stock price as shown:

$$120,000 \text{ X} = \$1,600,000$$
$$\text{X} = \$13.3333 \text{ (round to } \$13.33)$$

Here's the key to solving for the new stock price after a stock dividend: Remember that the total value of all the firm's stock remains the same as it was before the stock dividend. There may be a positive effect on the stock price due to expectations about the cash dividend. We explore this effect later in the chapter.

Stock Splits When an increase of more than 25 percent in the number of shares of common stock outstanding is desired, a corporation generally declares a *stock split* instead of a stock dividend. The firm's motivation for declaring a stock split is generally different from that for a stock dividend. A stock dividend appears to give stockholders something in place of a cash dividend. A **stock split** is an attempt to bring the firm's stock price into what management perceives to be a more popular trading range.

Stock splits are usually expressed as ratios. A *4–1 (four-for-one) stock split,* for example, indicates that new shares are issued such that there are four shares after the split for every one share before the split. In a 3–2 split, there would be three shares after the split for every two outstanding before the split.

TABLE 16-5

BB Corporation Stock Price Adjustment in Response to a 20% Stock Dividend

	Number of Shares Outstanding	×	Price per Share	=	Total Value of the Firm's Stock
Before Stock Dividend	100,000	×	$16	=	$1,600,000
After Stock Dividend	120,000	×	?	=	$1,600,000

The accounting treatment for a stock split is simpler than for a stock dividend. Table 16-6 shows how the BB Corporation would account for a 4–1 stock split.

Note in Table 16-6 that only the common stock entry in the equity section of the balance sheet is affected by the stock split. The entry indicates that after the stock split, there are four times as many shares of common stock outstanding as before the split. It also indicates that the par value of each share is one-fourth the value it was before (100,000 shares × 4 = 400,000 shares and $1 par value ÷ 4 = $.25). The total dollar value of $100,000 remains the same. Capital in excess of par and retained earnings accounts are completely unaffected by a stock split.

Adjustment of a Stock's Market Price after a Stock Split. As with a stock dividend, investors in the stock market recognize that a stock split simply increases the number of shares of a firm, but does not otherwise affect the total value of the firm. Therefore, the market price of the firm's stock will adjust accordingly following a stock split. Table 16-7 illustrates the stock price adjustment for the BB Corporation in response to the 4–1 stock split.

We see in Table 16-7 that just as with the stock dividend, no economically significant event has occurred. The common stock ownership "pie" has been cut into four times as many pieces as before, but the size of the pie is the same. As Table 16-7 shows, after the 4–1 split, each share will trade in the market at approximately one-fourth the price it commanded before the split.

Sometimes the cash dividend is increased at the same time as a stock split or a stock dividend. In our 4–1 stock split example, if the cash dividend per share is decreased by less than three-fourths, stockholder cash dividends received would increase. A $1 per share cash dividend, for instance, could be cut to $0.30 instead of the $0.25 value that would leave total cash dividends unchanged. The market sometimes anticipates such an increase in the cash dividend, leading to a possible increase in the total market value of the common stock.

The Rationale for Stock Splits. One of the most famous stock splits is the Berkshire Hathaway stock split of 1996. Before the split, Berkshire Hathaway shares were selling for $34,000 a share. Because the shares soared in value from $18 per share in 1965 to $34,000 in 1996, institutional investors began to offer—for a fee—thousand-dollar securities based on Berkshire stock. To halt others from making money off of the company's stock, Berkshire decided to offer some Baby Berkshire stock shares in a 30–1 stock split.[6]

TABLE 16-6

BB Corp. Capital Account as of Dec. 31, 1999 (before a 4–1 stock split)

Common Stock (100,000 shares, $1 par)	$ 100,000
Capital in Excess of Par	$ 900,000
Retained Earnings	$5,000,000
Total Common Stock Equity	$6,000,000

BB Corp. Capital Account as of Jan. 1, 2000 (after the 4–1 stock split)

Common Stock (400,000 shares, $.25 par)	$ 100,000
Capital in Excess of Par	$ 900,000
Retained Earnings	$5,000,000
Total Common Stock Equity	$6,000,000

---

[6]Allan Sloan, "Stop Kicking Yourself," *Newsweek* (April 8, 1996): 47.

TABLE 16-7

BB Corporation Stock Price Adjustment Due to a 4–1 Stock Split

	Number of Shares Outstanding	×	Price per Share	=	Total Value of the Firm's Stock
Before Stock Dividend	100,000	×	$16	=	$1,600,000
After Stock Dividend	400,000	×	$4	=	$1,600,000

Unlike Berkshire, which split some of its stock to prevent nonowners from dealing in Berkshire securities, many companies split stock to increase its value. Managers believe that as the market price per share of their common stock increases over time, it gets too expensive for some investors. Management perceives that a stock split will decrease the price per share, thereby increasing the number of potential investors who can afford to buy it. More potential investors might create additional buying pressure that results in an increase in stock price. This argument is less persuasive, however, as the percentage of stock ownership and trading activity by institutional investors (mutual funds, pension funds, insurance companies, and so on) continues to increase. These investors can afford to pay a very high price per share. Managers, however, continue to use stock splits to adjust the stock price to a lower level to make it more affordable.

What's Next

In this chapter we looked at how firms make their dividend decisions. This chapter ends part 3 on long-term financial management decisions. Part 4 focuses on short-term financial management decisions. In chapters 17–20 we will examine how firms manage the three primary current assets: cash, inventory, and accounts receivable.

Summary

1. **Explain the importance of and identify the factors that influence the dividend decision.**

Stock prices often change dramatically when a dividend change is announced, indicating that the market believes dividends affect value. The dividend decision, then, must be carefully planned and implemented. Factors that influence dividend decisions are a company's need for funds, its future financial prospects, stockholder preferences and expectations, and the firm's contractual obligations.

Although dividends are often discussed in the context of a firm's earnings, dividends are paid in cash. As a result, a firm's cash flow is a crucial factor affecting its dividend policy.

2. **Compare the major dividend theories.**

The major dividend theories that help guide dividend policy include the residual theory of dividends, the clientele theory, the signaling theory, the bird-in-the-hand theory, and the Modigliani and Miller theory. The residual theory posits that the amount of dividends matter less than the amount of earnings retained. If a firm enacts a residual policy, its dividend payments are likely to be unpredictable and erratic.

The clientele dividend theory assumes that one of the key reasons investors are attracted to a particular company is its dividend policy. Under this theory, it doesn't matter what the dividend policy is so long as the firm sticks to it.

The signaling dividend theory is based on the premise that management knows more about the future finances of the firm than do stockholders, so dividends signal the firm's future prospects. A dividend decrease signals an expected downturn in earnings; a dividend increase signals a positive future is expected. Managers who believe the signaling theory will be conscious of the message their dividend decision may send to investors.

The bird-in-the-hand theory assumes that stockholders prefer to receive dividends instead of having earnings reinvested in the firm on their behalf. If correct, stocks of companies that pay relatively high dividends will be more popular and, therefore, will have relatively higher stock prices than stocks of companies that reinvest their earnings. The Modigliani and Miller theory claims that earnings from assets, rather than dividend policy, affect firm value, so dividend policy is irrelevant.

3. Describe how firms make dividend payments.

Dividends are declared by the corporate board of directors. Stockholders on the date of record are entitled to the declared dividend. Investors who buy the stock before the ex-dividend date (two days before the date of record) will receive the dividend declared by the board. Investors who buy the stock on or after the ex-dividend date have bought the stock too late to get the dividend.

4. Identify alternatives to paying cash dividends.

Corporations often award stockholders stock dividends (additional shares of stock) and stock splits instead of cash dividends. Both stock splits and stock dividends increase the number of outstanding shares and decrease the price per share. Although neither of these actions is a real substitute for a cash dividend, many investors perceive these events as good news even though earnings do not increase nor does risk decrease.

You can access your FREE PHLIP/CW On-Line Study Guide, Current Events, and Spreadsheet Problems templates, which may correspond to this chapter either through the Prentice Hall Finance Center CD–ROM or by directly going to: **http://www.prenhall.com/gallagher.**

PHLIP/CW — *The Prentice Hall Learning on the Internet Partnership–Companion Web Site*

Equations Introduced in This Chapter

Equation 16-1. The Dividend Payout Ratio:

$$\text{Dividend Payout Ratio} = \frac{\text{Dividends Paid}}{\text{Net Income}}$$

Self-Test

ST-1. A company pays $4 per share in dividends and has 100,000 shares outstanding. The company has net income of $1 million. Its market price per share of $100. What is its dividend payout ratio?

ST-2. Is the amount of the dividend to be paid the primary focus of the board of directors if the board is guided by the residual theory of dividends? Explain.

ST-3. Does a company following the clientele theory of dividends pay high or low dividends? Explain.

ST-4. When paying dividends, what is the date of record?

ST-5. If a company has a $2 par value, what would be the accounting effect of a 4–1 stock split?

Review Questions

1. Explain the role of cash and of earnings when a corporation is deciding how much, if any, cash dividends to pay to common stockholders.
2. Are there any legal factors that could restrict a corporation in its attempt to pay cash dividends to common stockholders? Explain.
3. What are some of the factors that common stockholders consider when deciding how much, if any, cash dividends they desire from the corporation in which they have invested?
4. What is the Modigliani and Miller theory of dividends? Explain.
5. Do you believe an increased common stock cash dividend can send a signal to the common stockholders? If so, what signal might it send?
6. Explain the bird-in-the-hand theory of cash dividends.
7. What is the effect of stock (not cash) dividends and stock splits on the market price of common stock? Why do corporations declare stock splits and stock dividends?

Build Your Communication Skills

CS-1. Find a company that has recently cut its dividend. Write a report on the market's reaction to this decision and the timing of that reaction. What are some possible explanations for this market reaction?

CS-2. Many people perceive a stock split as a very positive event. Find a company that has gone through a stock split. Write a short report on the market's reaction to the stock split around the time it was announced and at the time the split actually occurred. Were they different? Were they what you would expect? Explain.

Problems

16-1. After discussion with the board of directors of the company, Lionel Mandrake, founder and chairman of Mandrake, Inc., decided to retain $600 million from its net income of $1 billion. Calculate the payout ratio.   *Payout Ratio*

16-2. The net income of Harold Bissonette Resorts, Inc. was $50 million this past year. The company decided to have a 40 percent payout ratio. How much was paid in dividends and how much was added to the retained earnings?   *Payout Ratio*

16-3. Hannah Brown International maintains a dividend policy with a constant payout ratio of 30 percent. In the last three years, the company had the following earnings.   *Constant Payout Ratio and Retained Earnings*

	Year 1	Year 2	Year 3
Net Income ($ million)	30	20	25

What is the total addition to retained earnings over the last three years?

Constant Dollar Dividend Policy

16-4. Use the same data given in problem 16-3. Now, if the company followed a constant dollar dividend policy and paid $10 million in dividends each year, compute the dividend payout ratios for each year and the total addition to retained earning over the last three years.

Residual Theory of Dividend Policy

16-5. Eliza Doolittle, the chief financial officer of East West Communications Corporation, has identified $14 million worth of new capital projects that the company should invest in next year. The optimal capital structure for the company is 40 percent debt and 60 percent equity. If the expected earnings for this year is $10 million, what amount of dividend should she recommend according to residual theory?

Residual Theory of Dividend Policy

16-6. Use the same data given in problem 16-5. Now, what would be the amount of dividend that could be paid if East West's net income for this year was:

a. $16 million?
b. $6 million?

Stock Dividend

16-7. Jan Brady, chief accountant of Mulberry Silk Products, is trying to work out the feasibility of a 20 percent stock dividend. The equity section of the balance sheet follows:

	($ 000s)
Common Stock (2 million shares, $1 par)	2,000
Capital in Excess of Par	8,000
Retained Earnings	10,000
Total Common Equity	20,000

The current market price of the company's stock is $31 per share. Is it possible to pay a 20 percent stock dividend? Is it possible to pay a 10 percent stock dividend? Explain.

Stock Dividend

16-8. Use the same data given in problem 16-7. After payment of a 10 percent stock dividend, what will be the expected market price of the stock? Also, show how the equity section of the balance sheet will change.

Stock Split

16-9. Wesley Crusher, chief accountant of Blue Sky Cruise Lines, is trying to figure out the effect of a 3–1 stock split. The equity section of the balance sheet follows:

	($ 000s)
Common Stock (3 million shares, $1.00 par)	3,000
Capital in Excess of Par	7,000
Retained Earnings	10,000
Total Common Equity	20,000

The current market price of the company's stock is $33 per share. If Blue Sky Cruise Lines decided to have a 3–1 stock split, how would the equity section change after the split? What would be the stock's market price?

Stock Split

16-10. Use the same information given in problem 16-9. If Blue Sky Cruise Lines' net income is $800,000, what is the earnings per share before and after the 3–1 stock split? Will there be any change in the price to earnings ratio before and after the stock split?

16-11. Sumner Outdoor Equipment Company has decided to go for a 5–1 stock split. The common shareholders received a dividend of $1.33 per share after the split.

a. If the dollar amount of dividends paid after the split is same as that paid last year before the split, what was the dividend per share last year?
b. If the dollar amount of dividends paid after the split is 10 percent higher than what was paid last year before the split, what was the dividend per share last year?

Challenge Problem

16-12. Market price of Linden Landscaping Company's stock is $30 the day before the stock goes ex-dividend. The earnings of the company are $10 million, and the company follows a dividend policy with constant payout ratio of 40 percent. There are 1 million shares of common stock outstanding. What would be the new ex-dividend price of the stock?

16-13. In its strategic plan for the next five years, Springfield Manufacturing Company has projected the following net income and capital investments (figures in $000s):

Long-Term Dividend (Residual Theory)

Year	Net Income	Investments
2001	1,000	800
2002	1,100	1,000
2003	1,200	2,000
2004	1,300	800
2005	1,400	1,000

The capital structure the company wishes to maintain is 40 percent debt and 60 percent equity. There are currently 500,000 shares of common stock outstanding.

If you own 500 shares of common stock, calculate the amounts you would receive in dividends over the next five years (2001 to 2005), assuming that the company uses the residual dividend theory each year to determine the dividend to be paid to its common stock holders.

16-14. Use the same data given in problem 16-13. Now assume that the company plans to issue 100,000 new shares of common stock in the year 2003 at $6 per share ($1 par plus $5 capital in excess of par). What will be the dividend that you would receive in the years 2003, 2004, and 2005 assuming your common stock holding remains same at 500 shares?

Long-Term Dividend (Residual Theory)

Comprehensive Problem

16-15. The equity section of the balance sheet of Cafe Vienna is given next:

	($ 000s)
Common Stock (500,000 shares, $3 par)	1,500
Capital in Excess of Par	3,500
Retained Earnings	5,000
Total Common Equity	10,000

The company earned a net income of $3 million this year. Historically, the company has paid dividends with a constant payout ratio of 50 percent. The stock will sell at $47 after the ex-dividend date.

William Riker, the vice president of finance for Cafe Vienna, is considering all possible ways to increase the company's earning per share (EPS). One possibility he is weighing is to buy back some of the company's outstanding shares of common stock from the market using all its net income earned this year without paying any dividend to its common stockholders.

a. Determine the repurchase price of the common stock.
b. Calculate the number of shares that could be repurchased using this year's net income.
c. Show the changes in the equity section of the balance sheet after the repurchase.
d. If net income next year is expected to be $4 million, what would be the EPS next year with and without the repurchase?
e. If you own 50 shares of the company's common stock, would you like the company's decision of buying back the stock instead of paying a dividend?

Answers to Self-Test

ST-1. Dividend Payout Ratio = Dividends Paid ÷ Net Income = ($4/share × 100,000 shares) ÷ $1,000,000 = $400,000 ÷ $1,000,000 = .40 = 40%

ST-2. No. According to the residual theory of dividends, the amount of earnings that should be retained is determined first. Whatever amount is not retained is paid out in dividends.

ST-3. A company following the clientele theory of dividends might have either high or low dividend payments. If the stockholders preferred high dividends, that is what would be paid. If they preferred low dividends, that would be the policy.

ST-4. The date of record is the date on which a company checks its stockholder records. Investors listed in the records on that date are entitled to receive the dividend that was recently declared.

ST-5. $2 original par value ÷ 4 (4 for 1 split) = $.50 new par value

PART 4:
SHORT-TERM FINANCIAL MANAGEMENT DECISIONS

CHAPTER 17

WORKING CAPITAL POLICY

Ready money is Aladdin's lamp.
— Lord Byron

WORKING CAPITAL AS AN ENDANGERED SPECIES

It's one thing to have a thriving business; it's another to have cash on hand to pay those daily bills. Tom Buschman, owner and CEO of a $3 million manufacturer of parts for paper machinery, thought he had a lot of cash. So he decided to expand his production facilities. Whoops! That year his company computed its taxes at the last minute and discovered it owed the IRS $15,000. The company didn't have the cash because it had spent the money on the expansion. To make matters worse, its biggest customer had filed for bankruptcy.

"I was really scared. We didn't have a credit line. I didn't know how we'd pay off that tax bill." Buschman's company was saved by a strong flow of orders, but he learned a hard lesson about having enough working capital available to meet daily expenses and emergencies.

For Roxanne Coady, owner of a small bookstore, the problem was small items that were eating too deeply into her working capital, combined with low inventory turnover that kept her cash flow down. The store finally showed a profit when she cut back on such items as the $2,500 annual expense for shopping bags and increased inventory turnover by 50 percent.

CHAPTER OBJECTIVES

After reading this chapter, you should be able to:

■ Explain the importance of managing working capital.

■ Discuss how the trade-off between liquidity and profitability affects a firm's current asset management policy.

■ Describe how a firm reaches an optimal level of current assets.

■ Discuss the effects of the three approaches to working capital financing policy.

Roxanne Coady and Tom Buschman both learned a hard lesson: Working capital must be there when you need it.

Sources: Jill Andresky Fraser, "The Tax-Wise Cash Monitor," *Entrepreneur Magazine* (January 1995): 75; Roxanne Coady, "The Cobbler's Shoes," *Inc. Magazine* (January 1996): 21; Brent Bowers, "This Store Is a Hit But Somehow Cash Flow Is Missing," *The Wall Street Journal* (April 13, 1993): B2.

Chapter Overview

In this chapter we examine *working capital policy*—the management of a firm's current assets and its financing. First, we'll see why firms manage working capital carefully, why they accumulate it, and how to classify current assets. We then investigate what determines a firm's working capital policy, and look at different types of policies.

Managing Working Capital

Working capital refers to a firm's current assets. By "current" we mean those assets that the firm expects to convert into cash within a year. Current assets include *cash*; *inventory*, which generates cash when the items are sold; and *accounts receivable*, which produces cash when customers pay off their credit accounts. Current assets are considered liquid because they can be transformed into cash in a relatively short time.

Net working capital is the firm's current assets minus current liabilities. Current liabilities are business obligations (i.e., debts) that the firm plans to pay off or otherwise satisfy within a year. Examples include *accounts payable*—bills due soon—and *notes payable*—loans due to be paid in less than a year.

Table 17-1 shows the working capital and net working capital for Green World Lawn Care Products Company, which manufactures lawn and gardening products and sells them to retailers.

In Table 17-1 we see that Green World has $13,000 in current assets (working capital), $11,000 in current liabilities, and $2,000 in net working capital ($13,000 − $11,000 = $2,000). Net working capital is important to firms. It represents the amount of current assets remaining if they were liquidated to pay the company's short-term debts.

Working capital policy is the firm's policy about its working capital level and how its working capital should be financed. For instance, a firm needs to make decisions about how much to keep in its cash account, what level of inventory to maintain, and how much to allow accounts receivable to build up. The firm must also decide whether to finance current assets with short-term funds, long-term funds, or

TABLE 17-1

Green World Lawn Care Products Company Balance Sheet, as of December 31, 1999

Cash	$ 2,000	Accounts Payable		$ 7,000
Accounts Receivable	1,000	Notes Payable		4,000
Inventory	10,000	Total Current Liabilities		11,000
Total Current Assets	13,000	Other Liabilities		7,000
Other Assets	32,000	Common Stock		27,000
Total Assets	$45,000	Total Liabilities and Equity		$45,000

some mixture of the two. Together, the level and financing decisions make up the firm's working capital policy.

WHY BUSINESSES ACCUMULATE WORKING CAPITAL

Why do firms accumulate working capital, and why does its level vary over time? In this section we examine the answers to these two questions.

Fluctuating Current Assets

Many factors affect a firm's working capital policy. For instance, a service firm may require a different level of current assets than a manufacturing firm. Or a business like Jason's Popcorn Wagon business (from chapter 13), which makes and sells popcorn during the summer months only, has different working capital needs than a firm that makes products year round.

To illustrate the principles of working capital policy, we focus on a manufacturing firm that has level production—that is, it produces the same amount of product every month, year round. However, its sales are seasonal—the firm sells more in certain time periods than in others. Many businesses are seasonal. (For instance, a swimwear manufacturer may sell many more swimsuits at the start of summer than it does in other months. A lawn care products manufacturer will probably have more sales at the start of the gardening season than it will in other months.)

If a business has level production but not level sales, inventory increases when production exceeds sales. Inventory then falls when sales exceed production. The firm's other current assets may fluctuate during the year as well. Accounts receivable, for example, will rise when new credit sales exceed customer payments, and will fall when customer payments exceed new credit sales. Cash will accumulate as sales revenues are collected and will decline when bills are paid. Thus, the current assets of the business will fluctuate over time.

Permanent and Temporary Current Assets

Although the level of current assets in the firm may fluctuate, it rarely reaches zero. The firm will nearly always have some cash on hand, hold some inventory in stock, and be owed some amount of money. Current assets thus reach various temporary levels, but will rarely fall below some minimal permanent level. This effect is illustrated for Green World Lawn Care Products Company in Figure 17-1.

Figure 17-1 shows three categories of business assets that affect a firm's working capital policy:

1. **Temporary current assets** represent the level of inventory, cash, and accounts receivable that fluctuate seasonally.
2. **Permanent current assets** represent the base level of inventory, cash, and accounts receivable, which tends to be maintained.
3. **Fixed assets** represent the land, buildings, equipment, and other assets that would not normally be sold or otherwise disposed of for a long period of time.

Permanent current assets tend to build up on a firm's balance sheet year after year. Cash collections increase as the business grows, accounts receivable grow as the list of credit customers lengthens, inventory on hand rises as new facilities are opened, and so on. Figure 17-2 shows how Green World Lawn Care Products' current assets might vary over several years.

The term **permanent current assets** *may sound like an oxymoron (like "jumbo shrimp"), but it's not. A portion of a firm's current assets are likely to remain on a firm's balance sheet indefinitely.*

Figure 17-1 **The Variation in Current Assets over Time for Green World Lawn Care Products**

Figure 17-1 shows how the current assets of a company tend to fluctuate over time, but never fall below a permanent level of current assets.

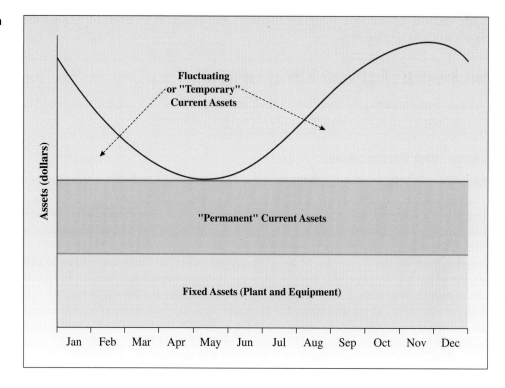

As Figure 17-2 shows, businesses have two tasks: First, they must contend with current assets that fluctuate through their business cycle. Second, they must manage permanent current asset growth due to long-term business growth over time. In the sections that follow, we discuss how firms do this.

Figure 17-2 **The Variation in Current Assets over Several Years for Green World Lawn Care Products**

Figure 17-2 shows how a typical firm's current assets tend to build up from year to year, while fluctuating within each year. The effect occurs because as firms grow, they accumulate more cash, accounts receivable, and inventory over time.

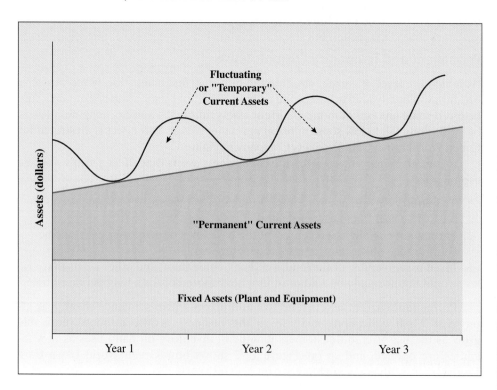

Liquidity versus Profitability

Lenders would like a company to have a large excess of current assets over current liabilities. But the owners don't necessarily feel the same way. Think about it. Current assets—in the form of cash, accounts receivable, and inventory—do not earn the firm a very high return. Cash is usually held in a commercial bank checking account that pays no interest.[1] Accounts receivable earns no return because it represents money that customers owe to the firm that the firm doesn't have yet. Inventory earns no return until it's sold. (Inventory being held by the firm is just material sitting in a warehouse, earning nothing.) These assets have the advantage of being liquid, but holding them is not very profitable.

Now consider the company's noncurrent assets—its land, buildings, machinery, equipment, and long-term investments. These assets can earn a substantial return. The company's land, buildings, machinery, and equipment are used to turn raw material into products that can be sold for a profit. Long-term investments (such as investments in subsidiaries) generally produce greater returns than current assets. These noncurrent assets may be profitable, but they are usually not very liquid. Lenders are reluctant to let firms use them for collateral (protection in the event of a default) for short-term loans. Why? Because lenders will have to spend more time and expense to sell noncurrent assets if firms default on their loans and the lenders become the asset owners. As a result, lenders prefer that firms use liquid assets as loan collateral.

Firms, then, are faced with a trade-off in their working capital management policy. At one extreme they can seek liquidity, holding a lot of cash and other current assets in case cash is needed soon. At the other extreme they can seek profitability, holding a low level of current assets and investing primarily in long-term, high-return-producing assets.

In practice, no firm would actually choose either of these extreme positions. Instead, managers seek a balance between liquidity and profitability that reflects their desire for profit and their need for liquidity.

Establishing the Optimal Level of Current Assets

The search for a balance between liquidity and profitability serves as a general guide for financial managers looking for an optimal level of current assets. However, the level managers eventually achieve is actually a result of their efforts to maintain optimal levels of each of the *components* of the current asset group. In other words, a firm's optimal level of current assets is reached when the optimal level of cash, of inventory, and of accounts receivable[2] is achieved. Each asset

---

[1] Business checking accounts almost never pay interest.
[2] Along with any other current assets the firm possesses, of course. In this book we concentrate on these three major categories of current assets.

account is managed separately, and the combined results produce the actual level of current assets. Here's a description of the attempt to find the optimal level for each current asset account:

- *Cash:* Managers try to keep just enough cash on hand to conduct day-to-day business, while investing extra amounts in short-term marketable securities. We discuss cash in detail in chapter 18.
- *Inventory:* Managers seek the level that reduces lost sales due to lack of inventory, while at the same time holding down inventory costs. We discuss inventory management in chapter 19.
- *Accounts receivable:* Firms want to enhance sales but hold down bad debt and collection expenses through sound credit policies. We discuss credit policies in chapter 19.

Once financial managers set policies to attain the optimal level of current assets, they must turn their attention to the flip side of working capital management: managing current liabilities.

Managing Current Liabilities: Risk and Return

A firm's current asset fluctuations and any long-term build-up of current assets (its working capital) must, of course, be financed. The question facing financial managers is whether to obtain the financing from short-term borrowing, long-term borrowing, contributions from the owners (equity financing), or some mixture of all three.

Take Note

In our discussion we assume that a firm has all available financing alternatives options. In practice, however, some firms may have limited financing options. For example, small firms usually have limited access to long-term capital markets.

As is so often the case, the choice of the firm's working capital financing blend depends on managers' desire for profit versus their degree of risk aversion. Short-term debt financing is generally less expensive than long-term debt and always less expensive than equity financing. Recall from chapter 2 that short-term interest rates are usually lower than long-term interest rates. Short-term loans, however, are more risky because a firm may not have enough cash to repay the loans (due to cash flow fluctuations) or interest rates may increase and increase the cost of short-term funds as loans are renewed. Long-term debt and/or equity financing is less risky from the firm's perspective because it puts repayment off (forever, in the case of stock) and locks in an interest rate for a long time period.

The balance between the risk and return of financing options depends on the firm, its financial managers, and its financing approaches. We discuss several financing approaches next.

Three Working Capital Financing Approaches

The three primary working capital financing approaches are the *aggressive* approach, the *conservative* approach, and the *moderate* approach. A firm that takes an aggressive approach uses more short-term financing to finance current assets. Firm risk increases, due to the risk of fluctuating interest rates, but the potential for higher returns increases because of the generally low-cost financing. A firm that implements the conservative approach avoids short-term financing to reduce risk but decreases the potential for maximum value creation because of the high cost of long-term debt and equity financing. The moderate approach tries to balance risk and return concerns.

The Aggressive Approach

We know that short-term interest rates are normally lower than long-term interest rates. We also know that borrowing short term is riskier than borrowing long term because the loan must be paid off or refinanced sooner rather than later.

The **aggressive working capital financing approach** involves the use of short-term debt to finance at least the firm's temporary assets, some or all of its permanent current assets, and possibly some of its long-term fixed assets. The aggressive approach is shown graphically in Figure 17-3.

If we compare current assets and current liabilities in Figure 17-3, we see that all the firm's temporary current assets and most of its permanent current assets are being financed with short-term debt (the current liabilities). As a result, the firm has very little *net* working capital. Depending on the nature of the firm's business, this small amount of net working capital can be risky. There isn't much cushion between the value of liquid assets and the amount of debt due in the short term.

Firms may be more aggressive than the firm depicted in Figure 17-3. If the firm's managers financed *all* working capital from short-term debt, then current assets would equal current liabilities and the firm would have zero net working capital—no cushion at all. Managers may go even further and finance a portion of the firm's long-term assets (plant and equipment) with short-term debt, creating a *negative* net working capital. However, such an approach is *very* risky. (Think what would happen to a firm using that approach if short-term interest rates rose unexpectedly.)

What tempts financial managers to take the aggressive approach and use a relatively large amount of short-term debt for working capital financing? Usually lower interest rates tempt them. Managers will take a risk if the promise of return is high enough to justify it.

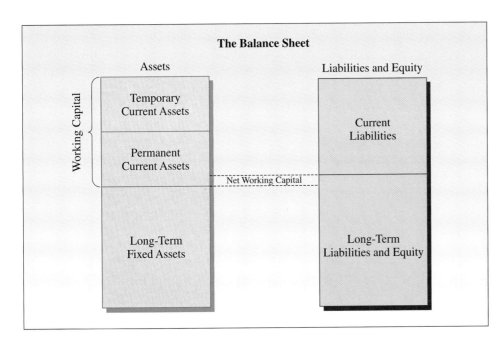

Figure 17-3 The Aggressive Working Capital Financing Approach

Figure 17-3 shows the firm's assets on the left, and liabilities and equity on the right. Subtracting current liabilities from current assets shows that the firm's working capital financing strategy is to finance nearly all current assets with current liabilities, resulting in a small amount of net working capital.

The Conservative Approach

Borrowing long term is considered less risky than borrowing short term. This is because the borrower has a longer time to use the loan proceeds before repayment is due. Furthermore, if interest rates should go up during the period of the loan, the long-term borrower has another advantage. The long-term borrower has locked in a fixed interest rate and may end up paying less total interest than the short-term borrower, who must renew the loan each time it comes due—at a new, higher interest rate. If market rates fall, the long-term borrower can usually refinance.

The **conservative working capital financing approach** involves the use of long-term debt and equity to finance all long-term fixed assets and permanent current assets, in addition to some part of temporary current assets. The conservative approach is shown graphically in Figure 17-4.

Compare current assets to current liabilities in Figure 17-4. Note that all the firm's permanent current assets and most of its temporary current assets are being financed with long-term debt or equity. As a result, current assets exceed current liabilities by a wide margin and the firm has a large amount of *net* working capital. Having a large amount of net working capital is a relatively low-risk position because the firm has many assets that could be liquidated to satisfy short-term debts.

A financial manager who applies an ultra-conservative approach would use cash from the owners to finance all asset financing needs (high cash balance supported by equity), and incur no debt. By using only equity capital, the firm would also have the maximum amount of net working capital possible because it would have no current liabilities.

The safety of the conservative approach has a cost. Long-term financing is generally more expensive than short-term financing. So relying on long-term debt and equity sources to finance working capital consumes funds that could otherwise be put to more productive use.

Figure 17-4 **The Conservative Working Capital Financing Approach**

The relative size of the current asset and current liability accounts in Figure 17-4 reveals the firm's working capital financing strategy. The figure shows that the firm is financing nearly all current assets with long-term liabilities and equity, resulting in a high level of net working capital.

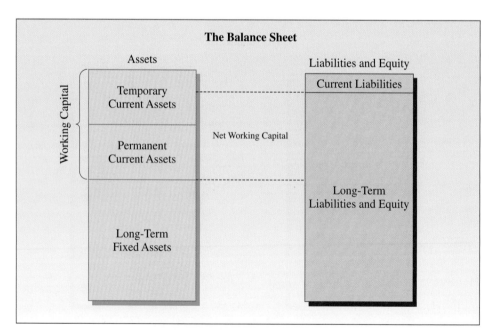

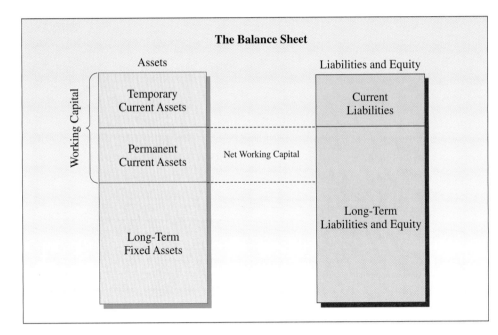

Figure 17-5 **The Moderate Working Capital Financing Approach**

In Figure 17-5 we see that the firm's approach to working capital financing policy is to finance its permanent current assets with long-term debt and equity, and its temporary current assets with current liabilities. This results in a moderate level of net working capital.

The Moderate Approach

An accounting concept known as the *matching principle* states that the cost of an asset should be recognized over the length of time that the asset provides revenue, or benefit, to the business.

The concept of the matching principle can be applied to define a moderate position between the aggressive and the conservative working capital financing approaches. According to the matching principle, temporary current assets that are only going to be on the balance sheet for a short time should be financed with short-term debt—that is, current liabilities. Permanent current assets and long-term fixed assets that are going to be on the balance sheet for a long time should be financed from long-term debt and equity sources. The **moderate working capital financing approach** is shown in Figure 17-5.

If we look at current assets and current liabilities in Figure 17-5, we see that the firm has matched its short-term temporary current assets to its current liabilities. It has also matched its long-term permanent current assets and fixed assets to its long-term financing sources. This policy gives the firm a moderate amount of net working capital. It calls for a relatively moderate amount of risk balanced by a relatively moderate amount of expected return.

Now that we have described three working capital financing policies, we turn to an analysis of the effect of such policies on a firm.

Working Capital Financing and Financial Ratios

The use of *ratio analysis* highlights how the three approaches to working capital financing policy can affect the risk and return potential of a firm. In Table 17-2 we compare selected financial ratios for three different firms that differ only in the manner in which they finance their working capital. Firm A takes the aggressive approach, Firm C takes the conservative approach, and Firm M takes the moderate approach.

TABLE 17-2

Ratio Analysis of Approaches to Working Capital Financing Policy (in thousands)

DATA AS OF THE END OF THE LAST FISCAL YEAR	FIRM A AGGRESSIVE	FIRM C CONSERVATIVE	FIRM M MODERATE
Temporary Current Assets	$ 200	$ 200	$ 200
Permanent Current Assets	$ 400	$ 400	$ 400
Fixed Assets	$ 600	$ 600	$ 600
Total Assets	$1,200	$1,200	$1,200
Current Liabilities	$ 300	$ 100	$ 200
Long-Term Debt	$ 300	$ 500	$ 400
Stockholders' Equity	$ 600	$ 600	$ 600
Total Liabilities and Equity	$1,200	$1,200	$1,200
Net Income for the Year	$ 126	$ 114	$ 120
Net Working Capital	$ 300	$ 500	$ 400
Current Ratio	2.0	6.0	3.0
Total Debt to Total Assets Ratio	50%	50%	50.0%
Return on Equity	21%	19%	20.0%

TABLE 17-3

The Three Approaches to Working Capital Financing Policy—Cost and Risk Factors

	AGGRESSIVE	CONSERVATIVE	MODERATE
Cost	Low	High	In Between
Risk	High	Low	In Between

Notice in Table 17-2 that Firm A, which follows an aggressive financing approach, has the highest net income, smallest amount of net working capital, the lowest current ratio, and the highest return on stockholders' equity of any of the three firms. This is consistent with the relationship between risk and return (the more risk a firm takes, the more return it earns). There is no guarantee, of course, that net income will be positive.

Firm C, which follows the conservative financing approach, has the lowest net income, the largest amount of net working capital, the highest current ratio, and the lowest return on stockholders' equity of the three firms. This reflects its relatively lower risk and lower return potential.

Firm M, which follows the moderate approach of matching its short-term temporary current assets to its current liabilities is, of course, in a position between the other two.

Table 17-3 summarizes the cost and risk considerations of the aggressive, conservative, and moderate approaches to working capital financing.

In the real world, of course, each firm must decide on its balance of financing sources and its approach to working capital management based on its particular industry and the firm's risk and return strategy.

WHAT'S NEXT

In this chapter we examined the general working capital policy of a firm. In chapter 18 we will look at cash. Accounts receivable and inventory will follow in chapter 19, and short-term financing in chapter 20.

FINANCE AT WORK

SALES·RETAIL·SPORTS·MEDIA TECHNOLOGY·PUBLIC RELATIONS·PRODUCTION·EXPORTS

Interview with Michael Coleman, Vice President of Tek Soft

Michael Coleman

Michael Coleman, an engineering major in college, now serves as the executive vice president of Tek Soft, a leading international technical software developer for large companies. Before joining Tek Soft, Michael was a marketing manager for Doefer Engineering, a capital-intensive company that designed and built machines for industry. His comments show how working capital policies and working capital requirements for new, information-based companies differ greatly from the capital requirements of traditional manufacturers.

Q. How important is managing working capital for a manufacturing company?

Very important. In manufacturing, working capital is a huge factor because a substantial portion of your costs can be directly tied to working capital, especially inventory costs. If you only have a 10 percent profit margin, the management of working capital becomes a primary concern. If you mismanage it, your 10 percent profit margin can easily be erased by changing interest rates, paying off loans at the wrong time, or other basic working capital management problems. You can't even deliver a product to the client unless you have enough working capital. So you must manage it well.

For example, if you are building a new machine for a client and you receive payments in installments, those payments don't cover the money you use to build the machine. You have to pay a lot of money to buy raw materials—machine bases, metal, stainless steel, copper, and wire. Then the client company you're building the machine for sometimes holds off paying for as long as possible, so you don't receive cash quickly from the sale even though you've paid cash to build the machine.

So a manufacturer must watch payment schedules carefully to ensure that clients pay within a reasonable time limit, and must also watch how much to build in advance given the cost of the inventory items. The firm also has to be prepared for the cash from a sale coming later than the time of the actual sale. All these factors affect working capital management.

Q. How important is managing working capital to a software development company?

It's not as important as it is to a manufacturing company. In the software development business, the cost of our company's current assets is about $80,000 a year. We have about $7 million in retail sales a year.

We don't need to have huge capital investments in raw materials because disks and the paper for software manuals are relatively inexpensive. We are working mainly with people's ideas. So working capital levels are low because we take those ideas, a disk worth about 90 cents, a couple of manuals, and we sell them for $10,000 to $12,000 apiece.

Q. Is it difficult to finance your current assets in the software business?

Our financing options are limited. It's tough to finance our assets in the software development business because the only asset we really have is our source code—the unique way we put together the binary code to make our products. If we go to a bank and say, "We have a unique way of putting a binary code together, lend us money," the bank would probably say, "You're nuts. We can't verify that your source code has any value."

A machine builder, on the other hand, has assets that the bank can see. It has buildings, machine tools, inventory, work in progress, and a loyal customer base. These assets are easier for lenders to value than a software source code.

Source: Interview with Michael Coleman.

Summary

1. Explain the importance of managing working capital.

A firm's current assets—such as cash, inventory, and accounts receivable—are referred to as working capital. Managing the levels and financing of working capital effectively is necessary to keep the firm's costs and risk under control while maintaining a firm's returns and cash flow over the long term.

Firms accumulate working capital because of fluctuations in sales, production, and cash or credit payments. For instance, cash accumulates as accounts are collected and declines when bills are paid. Inventory builds when production exceeds sales and falls when sales exceed production. Accounts receivable rises as credit sales are made and falls when customers pay off their accounts. The combined effect of changes in each current asset account causes working capital to fluctuate. Furthermore, working capital will gradually build up over time unless the firm takes some concrete action to either reinvest the funds in long-term assets or distribute them to the firm's owners.

Temporary current assets represents the level of current assets that fluctuates, and permanent current assets represents the level of current assets a firm keeps regardless of periodic fluctuations.

2. Discuss how the trade-off between liquidity and profitability affects a firm's current asset management policy.

Current asset management involves a trade-off between the need for liquidity and the desire for profitability. The more current assets a firm holds, the more liquid the firm because the assets can be converted to cash relatively quickly. However, tying up funds to sustain a certain level of current assets prevents the funds from being invested in long-term, high-return producing assets.

3. Describe how a firm reaches an optimal level of current assets.

Firms reach an optimal level of current assets when the optimal level for each individual current asset account (mainly cash, inventory, and accounts receivable) is achieved. Separate techniques exist for managing each of the current asset accounts. The techniques for managing each current asset account are described in chapters 18 and 19.

4. Discuss the effects of the three approaches to working capital financing policy.

Short-term interest rates are usually lower than long-term interest rates, so financing working capital with short-term debt generally lowers the firm's financing costs. However, using short-term debt increases the risk that cash won't be available to pay the loans back, and that the firm may have to renew its loans at higher interest rates. Relying on long-term debt and/or equity sources to finance working capital decreases risk because firms repay such obligations in the long term, and firms lock in an interest rate. A firm's return suffers, however, because long-term financing costs are normally higher than short-term costs.

The aggressive approach to working capital policy consists of financing all temporary current assets, and some or all long-term permanent current assets, and possibly a portion of fixed long term assets with short-term debt. The conservative approach to working capital policy consists of financing all permanent current assets and some short-term temporary current assets with long-term debt and/or equity financing. The moderate approach consists of financing temporary current assets with short-term debt, and financing long-term permanent current assets with long-term debt and/or equity.

Companion Website
content by Active Learning Technologies

You can access your FREE PHLIP/CW On–Line Study Guide, Current Events, and Spreadsheet Problems templates, which may correspond to this chapter either through the Prentice Hall Finance Center CD–ROM or by directly going to: **http://www.prenhall.com/gallagher**.

PHLIP/CW — *The Prentice Hall Learning on the Internet Partnership–Companion Web Site*

Self-Test

ST-1. Explain the liquidity–profitability trade-off associated with working capital management.

ST-2. How is the optimal level of working capital established?

ST-3. Using the following financial data, draw a diagram showing the company's temporary current assets, permanent current assets, and long-term fixed assets during this last year.

Selected Financial Data for Past Fiscal Year (in thousands)

	Jan 1	Mar 31	Jun 30	Oct 31	Dec 31
Total Assets	$580	$480	$280	$400	$580
Fixed Assets	$100	$100	$100	$100	$100

ST-4. As of today a company has $100,000 of temporary current assets, $50,000 of permanent current assets, and $80,000 of long-term fixed assets. If the company follows the moderate approach to working capital financing, how much of its assets will be financed with short-term debt (current liabilities) and how much will be financed with long-term debt and/or equity as of today?

Review Questions

1. What is working capital?
2. What is the primary advantage to a corporation of investing some of its funds in working capital?
3. Can a corporation have too much working capital? Explain.
4. Explain how a firm determines the optimal level of current assets.
5. What are the risks associated with using a large amount of short-term financing for working capital?
6. What is the matching principle of working capital financing? What are the benefits of following this principle?
7. What are the advantages and disadvantages of the aggressive working capital financing approach?
8. What is the most conservative type of working capital financing plan a company could implement? Explain.

Build Your Communication Skills

CS-1. Research a business you find interesting. Prepare a short report outlining at least two ways that company could increase its liquidity. Consider both asset management and liability management.

CS-2. Choose any company for which you can locate balance sheet data for the last five years. Plot on a graph the company's total assets for each year and connect the data points with a line. Next, plot the company's net fixed assets for each year and connect the data points with a line. Finally, sketch a line that just touches the low points reached by the total asset line across the graph.

You should now have a graph similar to Figure 17-2. Label the appropriate areas of your graph as temporary current assets, permanent current assets, and fixed assets.

Refer to the company's latest balance sheet and note the amount of current liabilities and long-term debt and equity shown. Compare these amounts with the ending amount of current and fixed assets shown on your graph. The results will reveal the company's working capital financing policy. Explain your findings to the class.

Problems

Assessing Liquidity

17-1. Consider the following two companies:

	Company A	Company B
Cash	$1,000	$ 80
Accounts Receivable	400	880
Net Fixed Assets	1,500	1,620
	$2,900	$2,580
Accounts Payable	$ 900	$ 600
Long-Term Debt	800	1,100
Common Equity	1,200	880
	$2,900	$2,580

Which of the two firms is the more liquid? Why?

Current Assets, Current Liabilities, and Net Working Capital

17-2. Capt. Louis Renault's Hikewell Outdoor Equipment Company has the following balance sheet accounts as of the end of last year. One-half current assets are permanent and one-half are temporary.

Cash	$ 30,000	Accounts Payable	$100,000
Accounts Receivable	15,000	Notes Payable	60,000
Inventory	130,000	Long-Term Debt	90,000
Fixed Assets	500,000	Common Equity	425,000
	$675,000		$675,000

a. What is the amount of the company's current assets, its working capital?
b. What is the amount of the company's current liabilities?
c. What is the amount of the company's net working capital?
d. What percentage of current assets is financed by current liabilities? Would you consider this an aggressive approach or a conservative approach?

Conservative Working Capital Financing Approach

17-3. Alexander Sebastian, the finance manager of Hikewell Outdoor Equipment Company of problem 17-2, thinks the way the company is financing its current assets is too risky. By the end of next year, he would like the *pro forma* balance sheet to look as follows:

Cash	$ 30,000	Accounts Payable	$ 30,000
Accounts Receivable	15,000	Notes Payable	20,000
Inventory	130,000	Long-Term Debt	150,000
Fixed Assets	500,000	Common Equity	475,000
	$675,000		$675,000

a. What is the amount of the company's projected current assets, its working capital?
b. What is the amount of the company's projected current liabilities?
c. What is the amount of the company's projected net working capital?
d. What percentage of current assets is projected to be financed by current liabilities? Would you consider this an aggressive approach or a conservative approach?

17-4. Consider the following balance sheet for Lulu Belle's Killer Guard Dogs, Inc.:

Assessing Working Capital Policy

Assets		Liabilities and Equity	
Cash	$ 50	Accounts Payable	$ 80
Marketable Securities	0	Short-Term Debt	90
Accounts Receivable	40	Long-Term Debt	210
Inventories	70	Common Equity	310
Net Fixed Assets	530		$690
	$690		

a. How much working capital does Lulu Belle have?
b. How much net working capital does Lulu Belle have?
c. What working capital financing policy (aggressive, moderate, or conservative) is Lulu Belle following?
d. Explain what actions Lulu Belle could take to increase the company's liquidity.

17-5. Marian Pardoo, the chief financial officer of Envirosafe Chemical Company, believes in a moderate approach of financing following the matching principle. Some of the projected balance sheet accounts of the company for the end of next year follow:

Moderate Working Capital Financing Approach

	Current and Fixed Assets	Permanent Current Assets
Cash	$ 30,000	$15,000
Accounts Receivable	15,000	5,000
Inventory	130,000	80,000
Fixed Assets	500,000	
Total Assets	$675,000	

Liabilities	
Accounts Payable	$ 20,000
Short-Term Debt	?
Long-Term Debt	?
Common Equity	450,000
Total Liabilities and Equity	$675,000

How much should Marian finance by short-term debt and long-term debt to conform to the matching principle?

17-6. Use the same data given in problem 17-5. Marian's boss, Ann Lowell, is the vice president of finance of Envirosafe Chemical Company and she expects interest rates to decrease in the future and, hence, would like to follow a very aggressive policy using a large amount of short-term debt and a small amount of long-term debt. She would also like to decrease net

Aggressive Working Capital Financing Approach

working capital to $25,000. How much should Ann finance by short-term debt and how much by long-term debt to conform to her aggressive approach?

Different Working Capital Financing Approaches

17-7. Comparative data at the end of this past year for three firms following aggressive, moderate, and conservative approaches to working capital policy follow (in thousands of dollars):

	Aggressive	Moderate	Conservative
Temporary Current Assets	$ 75	$ 75	$ 75
Permanent Current Assets	100	100	100
Fixed Assets	500	500	500
Total Assets	675	675	675
Current Liabilities	160	75	50
Long-Term Debt	90	150	150
Stockholders' Equity	425	450	475
Net Income	70	70	70

Calculate, compare, and comment on the current ratios, total debt to asset ratios, and returns on equity of the three firms.

Challenge Problem

17-8. Greenplanet Recycling Company is considering buying an additional facility at a cost of $500,000. The facility will have an economic life of five years. The company's financial officer, Karen Holmes, can finance the project by:

a. a five-year loan at an annual interest rate of 13 percent
b. a one-year loan rolled over each year for five years

Compare the total interest expenses for both the preceding alternatives under the following assumptions, and calculate the savings in the interest expenses by choosing one of two alternatives:

(i) the one-year loan has a constant interest rate of 11 percent per year over the next five years.
(ii) the one-year loan has an annual interest rate of 11 percent in the first two years, 14 percent in the third and fourth years, and 16 percent in the fifth year.

The Matching Principle

17-9. To analyze your company's working capital financing policy, you have gathered the following balance sheet data for the past 12 months (in thousands):

Date	Total Assets	Fixed Assets
Jan 31	$45	$14
Feb 28	46	14
Mar 31	34	14
Apr 30	48	14
May 31	40	14
Jun 30	30	14
Jul 31	28	14
Aug 31	39	14
Sep 30	45	14
Oct 31	39	14
Nov 30	52	14
Dec 31	50	14

Plot these data on a trendline graph. Indicate the amount of your firm's current liabilities each month if you follow the matching principle.

Comprehensive Problem

17-10. Milton Warden, the finance manager of WinHeart Gift Company, is analyzing past data on the firm's fixed assets, permanent current assets, and temporary current assets for each month over the last five years. The company maintains level production but its sales are seasonal. He found that the monthly level of the three types of assets over the last five years can be closely approximated by the following patterns (in thousands of dollars):

- Fixed assets remained constant at 39 each month over the last five years.
- Permanent current assets were equal to 2 in January of year 1 and had grown 0.16 per month each month over the last five years.
- Temporary current assets followed the same pattern each year, starting at 0 in January, then each year they increased by 1 monthly until July and reduced by 1 monthly until they reached 0 again in January of the next year.
  a. Plot these data on a graph similar to the one shown in Figure 17-2.
  b. Calculate and identify on the graph the level of temporary current assets, permanent current assets, and fixed assets in:
     (i) the month of September of year 4
     (ii) the month of August of year 5
  c. Now calculate the levels of current liabilities in those months if the company followed:
     (i) an aggressive working capital financing approach
     (ii) a moderate working capital financing approach
     (iii) a conservative working capital financing approach

Assume that the stockholders' equity remained constant at 20 over those five years.

17-11. Buddy Love, a finance officer of Christmas Tree Ornaments and Gifts Company, is analyzing past data on the fixed assets, permanent current assets, and temporary current assets of the company each month over the last several years. The company maintains level production but its sales peak at the end of the year. He found that the monthly level of the three types of assets over the last five years can be closely approximated by the following equations (in thousands of dollars):

Comprehensive Problem

- Fixed assets were equal to 55 and remained constant each month over the last five years.
- Permanent current assets were equal to 10 in January of year 1, and had grown 0.30 per month each month over the last five years.
- Temporary current assets = $m(m-1)/4$, where m = 1, 2, . . . , 12 for Jan, Feb, . . . , Dec, respectively.
  a. Plot these data on a graph similar to the one shown in Figure 17-2.
  b. Calculate and identify on your graph the level of temporary current assets, permanent current assets, and fixed assets at the following times:
     (i) the month of September of year 2
     (ii) the month of October of year 4
     (iii) year 5's minimum and maximum levels of total assets and the months those levels occurred
  c. Now calculate the levels of current liabilities in the months described in part b of this problem if the company followed:
     (i) an aggressive working capital financing approach
     (ii) a moderate working capital financing approach
     (iii) a conservative working capital financing approach

Assume that the stockholders' equity remained constant at 30 over those five years.

Answers to Self-Test

ST-1. Working capital (i.e., current assets in the form of cash, accounts receivable, inventory, and so on) can normally be exchanged for cash, or liquidated, in a relatively short time. Therefore, the more working capital a company maintains, the easier it is to raise cash quickly. However, maintaining working capital costs money and it ties up funds that could otherwise be used to invest in long-term income-producing assets, so that profits and returns suffer. For this reason, we say that managing working capital involves balancing liquidity and profitability.

ST-2. The optimal level of working capital is achieved when the optimal levels of each current asset account—cash, inventory, accounts receivable, and any others—are reached. Each current asset category is managed separately and the combined results produce the optimal level of current assets.

ST-3.

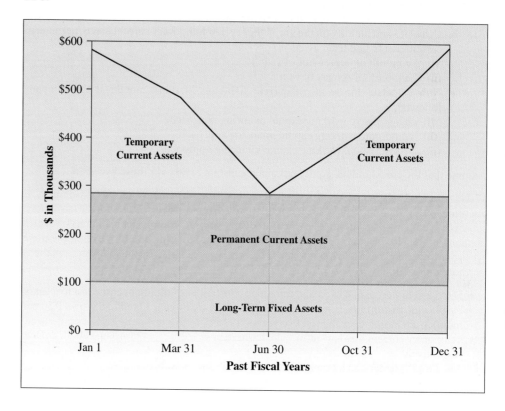

ST-4. Following the moderate approach, the company will finance its temporary current assets with short-term debt and the rest of its assets with long-term debt and equity. Therefore, as of today, the company will have on its balance sheet:

Current Liabilities	$100,000
Long-Term Debt and Equity	$130,000

CHAPTER 18

MANAGING CASH

CHAPTER OBJECTIVES

After reading this chapter, you should be able to:

- List the factors that affect a company's desired minimum cash balance.

- List the factors that affect a company's maximum cash balance.

- Apply the Miller–Orr model to establish a target optimal cash balance.

- Prepare a cash budget.

- Explain how firms manage their cash inflows and outflows to maximize value.

I've never been poor, but I've been broke.
—Mike Todd

CASH FLOW—THE BUSINESSES' LIFEBLOOD

Starting a small business requires more than a good idea and a truckload of energy to back it up. There are disciplines to follow, which could make the difference between riches and bankruptcy.

According to the small business spokesman for the Australian Society of Certified Practicing Accountants, Mr. Greg Hayes, one of the most fundamental yet overlooked disciplines in the running of a business is cash flow and how it is managed.

Mr. Hayes said a quarter of all bankruptcies cited cash flow problems as the cause of the business failure, and time and again accountants found themselves dealing with owners of failing businesses who did not understand the difference between cash flow and profit.

"If people don't know what cash flow is, then they are unlikely to know why it's important, let alone be able to manage it."

He said cash flow was the cycle of money moving in and out of the business. It was distinct from profit and loss statements, which were based on accounting standards and were generally historical views of a business.

For instance, a company that invoiced $100,000 on June 30 could—under accounting standards—declare a profit for that year. In terms of cash flow management, the invoice may not be paid for 60 days—or may never be paid, even though profit/loss had already counted it.

Mr. Hayes said the great majority of . . . small business owners had no idea about cash flow in general or how their own cash cycles operated.

"When you speak to so many small business owners, and you ask about their cash management policies, most of them just stare blankly. Others pull out their bank statements and they think they're on top of things."

Small businesses whose profit/loss statements suggested they were not in too bad shape could actually be failing. The problem—cash flow—had been distorted by oversights and ignorance. The most common distorters of cash flow data were creditor and debtor accounts. A business owner's bank statements could show $200,000 in the bank, which tempted them to spend cash on plant and equipment.

"They get a lot of money in the bank and they forget that there are suppliers to pay."

Suddenly, the creditors' reminder notices were arriving but the money had been spent. Businesses then went into overdraft to pay suppliers and wages. Once in the long-term overdraft trap, small businesses often never left it and were forced to start paying taxes from their overdraft accounts.

A proper policy for debtors was also essential so they didn't delay payment too long, thus distorting cash flow. A good policy was to not have any one client accounting for more than 10 percent of the business's income and not to deal with a business whose credit worthiness was uncertain, no matter how big the contract.

Source: Mark Abernethy, "Cash Flow—The SME's Lifeblood," *Australian Financial Review* (October 20, 1998): 35.

Chapter Overview

In this chapter we look at how firms manage cash. Cash flow management can differ greatly from firm to firm, as shown in the chapter opening. Here we start by exploring factors that affect a company's optimal cash balance and learn how to estimate the optimal balance. Then we examine how firms forecast their cash needs, develop a cash budget, and manage cash inflows and outflows.

Cash Management Concepts

> **Take Note**
>
> *In this chapter the term* **cash** *refers to dollar amounts in the firm's checking account at a commercial bank, and to coin and currency in the cash drawer. Interest is not generally paid on business checking accounts and, of course, coin and currency earn no interest when in the cash drawer.*

Whether they work in a large multinational corporation or a small business, financial managers need to know how much cash to keep on hand. Cash management may sound simple. Shouldn't businesses accumulate as much cash as possible? It's not that easy. Recall from chapter 17 that cash earns no return for the business owners. In fact, a business that accumulated as much cash as possible and did not invest in any assets would fail because it would earn no return for the stockholders.[1] Cash, then, should

---

[1] In fact, because there would be no investment in earning assets, such a business would never accumulate more cash than was originally contributed by the founders.

not be obtained for its own sake. Rather, it should be considered the "grease" that enables the machinery of the firm to run. Cash management is the process of controlling how much of this grease is needed and where and when to apply it.

DETERMINING THE OPTIMAL CASH BALANCE

To determine how much cash a firm should keep on hand, financial managers must:

- Maintain enough in the cash account to make payments when needed (minimum balance).
- Keep just the needed amount in the cash account so that the firm can invest excess funds and earn returns (maximum balance).

Let's examine the factors that affect the minimum cash balance and the maximum cash balance.

The Minimum Cash Balance

The size of a firm's minimum cash balance depends on three factors: (1) how quickly and cheaply a firm can raise cash when needed; (2) how accurately the firm's managers can predict when cash payment requirements will occur; and (3) how much precautionary cash the firm's managers want to keep to safeguard against emergencies. The effect of these three factors on the minimum cash balance is shown in Figure 18-1. We examine the three factors that affect a firm's minimum cash balance in the following sections.

Raising Cash Quickly When Needed If a firm's managers could obtain cash instantly whenever they needed it, at zero cost, they wouldn't need to maintain any balance in the cash account at all. All the firm's funds could be invested in short-term income-producing securities as soon as received. In the real world, of course, neither firms nor anyone else can borrow or sell assets to raise all the cash they want anytime, instantly, at zero cost. In practice, obtaining cash usually takes time and has a positive cost. Therefore, businesses maintain at least some cash in their checking accounts.

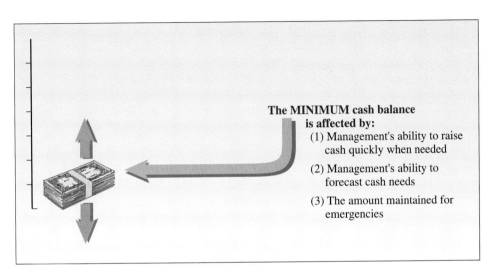

Figure 18-1 Factors Affecting the Minimum Cash Balance

Figure 18-1 shows the three factors that determine the minimum amount a firm will keep in its cash account.

Chapter 18 *Managing Cash* **441**

The question is, how much cash is enough? The answer is, only experience will tell. The more difficult or expensive it is to get cash when needed, the more a firm needs to keep in its checking account. At most—if cash were very difficult to obtain because the firm had no liquid assets, or if short-term interest rates were very high—the firm would want to keep enough cash on hand to cover all foreseeable needs until the next time the firm expects to receive more cash.

Predicting Cash Needs Cash flows can be volatile because of the business environment or the risk of the business. For instance, in a weak economy, people and firms pay bills more slowly. So even though sales may be strong, a firm may not have much cash. Also, in an economy with interest rate fluctuations or inflation, cash flow needs can vary suddenly because of economic factors. To protect against an uncertain business environment, firms may maintain extra cash to cover cash needs.

Similarly, the cash flows of a startup or high-risk business may vary because the company grows at uneven, often unpredictable rates. Managers, then, may have a tough time estimating the firm's cash needs with certainty. Such firms often keep extra cash to act as a buffer against cash flow volatility.

How much extra cash a firm keeps in its coffers to protect against uncertainty depends on two factors: how difficult and expensive it is to raise cash when needed, and how volatile the firm's cash flow patterns are.

Coping with Emergencies Most cash payments are expected and planned. But unforeseen emergencies may occur: storms, fires, strikes, riots, and most often, failure of business plans to materialize. These emergencies can cause unexpected, sometimes large, drains on a firm's cash. Insurance can help, but there is no substitute for having cash ready when you need it. Managers, then, assess the likelihood of potential emergencies, and how quickly and easily cash can be obtained in case of an emergency. They adjust their cash balances accordingly. The more *risk averse* managers are, the more precautionary cash they try to keep on hand for emergencies.

The Maximum Cash Balance

Suppose that a firm's managers decide they wish to keep at least $20,000 in the firm's cash account. The next question is, how much should be allowed to accumulate in the cash account before the excess is withdrawn and invested in something that produces a return? If the balance in the cash account is $20,001, for example, should a dollar be withdrawn and invested? Should $30,000 be allowed to accumulate before any is withdrawn and invested? The answer depends on three factors: (1) the available investment opportunities, (2) the expected return from these opportunities, and (3) the transaction cost of withdrawing cash and making an investment. The factors that affect a firm's maximum cash balance are summarized in Figure 18-2. We describe the three factors affecting the maximum cash balance in detail next.

We discussed types of short-term securities in chapter 2, pages 25–27.

Available Investment Opportunities All businesses have at least a few (and some have many) alternative short-term income-producing investments in which they could invest their cash. These range from money market mutual funds and CDs, to Eurodollars and commercial paper. The more opportunities a firm has, the sooner it will invest rather than allow cash to simply accumulate in the firm's checking account.

Expected Return on Investments The potential return on investments is just as important as the number of investments. If the expected return is relatively high,

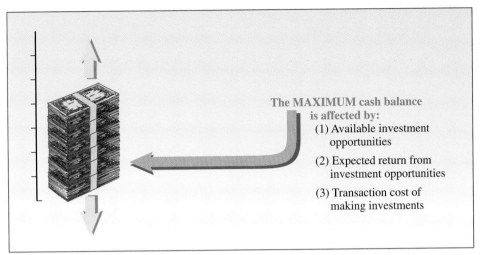

Figure 18-2 Factors Affecting the Maximum Cash Balance

Figure 18-2 shows the three factors that affect a firm's maximum cash balance.

firms will be quick to invest excess cash. If the expected return is relatively low, however, firms might let more cash accumulate before investing.

Transaction Cost of Making Investments Investing has costs. For instance, when you deposit money in a savings account, someone must search for information about and arrange for the transfer of the funds to the savings account. The search and implementation efforts take time. And that time has a cost.

Monetary and other costs of transferring cash into an investment are **transaction costs**—the costs associated with the transaction. Managers are interested in transaction costs because if the potential income from an investment does not exceed the cost of making the investment, then the investment is not worthwhile. Transaction costs also affect the frequency of a firm's investments. If transaction costs are relatively low, the firm will invest often, and will let only a small amount of excess cash accumulate in the cash account. Conversely, if transaction costs are relatively high, the firm will make fewer investments, letting a larger amount of cash accumulate in the meantime.

In this section we have seen that firms determine some minimum and maximum amount to keep in their cash accounts. The minimum amount is based on how quickly and cheaply firms can raise cash when needed, how accurately cash needs can be predicted, and how much precautionary cash a firm keeps for emergencies. The maximum amount depends on available investment opportunities, the expected returns from the investments, and the transaction costs of withdrawing the cash and making the investment.

The pattern of lottery ticket sales illustrates the principle of expected return on investments. When the jackpot is relatively low, ticket sales are sluggish. When the jackpot is relatively high, ticket sales increase.

Determining the Optimal Cash Balance

Financial theorists have developed mathematical models to help firms find an optimal "target" cash balance, between the minimum and maximum limits, that balances liquidity and profitability concerns. In the following sections, we discuss one of these models, the Miller–Orr model.

The Miller–Orr Cash Management Model In 1966 Merton Miller and Daniel Orr developed a cash management model that solves for an optimal target cash

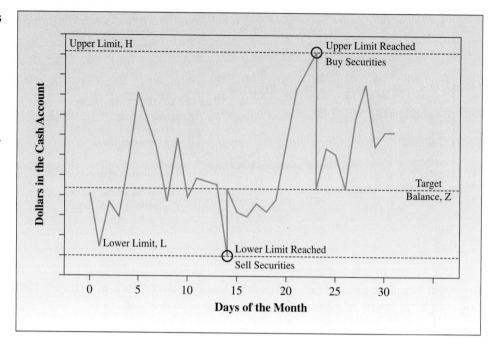

Figure 18-3 Cash Balances in a Typical Month Using the Miller–Orr Model

Figure 18-3 shows a firm's fluctuating cash flows, and its upper, lower, and optimal cash balances according to the Miller–Orr model.

balance about which the cash balance fluctuates until it reaches an upper or lower limit.[2] If the upper limit is reached, investment securities are bought, bringing the cash balance down to the target again. If the lower limit is reached, investment securities are sold, bringing the cash balance up to the target. Figure 18-3 shows the operation of the model.

The formula for the target cash balance Z shown in Figure 18-3 follows:

Miller–Orr Model
Formula for the Target Cash Balance (Z)

$$Z = \sqrt[3]{\frac{3 \times TC \times V}{4 \times r}} + L \tag{18-1}$$

where: TC = The transaction cost of buying or selling short-term investment securities
V = The variance of net daily cash flows
r = The daily rate of return on short-term investment securities
L = The lower limit to be maintained in the cash account

Notice in Figure 18-3 that the target cash balance, Z, is one-third of the way between the lower limit, L, and the upper limit, H. The Miller–Orr formula for the upper limit, H, is as follows:

Miller–Orr Model
Formula for the Upper Limit for the Cash Account

$$H = 3Z - 2L \tag{18-2}$$

---

[2]Merton Miller and Daniel Orr, "A Model of the Demand for Money by Firms," *Quarterly Journal of Economics* (August 1966): 413–35.

FINANCE AT WORK

SALES·RETAIL·SPORTS·MEDIA TECHNOLOGY·PUBLIC RELATIONS·PRODUCTION·EXPORTS

Karen Noble, Professional Golfer

Karen Noble

Karen Noble is a professional golfer and has been on the LPGA tour for five years. Her cash winnings have ranged from $20,000 one year to $115,000 another. With her income varying so greatly from year to year, Karen has taken several steps to ensure that she can forecast her cash needs and pay her expenses for a year of golf tournaments.

Q. How have you decreased the volatility of your cash flow?

Unlike many of the golfers on the LPGA tour, I have sponsors who advance me money at the beginning of the year. In exchange for the advance, I agree to repay my sponsors from my yearly winnings. That way, I know how much minimum cash I have for the year.

I also have one corporate sponsor that is a flat-out sponsorship. I don't have to pay them back. They pay me to advertise their company logo on my clothing and on my golf bag.

Q. How do you budget enough cash flow to cover the expenses of being on tour?

I construct my cash budget by first estimating my expenses. I start with the number of tournaments I'll play in for a year. I usually participate in 25 to 30 events. Once I've mapped out how many tournaments I plan to play in, I estimate how much it costs me to play in each tournament. Right now, tournament expenses range from $800 to $1,200 a week, so it averages out to about $1,000 a week. Once I have budgeted my expenses, I recruit sponsors, show them my estimated budget, and through them, I raise enough cash to cover the year's expenses.

Q. How do you estimate your cash flow?

Experience helps me estimate my cash needs. When I was just starting out, it was tough planning ahead. The biggest mistake I made was not realizing the amount of taxes I would have to pay. The year I won $115,000 I thought I could bank some of that for expenses down the road. But taxes took away the percentage I thought I could bank.

Now, after several years on the tour, I can gauge and budget for my cash needs. I have learned to minimize my expenses whenever possible. For instance, if I have to fly to a tournament, instead of drive, I try to use courtesy transportation instead of renting a car. And I don't necessarily stay in hotels. Often I can stay in private housing arranged by the tournament managers. Also, I don't always hire a caddy who tours with me. The average tour caddy costs between $400 and $450 a week. Instead, I might hire a local caddy who knows the golf course for about half the cost. One item I don't budget for is the cost of my golf equipment because the manufacturers provide it for free.

I then budget other large cash expenditures. Health insurance is one example. My computer upkeep and supplies are other examples. Without my laptop computer and printer, I can't keep track of my budget.

Q. What would be your ideal cash flow pattern?

To have a steady, reliable stream of cash. In my business, a complete corporate sponsorship provides that. A large company like AT&T signs you to a contract to represent them for a number of years. With a full sponsorship, I wouldn't have to spend the time and energy recruiting other sponsors each year or depend on the up-and-down earnings on the golf course.

Source: Interview with Karen Noble.

In the Miller–Orr model, the lower limit, L, is set by management according to the minimum cash balance concerns discussed earlier.

To illustrate how the Miller–Orr model works, assume that short-term investment securities are yielding 4 percent per year, and that it costs the firm $30 each time it buys or sells investment securities. Now assume that the firm's cash inflows and outflows occur irregularly, and the variance of the daily net cash flows has been found to be $90,846. Management wants to keep at least $10,000 in the cash account for emergencies, so L = $10,000. Under these circumstances, the firm will have the following target cash balance, according to Equation 18-1:

$$Z = \sqrt[3]{\frac{3 \times 30 \times \$90{,}846}{4 \times (.04/365)}} + \$10{,}000$$

$$= \sqrt[3]{\frac{\$8{,}176{,}140}{.00043836}} + \$10{,}000$$

$$= \$2{,}652 + \$10{,}000$$

$$= \$12{,}652$$

With a 4 percent annual return (converted to a daily figure), a lower limit of $10,000, transaction costs of $30, and a variance of $90,846, we see that the firm's target cash balance, Z, is $12,652.

According to Equation 18-2, the firm's upper limit for the target cash balance will be:

$$H = (3 \times \$12{,}652) - (2 \times \$10{,}000)$$

$$= \$37{,}956 - \$20{,}000$$

$$= \$17{,}956$$

According to the Miller–Orr model, then, the firm in this example will seek to maintain $12,652 in its cash account. If the cash balance increases to $17,956, the firm will buy $5,304 worth of investment securities to return the balance to $12,652. If the cash balance falls to $10,000, the firm will sell $2,652 worth of investment securities to raise the cash balance to $12,652. By finding the optimal cash balance, the firm seeks to accommodate its cash needs, given the volatility of cash inflows and outflows, and maximize its investment opportunities. We see, then, that the Miller–Orr model can help firms find their optimal cash balance.

Now that we have examined factors that affect a firm's minimum, maximum, and optimal cash balances, and described how a firm may find its target cash balance, we look at how a firm can forecast its cash needs next.

FORECASTING CASH NEEDS

Financial managers must frequently provide detailed estimates of their firm's future cash needs. The primary purpose of such estimates is to identify when excess cash will be available and when outside financing will be required to make up cash shortages.

Financial managers cannot base future cash estimates on *pro forma* income statements for the appropriate time period. Why? Because income and expenses are not always received and paid for in cash. Remember, if a firm is using an accrual-based accounting system, revenues and expenses may be recognized in one accounting period, but cash may not change hands until another period.

One technique for estimating true future cash needs is to develop a cash budget. A **cash budget** is a detailed budget plan that shows where cash is expected to come from and where it is expected to go during a given period of time.

The best way to learn about cash budgets is to practice creating one. In the following sections we will develop a cash budget for Bulldog Batteries that shows detailed cash flows from month to month throughout the upcoming year.

Developing a Cash Budget

The first step in developing a monthly cash budget is to identify sales revenue for each month of the period covered by the budget. Assume that it is the end of December 1999 and that Bulldog Batteries' 2000 sales are expected to occur as follows:

	SALES
Nov '99 (reference)	$ 13,441
Dec '99 (reference)	13,029
Jan '00	12,945
Feb '00	14,794
Mar '00	16,643
Apr '00	18,492
May '00	20,341
Jun '00	22,191
Jul '00	24,040
Aug '00	22,191
Sep '00	20,341
Oct '00	18,492
Nov '00	16,643
Dec '00	14,794
2000 Total	$221,907

Next, assume all of Bulldog's sales are on credit, so no cash is received immediately when a sale is made. Experience from past sales reveals that 30 percent of Bulldog's customers will pay off their accounts in the month of sale, 60 percent will pay off their accounts in the month following the sale, and the remaining 10 percent of the customers will pay off their accounts in the second month following the sale. Given this payment pattern, Bulldog's actual cash collections on sales throughout the year will follow the pattern shown in Table 18-1 on page 448.

In Table 18-1 we computed January's cash collections as follows:

30% of January's 2000's Sales:	$.30 \times \$12,945 =$	$ 3,884
+ 60% of December 1999's Sales:	$.60 \times \$13,029 =$	7,817
+ 10% of November 1999's Sales:	$.10 \times \$13,441 =$	1,344
= Total Collections in January 2000:		$13,045

Collections for the other months are computed similarly.

The next step in developing the cash budget is to turn to cash outflows. Assume that Bulldog Batteries' cost of materials is 27 percent of sales. Bulldog manufactures batteries expected to be sold in February one month ahead of time in January, and orders all the materials it needs for January's production schedule one month ahead of time in December. This schedule repeats for each month of the year. Bulldog makes all its purchases on credit and pays for them in cash during the month following the purchase. Therefore, December's purchase orders will be paid for in January, and so on. The situation is summarized in Figure 18-4 on page 449.

If Bulldog follows the production schedule illustrated in Figure 18-4 throughout 2000, its cash outflows for materials purchases will be as shown in Table 18-2 on page 448. (Materials purchases for November and December of 2000 are based on

Take Note

If Bulldog's managers expected other cash receipts during 2000, they would add them to sales collections in the appropriate month to obtain the total cash inflows for each month, as shown in Table 18-1.

TABLE 18-1 Bulldog's Actual Cash Collections

	1999		2000											
	Nov	Dec	Jan	Feb	Mar	Apr	May	Jun	Jul	Aug	Sep	Oct	Nov	Dec
Cash Inflows														
Sales (reference only; not a cash flow)	$13,441	$13,029	$12,945	$14,794	$16,643	$18,492	$20,341	$22,191	$24,040	$22,191	$20,341	$18,492	$16,643	$14,794
Cash collections on sales:														
30% in month of sale			$ 3,884	$ 4,438	$ 4,993	$ 5,548	$ 6,102	$ 6,657	$ 7,212	$ 6,657	$ 6,102	$ 5,548	$ 4,993	$ 4,438
60% in first month after sale			7,817	7,767	8,876	9,986	11,095	12,205	13,315	14,424	13,315	12,205	11,095	9,986
10% in second month after sale			1,344	1,303	1,295	1,479	1,664	1,849	2,034	2,219	2,404	2,219	2,034	1,849
Total collections			$13,045	$13,508	$15,164	$17,013	$18,862	$20,711	$22,561	$23,300	$21,821	$19,971	$18,122	$16,273
Other cash receipts			0	0	0	0	0	0	0	0	0	0	0	0
Total Cash Inflows			$13,045	$13,508	$15,164	$17,013	$18,862	$20,711	$22,561	$23,300	$21,821	$19,971	$18,122	$16,273

TABLE 18-2 Bulldog's Cash Outflows for Sales and Materials Purchased

	1999		2000											
	Nov	Dec	Jan	Feb	Mar	Apr	May	Jun	Jul	Aug	Sep	Oct	Nov	Dec
Sales (reference only; not a cash flow)	$13,441	$13,029	$12,945	$14,794	$16,643	$18,492	$20,341	$22,191	$24,040	$22,191	$20,341	$18,492	$16,643	$14,794
Materials purchases (27% of sales two months ahead—reference only, not a cash flow)		$ 3,994	$ 4,494	$ 4,993	$ 5,492	$ 5,992	$ 6,491	$ 5,992	$ 5,492	$ 4,993	$ 4,494	$ 3,994	$ 3,872	$ 3,994
Payments for materials purchases: 100% in month after purchase			$ 3,994	$ 4,494	$ 4,993	$ 5,492	$ 5,992	$ 6,491	$ 5,992	$ 5,492	$ 4,993	$ 4,494	$ 3,994	$ 3,872

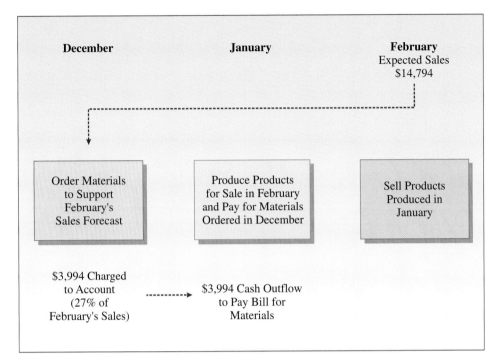

Figure 18-4 Timing for Bulldog Batteries' Cash Payments for Purchases

Figure 18-4 shows Bulldog Batteries' production and cash flow schedule.

sales forecasts for January and February 2001 of $14,342,000 and $14,794,000, respectively.)

For the sake of simplicity, assume Bulldog's remaining cash outflows are all direct expenses paid for in the month incurred as follows:

- Production expenses other than purchases are equal to purchases. (Bulldog's production costs are split evenly between materials cost and other production costs.)
- Sales and marketing expenses are 18.025 percent of sales each month.
- General and administrative expenses are $903,000 each month.
- Interest expense is expected to be $2,971 for the year. We assume that expense will be paid all at one time in December 2000.
- Bulldog's expected income tax bill for 2000 is $19,980. The bill will be paid in four installments in April, June, September, and December.
- Bulldog expects to declare four quarterly dividends of $6,098 in 2000. These will be paid in March, June, September, and December.

Bulldog's total cash outflows, including the preceding expenses and payments for materials purchases, are shown in Table 18-3 on page 450.

After all the cash inflows and outflows for each month are accounted for (as we have done, see Tables 18-1, 18-2, and 18-3), the next step is to summarize the *net* gain (or loss) for each month in 2000. Table 18-4 on page 450 contains the summary.

The final step in developing the cash budget is to summarize the effects of the monthly net cash flows on monthly cash balances and list any external financing required. The procedure for January 2000 is shown in Table 18-5 on page 451. We assume a cash balance of $65,313 at the beginning of January, a desired target cash balance of $65,000, and short-term loans of $302 outstanding at the beginning of the month.

TABLE 18-3 Bulldog's Total Cash Outflows

	1999	2000											
	Dec	Jan	Feb	Mar	Apr	May	Jun	Jul	Aug	Sep	Oct	Nov	Dec
Cash Inflows													
Materials purchases (reference only; not a cash flow)	$3,994	$4,494	$4,993	$5,492	$5,992	$6,491	$5,992	$5,492	$4,993	$4,494	$3,994	$3,872	$3,994
Payments for materials purchases: 100% in month after purchase		$3,994	$4,494	$4,993	$5,492	$5,992	$6,491	$5,992	$5,492	$4,993	$4,494	$3,994	$3,872
Other cash payments:													
Production costs other than purchases		3,994	4,494	4,993	5,492	5,992	6,491	5,992	5,492	4,993	4,494	3,994	3,872
Selling and marketing expenses		2,333	2,667	3,000	3,333	3,666	4,000	4,333	4,000	3,667	3,333	3,000	2,667
General and administrative expenses		903	903	903	903	903	903	903	903	903	903	903	903
Interest payments													2,971
Tax payments					4,995		4,995			4,995			4,995
Divided payments				6,098			6,098			6,098			6,098
Total Cash Outflows		$11,225	$12,557	$19,987	$20,215	$16,553	$28,978	$17,219	$15,887	$25,649	$13,223	$11,892	$25,378

TABLE 18-4 Bulldog's Cash Inflows and Outflows and Net Cash Flows

	2000											
	Jan	Feb	Mar	Apr	May	Jun	Jul	Aug	Sep	Oct	Nov	Dec
Total Cash Inflows	$13,045	$13,508	$15,164	$17,013	$18,862	$20,711	$22,561	$23,300	$21,821	$19,971	$18,122	$16,273
Total Cash Outflows	11,225	12,557	19,987	20,215	16,553	28,978	17,219	15,887	25,649	13,223	11,892	25,378
Net Cash Gain (Loss)	$1,820	$951	$(4,823)	$(3,203)	$2,309	$(8,267)	$5,341	$7,413	$(3,828)	$6,748	$6,230	$(9,105)

TABLE 18-5

Cash Flow and Financing Requirements Summary, January 2000

Cash Flow Summary:	
1. Cash balance at start of month	$65,313
2. Net cash gain (loss) during month	1,820
3. Cash balance at end of month before financing	67,133
4. Minimum cash balance desired	65,000
5. Surplus cash (deficit) (line 3 − line 4)	$ 2,133
External Financing Summary:	
6. External financing balance at start of month	$ 302
7. New financing required (negative amount on line 5)	0
8. Financing repayments (positive amount on line 5*)	302
9. External financing balance at end of month	0
10. Cash balance at end of month after financing (balance on line 3 + new financing on line 7 − repayments on line 8)	$66,831

*If the positive amount on line 5 exceeds the external financing balance on line 6, enter the external financing balance on line 6.

The procedure is repeated for February using January's cash balance at the end of the month after financing as the starting cash balance for February. Table 18-6 on page 452 contains the 12-month summary for Bulldog Batteries.

After filling out the cash budget through December 2000, Bulldog's managers can see that the firm will have surplus cash in January and February, but external financing will be needed in March, April, June, and September.

According to the budget, the loans can be fully paid off by October, and $66,598 will be in the cash account at the end of the year. Armed with this information, Bulldog's managers would approach their banker to establish a line of credit with a limit higher than $11,201—the largest anticipated loan balance.

For your convenience, Bulldog's complete cash budget for 2000 is shown in Table 18-7 on page 453. We have described how to forecast a firm's short-term cash needs by constructing a cash budget. Next, we explore ways to manage a firm's short-term cash flow.

> **Take Note**
>
> *Since the financing is only needed for a short time, it is typically obtained from a line of credit or short-term notes.*

Managing the Cash Flowing In and Out of the Firm

People who manage a firm's cash should focus on four objectives: (1) to increase the flow of cash into the business; (2) to decrease the flow of cash out of the business; (3) to receive cash as quickly as possible; and (4) to pay cash out as slowly as possible, without missing the due date. This gives you more time to put cash to work earning a return. These four objectives principles are displayed in Figure 18-5 on page 454. In the sections that follow, we discuss ways to accomplish the four objectives of cash flow management.

Increasing Cash Inflows

There are really only two ways to increase the amount of cash flowing into a business during any given time period. First, the firm can do more of whatever it is that makes money—that is, a manufacturing business can sell more products or a service business can serve more people. Of course, when sales increase, costs increase too. Hopefully, however, the sales increase will be bigger than the cost increase. Second,

TABLE 18-6 Bulldog's 12-Month Summary of Cash Flow and Financing Requirements

	Dec	Jan	Feb	Mar	Apr	May	Jun	Jul.	Aug	Sep	Oct	Nov	Dec
Cash Flow Summary:							**2000**						
1. Cash balance at start of month		$65,313	$66,831	$67,782	$65,000	$65,000	$65,000	$65,000	$65,000	$66,554	$65,000	$69,473	$75,704
2. Net cash gain (loss) during month		1,820	951	(4,823)	(3,203)	2,309	(8,267)	5,341	7,413	(3,828)	6,748	6,230	(9,105)
3. Cash balance at end of month before financing (line 1 + line 2)		$67,133	$67,782	$62,959	$61,797	$67,309	$56,733	$70,341	$72,413	$62,725	$71,748	$75,704	$66,598
4. Minimum cash balance desired		65,000	65,000	65,000	65,000	65,000	65,000	65,000	65,000	65,000	65,000	65,000	65,000
5. Surplus cash (deficit) (line 3 − line 4)		$ 2,133	$ 2,782	$(2,041)	$(3,203)	$ 2,309	$(8,267)	$ 5,341	$ 7,413	$(2,275)	$ 6,748	$10,704	$ 1,598
External Financing Summary:													
6. External financing balance at start of month		$ 302	$ 0	$ 0	$ 2,041	$ 5,243	$ 2,934	$11,201	$ 5,860	$ 0	$ 2,275	$ 0	$ 0
7. New financing required (negative amount from line 5)		0	0	2,041	3,203	0	8,267	0	0	2,275	0	0	0
8. Financing repayments (positive amount from line 5 up to the amount on line 6)		302	0	0	0	2,309	0	5,341	5,860	0	2,275	0	0
9. External financing balance at end of month	$302	0	0	2,041	5,243	2,934	11,201	5,860	0	2,275	0	0	0
10. Cash balance at end of month after financing (line 3 + line 7 − line 8)		$66,831	$67,782	$65,000	$65,000	$65,000	$65,000	$65,000	$66,554	$65,000	$69,473	$75,704	$66,598

	1999		2000											
	Nov	Dec	Jan	Feb	Mar	Apr	May	Jun	Jul	Aug	Sep	Oct	Nov	Dec
Cash Inflows:														
Sales (reference only; not a cash flow)	$13,441	$13,029	$12,945	$14,794	$16,643	$18,492	$20,341	$22,191	$24,040	$22,191	$20,341	$18,492	$16,643	$14,794
Cash collections on sales:														
30% in month of sale			$ 3,884	$ 4,438	$ 4,993	$ 5,548	$ 6,102	$ 6,657	$ 7,212	$ 6,657	$ 6,102	$ 5,548	$ 4,993	$ 4,438
60% in first month after sale			7,817	7,767	8,876	9,986	11,095	12,205	13,315	14,424	13,315	12,205	11,095	9,986
10% in second month after sale			1,344	1,303	1,295	1,479	1,664	1,849	2,034	2,219	2,404	2,219	2,034	1,849
Total collections			$13,045	$13,508	$15,164	$17,013	$18,862	$20,711	$22,561	$23,300	$21,821	$19,971	$18,122	$16,273
Other cash receipts			0	0	0	0	0	0	0	0	0	0	0	0
Total Cash Inflows			$13,045	$13,508	$15,164	$17,013	$18,862	$20,711	$22,561	$23,300	$21,821	$19,971	$18,122	$16,273
Cash Outflows:														
Materials purchases (reference only; not a cash flow)		$ 3,994	$ 4,494	$ 4,993	$ 5,492	$ 5,992	$ 6,491	$ 5,992	$ 5,492	$ 4,993	$ 4,494	$ 3,994	$ 3,872	$ 3,994
Payments for materials purchases:														
100% in month after purchase			$ 3,994	$ 4,494	$ 4,993	$ 5,492	$ 5,992	$ 6,491	$ 5,992	$ 5,492	$ 4,993	$ 4,494	$ 3,994	$ 3,872
Other cash payments:														
Production costs other than purchases			3,994	4,494	4,993	5,492	5,992	6,491	5,992	5,492	4,993	4,494	3,994	3,872
Selling and marketing expenses			2,333	2,667	3,000	3,333	3,666	4,000	4,333	4,000	3,667	3,333	3,000	2,667
General and administrative expenses			903	903	903	903	903	903	903	903	903	903	903	903
Interest payments														2,971
Tax payments						4,995		4,995			4,995			4,995
Dividend payments					6,098			6,098			6,098			6,098
Total Cash Outflows			$11,225	$12,557	$19,987	$20,215	$16,553	$28,978	$17,219	$15,887	$25,649	$13,223	$11,892	$25,378
Net Cash Gain (Loss)			$ 1,820	$ 951	$(4,823)	$(3,203)	$ 2,309	$(8,267)	$ 5,341	$ 7,413	$(3,828)	$ 6,748	$ 6,230	$(9,105)
Cash Flow Summary:														
1. Cash balance at start of month			$65,313	$66,831	$67,782	$65,000	$65,000	$65,000	$65,000	$65,000	$66,554	$65,000	$69,473	$75,704
2. Net cash gain (loss) during month			1,820	951	(4,823)	(3,203)	2,309	(8,267)	5,341	7,413	(3,828)	6,748	6,230	(9,105)
3. Cash balance at end of month before financing (line 1 + line 2)			$67,133	$67,782	$62,959	$61,797	$67,309	$56,733	$70,341	$72,413	$62,725	$71,748	$75,704	$66,598
4. Minimum cash balance desired			65,000	65,000	65,000	65,000	65,000	65,000	65,000	65,000	65,000	65,000	65,000	65,000
5. Surplus cash (deficit) (line 3 – line 4)			$ 2,133	$ 2,782	$(2,041)	$(3,203)	$ 2,309	$(8,267)	$ 5,341	$ 7,413	$(2,275)	$ 6,748	$10,704	$ 1,598
External Financing Summary:														
6. External financing balance at start of month			$ 302	$ 0	$ 0	$ 2,041	$ 5,243	$ 2,934	$11,201	$ 5,860	$ 0	$ 2,275	$ 0	$ 0
7. New financing required (negative amount from line 5)			0	0	2,041	3,203	0	8,267	0	0	2,275	0	0	0
8. Financing repayments (positive amount from line 5 up to the amount on line 6)			302	0	0	0	2,309	0	5,341	5,860	0	2,275	0	0
9. External financing balance at end of month	$ 302		0	0	2,041	5,243	2,934	11,201	5,860	0	2,275	0	0	0
10. Cash balance at end of month after financing (line 3 + line 7 – line 8)			$66,831	$67,782	$65,000	$65,000	$65,000	$65,000	$65,000	$66,554	$65,000	$69,473	$75,704	$66,598

TABLE 18-7 Bulldog's Complete Cash Budget

Figure 18-5 **Managing the Cash Flowing In and Out of the Firm**

Figure 18-5 shows the four objectives of cash flow management.

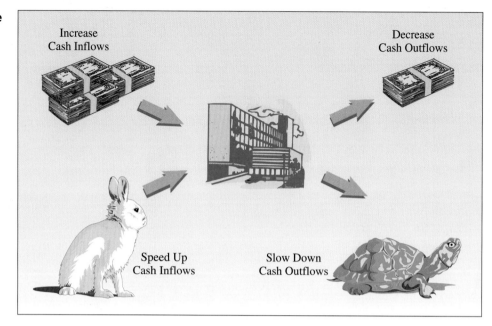

firms can increase the return that the company's assets are earning—that is, firms can find ways to produce more money with the same amount of assets.

Decreasing Cash Outflows

Managers can also increase the *net* amount of cash flowing into their firms during any given time period by decreasing the amount of cash flowing out. How? By cutting costs.

Finding the proper balance between short-term and long-term considerations is a subject that business managers must grapple with every day. There are no easy answers.

A less obvious way to decrease cash outflows from the business is to decrease the *risk* of doing business. Risk in business equates to uncertainty, and a business that faces a lot of uncertainty must keep a lot of cash on hand to deal with unexpected events. If a firm could somehow reduce the degree of risk of doing business, the number of unexpected events would drop, and the amount needed in the cash account could be reduced.

Be wary of cost-cutting measures that hurt the business in the long term. For example, a drug company might boost profits now by cutting research and development spending (an expense). But that action will rob the company of new products that would generate extra value later.

Speeding Up Cash Inflows

Most business managers would agree that, other things being equal, they would rather have cash earlier than later. This makes sense. The earlier a firm receives cash, the earlier it can put it to work earning a return. Accordingly, managers try to figure out how to speed the flow of cash into their firms.

Collecting funds from the firm's customers more quickly speeds cash inflow. The ideal situation, from a business firm's point of view, would be for all customers to pay for the products or services they buy immediately. However, the realities of the marketplace demand that credit be extended.[3] Given that credit often must be extended, the business firm's goal is to encourage customers to pay off their accounts as quickly as possible. The firm might even offer customers a *discount* if they pay their bills early, say, within 10 days. This technique works and the firm's managers hope that the return they can earn by getting the cash early outweighs the amount lost through the discount.[4]

Another way to speed up cash inflows is to make use of computerized fund transfers wherever possible. An **electronic funds transfer** is the act of crediting one account and debiting another automatically by a computer system. Electronic transfer is much faster than checks, which can take over a week to mail and clear.

Another method of speeding cash collections is a lockbox system. A **lockbox** system allows customers to send checks to a nearby post office box. The firm arranges for these funds to be deposited in a bank in or near the customer's town for electronic transfer to the receiving firm's account. Here's how a lockbox system works. A San Francisco-based business that has customers nationwide rents post office boxes in major cities all around the country. The firm directs its customers to send payments for their bills to the post office box in the city nearest them. The firm arranges for a bank in each city to pick up the mail from the post office box at least once a day, and to deposit any payments received in the firm's account at the bank. From that point on, the funds are immediately available for the firm's use, either from the individual banks directly or by having the banks electronically transfer the funds to the firm's bank in San Francisco. By using the lockbox system, the firm can receive cash two to five days faster than if customers mailed all their payments to San Francisco.[5]

Slowing Down Cash Outflows

Just as speeding up the flow of cash into the firm gives managers more time to earn a return on the cash, so does slowing down the flow of cash out of the firm. Either way, the idea is to increase the amount of time that the firm has possession of the cash. One obvious way to slow down cash outflows is to delay paying bills as long as possible. However, the firm must take great care not to overstep the bounds of good sense and fair play in applying this principle. Imagine what would happen if a firm didn't pay its employees on time or delayed paying suppliers. Its business operations would suffer or it might not stay in business at all. Firms shouldn't pay bills that are due at the end of the month on the first of the month, but neither should they take unfair advantage of creditors by making late payments.

---

[3] A quick example illustrates why businesses must grant credit: Suppose two firms in town both sold office supplies. One will let you, a business owner, order supplies over the phone anytime you want, and you are billed once a month. The other demands that you go to the store and pay cash each time you want to buy supplies. How long do you think the second store will be able to stay in business?

[4] This subject is discussed more fully in chapter 19.

[5] Before implementing such a system, of course, the firm would have to evaluate whether the extra amount that could be earned on the funds collected two to five days early is greater than the cost of maintaining the lockbox system.

What's Next

This chapter examined how firms manage cash and the factors that determine how large a firm's cash balance will be. In chapter 19 we study another component of working capital management: accounts receivable and inventory management.

Summary

1. List the factors that affect a company's desired minimum cash balance.

Having a minimum balance ensures that enough money is maintained in the cash account to make payments when needed. The base level to be maintained is affected by: (1) how quickly and cheaply the firm can raise cash when needed; (2) how accurately the firm can predict cash needs; and (3) how much extra the firm wants to keep in the cash account for emergencies.

2. List the factors that affect a company's maximum cash balance.

Having a maximum balance ensures that firms limit the cash balance so that the firm invests and earns a return on as much cash as possible. The maximum amount to be maintained is affected by three factors: (1) available investment opportunities; (2) expected returns from these opportunities; and (3) the transaction costs of withdrawing the cash and making the investments.

3. Apply the Miller–Orr model to establish a target optimal cash balance.

The Miller–Orr model recognizes that a firm's cash balance might fluctuate up and down in an irregular fashion over time. The model solves for an optimal target cash balance about which the cash balance fluctuates until it reaches an upper or lower limit. If the upper limit is reached, short-term investment securities are bought, bringing the cash balance down to the target again. If the lower limit is reached, short-term investment securities are sold, bringing the cash balance up to the target.

4. Prepare a cash budget.

Managers use a cash budget to estimate detailed cash needs for future periods. Cash budgets are necessary because *pro forma* income statements and balance sheets do not indicate the actual flow of cash in and out of the firm. A monthly cash budget shows detailed cash flows from month to month throughout the year, and how much over or short the firm's cash account will be at the end of each month. By using a cash budget, managers can estimate when it will be necessary to obtain short-term loans from their bank.

5. Explain how firms manage their cash inflows and outflows to maximize value.

The four objectives of cash management are to increase the flow of cash into the business, to decrease the flow of cash out of the business, to receive cash more quickly, and to pay cash out more slowly.

Ways to increase the flow of cash into the firm include selling more products, serving more customers, and increasing the return earned by the firm's assets. Ways to reduce the flow of cash out of the business include cutting costs and decreasing the degree of risk in the business. Ways to speed up the flow of cash into the business include helping customers pay off their credit accounts more quickly and using electronic funds transfer or lockbox techniques. Ways to slow down the flow of cash out of the firm include taking advantage of credit terms whenever possible.

Companion Website
content by Active Learning Technologies

You can access your FREE PHLIP/CW On–Line Study Guide, Current Events, and Spreadsheet Problem templates, which may correspond to this chapter either through the Prentice Hall Finance Center CD–ROM or by directly going to: **http://www.prenhall.com/gallagher.**

PHLIP/CW — *The Prentice Hall Learning on the Internet Partnership–Companion Web Site*

Equations Introduced in This Chapter

Equation 18-1. Formula for Miller–Orr Model Target Cash Balance:

$$Z = \sqrt[3]{\frac{3 \times TC \times V}{4 \times r}} + L$$

where: TC = The transaction cost of buying or selling short-term investment securities
V = The variance of net daily cash flows
r = The daily rate of return on short-term investment securities
L = The lower limit to be maintained in the cash account.

Equation 18-2. Formula for the Upper Limit in the Miller–Orr Model:

$$H = 3Z - 2L$$

Self Test

ST-1. Assume that short-term investment securities are yielding 5 percent, and it costs a firm $20 each time it buys or sells investment securities. The variance of the firm's daily net cash flows has been found to be $20,000. Management wants to keep at least $1,000 in the cash account for emergencies. Given these conditions, what is the firm's target cash balance?

ST-2. What is the maximum amount the firm in question ST-1 will let accumulate in its cash account before investing excess cash in marketable securities?

ST-3. Continuing with the firm in ST-1 and ST-2, how much will the firm invest in marketable securities if the upper limit in ST-2 is reached?

ST-4. Assume a company has $1,000 in its cash account at the beginning of a month and short-term loan balances of $3,500. If the company experiences a $14,000 net cash inflow during the month and its desired target cash balance is $3,000, how much of the outstanding loans can be paid off this month?

Review Questions

1. What are the primary reasons that companies hold cash?
2. Explain the factors affecting the choice of a minimum cash balance amount.
3. What are the negative consequences of a company holding too much cash?
4. Explain the factors affecting the choice of a maximum cash balance amount.

5. What is the difference between *pro forma* financial statements and a cash budget? Explain why *pro forma* financial statements are not used to forecast cash needs.
6. What are the benefits of "collecting early" and how do companies attempt to do this?
7. What are the benefits of "paying late" (but not too late) and how do companies attempt to do this?
8. Refer to the Bulldog battery company's cash budget in Table 18-7. Explain why the company would probably not issue $1 million worth of new common stock in January to avoid all short-term borrowing during the year.

Build Your Communication Skills

CS-1. Assume the company you work for has $10,000 in its cash account on January 1, 2000 and short-term loans outstanding in the amount of $15,000. Net cash flows for January, February, and March 2000 are forecasted to be +$2,000, −$5,000, and +$8,000, respectively. Prepare a report for your company's CEO showing any new borrowing required and any possible debt payoffs that may be made during the three-month period. Your company's minimum desired cash balance is $5,000. (Hint: Use the format in Table 18-5.)

CS-2. Assume you work for an electric utility that serves several western states and your boss is concerned that there is too much delay between the time the company's customers pay their bills and the time the cash is actually deposited in the corporation's cash account. Prepare a briefing for your boss outlining a method for streamlining cash collection procedures in such a way that the delay in making cash available is minimized.

Problems

Miller–Orr Model

18-1. Your company wants to have a minimum cash balance of $3,000 and an upper limit cash balance equal to $9,000. What would be your target cash balance?

Miller–Orr Model

18-2. Selena Rogers, the financial analyst of Keep-Fit Health Equipment Company, has a short-term investment yield of 3 percent and transaction cost of $40 per transaction. The cash inflows and outflows have traditionally been irregular with a variance of daily net cash flows equal to $39,000. Management of the company wants a minimum cash balance of $2,200. Calculate:

a. the target cash balance
b. the upper limit of cash balance.

Miller–Orr Model

18-3. Will Clark, the financial analyst of Get-Fit Health Equipment Company, a competitor of Keep-Fit of problem 18-2, has the same yield and transaction cost. However, the variance of daily net cash flows equal to $52,000. Management of the company wants a minimum cash balance of $3,900. Calculate:

a. the target cash balance
b. the upper limit of cash balance.

Cash Inflow

18-4. Marion Crane, a financial analyst of Lifelong Appliances Company, is trying to develop a cash budget for each month of the year 2000. The sales are expected to occur as follows:

Month	Sales (in thousand dollars)
Nov '99 (reference)	$131
Dec '99 (reference)	129
Jan '00	126
Feb '00	133
Mar '00	139
Apr '00	143
May '00	191
Jun '00	226
Jul '00	242
Aug '00	224
Sep '00	184
Oct '00	173
Nov '00	166
Dec '00	143
Jan '01 (reference)	136
Feb '01 (reference)	139

Assume all of Lifelong's sales are on credit, so no cash is received immediately when a sale is made. It is expected that 20 percent of Lifelong's customers will pay off their accounts in the month of sale, 70 percent will pay off their accounts in the month following the sale, and the remaining 10 percent of the customers will pay off their accounts in the second month following the sale. Given this payment pattern, help Marion in calculating Lifelong's actual monthly cash collections throughout the year 2000.

18-5. Use the same sales data given in problem 18-4. To improve the cash collections, Marion has decided to undertake stricter credit terms. With this change she expects that 40 percent of Lifelong's customers will pay off their accounts in the month of sale, 55 percent will pay off their accounts in the month following the sale, and the remaining 5 percent of the customers will pay off their accounts in the second month following the sale. Given this payment pattern, what would be Lifelong's actual monthly cash collections throughout the year 2000?

Cash Inflow

18-6. Use the same sales data given in problem 18-4. Assume that Lifelong's cost of materials is 30 percent of sales. Appliances that are expected to be sold in February will be manufactured one month ahead of time in January, and all the materials needed for January's production schedule will be ordered one month ahead of time in December. This schedule repeats for each month of the year. Lifelong makes all of its purchases on credit and pays for them in cash during the month following the purchase. That is, December's purchase orders are paid for in January, and so on. Assume Lifelong's remaining cash outflows are all direct expenses paid for in the month incurred as follows:

Challenge Problem

- Production expenses other than purchases are equal to 80 percent of purchases.
- Sales and marketing expenses are 19 percent of sales each month.
- General and administrative expenses are $11,000 each month.
- Interest expense is expected to be $31,000 for the year. Assume it will be paid all at once in December 2000.
- Lifelong's income tax bill for 2000 is expected to be $100,000. The bill will be paid in four equal installments in April, June, September, and December.
- Two semiannual dividends of $50,000 each are expected to be declared in 2000. These will be paid in June and December.

Calculate Lifelong's total cash outflows, including the preceding expenses and payments for materials purchases.

Cash Outflow

18-7. Use the same data given in problem 18-6, except for the payment schedule for materials purchased. Now assume that Lifelong pays for the material purchased in the following manner: 30 percent is paid in cash in the month of purchase, and the remaining 70 percent is paid in cash during the month following the purchase. That is, 30 percent of December's purchase orders are paid for in December, and the balance of 70 percent is paid in January, and so on. With this change, calculate Lifelong's total cash outflows, including the preceding expenses, dividends, and payments for materials purchases.

Comprehensive Problem

18-8. Rose Sayer, a financial analyst of Fit-and-Forget Fittings Company, is trying to develop a cash budget for each month of the year 2000. The sales are expected to occur as follows:

Month	Sales (in thousand dollars)
Nov '99 (reference)	$2,266
Dec '99 (reference)	2,230
Jan '00	2,116
Feb '00	2,300
Mar '00	2,402
Apr '00	2,420
May '00	3,390
Jun '00	3,909
Jul '00	4,164
Aug '00	3,933
Sep '00	3,163
Oct '00	2,912
Nov '00	2,886
Dec '00	2,424
Jan '01 (reference)	2,353
Feb '01 (reference)	2,442

Assume all of Fit-and-Forget's sales are on credit, so no cash is received immediately when a sale is made. It is expected that 30 percent of Fit-and-Forget's customers will pay off their accounts in the month of sale, 65 percent will pay off their accounts in the month following the sale, and the remaining 5 percent of the customers will pay off their accounts in the second month following the sale.

Assume that Fit-and-Forget's cost of materials is 20 percent of sales. Fit-and-Forget manufactures fittings expected to be sold in February one month ahead of time, in January. They order all the materials they need for January's production schedule one month ahead of time, in December. This schedule repeats for each month of the year. Fit-and-Forget makes all purchases on credit, and pays for the material purchased in the following manner: 20 percent is paid in cash in the month of ordering, and the balance of 80 percent is paid in cash during the month following the purchase. That is, 20 percent of December's purchase orders are paid for in December, and the balance of 80 percent is paid in January, and so on. Assume Fit-and-Forget's remaining cash outflows are all direct expenses paid for in the month incurred as follows:

- Production expenses other than purchases are equal to 14 percent of purchases.
- Sales and marketing expenses are 16 percent of sales each month.
- General and administrative expenses are $180,000 each month.
- Interest expense is expected to be $500,000 for the year. Assume it will be paid all at once in December 2000.

- Fit-and-Forget's income tax bill for 2000 is expected to be $1,600,000. The bill will be paid in four equal installments in April, June, September, and December.
- Two semiannual dividends of $855,000 each are expected to be declared in 2000. These will be paid in June and December.

Assuming a cash balance of $1,133,000 at the beginning of January, a desired target cash balance of $1,110,000, and short-term loans of $50,000 outstanding at the beginning of the month, calculate total cash inflows, total cash outflows, net cash gain (loss), cash flow summary, and external financing (if any) summary in the same format as in Table 18-7.

Answers to Self-Test

ST-1. Per Equation 18-1, the firm's target cash balance per the Miller–Orr model is:

$$Z = \sqrt[3]{\frac{3 \times 20 \times \$20,000}{4 \times (.05/365)}} + \$1,000$$

$$= \sqrt[3]{\frac{\$1,200,000}{.000547945}} + \$1,000$$

$$= \$1,299 + \$1,000$$

$$= \$2,299$$

ST-2. Per Equation 18-2, the upper limit will be:

$$H = (3 \times \$2,299) - (2 \times \$1,000)$$

$$= \$6,897 - \$2,000$$

$$= \$4,897$$

ST-3. If the upper limit is reached the firm will invest the amount necessary to bring the cash balance back down to the target balance level. This amount is:

$$\$4,897 - \$2,299 = \$2,598$$

ST-4. A summary of the company's cash flows and external financing is shown next:

1. Cash balance at start of month	$ 1,000
2. Net cash gain (loss) during month	14,000
3. Cash balance at end of month before financing	15,000
4. Minimum cash balance desired	3,000
5. Surplus cash (deficit) (line 3 − line 4)	12,000
External Financing Summary:	
6. External financing balance at start of month	$ 3,500
7. New financing required (negative amount on line 5)	0
8. Financing repayments (the entire loan balance)	3,500
9. External financing balance at end of month	0
10. Cash balance at end of month after financing (balance on line 3 + new financing on line 7 − repayments on line 8)	$11,500

As shown on line 8 of the preceding summary, the company can pay off the entire $3,500 short-term loan balance this month.

CHAPTER 19

ACCOUNTS RECEIVABLE AND INVENTORY

Everybody likes a kidder, but nobody lends him money.
—Arthur Miller

ON-DEMAND PUBLISHING SOLVES THE PROBLEM OF INVENTORY COSTS

Phil Theibert decided to start a small publishing business called Ascendix Publishing. With desktop publishing available, he thought it would be fairly easy. Typesetting and designing Ascendix's first book on the computer, *Kids Cook Microwave,* was easy. Then the hard decisions began. It was one thing to have a book on the computer screen. It was another to have it printed and marketed—especially on Ascendix's limited budget of $10,000.

Phil discovered that he could get the book printed for about $2 a book. Yet to get this low price, he had to commit to print 4,000 books. The cost of printing, then, would be $8,000. To complicate matters, he would have to store the books somewhere, plus pay tax on the books in inventory. This storage and tax would cost an additional $500.

Thus, $8,500 of Ascendix's limited budget would be spent on production costs to produce an inventory of 4,000 books that required additional money to store. That only left $1,500 for a marketing campaign.

Obviously, production costs and inventory would eat too much of Ascendix's cash to market the book aggressively. Phil investigated further and found out that many books could be printed "on demand." The text of his book could be stored in

CHAPTER OBJECTIVES

After reading this chapter, you should be able to:

■ Describe how and why firms must manage accounts receivable and inventory as investments.

■ Compute the optimal levels of accounts receivable and inventory.

■ Describe alternative inventory management approaches.

■ Explain how firms make credit decisions and create collection policies.

a digital copy machine, then printed when orders came in. The cost of each book would be $3 compared to $2, but almost all of Phil's $10,000 budget could be spent on a direct-marketing campaign to target potential buyers. Then, when the orders came in, the books could be printed and sent out.

This "just-in-time" printing helped Ascendix as it struggled to launch its first book and let Phil use the cash where it was more important—to get buyers for the book. The decision Ascendix faced, to carry inventory or not carry inventory, and the expenses associated with that decision, are the same problems big and small businesses across the nation face every day.

Source: Interview with Phil Theibert.

Chapter Overview

A key component of working capital policy is managing accounts receivable and inventory. In this chapter we see that accounts receivable and inventory are necessary investments that affect a firm's profitability and liquidity. Then we investi-

Source: *From The Wall Street Journal—Permission, Cartoon Features Syndicate.*

gate how financial managers determine the optimal level of these current assets. Finally, we examine inventory management techniques and collection policies.

Why Firms Accumulate Accounts Receivable and Inventory

As we saw in chapter 4, accounts receivable represent money that customers owe to the firm because they have purchased goods or services on credit. Accounts receivable, therefore, are assets that have value.[1] Nonetheless, any time a firm accumulates accounts receivable, it suffers opportunity costs because it is unable to invest or otherwise use the money owed until customers pay. A firm may also incur a direct cost when it grants credit because some customers may not pay their bills at all. The ideal situation, from a firm's point of view, is to have customers pay cash at the time of the purchase.[2]

In the real world, of course, it's unrealistic to expect customers always to pay cash for products and services. Who would buy lumber from a firm that insisted that all its customers pay cash if other lumber companies offered credit? Like it or not, for most firms, granting credit is an essential business practice. The real question managers must answer is, how much credit should the firm grant and to whom? Offering more credit enhances sales but also increases costs. At some point the cost of granting credit outweighs the benefits. Financial managers must manage accounts receivable carefully to make sure this asset adds to, rather than detracts from, a firm's value.

The situation with inventory is similar. As Phil Theibert's experience from the chapter opener shows, inventory is costly to accumulate and maintain, so firms generally want to hold as few products in inventory as possible. For most firms, however, operating without any inventory is impractical—can you imagine a grocery store with no food on display? Most firms that sell products accumulate some inventory. Financial managers must find the best level of inventory. They do this by weighing the risk of losing sales due to unavailable products against the cost savings produced by reducing inventory.

Accounts receivable and inventory are investments because both tie up funds and have opportunity costs, but can add to the firm's value. Be careful not to be confused by the term *investment*. Investment usually implies something desirable—a long-term venture specifically planned and implemented for profit. Instead, accounts receivable and inventory may be viewed as necessary evils. Most firms need accounts receivable and inventory to do business, but less is generally better. Managing accounts receivable and inventory, then, ought to be done with an eye toward reducing these assets to the lowest level possible consistent with the firm's goal of maximizing value.

How Accounts Receivable and Inventory Affect Profitability and Liquidity

Holding different levels of accounts receivable and inventory can affect a company's profitability and liquidity. To illustrate, consider Firms A and B in Table 19-1.

---

[1] In fact, firms sometimes sell their cutomers' "IOUs" to other businesses for cash. This process, known as *factoring* accounts receivable, is discussed later in this chapter.
[2] Individuals are the same way. For example, if you sell your bicycle to another student, which would you prefer: to be paid in cash at the time of sale, or to let the buyer pay you a little bit each month?

TABLE 19-1

Comparison of Accounts Receivable and Inventory Policies, Selected Financial Data for Firms A and B (in 000's)

Data as of December 31	FIRM A Sells Products for Cash and Holds No Inventory	FIRM B Sells Products for Credit and Accumulates Inventory
Cash	$ 100	$ 100
Accounts Receivable	0	200
Inventory	0	400
Fixed Assets	500	500
Total Assets	$ 600	$1,200
Current Liabilities	$ 50	$ 100
Long-Term Debt	200	400
Stockholders' Equity	350	700
Total Liabilities and Equity	$ 600	$1,200
Sales	$2,433	$2,433
Expenses	2,383	2,383
Net Income for the Year	$ 50	$ 50
Current Ratio	2.0	7.0
Quick Ratio	2.0	3.0
Return on Equity	14.3%	7.1%

Firm A sells all its products for cash and keeps no inventory. Firm B gives its customers 30 days to pay and maintains a large product inventory. Assuming every other factor is equal, including the firms' capital structures, Firm A can earn over twice the return on its stockholders' equity as Firm B simply by eliminating accounts receivable and inventory (and any associated current liabilities and long-term debt).

The comparison between Firms A and B in Table 19-1 illustrates the liquidity, profitability, and risk of each firm. Observe that although Firm A is more *profitable* than Firm B as measured by return on equity (14.3 percent versus 7.1 percent), it is much less *liquid,* as measured by the current ratio[3] (2.0 versus 7.0). If the managers of Firm A needed to raise more than $100,000 cash in a hurry, they would have no recourse but to seek outside financing or sell some of their fixed assets. The managers of Firm B, however, could collect cash from customers, draw down inventory, or both.

However, Firm B's business practice of accumulating inventory adds risk to the firm if the inventory is hard to liquidate. Some inventory may not be sold, or it may be sold for a low value. Note how when using the quick ratio[4] to compare firm liquidity, we see that Firm A's quick ratio is the same as its current ratio (2.0), but Firm B's quick ratio is 3.0 compared to its current ratio of 7.0. When using the quick ratio, then, we see that Firm A is less liquid than Firm B (2.0 versus 3.0), but how much less depends on the liquidity of the inventory.

---

[3] The Current Ratio = $\dfrac{\text{Current Assets}}{\text{Current Liabilities}}$

[4] The Quick Ratio = $\dfrac{\text{Current Assets Less Inventory}}{\text{Current Liabilities}}$

Most firms accumulate some accounts receivable and inventory. Because these current assets can affect the profitability and liquidity of the firm, financial managers try to find the amounts of both assets that maximize firm value. In the following section, we examine how to find the optimal level of these current assets.

We discussed how to calculate return on equity, the current ratio, and quick ratio in chapter 5, pages 89–91.

Finding Optimal Levels of Accounts Receivable and Inventory

The conclusions drawn from Table 19-1 assumed that sales for both firms were $2,433,000. That may not be a reasonable assumption. Firm B may have greater sales than Firm A because it grants credit and has inventory immediately available for purchase. Depending on how well customers respond to Firm B's decision to grant credit and maintain inventory, the resulting increase in sales and net income might boost Firm B's return on equity beyond that of Firm A.

However, bad debts and inventory costs are likely to drive up Firm B's expenses, possibly causing its net income to fall. The net result could be a *decrease* in Firm B's return on equity.

Conflicting forces make it more difficult to assess the situation. For accounts receivable, the forces are sales that increase as more generous credit terms are offered *versus* costs that increase with collections, bad debts, and opportunity costs from foregone investments. For inventory, the conflicting forces are sales that increase as more products are made available *versus* storage costs that increase as more inventory is accumulated.

In the following sections, we discuss how to find a balance between these conflicting forces in order to determine the optimal levels of accounts receivable and inventory.

The Optimal Level of Accounts Receivable

To find the best level of accounts receivable, financial managers must review the firm's credit policies, any proposed changes in those policies, and the incremental cash flows of each proposed credit policy. We must then compute the net present value of each policy.

Credit Policy A firm's credit terms and credit standards make up the firm's **credit policy.** Remember, accounts receivable are created by customers taking advantage of the firm's credit terms. These terms generally offer a discount to credit customers who pay off their accounts within a short time, and specify a maximum number of days that credit customers have to pay off the total amount of their accounts. An example of such terms is "2/10, n30." This means that credit customers will receive a 2 percent discount if they pay off their accounts within 10 days of the invoice date, and the full net amount is due within 30 days if the discount is not taken.

For example, suppose you purchase $1,000 worth of camping supplies on credit July 1. If you pay the bill before July 11, you receive a 2 percent discount and the equipment will only cost you $980. If you pay the bill between July 11 and July 31, you'll owe the full amount—$1,000. The bill is past due if you don't pay it by July 31.

Some firms offer credit without a discount feature, simply giving their customers so many days to pay. An example would be "n90"—net 90, pay the invoice amount within 90 days.

TABLE 19-2

The Effects of Tightening Credit Policy

ACTION	EFFECT ON ACCOUNTS RECEIVABLE
Raise standards for granting credit	Fewer credit customers Fewer people owing money to the firm at any given time Accounts receivable goes down
Shorten net due period	Accounts paid off sooner Less money owed to the firm at any given time Accounts receivable goes down
Reduce discount percent	Fewer credit customers, *but* Some customers who previously took discount now do not Net effect on accounts receivable depends on whether the number of customers lost will be more than the number of those now forgoing the discount and paying the higher net amount
Shorten discount period	Same as above; some credit customers will leave, others will forgo the discount and pay the higher net amount. Net effect on accounts receivable is indeterminate

Individual customers receive credit from the firm if they meet the firm's **credit standards** for character, payment history, and so on.[5] Taken together, the credit terms and credit standards comprise the firm's credit policy.

A firm that wishes to change its level of accounts receivable does so by changing its credit policy. *Relaxing* the credit policy—by adopting less stringent credit standards or extending the net due period—will tend to cause accounts receivable to increase. *Tightening* the credit policy—by adopting more stringent credit standards or shortening the net due period—will tend to cause accounts receivable to decrease. The discount percent or time period could also be changed. This may either increase or decrease accounts receivable, depending on the reaction of customers to competing influences. Table 19-2 summarizes the effects of tightening credit policy on accounts receivable.

Analyzing Accounts Receivable Levels   To decide what level of accounts receivable is best for the firm, we follow this three-step process:

1. Develop *pro forma* financial statements for each credit policy under consideration.
2. Use the *pro forma* financial statements to estimate the incremental cash flows of the proposed credit policy and compare them to the current policy cash flows.
3. Use the incremental cash flows and calculate the net present value (NPV) of each policy change proposal. Select the credit policy with the highest NPV.

To demonstrate how the three-step credit policy evaluation works, let's analyze a proposed credit policy for a fictional firm, Bulldog Batteries. Assume Bulldog Batteries currently offers credit terms of 2/10, n30, and Jackie Russell, the vice president for marketing, thinks the terms should be changed to 2/10, n40. Doing so, she says, will result in a 10 percent increase in sales, but only a small increase in bad debts. Should Bulldog make the change?

Take Note

In real life, the firm's credit policy may be limited by marketplace constraints. A new, small firm attempting to sell to large, well-established customers may have to offer credit terms that match those of competitors in the industry.

---

[5]We'll discuss credit standards more fully later in this chapter.

For the sake of simplicity, we make the following additional assumptions:

- Ms. Russell is correct that sales will increase by 10 percent if the new credit policy is implemented.
- Cost of goods sold and other operating expenses on the firm's income statement, and all current accounts on the balance sheet, will vary directly with sales. So each of these accounts will increase by 10 percent with the change in credit policy.
- All of Bulldog's sales are made on credit.

Suppose further that Ms. Russell produced the following data on Bulldog's customers' historical payment patterns:

- 45 percent of Bulldog's customers take advantage of the discount and pay off their accounts in 10 days.
- 53 percent of Bulldog's customers forgo the discount, but pay off their accounts in 30 days.
- The remaining 2 percent of Bulldog's customers pay off their accounts in 100 days.[6]

Ms. Russell expects that under the new credit policy:

- 45 percent of Bulldog's customers will still take advantage of the discount and pay off their accounts in 10 days.
- 52 percent of Bulldog's customers will forgo the discount and pay off their accounts in 40 days.
- 3 percent of Bulldog's customers will pay off their accounts in 100 days.

With this information, we calculate the weighted average of the customers' payment periods (average collection period, or ACP) under the old and new credit policies. We weight the averages for each scenario by multiplying the percentage of customers who pay times the number of days they take to pay. Then we total each scenario result, as follows:

Under the old credit policy:

$$ACP = (.45 \times 10 \text{ days}) + (.53 \times 30 \text{ days}) + (.02 \times 100 \text{ days})$$
$$= 22.4 \text{ days}.$$

Under the new credit policy:

$$ACP = (.45 \times 10 \text{ days}) + (.52 \times 40 \text{ days}) + (.03 \times 100)$$
$$= 28.3 \text{ days}.$$

Based on Ms. Russell's information, we know that under the current credit policy, bad debt expenses are 2 percent of sales; under the new policy, they will climb to 3 percent of sales. Bulldog's CFO has informed us that any increases in current assets resulting from the policy change will be financed from short-term notes at an interest rate of 6 percent. The CFO also tells us that the firm's effective tax rate is 40 percent, and that the cost of capital is 10 percent. The long-term interest rate is 8 percent.

---

[6]In real life, many of these customers will never pay off their accounts, creating bad debts. Assume here, however, that they do pay off their accounts eventually. If we do not make that assumption, our mathematical average will include a certain percentage of customers taking an infinite amount of time to pay. As a result, we get an infinite average collection period (ACP), and that won't provide usable information.

TABLE 19-3

Incremental Cash Flows Associated with Changing Bulldog Batteries' Credit Policy from 2/10, n30 to 2/10, n40

1. Net Incremental Cash Outflow at Time Zero (t_0)	
External Financing Required from the Projected Balance Sheet	$ 9,352
2. Incremental Cash Flows Occurring in the Future	
Incremental Cash Inflow:	
Increase in Sales	$20,173
Incremental Cash Outflows:	
Increase in Cost of Goods Sold	$10,728
Increase in Bad Debt Expenses	2,623
Increase in Other Operating Expenses	4,323
Increase in Interest Expense	2
Increase in Taxes	999
Total Incremental Cash Outflows:	$18,675
Net Incremental Cash Flows Occurring from Time t_1 through Infinity: $20,173 − $18,675 =	$ 1,498 per year

Step 1: Develop the *Pro Forma* Financial Statements The first step is to develop the *pro forma* financial statements that reflect the effects of the proposed credit policy change. We begin by reviewing the firm's current income statement and balance sheet, and then we create new statements that incorporate the changes. The statements for Bulldog Batteries are shown in Figure 19-1. The left-hand column shows Bulldog's financial statements before the credit policy change, the middle column shows them after the change, and the right-hand column shows how the new figures were calculated, given our assumptions.

Step 2: Compute the Incremental Cash Flows Now it's time to compute the incremental cash flows that occur as a result of the credit policy change. Table 19-3 contains these cash flows. Table 19-3 shows that Bulldog's initial investment cash flow is $9,352 and its net incremental cash flows from t_1 through infinity are $1,498 per year.

Step 3: Compute the NPV of the Credit Policy Change Now that we have all the incremental cash flows, we can calculate the NPV of the proposed credit terms change. We learned in chapter 10 that NPV is calculated by summing the present value (PV) of all a project's projected cash flows and then subtracting the amount of the initial investment.[7] In our example, we have a net incremental cash outflow (the initial investment) that occurs at time t_0 of $9,354, followed by net incremental cash inflows occurring from time t_1 through infinity of $1,498 per year. The PV of the $1,498 per year from time t_1 through infinity can be found using the formula in chapter 8 for the present valve of a perpetuity:

$$\text{PVP} = \text{PMT} \times \left(\frac{1}{k}\right) \qquad (8\text{-}5)$$

where: PMT = the cash flow per period
           k = the required rate of return

---

[7] See Equation 10-1a in chapter 10.

Figure 19-1 Bulldog Batteries Financial Statements

Figure 19-1 shows the pro forma financial statements for Bulldog Batteries with the current collection policy and with the proposed collection policy.

INCOME STATEMENTS

	With Old Credit Terms: 2/10, n30	With New Credit Terms: 2/10, n40	Remarks
Sales (all on credit)	$201,734	$221,907	10% increase assumed
Cost of Goods Sold	107,280	118,008	Increase in proportion with sales (10%)
Gross Profit	94,454	103,899	
Bad Debt Expenses	4,035	6,657	Old: 2% of sales / New: 3% of sales
Other Operating Expenses	43,229	47,552	Increase in proportion with sales (10%)
Operating Income	47,190	49,690	
Interest Expense	1,221	1,223	Notes Payable × .06 + LTD × .08
Before-Tax Income	45,969	48,468	
Income Taxes (tax rate = 40%)	18,388	19,387	
Net Income	$ 27,581	$ 29,081	

BALANCE SHEETS, AS OF DECEMBER 31

	With Old Credit Terms: 2/10, n30	With New Credit Terms: 2/10, n40	Remarks
Assets			
Current Assets:			
Cash and Marketable Securities	$ 65,313	$ 71,844	Increase in proportion with sales (10%)
Accounts Receivable	12,380	17,205	See note 1
Inventory	21,453	23,598	Increase in proportion with sales (10%)
Total Current Assets	$ 99,146	$112,647	Increase in proportion with sales (10%)
Prop, Plant, and Equipment, Net	92,983	92,983	No change
Total Assets	$192,129	$205,630	
Liabilities and Equity			
Current Liabilities:			
Accounts Payable	$ 26,186	$ 28,805	Increase in proportion with sales (10%)
Notes Payable	302	332	Increase in proportion with sales (10%)
Total Current Liabilities	$ 26,488	$ 29,137	
Long-Term Debt	15,034	15,034	No change
Total Liabilities	$ 41,522	$ 44,171	
Common Stock	35,000	35,000	No change
Capital in Excess of Par	32,100	32,100	No change
Retained Earnings	83,507	85,007	Old RE + 1,500 net income difference
Total Stockholders' Equity	$150,607	$152,107	
Total Liabilities and Equity	$192,129	$196,278	
Additional Funds Needed		$ 9,352	See notes 2 and 3
		$205,630	

Note 1: Accounts Receivable (AR) = Credit sales per day × ACP
Under the old credit policy: AR = ($201,734/365) × 22.4 = $12,380
Under the new credit policy: AR = ($221,907/365) × 28.3 = $17,205
Note 2: $9,352 is the amount of additional financing needed (AFN) to balance the balance sheet. It is the amount that must be obtained from outside sources to undertake the proposed credit policy change. Therefore, $9,352 may be viewed as the net investment required at time zero for the project.
Note 3: If $9,352 is borrowed to make up for AFN, Bulldog will incur new interest charges. If included in the income statement, these will reduce the net income and retained earnings—throwing the balance sheet off balance again and changing the amount of AFN. If the problem is solved using an electronic spreadsheet, the financial statements can be recast several times until the additional interest expense becomes negligible. Here, however, we will use the original interest rate to simplify the calculations.

According to our assumptions, Bulldog Batteries' cost of capital is 10 percent. Applying Equation 8-5, we find the PV of an endless stream of payments of $1,499 discounted at 10 percent as follows:

$$PVP = PMT \times \left(\frac{1}{k}\right)$$

$$= \$1,498 \times \left(\frac{1}{.10}\right)$$

$$= \$14,980$$

We see that the present value of the $1,498 perpetuity with a 10 percent required rate of return is $14,980.

To complete the NPV calculation, we now need to subtract the $9,352 net cash outflow that occurs at time t_0—the initial investment—from the present value of the $14,980 perpetuity, as shown next:

$$NPV = \$14,980 - \$9,352$$

$$= \$5,628$$

We find that the net present value of the credit policy change is $5,628. Since the NPV is positive, the credit terms change proposal should be accepted. Doing so will increase the value of Bulldog Batteries by $5,628.[8]

Any credit policy change proposal can be evaluated using this framework. Managers may try any number of discount amounts, discount periods, and net due periods, until they discover that combination with the greatest NPV.

Take Note

Remember that this analysis depends on the accuracy of our assumptions. In this case, we assumed that sales would increase 10 percent, that bad debts would increase to 3 percent of sales, and that customers would pay according to the pattern described. If these assumptions are invalid, the credit policy analysis will also be invalid.

The Optimal Level of Inventory

Firms may be able to stimulate sales by maintaining more inventory, but they may drive up costs as well. The financial manager's task is to figure out what level of inventory produces the greatest benefit to the firm. Financial managers first estimate the costs that are associated with inventory.

The Costs of Maintaining Inventory The two main costs associated with inventory are carrying costs and ordering costs. *Carrying costs* are those costs associated with keeping inventory on hand—warehouse rent, insurance, security expenses, utility bills, and so on. Carrying costs are generally expressed in dollars per unit per year.

---

[8]Given all the information presented so far, we could also calculate the internal rate of return (IRR) of the credit terms proposal. We simply solve the PV of a perpetuity formula for k, which represents the IRR:

$$k, \text{ or } IRR, = PMT/PVP$$
$$\text{In our example: } IRR = \$1,498/\$9,352$$
$$= .1602, \text{ or } 16.02\%$$

Because 16.02 percent exceeds Bulldog's cost of capital of 10 percent, the credit terms change proposal should be accepted.

Ordering costs are those costs incurred each time an order for inventory materials is placed—clerical expense, telephone calls, management time, and so on. Ordering costs tend to be fixed no matter what the size of the order, so they are generally expressed in dollars per order.

Although you would think financial managers would like to minimize these two costs, it's not so easy. Carrying costs tend to rise as the level of inventory rises, but ordering costs tend to fall as inventory rises (because less ordering is necessary). Firms that minimize carrying costs by keeping no inventory have to order materials every time they want to produce an item, so they actually maximize ordering costs. Likewise, firms that minimize ordering costs by ordering all materials at once have sky-high carrying costs.

Complicating the situation is the possibility that a larger stock of inventory might increase sales. More inventory on display means more opportunities to catch the customer's eye, and more inventory on hand means fewer sales lost due to not having the correct size or model available. The fact that more inventory might translate into more sales means the lowest cost level of inventory might not be the optimal level of inventory. To find the optimal level, managers have to balance the costs and benefits of various inventory levels.

Analyzing Inventory Levels To maximize the value created from the firm's investment in inventory, use a three-step process similar to the one used to find accounts receivable levels. First, generate *pro forma* income statements and balance sheets for each proposed inventory level. Second, observe the incremental cash flows that occur with the change. Third, compute the NPV of the incremental cash flows. The following example illustrates the three-step process.

Dealin' Dan, the owner of Cream Puff Used Cars, wants to determine the optimal number of cars to display on his lot. Dan knows that increasing the number of cars on display will probably cause sales to increase, but would also increase his inventory carrying costs. Dan also knows that decreasing the number of cars on display will save him inventory costs but might also cost him sales. As a result, Dan is not sure how a change in his planned average inventory level from 32 cars to 48 cars will affect the value of his firm.

We'll make the following assumptions about Cream Puff's financial condition:

1. Cream Puff's inventory ordering costs are $100 per order. (Each time Dealin' Dan takes action to obtain cars, whether from other dealers, or from trade-ins, he incurs $100 in processing costs.)
2. Inventory carrying costs are $500 per car per year.
3. Because Dealin' Dan does not expect to keep cars on the lot more than a few weeks, he finances all the firm's inventory with short-term debt. The short-term interest rate available to Cream Puff is 6 percent.
4. Cream Puff pays $5,000, on average, for each car it purchases for resale. The firm's average selling price per car is $6,000.
5. Cream Puff displays, on average, about 32 cars on its lot. Sales occur regularly throughout the year, and Dan expects to sell 200 cars this year.
6. Based on his business experience, Dan believes that the relationship between inventory and Cream Puff's car sales is direct, as shown in Figure 19-2. According to the graph, an increase in inventory from 32 to 48 cars should produce an increase in the number of cars sold per year from 200 to 232.

Figure 19-2 Cream Puff Used Cars Inventory versus Number of Cars Sold

Figure 19-2 shows the direct relationship between inventory and sales. As inventory increases, sales increase; as inventory decreases, sales decrease.

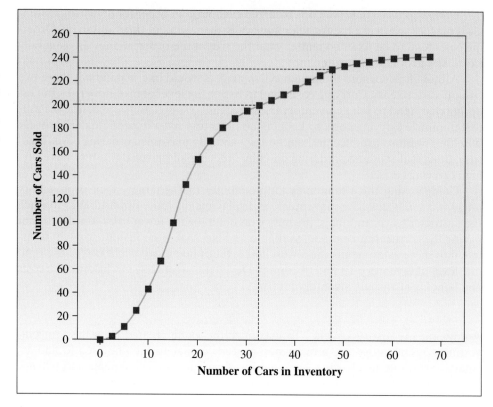

Dan uses the *economic order quantity* (EOQ) model to compute the number of cars to order from wholesale dealers when he replenishes his inventory.[9] According to the EOQ model, the optimal order size follows:

Formula for Economic Order Quantity Model

$$OQ = \sqrt{\frac{2 \times S \times OC}{CC}} \qquad (19\text{-}1)$$

where: OQ = Order quantity
S = Annual sales volume in units
OC = Ordering costs, per order
CC = Carrying costs, per unit per year

We may use the model to see what order size Cream Puff should have. We know that sales are 200 cars per year, ordering costs are $100 per order, and the carrying costs are $500 per year. According to Equation 19-1, the ordering quantity for Cream Puff is as follows:

$$OQ = \sqrt{\frac{2 \times 200 \times 100}{500}}$$
$$= \sqrt{80}$$
$$= 8.94 \text{ (round to 9)}$$

---

[9]The EOQ model computes the inventory order size that, if certain other conditions are met, will minimize total inventory costs for the year. For a complete discussion of the model, refer to a production management text.

We find that the ordering quantity that minimizes inventory costs is nine cars per order. Cream Puff's sales occur regularly, so Dan orders nine replacement cars at even intervals throughout the year. The sales forecast calls for 200 cars to be sold this year, so the number of orders for replacement cars will be 200/9 = 22.22 (round to 22).[10] Each order for replacement cars costs $100, so Cream Puff's total ordering cost—assuming 22 orders are made—is $2,200.

Under the current inventory policy, Cream Puff's average inventory level is 32 cars. Carrying costs are $500 per car per year, so total carrying costs are 32 × $500 = $16,000. We sum the ordering and carrying costs to find the total inventory costs for Cream Puff. The total costs are $2,200 + $16,000 = $18,200.

We assume in our example that other operating expenses on Cream Puff's income statement and all current accounts on the balance sheet vary directly with sales, so each of these accounts will increase (or decrease) proportionally with sales. We also assume that the interest rate on short-term debt is 6 percent and the rate on long-term debt is 8 percent. Finally, assume that Cream Puff's effective tax rate is 40 percent, and the firm's cost of capital is 10 percent.

Now we're ready to apply the three-step process to determine the optimal level of inventory for Cream Puff Used Cars.

Step 1: Develop *Pro Forma* Financial Statements. Let's see how Dealin' Dan's proposed inventory change from 32 cars to 48 cars will affect the business. The left-hand column in Figure 19-3 shows Cream Puff's projected 2000 income statement and balance sheet. It also shows selected financial ratios, given our assumptions, current inventory policy of 32 cars, and the total inventory costs of $18,200. The right-hand column contains revised statements and ratios assuming average inventory is raised to 48 cars. The remarks column explains how the various numbers were computed.

If Dealin' Dan's assumptions are correct, increasing the number of cars on display at the Cream Puff used car lot to 48 will produce a $192,000 increase in sales (32 extra cars). Inventory costs and other expenses will increase too, of course, but only by a total of $186,534, so profits should rise by $5,466. On the surface it looks like Dan should go ahead with the inventory change, doesn't it? Perhaps, but let's see how the change will affect the value of Dan's firm by computing the incremental cash flows associated with the change and the NPV of those incremental cash flows.

Step 2: Compute the Incremental Cash Flows. Drawing the necessary information from the *pro forma* statements shown in Figure 19-3, Table 19-4 lists the incremental cash flows of changing the average inventory level from 32 cars to 48 cars.

We see in Table 19-4 that the initial investment cash flows of changing Dealin' Dan's inventory policy are $72,294. We also find that the net cash flows associated with changing the average inventory level are $5,466 per year.

Step 3: Compute the NPV of the Inventory Policy Change. We are ready now to calculate the NPV of the inventory policy change just as we did for accounts receivable. First, we compute the present value of the net cash inflows occurring from time

---

[10]Astute readers will note that after rounding, 22 × 9 = 198. So Dealin' Dan will not actually order the exact number of cars he expects to sell this year. However, this small discrepancy will not materially affect our analysis.

Figure 19-3 Dealin' Dan's Cream Puff Used Cars Inventory Analysis

Figure 19-3 shows the pro forma financial statements with Cream Puff's current average inventory policy of 32 cars and with the new inventory policy of 48 cars.

INCOME STATEMENTS

	Current Inventory Policy	New Inventory Policy	Remarks
Sales	$1,200,000	$1,392,000	32-car increase × $6,000 each
Cost of Goods Sold	1,000,000	1,160,000	32-car increase × $5,000 each
Gross Profit	200,000	232,000	
Inventory Costs	18,200	26,300	See Note 1
Other Operating Expenses	85,000	98,600	16% increase (same as sales)
Operating Income	96,800	107,100	
Interest Expense	11,040	12,230	CL × .06 + LTD × .08
Before-Tax Income	$ 85,760	$ 94,870	
Income Taxes (tax rate = 40%)	34,304	37,948	
Net Income	$ 51,456	$ 56,922	

BALANCE SHEETS, AS OF DECEMBER 31

	Current Inventory Policy	New Inventory Policy	Remarks
Assets			
Current Assets:			
Cash and Securities	$ 47,000	$ 54,520	Change in proportion with sales (16% increase)
Accounts Receivable	63,000	73,080	Change in proportion with sales (16% increase)
Inventory	160,000	240,000	16-car increase (48 − 32) × $5,000 each
Total Current Assets	$ 270,000	$ 367,600	
Prop, Plant, and Equipment, Net	72,000	72,000	No change
Total Assets	$ 342,000	$ 439,600	
Liabilities and Equity			
Current Liabilities:			
Accounts Payable	$ 50,000	$ 58,000	Change in proportion with sales (16% increase)
Notes Payable	74,000	85,840	Change in proportion with sales (16% increase)
Total Current Liabilities	$ 124,000	$ 143,840	
Long-Term Debt	45,000	45,000	No change
Total Liabilities	$ 169,000	$ 188,840	
Common Stock	35,000	35,000	No change
Capital in Excess of Par	34,000	34,000	No change
Retained Earnings	104,000	109,466	Old RE + 5,466 net income difference
Total Stockholders' Equity	$ 173,000	$ 178,466	
Total Liabilities and Equity	$ 342,000	$ 367,306	
Additional Funds Needed		72,294	To be obtained from external sources
		$ 439,600	

Note 1: The new total inventory costs were computed as follows:

$$\text{New Order Quantity per the EOQ} = \sqrt{\frac{2 \times 232 \times \$100}{500}}$$

$$= 9.63 \text{ (round to 10)}$$

Number of orders this year = 232/10 = 23.2 (round to 23)
New total ordering cost = 23 × $100 = $2,300
New average inventory level = 48
New total carrying cost = 48 × $500 = $24,000
New total inventory cost = $2,300 + $24,000 = $26,300 $8,100 increase

TABLE 19-4

Incremental Cash Flows If Cream Puff Changes Inventory Level from 32 to 48 Cars

1. Incremental Cash Outflows at t_0	
Additional Funds Needed from the Projected Balance Sheet	$72,294
2. Incremental Cash Flows Occurring in the Future	
Incremental Cash Inflows:	
Increase in Sales	$192,000
Incremental Cash Outflows:	
Increase in Cost of Goods Sold	$160,000
Increase in Inventory Costs	8,100
Increase in Other Operating Expenses	13,600
Increase in Interest Expense	1,190
Increase in Taxes	3,644
Total Incremental Cash Outflows:	$186,534
Net Incremental Cash Flows Occurring from Time t_1 through Infinity: $192,000 − $186,534 =	$5,466 per year

t_1 through infinity from Table 19-4 ($5,466 per year) using Equation 8-5, the formula for the present value of a perpetuity:

$$PVP = PMT \times \left(\frac{1}{k}\right)$$

$$= \$5,466 \times \left(\frac{1}{.10}\right)$$

$$= \$54,660$$

We find that the present value of the net cash flow perpetuity is $54,660.

Next, we subtract the present value of the net cash outflows at t_0 (the initial investment) to find the NPV of the inventory policy change. From Table 19-4 we see the net cash outflow that occurs at t_0 is $72,294. So, the NPV of Dealin' Dan's Cream Puff Used Cars inventory change proposal is as follows:

$$NPV = \$54,660 - \$72,294$$

$$= \$(17,634)$$

We find that the net present value of the inventory change proposal is −$17,634. Because the NPV is negative, the inventory change proposal should be rejected. Accepting it would decrease the value of the Cream Puff Used Cars company by $17,634.[11]

In the preceding analysis, we used one possible inventory level and observed the effect on the value of the Cream Puff firm. By repeating this procedure a number of times, we could eventually find one inventory level, or a range of levels, at which the firm's value was maximized. We would then have found the true optimal level of inventory for the firm.

---

[11]The internal rate of return (IRR) of the inventory proposal is:

$$k, \text{ or IRR,} = PMT / PV \text{ (initial investment)}$$

$$IRR = 5,466 / 72,294$$

$$= .0756, \text{ or } 7.56\%$$

Because 7.56 percent is less than Dealin' Dan's cost of capital of 10 percent, the inventory change proposal should be rejected.

Inventory Management Approaches

Managing inventory is more than just determining the optimal level of items to keep on hand. Remember, the task is to hold down inventory costs without sacrificing sales too much. Techniques for doing this abound, but two approaches deserve special mention: the ABC classification system and the just-in-time (JIT) system.

The ABC Inventory Classification System

The **ABC system** of inventory classification is a tool used to lower inventory carrying costs. The system classifies inventory according to value. In many firms, inventory items may range in value from relatively expensive to relatively cheap. Generally, firms have fewer expensive items and more inexpensive items. In such a situation, it doesn't make sense to use one inventory control system to manage all inventory items because the firm would waste a lot of time and effort monitoring the relatively cheap items. For example, imagine the inventory system of a bicycle store. Its inventory would probably include several custom-designed racing bicycles; standard 10-speed and mountain bikes; and cycling helmets, water bottles, and other cycling equipment. Wouldn't it waste time and effort to assign serial numbers to all items in inventory and keep them all locked up in glass cases?

Under the ABC system, firm managers classify the relatively few, very expensive items as group A, the larger number of less expensive items as group B, and the rest of the relatively cheap items as group C.[12] Then different inventory control systems are designed for each group, appropriate for the value of that group. For example, the owner of a bicycle shop might assign custom-designed racing bikes to group A, less expensive standard bicycles to group B, and the rest of the inventory to group C. Then the owner could apply inventory control techniques appropriate for each group, as follows:

- *Group A:* Assign serial numbers to each item. Keep in secure storage. Check inventory daily. Keep fixed number on display, ordering replacements as each is sold.
- *Group B:* Assign serial numbers to each item. Keep in secure storage. Check inventory monthly. Manage levels of each type per the EOQ model.
- *Group C:* Check inventory annually. Reorder when visual checks of shelves indicate need.

This technique allows the bicycle store owner to concentrate his or her time and effort on those items that deserve it. Unnecessary carrying costs on the rest of the inventory items are thus avoided.

Just-in-Time Inventory Control (JIT)

The just-in-time (JIT) inventory system, developed in Japan, is useful when storage space is limited and inventory carrying costs are high. As shown in the chapter opener, the system attempts to operate the firm on little or no inventory.

---

[12]Of course, the classifications are not limited to just A, B, and C. Depending on the firm's product lines, some companies might have D, E, and F as well. However, the guiding principle is the same.

Here is an example of how JIT works. A firm that makes kitchen cabinets needs wood, brass handles and knobs, screws, and varnish. All these items constitute the firm's raw materials inventory. On the one hand, the firm could order these raw materials once a month and keep them in storage areas until needed in the manufacturing process (incurring inventory carrying costs as a result). On the other hand, the firm might strike a deal with its raw materials suppliers to deliver just the number of items needed immediately upon request. The items would thus arrive just in time to be used. The firm would not need to store materials, and inventory carrying costs would be eliminated.[13]

In addition to lowering inventory carrying costs, just-in-time systems tend to force quality into the manufacturing process. Any defect in materials will force the entire production line to shut down until the firm can obtain replacement materials.

Carrying little or no inventory can have drawbacks, however. Suppliers that are late or produce poor-quality products jeopardize the firm's customer relations. Little or no inventory means that the business does not have a buffer when a work slowdown occurs due to illness, natural disaster, or a labor dispute.

For instance, General Motors had an inventory system similar to JIT. When 3,000 workers at two brake plants in Dayton, Ohio, went on strike in early 1996, the ripple effects were staggering. Without the brake parts in inventory, workers at other GM plants could not complete car assembly, so eventually 177,775 GM workers were idled. The labor dispute virtually shut down GM's North American operations.[14]

Recall that the JIT inventory system was developed in Japan, where labor unions do not exist. Clearly, firms that employ organized labor should carefully consider the risks of JIT, as evidenced by the GM strike.

MAKING CREDIT DECISIONS

Earlier in this chapter, we said that individual customers receive credit from the firm if they meet the firm's credit standards. Credit standards are those requirements each individual customer must satisfy in order to receive credit from the firm. They are tests, in other words, of a person's creditworthiness.

Firms often base their credit standards on the *Five Cs of Credit:* character, capacity, capital, collateral, and conditions.

1. *Character:* the borrower's *willingness to pay.* Lenders evaluate character by looking at borrowers' past payment patterns. A good payment record in the past implies willingness to pay debts in the future.
2. *Capacity:* the borrower's *ability to pay,* as indicated by forecasts of future cash flows. The more confidence a lender has that a borrower is going to receive cash in the future, the more willing the lender will be to grant credit now.
3. *Capital:* how much wealth a borrower has to fall back on, in case the expected future cash flows with which the borrower plans to pay debts don't materialize.

---

[13]You can see that, in effect, the requirement to store materials (and the attendant carrying costs) are passed on to the suppliers. Presumably, the suppliers would adopt such systems as well, until a closely coordinated chain from original suppliers to final customers developed.
[14]Rebecca Blumenstein and Nichole Christian, "Parts Dispute to Remain Despite GM–UAW Accord," *The Wall Street Journal* (March 25, 1996): A3; John Byrne, "Has Outsourcing Gone Too Far?" *Business Week* (April 1, 1996): 26.

Lenders feel more comfortable if borrowers have something they could liquidate if necessary to pay their debts.
4. *Collateral:* what the lender gets if capacity and capital fail, and the borrower defaults on a loan. Collateral is usually some form of tangible asset, such as the firm's inventory, buildings, manufacturing equipment, and so on that has been pledged as security by the borrower.
5. *Conditions:* the business conditions the borrower is expected to face. The more favorable business conditions appear to be for the borrower, the more willing lenders are to grant credit.

To evaluate potential credit customers (in terms of the Five Cs of Credit, or any other criteria), firms find some way to quantify how well the customers compare to the measurement criteria. Some firms use a method known as **credit scoring**. Credit scoring works by assigning points according to how well customers meet indicators of creditworthiness. For example, statisticians have determined that established businesses tend to pay their debts more faithfully than new businesses. So, a credit applicant might be awarded points for each year that the applicant's firm has been in business. A sample of a simplified credit score sheet is shown in Figure 19-4 for Wishful Thinking Company.

We cannot overemphasize the importance of investigating creditworthiness carefully before granting credit. Not doing so is a quick way to end up with lots of accounts receivable and no cash!

Collection Policies to Handle Bad Debts

Sometimes, despite precautions, a firm ends up with customers who don't pay their bills. You thought your firm did a good job of scoring customers. But now some of them are not paying their bills on time, and a few haven't paid at all. Now what?

Slow or no payment is bound to happen to any credit-granting business. Firms establish a **collection policy** to cope with the problem. For instance, what do you do if one of your long-time customers fails to pay a bill on the due date? Send it to a collection agency the next day? Ignore the situation and hope for the best? Send a reminder notice? Start charging interest? It helps to have a collection policy in place. That way, both the firm and its customers know what to expect once credit has been granted.

No one collection policy will be best for all firms and for all customers. The best policy depends on the business situation, the firm's tolerance for abuse, and the relationship it has with customers. However, most firms consider one or more of the following collection policies:

- *Send reminder letters.* Send one or more letters, each one becoming less friendly in tone. Certainly, the first letter should not sound threatening. (How often have you simply misplaced a bill and not realized it until you received a reminder notice?)
- *Make telephone calls.* If gentle reminders in the mail don't produce results, call the customer to see what the problem is. If there is a good reason why the customer hasn't paid the bill, you may choose to take no action or make accommodating arrangements. Make sure that any alternative payment plan is specific so that the firm can follow up early if the customer fails to pay again.

Criteria	Score
1. Length of time since missing a payment on any loan	_____
More than four years — 4 points	
Three to four years — 3 points	
Two to three years — 2 points	
One to two years — 1 point	
Less than one year — 0 points	
2. Length of time in business	_____
More than four years — 4 points	
Three to four years — 3 points	
Two to three years — 2 points	
One to two years — 1 point	
Less than one year — 0 points	
3. Net income	_____
More than $200,000 — 4 points	
$100,000 to $200,000 — 3 points	
$50,000 to $100,000 — 2 points	
$25,000 to $50,000 — 1 point	
Less than $25,000 — 0 points	
4. Net worth	_____
More than $1 million — 4 points	
$500,000 to $1 million — 3 points	
$100,000 to $500,000 — 2 points	
$50,000 to $100,000 — 1 point	
Less than $50,000 — 0 points	
5. Market value of tangible assets	_____
More than $1 million — 4 points	
$500,000 to $1 million — 3 points	
$100,000 to $500,000 — 2 points	
$50,000 to $100,000 — 1 point	
Less than $50,000 — 0 points	
6. Expected business growth rate in next five years	_____
More than 20 percent — 4 points	
15 to 20 percent — 3 points	
10 to 15 percent — 2 points	
5 to 10 percent — 1 point	
Less than 5 percent — 0 points	
Total Score:	_____
Approved for credit if score = 12 or more	

Figure 19-4 Sample Items on a Credit Scoring Worksheet for Wishful Thinking Company

Figure 19-4 is an example of items on a credit scoring sheet to determine an applicant's creditworthiness. No attempt is made here to determine the amount of credit that may be extended.

- *Hire collection agencies.* When all efforts to collect are unsuccessful, you may want to turn to professional collection agencies. This action should be used sparingly for two reasons. First, it will probably cost the firm any future business from this customer. Second, the price of the collection agency service may be very high, often 50 percent of the uncollected debt.
- *Sue the customer.* Legal action is a last resort. A lawsuit is even more expensive than using a collection agency, so firms should determine whether the court action is worth the trouble. Remember the "lawyers first" rule: Lawyers almost always get paid first.

FINANCE AT WORK

SALES·RETAIL·SPORTS·MEDIA TECHNOLOGY·PUBLIC RELATIONS·PRODUCTION·EXPORTS

Joann K. Jones, CEO of Capital Electric Supply

Joann K. Jones

"I ain't gonna have no broad tell me what to do." These words from a hostile employee greeted Joann K. Jones after she took charge of her husband's company when he was diagnosed with Alzheimer's. Joann, whose previous years were spent taking care of the home front, had to become a CEO almost overnight. And she had to turn around Capital Electric Supply, a wholesale distributor, whose equity had dropped $2 million in two years, while sales had dropped by $8 million. Joann's efforts succeeded. One of the keys to the company's dramatic success was Joann's quick action in taking command of inventory and accounts receivable. Under Joann's guidance, Capital Electric's net income rose in four years from a $231,000 loss to a $300,000 profit.

Q. What is one basic premise everyone should understand about inventory?

Inventory costs money—lots of it. Inventory expenses include financing, storage, accounting, taxes, insurance, security, obsolescence, and shrinkage. Even if your inventory is just sitting on the shelf, you must cover those costs.

Q. What was Capital Electric's inventory level when you took over?

We had much too much inventory for our customers' demands. We also had too many items that moved slowly and held up our ability to collect money. For example, we kept five special electrical panels in stock, even though on average we only sold one a year. We had a great deal of money tied up in the other four. We also faced another problem. Our purchasing agent would buy the complete line that his favorite supplier sold, even though we had no demand for half the items.

Q. How did you correct your inventory problem?

First, we returned items to the vendor where possible to credit our account with that vendor. Doing so added more cash to the bottom line so we could pay our bills. Then we sold remaining excess inventory at heavily discounted rates. If we couldn't sell items, we donated them to charities. Once we got rid of items that didn't move quickly, we had the cash to stock quick-moving items. Now at least 60 percent of our inventory items move quickly—items like lock nuts, extension cords, light bulbs. These items are our bread and butter; they pay employees' salaries.

Because we are a relatively small company, we need to be smarter than our competitors. We have a strong relationship with our vendors and this allows us to obtain special items overnight, if needed. Those are items that our competitor has foolishly sitting on shelves collecting dust.

Q. How do you monitor your inventory and accounts receivable?

We take charge of our inventory with weekly "flash reports" that indicate gross sales, gross margin, administrative overhead, accounts receivable, and inventory turnover. These weekly reports are an important management tool. A business owner shouldn't wait until the middle of the following month to learn what happened this week or this month.

Q. How do you keep your accounts receivable at a low level?

Our accounts receivable are a goodwill gesture to our customers so they will come back more often. It's a way of extending credit to your customers for 30 days, so they can collect their money before they pay us. But we don't want to hold onto accounts receivable too long. Holding accounts receivable for too long affects our cash flow, which affects our ability to pay our bills. That's why we encourage our customers to pay early by offering them a 2 percent discount if they pay by the tenth of the following month.

Q. What do you view as the optimal level of accounts receivable and inventory?

Simply put, you want the right level of accounts receivable and inventory that not only satisfies your customers' needs, but that gives you a positive, timely cash flow.

Source: Interviews with Joann K. Jones.

- *Settle for a reduced amount.* A firm should keep in mind that trying too hard to collect from a customer may force the customer into bankruptcy. Once the client is in bankruptcy, the firm may not receive any money. In such a case, settling for a reduced amount, or a stretched-out payment schedule, may be the firm's best option.
- *Write off the bill as a loss.* In other words, forget it. This may be a firm's best alternative if the amount owed is relatively small or too costly to collect. Firms may have to write off all or part of the bill as a loss anyway, if efforts to collect are unsuccessful.
- *Sell accounts receivable to factors.* Selling accounts receivable to some other person or business is known as **factoring**. Businesses that make money by buying accounts receivable from other firms, at less than their face value, are called **factors**. Suppose your firm had 100 customers who owed you a total of $10,000. Rather than wait for the customers to pay, you might sell the "IOUs" to a factor for $9,000 in cash. The factor discounts the accounts by an amount that both generates a return and compensates for the risk that some customers won't pay. Your firm no longer has to manage the accounts, plus it has cash to put to use elsewhere.

A firm that frequently writes off uncollected amounts needs to tighten its standards for granting credit.

When granting credit to customers, it is best to remember the old saying, "An ounce of prevention is worth a pound of cure." In other words, crafting credit standards that avoid frequent collections can save a firm time and money.

What's Next

In this chapter we have explored how and why firms manage inventory and accounts receivable. We have examined methods for finding the optimal levels of both categories of current assets and have explored additional management techniques. In chapter 20 we'll look at short-term financing.

Summary

1. **Describe how and why firms must manage accounts receivable and inventory as investments.**

Granting credit and maintaining inventory are necessary business practices. Offering more credit and increasing inventory enhance sales. However, both accounts receivable and inventory tie up cash and incur opportunity costs—the cost of not having funds that could generate returns. The financial manager's task is to balance (1) the risk of losing sales due to not granting credit or having products available, against (2) the savings produced by collecting cash immediately or maintaining inventory at a reduced level.

2. **Compute the optimal levels of accounts receivable and inventory.**

The optimal level of accounts receivable or inventory may be found by using a three-step process: (1) create *pro forma* financial statements for alternative credit or inventory policies; (2) use the *pro forma* data to estimate the incremental cash flows associated with the proposed changes; and (3) compute the net present value of each alternative. When this process is complete, managers compare the NPVs of each alternative to see which policy produces the most favorable effect on the value of the firm.

3. Describe alternative inventory management approaches.

Two popular inventory management approaches are the ABC and just-in-time (JIT) inventory systems. The ABC system classifies inventory items into categories according to their relative value. Management's time, money, and effort can then be directed to those inventory items in the proportions that they deserve.

The JIT inventory system calls for close coordination between manufacturers and suppliers to ensure that parts and materials used in the manufacturing process arrive just in time to be used. If the coordination is close enough, raw materials inventories at the manufacturing firm can be eliminated.

4. Explain how firms make credit decisions and create credit policies.

Firms evaluate the creditworthiness of their customers by applying the following Five Cs of Credit:

- *Character*—a borrower's willingness to pay.
- *Capacity*—a borrower's ability to pay.
- *Capital*—how much wealth a borrower has to fall back on, in case the expected future cash flows with which the borrower plans to pay debts don't materialize.
- *Collateral*—what the lender gets if capacity and capital fail, and the borrower defaults on a loan.
- *Conditions*—the business conditions a borrower is expected to face.

To evaluate potential credit customers, firms use a credit scoring procedure that assigns numerical values to the various indicators of creditworthiness.

A firm's collection policy includes the actions the firm plans to take in the event that credit customers don't pay their bills on time. Policy actions include reminder letters, telephone calls, use of collection agencies, court action, settling for partial payment, factoring the accounts, and writing off the bills as a loss.

Companion Website
content by Active Learning Technologies

You can access your FREE PHLIP/CW On–Line Study Guide, Current Events, and Spreadsheet Problems templates, which may correspond to this chapter either through the Prentice Hall Finance Center CD-ROM or by directly going to: **http://www.prenhall.com/gallagher.**

PHLIP/CW — *The Prentice Hall Learning on the Internet Partnership–Companion Web Site*

Equations Introduced in This Chapter

Equation 19-1. The Economic Order Quantity for Inventory:

$$OQ = \sqrt{\frac{2 \times S \times OC}{CC}}$$

where: OQ = Order quantity
S = Annual sales volumes in units
OC = Ordering costs, per order
CC = Carrying costs, per unit per year

Self-Test

ST-1. Assume the Cash & Carry company is considering offering credit to its customers. Management estimates that if it does so, sales will increase 10 percent, expenses will increase 5 percent, accounts receivable will increase to $200,000, cash will decrease by $50,000, and current liabilities will increase to $300,000. No other accounts on the financial statements will be affected. Compute the company's return on equity (ROE) ratio and current ratio if it adopts the proposal. Base your calculations on Cash & Carry's latest financial statements shown next.

Selected Financial Data for Cash & Carry, Inc.
(in 000's)

	Prior to Granting Credit	After Granting Credit
Cash	$ 100	
Accounts Receivable	0	
Total Current Assets	$ 100	
Current Liabilities	$ 50	
Total Equity	$2,000	
Sales	$2,500	
Expenses	2,300	
Net Income for the Year	$ 200	

ST-2. Your firm sells inventory at an even rate throughout the year. Sales volume this year is expected to be 100,000 units. Inventory ordering costs are $30 per order and inventory carrying costs are $60 per unit per year. Given these conditions, what is the most economical inventory order quantity (EOQ)?

ST-3. Refer again to the Dealin' Dan's Cream Puff Used Cars example in the text. Continue with the assumptions given and compute the NPV of raising the company's average inventory level of cars to sixty. Assume the inventory change would cause unit sales to increase to 240 per year.

ST-4. Your firm uses the credit scoring worksheet in Figure 19-4 to evaluate potential credit customers. If an applicant has never missed a loan payment, has been in business for two and a half years, has net income of $75,000, has a net worth of $90,000, has tangible assets worth $120,000, and has an expected business growth rate of 12 percent a year, will the applicant be granted credit?

Review Questions

1. Accounts receivable are sometimes not collected. Why do companies extend trade credit when they could insist on cash for all sales?
2. Inventory is sometimes thought of as a necessary evil. Explain.
3. What are the primary variables being balanced in the EOQ inventory model? Explain.
4. What are the benefits of the JIT inventory control system?
5. What are the primary requirements for a successful JIT inventory control system?
6. Can a company have a default rate on its accounts receivable that is too low? Explain.
7. How does accounts receivable factoring work? What are the benefits to the two parties involved? What are the risks?

Build Your Communication Skills

CS-1. The three-step process of evaluating the NPV of a proposed credit policy or inventory level change can intimidate some managers. Prepare an oral presentation to explain how to measure the value of the credit policy change analysis presented in this chapter. Lead a study group through the process.

CS-2. Evaluate the creditworthiness of a business in your local community using the credit scoring worksheet in Figure 19-4. To collect information, you may want to research local business publications, conduct interviews, or contact the local chamber of commerce. Prepare a brief written report of the results and include your credit scoring worksheet as an exhibit.

Problems

Accounts Receivable, ACP

19-1. Compu-Chip Co. had an annual credit sales of $8,030,000 and an average collection period (ACP) of 22 days in 1999. What is the company's accounts receivable for the year? Assume 365 days in a year.

Accounts Receivable, ACP

19-2. If Compu-Chip Co. of problem 19-1 is expected to have annual credit sales of $7,600,000 and an average collection period of 26 days in 2000, what would be the company's accounts receivables? Do you think that the company relaxed or tightened its credit policy in 2000, compared to its policy in 1999?

Accounts Receivable, ACP, Credit Policy

19-3. Fitzgerald Company has credit terms of 2/15, n60. The historical payment patterns of its customers are as follows:

- 40 percent of customers pay in 15 days.
- 57 percent of customers pay in 60 days.
- 3 percent of customers pay in 100 days.

Annual sales of tools is $730,000. Assume there are 365 days in a year.

a. Calculate the average collection period (ACP).
b. What is the accounts receivable (AR) assuming all goods are sold on credit?

Accounts Receivable, ACP, Credit Policy

19-4. If Fitzgerald in problem 19-3 decides to adopt a more stringent credit terms of 2/10, n30, sales are expected to drop by 10 percent but the following improved payment pattern of its customers is expected:

- 40 percent of customers will pay in 10 days.
- 58 percent of customers will pay in 30 days.
- 2 percent of customers will pay in 100 days.

Calculate the new average collection period (ACP) and accounts receivable (AR) assuming all goods are sold on credit.

Effect of Change of Credit Policy

19-5. Elwood Blues, vice president of sales for East-West Trading, Inc., wants to change the firm's credit policy from 2/15, n40 to 2/15, n60, effective January 1, 2000. He is confident that the proposed relaxation will result in a 20 percent increase over the otherwise expected annual sales of $350,000 with the old policy. All sales are made on credit. The historical payment pattern under the present credit terms are as follows:

Under the old policy:

- 40 percent of the customers take advantage of the discount and pay in 15 days.

- 58 percent of the customers forgo the discount and pay in 40 days.
- The remaining 2 percent pay in 100 days.

Under the new credit policy, the payment pattern is expected to be as follows:

- 40 percent of the customers will take advantage of the discount and pay in 15 days.
- 57 percent of the customers will forgo the discount and pay in 60 days.
- The remaining 3 percent will pay in 100 days.

Bad debt expenses are expected to rise from 2 percent to 3 percent with the change in credit policy. Assume (i) any increase in the current assets will be financed by short-term notes at an interest rate of 7 percent; (ii) long-term interest rate is 10 percent; (iii) income tax rate is 40 percent; (iv) cost of capital for East-West is 11 percent; (v) cost of goods sold is 80 percent of sales; (vi) other operating expenses is $10,000 under the old policy.

The *pro forma* balance sheet items of the company under the old policy would be as follows:

Cash and Securities	$ 15,000	Accounts Payable	$14,918
Inventory	50,000	Notes Payable	35,000
Plant and Equipment	120,000	Long-Term Debt	30,000
Common Stock	25,000	Capital in Excess of Par	60,000
Retained Earnings	50,000		

Also assume that the cost of goods sold and other operating expenses in the income statement and all current asset and current liability items, except notes payable, vary directly with sales.

a. Calculate average collection periods and accounts receivable under the old and the new policies.
b. Develop *pro forma* income statements and balance sheets under the old and the new policies.
c. Calculate the incremental cash flows for the year 2000 and the subsequent years.
d. Advise Mr. Blues if he should adopt the new policy.

19-6. Use the same information given in problem 19-5 with the following changes. Mr. Blues asked Mr. Scott Hayward, the general manager of sales of East-West, to recheck the payment patterns and credit history of East-West's customers to be absolutely sure that the change in credit policy would indeed be beneficial to the company. A strict scrutiny by Mr. Hayward resulted in the following changes in expected payment pattern and bad debts:

Effect of Change of Credit Policy

Under the old policy:

- 40 percent of the customers take advantage of the discount and pay in 15 days.
- 58 percent of the customers forgo the discount and pay in 40 days.
- The remaining 2 percent pay in 100 days.

Under the new credit policy, the payment pattern is expected to be as follows:

- 30 percent of the customers will take advantage of the discount and pay in 15 days.
- 60 percent of the customers will forgo the discount and pay in 60 days.
- The remaining 10 percent will pay in 100 days.

Bad debt expenses are expected to rise from 2 percent to 4 percent with the change in credit policy. Under this changed scenario, is adoption of the new credit policy advisable?

19-7. Tom Jackson, the vice president of sales for A-Z Trading, Inc., wants to change the firm's credit policy from 3/10, n40 to 3/15, n30, effective January 1, 2000. He is confident that though the proposed tightening will result in a 10 percent decrease over the otherwise ex-

Challenge Problem

pected annual sales of $2 million with the old policy, it will increase profitability and value of the firm. All sales are made on credit. The historical payment pattern under the present credit terms are as follows:

Under the old policy:

- 30 percent of the customers take advantage of the discount and pay in 10 days.
- 60 percent of the customers forgo the discount and pay in 40 days.
- The remaining 10 percent pay in 100 days.

Under the new credit policy, the payment pattern is expected to be as follows:

- 42 percent of the customers will take advantage of the discount and pay in 15 days.
- 57 percent of the customers will forgo the discount and pay in 30 days.
- The remaining 1 percent will pay in 100 days.

Bad debt expenses are expected to decrease from 3 percent to 1 percent with the change in credit policy. Assume (i) any decrease in the current assets will be used to pay off short-term notes currently outstanding at an interest rate of 8 percent; (ii) long-term interest rate is 11 percent; (iii) income tax rate is 40 percent; (iv) cost of capital for A-Z is 13 percent; (v) cost of goods sold is 80 percent of sales; (vi) other operating expenses is $60,000 under the old policy.

The *pro forma* balance sheet items of the company under the old policy would be as follows:

Cash and Securities	$ 86,000	Accounts Payable	$ 85,000
Inventory	285,000	Notes Payable	200,000
Plant and Equipment	652,000	Long-Term Debt	171,000
Common Stock	143,000	Capital in Excess of Par	342,000
Retained Earnings	285,000		

Also assume that the cost of goods sold and other operating expenses in the income statement and all current account items in the balance sheet vary directly with sales.

a. Calculate average collection periods and accounts receivable under the old and the new policies.
b. Develop *pro forma* income statements and balance sheets under the old and the new policies.
c. Calculate the incremental cash flows for the year 2000 and the subsequent years.
d. Advise Mr. Jackson if he should adopt the new policy.

Economic Order Quantity

19-8. Windhome and Drake Co., a dealer in building products, has the following costs associated with its business in 1999:

Ordering Cost	$250 per Order
Carrying Cost	$300 per Unit per Year
Annual Sales	500 Units

Calculate the EOQ and the number or orders placed per year.

Economic Order Quantity

19-9. Use the same data for problem 19-8 except that the carrying cost is expected to increase by 10 percent in 2000 due to increase in rentals of warehouse space. Recalculate the EOQ, number of orders placed per year, and total ordering cost for 2000.

Credit Scoring

19-10. Mr. Danny Fisher's firm uses the credit scoring sheet in Figure 19-4 to evaluate the creditworthiness of its customers. An applicant missed a loan payment three and a half years back, has been in business for six years, has a net income of $143,000, a net worth of $1.5 million, a market value of $550,000 in tangible assets, and a business growth rate of 14 percent. Would Mr. Fisher approve credit to the applicant?

19-11. Mr. Homer Smith is the vice president of sales for Sunrise Corporation, which buys and sells mobile homes. He is sure that increasing the number of homes on display will cause sales to increase. He thinks that an increase in inventory effective January 1, 2000, from the present level of 60 units to 100 units will boost up sales from 350 units per year to 450 units per year. Assume ordering cost to be $200 per order, carrying cost to be $600 per unit per year, unit sales price to be $10,000, unit purchase price to be $8,000, and applicable income tax rate to be 40 percent. Also assume that any increase in the current assets will be financed by short-term notes at an interest rate of 7 percent, long-term interest rate to be 10 percent, cost of capital for Sunrise to be 11 percent, cost of goods sold to be 80 percent of sales, and other operating expenses to be $100,000 under the old policy. The *pro forma* balance sheet items of the company under the old policy would be as follows:

Effect of Change of Inventory Policy

Cash and Securities	$ 55,000	Accounts Payable	$100,000
Accounts Receivable	105,000	Notes Payable	95,000
Plant and Equipment	100,000	Long-Term Debt	65,000
Common Stock	60,000	Capital in Excess of Par	220,000
Retained Earnings	200,000		

Also assume that the cost of goods sold and other operating expenses in the income statement and all current asset and current liability items vary directly with sales.

a. Calculate the EOQ, number of orders issued per year, and inventory cost.
b. Develop *pro forma* income statements and balance sheets under the old and the new policies.
c. Calculate the incremental cash flows for the year 2000 and the subsequent years.
d. Advise Mr. Smith if he should adopt the new policy.

19-12. Use the same information given in problem 19-11 except that Ms. Judy Benjamin, the general manager of sales for Sunrise, thinks that increasing the inventory level from 60 to 90 will increase sales from 350 to 390 units per year. With other assumptions remaining the same as in problem 19-11, evaluate this change in the inventory policy.

Effect of Change of Inventory Policy

19-13. Ms. Terry McKay is the vice president of sales for Windermere Corporation, which sells hot air balloons. She is sure that increasing the number of balloons on display will cause sales to increase. She thinks that an increase in inventory effective January 1, 2000, from the present level will boost sales to higher levels in 2000 as shown:

Comprehensive Problem

	Inventory Level (Units)	Sales (Units)
Present	70	340
Future (2000)	(1) 80	375
	(2) 90	390
	(3) 100	400

Assume ordering costs are $160 per order, carrying costs are $400 per unit per year, unit sales price is $16,000, unit purchase price is $12,800, and applicable income tax rate is 40 percent. Also assume that any increase in the current assets will be financed by short-term notes at an interest rate of 7 percent, the long-term interest rate is 11 percent, cost of capital for Windermere is 13 percent, cost of goods sold is 80 percent of sales, and other operating expenses are $130,000 under the present policy. The *pro forma* balance sheet items of the company under the present policy would be as follows:

Cash and Securities	$ 65,000	Accounts Payable	$110,000
Accounts Receivable	114,000	Notes Payable	95,000
Plant and Equipment	113,000	Long-Term Debt	65,000
Common Stock	80,000	Capital in Excess of Par	320,000
Retained Earnings	518,000		

Also assume that the cost of goods sold and other operating expenses in the income statement and all current account items in the balance sheet vary directly with sales.

Find out for Ms. McKay what level of inventory maximizes value of the firm by doing the following, and comparing the net present values of the cash flows associated with each level of inventory:

a. Calculate the EOQ, number of orders issued per year, and inventory cost.
b. Develop *pro forma* income statements and balance sheets under the present and the future policies.
c. Calculate the incremental cash flows for the year 2000 and the subsequent years.
d. Advise Ms. McKay whether she should change the present inventory policy. If so, which inventory level should she adopt?

Answers to Self-Test

ST-1.

Selected Financial Data for Cash & Carry, Inc.
(in 000's)

	Prior to Granting Credit	After Granting Credit
Cash	$ 100	$ 50
Accounts Receivable	0	200
Total Current Assets	$ 100	$ 250
Current Liabilities	$ 50	$ 300
Total Equity	$2,000	$2,000
Sales	$2,500	$2,750
Expenses	2,300	2,415
Net Income for the Year	$ 200	$ 335
Current Ratio	$100/$50 = 2.0	$250/$300 = .833
Return on Equity	$200/$2,000 = .10	$335/$2,000 = .1675

ST-2. Per Equation 19-1, the EOQ model:

$$OQ = \sqrt{\frac{2 \times 100,000 \times 30}{60}}$$

$$= \sqrt{100,000}$$

$$= 316.228 \text{ (round to 316)}$$

ST-3. Use the three-step process to compute the NPV of raising the average inventory to 60 cars:

- *Step 1:* Create *pro forma* financial statements for the new inventory policy:

Dealin' Dan's Cream Puff Used Cars Inventory Analysis

Income Statements

	Current Inventory Policy	New Inventory Policy	Remarks
Sales	$ 1,200,000	$ 1,440,000	Unit sales × price each, from assumptions
Cost of Goods Sold	1,000,000	1,200,000	Unit sales × cost each, from assumptions
Gross Profit	200,000	240,000	
Inventory Costs	18,200	32,400	See Note 1
Other Operating Expenses	85,000	102,000	Increase in proportion with sales
Operating Income	96,800	105,600	
Interest Expense	11,040	12,528	ST & LT Debt × Costs of Debt
Before-Tax Income	85,760	93,072	
Income Taxes (rate = 40%)	34,304	37,229	
Net Income	$ 51,456	$ 55,843	

Balance Sheets, as of December 31

	Current Inventory Policy	New Inventory Policy	Remarks
Assets			
Current Assets:			
Cash and Securities	$ 47,000	$ 56,400	Change in proportion with sales
Accounts Receivable	63,000	75,600	Change in proportion with sales
Inventory	160,000	300,000	Average inventory × cost each, from assumptions
Total Current Assets	270,000	432,000	
Prop, Plant, and Equipment, Net	72,000	72,000	No change
Total Assets	$ 342,000	$ 504,000	
Liabilities and Equity			
Current Liabilities:			
Accounts Payable	$ 50,000	$ 60,000	Change in proportion with sales
Notes Payable	74,000	88,800	Change in proportion with sales
Total Current Liabilities	124,000	148,800	
Long-Term Debt	45,000	45,000	No change
Total Liabilities	169,000	193,800	
Common Stock	35,000	35,000	No change
Capital in Excess of Par	34,000	34,000	No change
Retained Earnings	104,000	108,387	Old RE + net income change
Total Stockholders' Equity	173,000	177,387	
Total Liabilities and Equity	$ 342,000	$ 371,187	
Additional Funds Needed		132,813	Obtained from external sources
		$ 504,000	

Note 1: The new total inventory costs were computed as follows:

$$\text{New Order Quantity per the EOQ} = \sqrt{\frac{2 \times 240 \times \$100}{500}}$$

$$= 9.8 \text{ (round to 10)}$$

Number of orders this year = 24/10 = 24
New total ordering cost = 24 × $100 = $2,400
New average inventory level = 60
New total carrying cost = 60 × $500 = $30,000
New total inventory cost = $2,400 + $30,000 = $32,400

- *Step 2:* Compute the incremental cash flows:

Incremental Cash Flows Associated with Changing Cream Puff Used Cars Inventory Level from 32 Cars to 60 Cars

1. Incremental Cash Outflows at Time Zero (t_0):

    Additional Funds Needed from the Projected Balance Sheet $132,813

2. Incremental Cash Flows Occurring in the Future

    Incremental Cash Inflows:

        Increase in Sales $240,000

    Incremental Cash Outflows:

Increase in Cost of Goods Sold	$200,000
Increase in Inventory Costs	14,200
Increase in Other Operating Expenses	17,000
Increase in Interest Expense	1,488
Increase in Taxes	2,925
Total Incremental Cash Outflows:	$235,613

    Net Incremental Cash Flows Occurring from Time t_1 through Infinity: $ 4,387 per year

- *Step 3:* Compute the NPV of the Inventory Policy Change.

PV of the net cash inflows occurring from time t_1 through infinity ($4,387 per year) using Equation 8-5:

$$PVP = PMT \times \left(\frac{1}{k}\right)$$

$$= \$4,387 \times \left(\frac{1}{.10}\right)$$

$$= \$43,870$$

Subtract the PV of the net cash outflows at t_0 ($132,813) to obtain the NPV of the inventory policy change:

$$NPV = \$43,870 - \$132,813$$

$$= (\$88,943)$$

Because the NPV is negative, the inventory change proposal should be rejected.

ST-4. The applicant's completed credit scoring worksheet from Figure 19-4 is shown next.

Credit Scoring Worksheet
for _____(applicant)_____

Criteria		Score
1. Length of time since missing a payment on any loan		4
More than four years	4 points	
Three to four years	3 points	
Two to three years	2 points	
One to two years	1 point	
Less than one year	0 points	
2. Length of time in business		2
More than four years	4 points	
Three to four years	3 points	
Two to three years	2 points	
One to two years	1 point	
Less than one year	0 points	
3. Net income		2
More than $200,000	4 points	
$100,000 to $200,000	3 points	
$50,000 to $100,000	2 points	
$25,000 to $50,000	1 point	
Less than $25,000	0 points	
4. Net worth		1
More than $1 million	4 points	
$500,000 to $1 million	3 points	
$100,000 to $500,000	2 points	
$50,000 to $100,000	1 point	
Less than $50,000	0 points	
5. Market value of tangible assets		2
More than $1 million	4 points	
$500,000 to $1 million	3 points	
$100,000 to $500,000	2 points	
$50,000 to $100,000	1 point	
Less than $50,000	0 points	
6. Expected business growth rate in next five years		2
More than 20 percent	4 points	
15 to 20 percent	3 points	
10 to 15 percent	2 points	
5 to 10 percent	1 point	
Less 5 percent	0 points	
	Total Score:	13
Approved for credit if score = 12 or more		
Result for applicant:	**Approved**	

20

CHAPTER

SHORT-TERM FINANCING

If you would know the value of money try to borrow some.
—Benjamin Franklin

SUCCESS STORY

Most financiers would rather go hungry than lend money to a family-run restaurant. But when the restaurant has been around for many years—owned by the same family throughout its history—some lenders might be willing to take another look.

When the owners of Julian's Restaurant in Richmond decided to open a second location in western Henrico County, the company borrowed nearly $1.5 million from the Business Loan Center. They used the proceeds from the SBA-backed loan to buy land and build an upscale facility in a high-growth spot on Three Chopt Road.

"We've been working on this idea for more than four years," says Robert Hickman, one of five owners of the restaurant. "We knew we wanted to expand to the west end. We knew we wanted to own the building for our new restaurant, and we knew we could do well enough to make it worthwhile."

Founded in 1929, Julian's has become one of the city's top medium-priced Italian restaurants. In 1947 the restaurant moved to a building on West Broad Street, where it became a city icon. It moved two doors down in 1988 and continued to

CHAPTER OBJECTIVES

After reading this chapter, you should be able to:

- Explain the need for short-term financing.

- List the advantages and disadvantages of short-term financing.

- Describe three types of short-term financing.

- Compute the cost of trade credit and commercial paper.

- Calculate the cost of a loan and explain how loan terms affect the effective interest rate.

- Describe how accounts receivable and inventory can be used as collateral for short-term loans.

prosper. With its expansion to the affluent suburbs of western Henrico, Julian's now employs about 100 people, and Hickman says the family hopes to open two or three more locations in the Richmond area in the coming years.

Matthew McGee, chief operating officer of the Business Loan Center, says his company liked Julian's because the restaurant's owners had collateral, a strong reputation and a clear image of what they wanted. In sharp contrast, many would-be borrowers approach the Business Loan Center with no idea of how much money they need, McGee says.

The Business Loan Center, one of the largest nonbank lenders in the nation, is owned by BLC Financial Services of New York, a publicly held company that was first listed on the American Stock Exchange in April 1998. With regional offices in Richmond, Panama City, Florida, and Wichita, Kansas, BLC offers financing to businesses throughout the United States, but McGee says the firm historically has lent to companies east of the Mississippi. BLC does more business in Virginia, he says, than in any other state.

Source: Bill Edwards, "Success Story," Virginia Business MagNet <http://www.virginia-business.com/electro/smfinanc/julians.html>.

Chapter Overview

The credit card is perhaps the best-known source of short-term financing. However, businesses use many other types of short-term financing to sustain their business operations. In this chapter we discuss the advantages and disadvantages of short-term financing, sources of that financing, and methods for calculating the cost of each source. We also show how loan terms can affect a loan's effective interest rate, and how accounts receivable and inventory can be used as short-term loan collateral.

The Need for Short-Term Financing

Businesses rely on short-term financing from external sources for two reasons. The first is growth—profits may simply not be high enough to keep up with the rate at which the company is buying new assets. Imagine a convenience store chain that wished to open one new store a month. If each new store cost $100,000, the company would have to be *very* profitable to be able to do this without obtaining external financing.

The second reason that businesses rely on external short-term financing is choice. Rather than waiting to save enough money from net profits to make their desired purchases, many firms would rather borrow the money at the outset and make their purchases on time. People make the same choices in their personal lives. For example, you could save a little money each month until you saved enough to buy a car with cash. This might take a long time, however, and in the meantime you would be without transportation. Alternatively, you could borrow the money to buy the car, and have it to drive around while you're paying off the loan. People—and businesses—often choose the latter alternative.

Clearly, the ability to obtain external financing is crucial for most businesses. Without it, most businesses could never even get started.

Short-Term Financing versus Long-Term Financing

Two factors influence the duration of external financing that businesses seek. The first, of course, is availability. A firm may want to take out a 10-year loan to finance its inventory purchases, but may find no one willing to make such a loan. In general, businesses can usually find financing for short time periods. It is more difficult to find long-term financing.

We discussed the liquidity–profitability trade-off in chapter 17, page 425.

The second factor influencing the length of time that firms finance for is the risk–return, or liquidity–profitability, trade-off discussed in chapter 17.

In the context of financing alternatives, here is how the trade-off works:

- *Short-term financing* is usually cheaper than long-term financing because short-term interest rates are normally lower than long-term interest rates.[1] Therefore, the desire for profitability (return) pushes firms toward short-term financing.
- *Long-term financing* is regarded as less risky than short-term financing for the borrower because the borrower locks in the agreed-on interest rate for a long period of time. No matter how interest rates change during the life of the loan, the borrower's interest costs are certain. Furthermore, the borrower does not have to incur the transaction costs of obtaining new financing every few months. So, the desire to avoid risk encourages firms to use long-term financing.

The length of time that firms finance for depends on whether they want "to eat well or to sleep well."[2] Returns generally increase as financing maturities grow shorter, but so does risk. Risk decreases as financing maturities grow longer, but so do returns. The blend of financing maturities that a firm selects reflects how aggressive or conservative the firm's managers are.

Figure 20-1 summarizes the factors that influence the sources of external financing. External financing can come from short-term or long-term sources. We discuss short-term financing sources next.

Short-Term Financing Alternatives

When most businesses need money for a short time—that is, for less than one year—they usually turn to two sources: short-term loans and trade credit (the process of delaying payments to suppliers). Large, well-established businesses may make use of a third financing source: commercial paper. In the sections that follow, we discuss the various aspects of obtaining money from these three sources.

Short-Term Loans from Banks and Other Institutions

Financial institutions offer businesses many types of short-term loans. No matter what the type of loan, however, the cost to a borrower is usually measured by the percent interest rate charged by the lender. The annual interest rate that reflects the dollars of interest paid divided by the dollars borrowed is the **effective interest rate.**

---

[1] Remember from chapter 2 that a normal yield curve is upward sloping.
[2] The phrase is adapted from a remark by J. Kenfield Morley, who said, "In investing money, the amount of interest you want should depend on whether you want to eat well or sleep well."

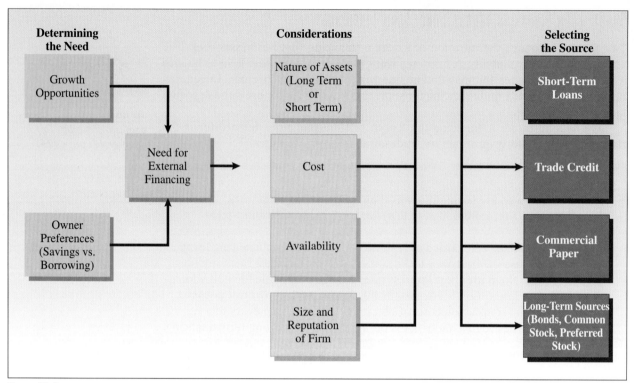

Figure 20-1 **External Financing Source-Selection Process**

This flow chart illustrates the external financing source-selection process. A firm first determines the need for external financing, and then considers several factors before selecting the short-term financing sources.

Often, the effective interest rate differs from the interest rate advertised by the bank, which is known as the **stated interest rate.**

Two common types of short-term loans are the *self-liquidating loan* and the *line of credit*. We examine these loan alternatives next. No matter what type of loan a firm uses, the firm must sign a promissory note. A **promissory note** is the legal instrument that the borrower signs and is the evidence of the lender's claim on the borrower.

Self-Liquidating Loans Many of the short-term loans obtained from banks are self-liquidating. A **self-liquidating loan** is one in which the proceeds of the loan are used to acquire assets that will generate enough cash to repay the loan. An example is a loan used to finance a seasonal increase in inventory, such as the purchase of swimwear to sell during the summer months. The sale of the inventory generates enough cash to repay the loan.

The Line of Credit As we now know, each time a firm borrows money from a bank, it signs a promissory note. However, a firm may have more than one promissory note outstanding at any one time. Indeed, a firm could have a substantial number of promissory notes outstanding, all with overlapping terms of payment. To keep loans under control, banks may specify the maximum total balance that firms may have in outstanding short-term loans. A **line of credit** is the borrowing limit a bank sets for a firm. A line of credit is an informal arrangement. The bank may change a firm's credit limit or withdraw it entirely at any time as business conditions change.

In contrast, a *revolving credit agreement* is a formal agreement between a bank and a borrower to extend credit to a firm up to a certain amount for some period of time (which may be for several years). The agreements are usually set forth in a written contract, and firms generally pay a fee for the revolving credit.

Trade Credit

When a company purchases materials, supplies, or services on credit instead of paying cash, that frees up funds to be used elsewhere, just as if the funds had been borrowed from a bank. **Trade credit** is the act of obtaining funds by delaying payment to suppliers. The longer a company delays paying for purchases, the more trade credit the firm is said to be using.

Even though trade credit is obtained by simply delaying payment to suppliers, it is not always free. Next, we explain the cost of trade credit and how to compute that cost so it can be compared to the cost of a bank loan.

Computing the Cost of Trade Credit
If a supplier charges a firm interest on credit balances, then computing the cost of trade credit is easy—simply read the interest rate charges on the supplier's account statements, much as we would read a credit card's interest charges.

Most wholesale suppliers, however, do not charge interest on credit balances. Instead, they simply give their customers so many days to pay, and offer them a discount on the amount of the purchase if they pay early. A typical example of such credit terms is 2/10, n30—if customers pay their bills within 10 days of the invoice date, they will receive a 2 percent discount; if not, the net amount of the bill is due within 30 days.[3] Figure 20-2 diagrams a purchaser's payment deadlines for a $100 purchase on credit terms of 2/10, n30.

We see from Figure 20-2 that if a firm takes the discount, it can obtain the use of $98 for up to 10 days without any cost. In this case the trade credit the firm receives is free. But suppose a firm doesn't take the discount? Look at Figure 20-2 again and think of the situation this way: Instead of paying $98 on the tenth day, the firm can pay $98 anytime during the next 20 days as long as it pays a "fee" of $2 for delaying payment. In essence, the firm is "borrowing" $98 for 20 days at a cost of $2. Assuming the firm pays its bill on day 30, we can compute the effective annual interest rate of the trade credit using the following equation:

---

[3]Credit terms of this type were introduced in chapter 19 from the point of view of the supplier granting the credit. Here, we discuss the terms from the purchaser's point of view.

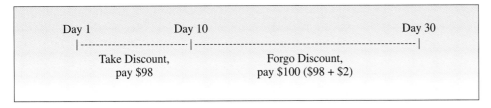

Figure 20-2 Payment Schedule with 2/10, n30 Credit Terms

Figure 20-2 shows the purchaser's payment schedule for a $100 purchase from a supplier who offers 2/10, n30 credit terms.

Trade Credit Effective Annual Interest Rate Formula

$$k = \left(1 + \frac{\text{Discount \%}}{100 - \text{Discount \%}}\right)^{\left(\frac{365}{\text{Days to Pay} - \text{Discount Period}}\right)} - 1 \qquad (20\text{-}1)$$

where: k = The cost of trade credit expressed as an effective annual interest rate

Discount % = The percentage discount being offered

Days to Pay = The time between the day of the credit purchase and the day the firm must pay its bill

Discount Period = The number of days in the discount period

The 365 in the equation represents the number of days in a year. We also multiply the result, k, by 100 to express it as a percentage.

In our example, the discount percentage is 2 percent, the total number of days to pay is thirty, and the number of days in the discount period is ten. We use Equation 20-1 to solve for k as follows:

$$k = \left(1 + \frac{2}{100 - 2}\right)^{\left(\frac{365}{30 - 10}\right)} - 1$$

$$= (1 + .020408)^{(18.25)} - 1$$

$$= (1.020408)^{(18.25)} - 1$$

$$= 1.4458 - 1$$

$$= .4458$$

$$\times 100 = 44.58\%$$

As the calculation shows, the firm's trade credit—the use of $98 for an additional 20 days—costs the firm an effective annual percentage rate of interest of nearly 45 percent! Why would any reasonable financial manager pay such high rates? Most reasonable financial managers wouldn't, unless very unfavorable circumstances forced them to or they didn't realize they were doing it.

Instead, because bank loan rates are usually much lower than 45 percent, most reasonable financial managers would borrow $98 from the bank and use it to pay the supplier on the tenth day to take advantage of the discount. Twenty days later, the financial manager would repay the loan plus the interest charges, which would be considerably less than $2. Either way, a firm can obtain the use of $98 for 20 days, but borrowing from a bank is usually the much cheaper alternative.

Commercial Paper

We described commercial paper in chapter 2, pages 25–26.

Firms can sell **commercial paper**—unsecured notes issued by large, very creditworthy firms for up to 270 days—to obtain cash. Selling commercial paper is usually a cheaper alternative to getting a short-term loan from a bank. Remember that only large, creditworthy corporations sell commercial paper because only they can attract investors who will lend them money for lower rates than banks charge for short-term loans.

Calculating the Cost of Commercial Paper Commercial paper is quoted on a *discount basis*. So, to compare the percent cost of a commercial paper issue to the percent cost of a bank loan, we first convert the commercial paper **discount yield** to an effective annual interest rate. We use the following three-step process to find this rate.

1. Compute the discount from face value using Equation 20-2, the formula for the dollar amount of the discount on a commercial paper note:

Dollar Amount of the Discount on a Commercial Paper Note

$$D = \frac{DY \times Par \times DTG}{360} \quad (20\text{-}2)$$

where: D = The dollar amount of the discount
DY = The discount yield
Par = The face value of the commercial paper issue; the amount to be paid at maturity
DTG = The days to go until maturity

2. Compute the price of the commercial paper issue by subtracting the discount (D) from Par, as shown in Equation 20-3:

Price of a Commercial Paper Note

$$Price = Par - D \quad (20\text{-}3)$$

3. Compute the effective annual interest rate using the following formula, Equation 20-4:

Effective Annual Interest Rate of a Commercial Paper Note

$$\text{Effective Annual Interest Rate} = \left(\frac{Par}{Price}\right)^{\frac{365}{DTG}} - 1 \quad (20\text{-}4)$$

To illustrate the three-step process, imagine you are a financial analyst at Texaco, Inc., and your commercial paper dealer[4] has informed you that she is willing to pay 3.3 percent discount yield for a $1 million issue of Texaco 90-day commercial paper notes. What effective annual interest rate does the 3.3 percent discount yield equate to?

Step 1: Compute the discount using Equation 20-2.

$$D = \frac{DY \times Par \times DTG}{360}$$

$$= \frac{.033 \times \$1,000,000 \times 90}{360}$$

$$= \frac{\$2,970,000}{360}$$

$$= \$8,250$$

We see that with a 3.3 percent discount rate, $1 million face value, and 90 days to go until maturity, the dollar amount of the discount on the commercial paper note is $8,250.

---

[4]Some firms sell commercial paper through dealers. Others, such as General Motors Acceptance Corporation (GMAC), sell it directly to the public.

Step 2: Compute the price using Equation 20-3.

$$\text{Price} = \text{Par} - D$$
$$= \$1,000,000 - \$8,250$$
$$= \$991,750$$

Our calculations show that the price of the 90-day commercial paper note with a face value of $1 million at a discount price of $8,250 is $991,750.

Step 3: Compute the effective annual interest rate using Equation 20-4.

$$\text{Effective Annual Interest Rate} = \left(\frac{\text{Par}}{\text{Price}}\right)^{\left(\frac{365}{\text{DTG}}\right)} - 1$$
$$= \left(\frac{\$1,000,000}{\$991,750}\right)^{\left(\frac{365}{90}\right)} - 1$$
$$= (1.00832)^{4.056} - 1$$
$$= 1.0342 - 1$$
$$= .0342$$
$$.0342 \times 100 = 3.42\%$$

Applying Equation 20-4, we find that the effective annual interest rate of a $1 million, 90-day commercial paper note with a price of $991,750 is 3.42 percent. Now, you can compare the 3.42 percent effective annual interest rate Texaco would pay for commercial paper to the various loan rates available, and choose the best deal.

In the next section, we examine the effect of loan terms on the effective interest rate.

How Loan Terms Affect the Effective Interest Rate of a Loan

The effective interest rate of a bank loan may not be the same as the stated interest rate advertised by the bank because of a lender's loan terms. In the following sections, we describe how to find the effective interest rate, and what terms affect the effective interest rate.

The Effective Interest Rate

Some loans have the same effective rate of interest as the stated rate of interest because the bank places no terms on the loan other than the amount of interest and the amount borrowed. In these cases finding the effective interest rate per period is straightforward. We divide the interest paid on the loan by the amount of money borrowed during the period of the loan (and afterwards multiply the result by 100 to obtain a percent). Equation 20-5 shows the effective interest rate formula:

Effective Interest Rate of a Loan

$$\text{Effective Interest Rate } k = \frac{\$ \text{ Interest You Pay}}{\$ \text{ You Get to Use}} \quad (20\text{-}5)$$

For example, suppose you borrow $10,000 from a bank for one year, and your promissory note specifies that you are to pay $1,000 in interest at the end of the year. We use Equation 20-5 to find the effective interest rate for the loan as follows:

$$k = \frac{\$1,000}{\$10,000}$$
$$= .10$$
$$\times 100 = 10\%$$

The calculations show that for a $10,000 loan with $1,000 in interest, the effective interest rate is 10 percent.

Effective interest rates are customarily expressed as annual rates. If a loan's maturity is for one year and there are no complicating factors, computing effective interest rates is quite simple, as we have just seen. Equation 20-5 gives the effective rate per period.

For many loans, however, things are not so simple. Lenders have a variety of terms and conditions that they apply to loans, and many of them affect the effective interest rate. Two of the more common loan terms, *discount loans* and *compensating balances*, are discussed next.

Discount Loans

Sometimes a lender's terms specify that interest is to be collected up front, at the time the loan is made, rather than at maturity. When this is the case, the loan is referred to as a **discount loan.** In a discount loan the amount the borrower actually receives is the principal amount borrowed minus the interest owed. So the amount the borrower may use is lower than if the loan were a standard loan with interest paid annually at year's end. As a result, the borrower's effective interest rate is higher than it would be for a standard loan.

Let's return to our earlier one-year, $10,000 loan example to see what happens if it were a discount loan. Instead of paying $1,000 in interest at the end of the year (the equivalent of an effective interest rate of 10 percent), the $1,000 in interest must be paid at the beginning of the year. According to Equation 20-5, the effective interest rate is as follows:

$$k = \frac{\$ \text{Interest You Pay}}{\$ \text{You Get to Use}}$$
$$= \frac{\$1,000}{\$10,000 - \$1,000}$$
$$= \frac{\$1,000}{\$9,000}$$
$$= .1111$$
$$\times 100 = 11.11\%$$

Note that by collecting the $1,000 interest on the loan at the start of the year, the effective rate of interest rose from 10 percent to 11.11 percent, solely because of the timing of the interest payment. The stated interest rate, then, is lower than the borrower's effective rate of interest.

Compensating Balances

Sometimes a lender's loan terms will specify that while a loan is outstanding the borrower must keep some minimum balance in a checking account at the lender's institution. The amount required is called a **compensating balance.** The lender would say that this minimum balance is its compensation for granting the borrower favorable loan terms (even though the terms may not be especially favorable). Because the borrower cannot allow the balance in the checking account to fall below the required minimum during the life of the loan, the borrower may not use these funds during the life of the loan. As a result, the borrower's effective interest rate is higher than it would be without a compensating balance requirement. This assumes that the borrower would not have kept the required compensating balance funds in the checking account if the loan were a standard loan.

Let's add a compensating balance requirement to our one-year, $10,000 loan with a year-end interest payment of $1,000. The stated rate of interest is 10 percent. Assume the bank requires a compensating balance of 12 percent of the amount borrowed in a checking account during the life of the loan. This compensating balance requirement would be referred to as a "12 percent compensating balance requirement." We quickly figure out that 12 percent of $10,000 is $1,200. Then we use Equation 20-5 to find the following effective interest rate:

$$k = \frac{\$ \text{ Interest You Pay}}{\$ \text{ You Get to Use}}$$

$$= \frac{\$1,000}{\$10,000 - \$1,200}$$

$$= \frac{\$1,000}{\$8,800}$$

$$= .1136$$

$$\times 100 = 11.36\%$$

We find that the effect of the bank's 12 percent compensating balance requirement is to raise the effective interest rate to the borrower by 1.36 percentage points. Instead of paying 10 percent, the borrower actually pays 11.36 percent. The effect of the compensating balance requirement is to increase the effective rate of interest, 11.36 percent, compared to the stated rate of interest of 10 percent.

Figure 20-3 shows how changing the terms of a one-year loan can affect the effective interest rate. The chart summarizes the effect of simple interest, discount interest, and compensating balances. Figure 20-3 demonstrates that loan terms such as discount interest and compensating balances reduce the amount the borrower gets to use, thus raising the effective interest rate.

Figure 20-3 How Changing the Terms of a Loan Can Affect the Effective Interest Rate

Figure 20-3 shows the impact of loan terms on the effective interest rate (the amount you pay to obtain the loan divided by the amount you get to use during the life of the loan).

Simple Interest Rate	$\frac{\$ \text{ interest you pay}}{\$ \text{ you get to use}}$	$\frac{\$1,000}{\$10,000}$	= .10, or 10%
Discount Interest	$\frac{\$ \text{ interest you pay}}{\$ \text{ you get to use}}$	$\frac{\$1,000}{\$10,000 - \$1,000 = \$9,000}$	= .1111, or 11.11%
Compensating Balance (12%)	$\frac{\$ \text{ interest you pay}}{\$ \text{ you get to use}}$	$\frac{\$1,000}{\$10,000 - \$1,200 = \$8,800}$	= .1136, or 11.36%

Loan Maturities Shorter Than One Year

Another term that affects the effective interest rate is a loan maturity that is less than one year. In such cases we modify Equation 20-5 to convert the effective interest rate of the loan that is for less than a year into an annual rate. We find the annual rate so that we can compare that rate with those from other lenders, almost all of which are expressed as annual rates. Annualizing the rates allows a comparison of apples to apples, rather than apples to oranges. An example demonstrates this point.

Suppose you are borrowing $10,000 for one *month* and paying $1,000 in interest at the end of the *month* (with no other conditions). The effective interest rate of this loan is 10 percent, according to Equation 20-5. However, remember that the rate is 10 percent *per month*. It would be inaccurate to say that the interest rate on this loan was the same as a 10 percent loan from another financial institution. Why? Because the 10 percent stated rate from the other institution is an annual rate and you are comparing it to a monthly rate. For one month, the other institution's stated rate would be 10 percent divided by 12 months equals .83 percent, which is considerably less than the 10 percent monthly interest on your loan.

Annualizing Interest Rates We can modify Equation 20-5 so that it annualizes interest rates that are not paid yearly. The modified formula, Equation 20-6, follows:

Effective Annual Interest Rate

$$\text{Effective Annual Interest Rate } k = \left(1 + \frac{\$ \text{ Interest You Pay}}{\$ \text{ You Get to Use}}\right)^{\binom{\text{Loan Periods}}{\text{in a Year}}} - 1 \quad (20\text{-}6)$$

We then multiply the result by 100 to find the percentage rate.

Now let's use Equation 20-6 to annualize the $10,000 loan with interest of $1,000 a month. Remember, the *monthly* interest rate for this loan is 10 percent, and there are 12 monthly loan periods in a year. The calculations follow:

$$k = \left(1 + \frac{\$1,000}{\$10,000}\right)^{(12)} - 1$$
$$= (1 + .10)^{(12)} - 1$$
$$= (1.10)^{(12)} - 1$$
$$= 3.1384 - 1$$
$$= 2.1384$$
$$\times 100 = 213.84\%$$

We find that the interest rate is more than 213 percent. Surely there is a cheaper alternative at another bank. Suppose you find one whose *stated* annual interest rate for a $10,000 one-month loan is 12 percent. What's the *effective* annual interest rate for this loan? In order to apply Equation 20-6 to find out, we first compute the dollar amount of interest to be paid:

- The stated rate for one year, or 12 months, is 12 percent, so the rate for one month is 12 percent/12 = 1 percent.[5]
- 1 percent of $10,000 is $10,000 × .01 = $100.
- So, the amount of "dollars you pay" to get this loan is $100.

---

[5] If the loan's term is one week, divide by 52. If it is one day, divide by 365, and so on.

Because the loan is for one month, we know that there are 12 loan periods in a year. Now we're ready to plug these numbers into Equation 20-6 as shown next:

$$k = \left(1 + \frac{\$ \text{ Interest You Pay}}{\$ \text{ You Get to Use}}\right)^{\left(\text{Loan Periods in a Year}\right)} - 1$$

$$= \left(1 + \frac{\$100}{\$10{,}000}\right)^{(12)} - 1$$

$$= (1 + .01)^{(12)} - 1$$

$$= (1.01)^{(12)} - 1$$

$$= 1.1268 - 1$$

$$= .1268$$

$$\times 100 = 12.68\%$$

The effective annual rate, 12.68 percent, is a little higher than the bank's stated rate of 12 percent because of the compounding effect of adding "interest on interest" for 12 months.

We have seen how discount loans, compensating balances, and loans that have maturities less than a year affect the effective annual interest rate. Next, we walk through an example of a loan with more than one complicating term.

A Comprehensive Example

Let's consider a loan that includes all the complicating factors discussed in the preceding sections. Suppose you want to borrow $5,000 for one week, and the bank's terms are 8 percent interest, collected on a discount basis, with a 10 percent compensating balance. What is the effective annual interest rate of this loan?

Computing the Interest Cost in Dollars The bank's stated rate of interest for one year, or 52 weeks, is 8 percent, so the rate for one week is 8 percent/52 = 0.1538 percent.

$$0.1538 \text{ percent of } \$5{,}000 \text{ is } \$5{,}000 \times .001538 = \$7.69$$

So, the amount of dollars in interest that you pay to obtain this loan is $7.69.

Computing the Net Amount Received Because this is a discount loan, the interest will be collected up front. That means $7.69 will be deducted from the $5,000 loan.

The loan also has a 10 percent compensating balance requirement, so 10 percent of the $5,000, or $500, must remain in a checking account at the bank, denying you the use of it during the life of the loan.

The net amount of money that you will get to use during the life of the loan is $5,000 − $7.69 − $500 = $4,492.31.

Computing the Effective Annual Interest Rate We use Equation 20-6, the formula for annualizing a loan with a term of interest payments less than a year, to find the effective annual interest rate for this loan. The calculations follow:

$$k = \left(1 + \frac{\$ \text{ Interest You Pay}}{\$ \text{ You Get to Use}}\right)^{\left(\text{Loan Periods in a Year}\right)} - 1$$

$$= \left(1 + \frac{\$7.69}{\$4{,}492.31}\right)^{(52)} - 1$$

$$= (1 + .001712)^{(52)} - 1$$

$$= (1.001712)^{(52)} - 1$$
$$= 1.093 - 1$$
$$= .093$$
$$\times 100 = 9.3\%$$

We see that the effective annual interest rate for a one-week, $5,000 discounted loan with an interest rate of 8 percent, and a 10 percent compensating balance requirement is 9.3 percent. The effective rate of interest, 9.3 percent, is higher than the 8 percent stated rate of interest.

Computing the Amount to Borrow

In the preceding comprehensive example, you tried to borrow $5,000. Presumably that was the amount you needed to use for a week. But, as shown, if the bank collected $7.69 in interest up front, and made you keep $500 in a checking account during the term of the loan, you would only receive $4,492.31. Clearly, given the bank's terms, you will have to borrow some amount greater than $5,000 to end up with the $5,000 you need. So the question is, how much do you have to borrow to walk out of the bank with $5,000?

We solve this question algebraically. Let X = the amount to borrow. Now, because the loan is a discount loan, the bank will collect one week's worth of interest, or (.08/52) times X at the beginning of the week. Furthermore, 10 percent of X must remain in a checking account at the bank as a compensating balance. When these two amounts are subtracted from X, the remainder must equal $5,000. Here is the equation describing the situation:

$$X - \left(\frac{.08}{52}\right)X - .10X = \$5,000$$

We then solve for X as follows:

$$X - \left(\frac{.08}{52}\right)X - .10X = \$5,000$$
$$X - .001538X - .10X = \$5,000$$
$$.8985X = \$5,000$$
$$X = \$5,564.83$$

We find that if you borrow $5,564.83 for one week at 8 percent, discount interest, with a 10 percent compensating balance requirement, you will leave the bank with $5,000.[6]

We have examined how loan terms can affect the effective interest rate. Now we turn to types of collateral that are used to secure short-term loans.

COLLATERAL FOR SHORT-TERM LOANS

The promissory note that specifies the terms of the loan often includes the type of **collateral** used to secure the loan.

Loans for which collateral is required are called secured loans. If no collateral is required, the loan is unsecured.

---

[6] In case you're wondering, the effective annual interest rate for this loan, per Equation 20-2, is 9.3 percent.

For secured short-term loans, lenders usually require that the assets pledged for collateral be short term in nature also. Lenders require short-term assets because they are generally more liquid than long-term assets and are easier to convert to cash if the borrower defaults on the loan. The major types of short-term assets used for short-term loan collateral are accounts receivable and inventory.

Accounts Receivable as Collateral

Accounts receivable are assets with value because they represent money owed to a firm. Because of their value, a lender might be willing to accept the accounts as collateral for a loan. If so, the borrowing firm may *pledge* its accounts receivable. The pledge is a promise that the firm will turn over the accounts receivable collateral to the lender in the case of default.

Loan agreements that use accounts receivable as collateral usually specify that the firm is responsible for the amount of the accounts receivable even if the firm's credit customers fail to pay. In short, the borrowing firm still has to pay even if its customers don't.

Lenders try to safeguard against accounts receivable that fluctuate so much that the value of the account becomes less than the value of the loan. Accounts receivable fluctuate because some credit customers may send in payments on their outstanding accounts receivable, others may make new charges, and some may be late with payments. If accounts receivable are pledged as short-term loan collateral, lenders usually require a loan payment plan that prevents the value of the accounts from dropping below the value of the loans. For instance, a bank may require a borrower to send payments received on pledged accounts to the lender to apply against the loan balance. Sending payments as received decreases the balance of the loan as the value of accounts receivable decreases, thereby protecting the lender.

Inventory as Collateral

Like accounts receivable, inventory represents assets that have value, and so can be used as collateral for loans. The practice of using inventory as collateral for a short-term loan is called **inventory financing.**

A major problem with inventory financing is valuing the inventory. If a borrowing firm defaults on a loan secured by inventory, the lender wants to know that the inventory can compensate for the remaining loan balance. To illustrate how important valuing inventory is, suppose you were a banker that lent a firm $200,000 for six months. As collateral, the firm put up its entire inventory of Alien Angels dolls, based on characters in a soon-to-be-released major motion picture. Unfortunately, the movie was a bust. The firm was unable to repay its loan, and you as the banker have ended up with 10,000 dolls no one wants. It is small comfort to you now that the firm said the angels were worth $20 each when they were offered as collateral.

To compensate for the difficulties in valuing inventory, lenders usually lend only a fraction of the stated value of the inventory. If the inventory consists of fairly standard items that can be resold easily, like 2 × 4s, then the lender might be willing to lend up to 80 percent of the inventory's stated value. In contrast, if the inventory consists of perishable or specialized items, like the Alien Angels in our example, then the lender will only lend a small fraction of their value, or might not be willing to accept them as collateral at all.

Financial Management and You

Easy Come, Easy Go: The Cost of Credit

Many people use credit cards to take care of their short-term business and personal financial needs. Because credit cards are easy to obtain, this makes them an easy-to-tap source of capital. Just how easy is it to get a credit card? Just for fun, John Galbreath filled out a credit card application on which he stated that he was 97 years old, had no income, no telephone, and no Social Security number. Plus, he wrote that he owed money to the Mafia. Back came a credit card.

Although it may be easy to get a credit card, it can be tough to find out how expensive that credit is. To help you assess the cost of the credit, start with these five items.

1. *The stated rate of interest:* Credit card interest rates vary. Compare interest rates before applying to make sure you select the lowest interest rate possible.
2. *The type of interest rate:* Some credit card companies may entice you with low interest rates, but those rates may not be fixed. Instead the rate may vary according to market rates, meaning that every quarter, or even every month, your interest rates may be adjusted upward or downward. Well over 60 percent of all cards in use have variable rates.
3. *When the rates change:* Variable-rate card issuers aren't legally required to tell customers of rate changes. Often, the only way holders of variable-rate cards can tell if their rate has changed is to look at the monthly statement, see what the rate is, and then compare it with previous months.
4. *The rates that apply to every type of credit:* Many cards offer additional credit options to cardholders, including cash advances. However, card issuers often charge higher interest rates on the cash advances than on regular credit purchases.
5. *The annual fee:* To lure you to sign up for a card, credit card issuers often waive the annual fee for use of the credit card. The following year, though, issuers charge the fee without prior notification. The annual fee can range from $25 to $100, and is part of the cost of the credit.

The point? Read the fine print of the credit card terms to learn how much the credit will cost you. If it costs too much, you should consider other types of short-term loans.

Sources: "Give the Man Credit," *Forbes* (August 14, 1995): 19; Albert B. Crenshaw, "Variable Rate Credit Cards: How to Avoid a Bumpy Ride," *Washington Post* (October 21, 1994): H1.

Inventory depletion is an additional concern for lenders who allow borrowers to use inventory as short-term loan collateral. The borrowing firm can sell the pledged inventory and use the cash received for other purposes, leaving the lender with nothing if the borrower defaults. This can happen when the lender has only a general claim, or *blanket lien,* on the borrower's inventory in the event of a default. Therefore, when inventory is used as collateral for a loan, the lender will often insist on some procedures to safeguard its interests.

One procedure to safeguard the interests of the lender is for the borrower to issue trust receipts to the lender. A *trust receipt* is a legal document in which specifically identified assets, inventory in this case, are pledged as collateral for the loan. Automobiles, railroad cars, and airplanes are often financed this way. The lender can make surprise visits to the borrower's business, checking to be sure that the pledged assets are on hand as they should be. There is often a unique identification number (a car's VIN, vehicle identification number, for example) on these assets.

Another procedure to control pledged inventory is to use a *public warehouse* where the inventory cannot be removed and sold without permission of the lender.

When the inventory is sold (with the lender's permission), the proceeds are sent to the lender and used to reduce the outstanding loan balance. While this arrangement gives the lender control, it is expensive for the borrowing firm. Usually, the borrowing firm must pay for the warehouse and seek the lender's permission each time it wants to sell some inventory.

We have seen that short-term secured loans generally have short-term liquid assets pledged as collateral, such as accounts receivable or inventory. Lenders often add loan terms to protect against problems such as fluctuating accounts receivable, and overvalued or depleted inventory.

What's Next

We discussed short-term financing in this chapter, the final chapter of part 4 of the text. We turn next to part 5, Finance in the Global Economy. In chapter 21, we discuss international finance.

Summary

1. **Explain the need for short-term financing.**

Firms rely on short-term financing from outside sources for two reasons:

- *Growth:* Profits simply may not be high enough to keep up with the rate at which they are buying new assets.
- *Choice:* Rather than save enough money to make desired purchases, many firms borrow money at the outset to make their purchases.

2. **List the advantages and disadvantages of short-term financing.**

Short-term financing is usually a cheaper option than long-term financing because of its generally lower interest rates. However, short-term financing is riskier than long-term financing because, unlike long-term financing, the loans come due soon, the lender may not be willing to renew financing on favorable terms, and short-term interest rates may rise unexpectedly.

3. **Describe three types of short-term financing.**

- *Loans from banks and other institutions:* When a bank or other institution agrees to lend money to a firm, the firm signs a promissory note that specifies the repayment terms. Two common types of short-term business loans are self-liquidating loans and a line of credit. A self-liquidating loan is a loan for an asset that will generate enough return to repay the loan balance. A line of credit is a maximum total balance that a bank sets for a firm's outstanding short-term loans.
- *Trade credit:* Trade credit is obtained by purchasing materials, supplies, or services on credit. By buying on credit, the firm has use of the funds during the time of the purchase until the account is paid.
- *Commercial paper:* Commercial paper consists of unsecured notes issued by large, creditworthy corporations for periods up to 270 days.

4. **Compute the cost of trade credit and commercial paper.**

The cost of trade credit is calculated by dividing the amount of the discount offered by the supplier by the amount the buyer owes. The result is annualized for comparison with other financing sources.

The cost of commercial paper is quoted as a discount yield. To compare the percent cost of a commercial paper issue to the percent cost of a bank loan, the commercial paper's discount yield must be converted to an effective annual interest rate.

5. **Calculate the cost of a loan and explain how loan terms affect the effective interest rate.**

The cost of a loan is normally measured by dividing the amount paid to obtain the loan by the amount the borrower gets to use during the life of the loan. The result is converted to a percentage. The stated interest rate on a loan is not always the same as the loan's effective annual interest rate. If the lender collects interest up front (a discount interest loan) or requires the borrowing firm to keep a fraction of the loan in an account at the lending institution (a compensating balance), then the amount of money the borrower gets to use is reduced. As a result, the effective rate of interest the borrower is paying is increased.

6. **Describe how accounts receivable and inventory can be used as collateral for short-term loans.**

Short-term loans are often secured by short-term, liquid assets, such as accounts receivable and inventory. When accounts receivable are used for collateral, the borrower pledges to turn over its accounts receivable to the lender if the borrower defaults. When inventory is used for collateral, the borrowing firm often sets aside the inventory that has been identified for collateral in a separate warehouse. When the inventory is sold, the cash received is forwarded to the lender in payment for the loan.

You can access your FREE PHLIP/CW On–Line Study Guide, Current Events, and Spreadsheet Problems templates, which may correspond to this chapter either through the Prentice Hall Finance Center CD–ROM or by directly going to: **http://www.prenhall.com/gallagher**.

PHLIP/CW — *The Prentice Hall Learning on the Internet Partnership–Companion Web Site*

Equations Introduced in This Chapter

Equation 20-1. Trade Credit Effective Annual Interest Rate Formula:

$$k = \left(1 + \frac{\text{Discount \%}}{100 - \text{Discount \%}}\right)\left(\frac{365}{\text{Days to Pay} - \text{Discount Period}}\right) - 1$$

where: k = The cost of trade credit expressed as an effective annual interest rate
discount % = The percent discount being offered
days to pay = The time between the day the firm makes the credit purchase and the day it pays its bill
discount period = The number of days in the discount period

Equation 20-2. Dollar Amount of the Discount on a Commercial Paper Note:

$$D = \frac{DY \times \text{Par} \times DTG}{360}$$

where: D = The dollar amount of the discount
DY = The discount yield
Par = The face value of the commercial paper issue; the amount to be paid at maturity
DTG = The days to go until maturity

Equation 20-3. Price of a Commercial Paper Note:

$$\text{Price} = \text{Par} - D$$

where: Par = The face value of the note at maturity
D = The dollar amount of the discount

Equation 20-4. Effective Annual Interest Rate of a Commercial Paper Note:

$$\text{Effective Annual Interest Rate} = \left(\frac{\text{Par}}{\text{Price}}\right)^{\left(\frac{365}{\text{DTG}}\right)} - 1$$

where: Par = The face value of the note at maturity
Price = The price of the note when purchased
DTG = The number of days until the note matures

Equation 20-5. Effective Interest Rate of a Loan:

$$\text{Effective Interest Rate } k = \frac{\$ \text{ Interest You Pay}}{\$ \text{ You Get to Use}}$$

Equation 20-6. Effective Annual Interest Rate:

$$\text{Effective Annual Interest Rate } k = \left(1 + \frac{\$ \text{ Interest You Pay}}{\$ \text{ You Get to Use}}\right)^{\left(\text{Loan Periods in a Year}\right)} - 1$$

Self-Test

ST-1. Your company's suppliers offer terms of 3/15, n40. What is the cost of forgoing the discount and delaying payment until the fortieth day?

ST-2. A commercial paper dealer is willing to pay 4 percent discount yield for a $1 million issue of Pennzoil 60-day commercial paper notes. To what effective annual interest rate does the 4 percent discount yield equate?

ST-3. A bank is willing to lend your company $20,000 for 6 months at 8 percent interest, with a 10 percent compensating balance. What is the effective annual interest rate of this loan?

ST-4. Using the loan terms from question ST-3, how much would your firm have to borrow in order to have $20,000 for use during the loan period?

Review Questions

1. Companies with rapidly growing levels of sales do not need to worry about raising funds from outside the firm. Do you agree or disagree with this statement? Explain.
2. Banks like to make short-term, self-liquidating loans to businesses. Why?
3. What are compensating balances and why do banks require them from some customers? Under what circumstances would banks be most likely to impose compensating balances?
4. What happens when a bank charges discount interest on a loan?
5. What is trustworthy collateral from the lender's perspective? Explain whether accounts receivable and inventory are trustworthy collateral.
6. Trade credit is free credit. Do you agree or disagree with this statement? Explain.
7. What are the pros and cons of commercial paper relative to bank loans for a company seeking short-term financing?

Build Your Communication Skills

CS-1. Your firm's request for a $50,000 loan for one month has been approved. The bank's terms are 10 percent annual discount interest with a 10 percent compensating balance. Prepare a one-page report for the CEO of your firm explaining how the effective interest rate of this loan is calculated.

CS-2. Imagine you are a loan officer for a bank. One of the town's businesses has applied for a loan of $200,000 for six months. The company has offered to put up the building in which its manufacturing operations are located as collateral for the loan. Local real estate agents estimate the building is worth at least $220,000. Write a letter to the company explaining why your bank does not wish to accept the building as collateral. Propose two alternative assets that your bank would accept.

Problems

20-1. Harold Hill is planning to borrow $20,000 for one year paying interest in the amount of $1,600 to a bank. Calculate the effective annual interest rate if the interest is paid:

Simple and Discount Loans

   a. at the end of the year.
   b. at the beginning of the year (discount loan).

20-2. Chad Gates is planning to borrow $40,000 for one year paying interest of $2,400 to a bank at the beginning of the year (discount loan). In addition, according to the terms of the loan, the bank requires Chad to keep 10 percent of the borrowed funds in a non-interest-bearing checking account at the bank during the life of the loan. Calculate the effective annual interest rate.

Loans with Compensating Balance

20-3. Ralph Bellamy is considering borrowing $20,000 for a year from a bank that has offered the following alternatives:

Challenge Problem

   a. an interest payment of $1,800 at the end of the year.
   b. an interest payment of 8 percent of $20,000 at the beginning of the year.
   c. an interest payment of 7.5 percent of $20,000 at the end of the year in addition to a compensating balance requirement of 10 percent.
      (i) Which alternative is best for Ralph from the effective interest rate point of view?
      (ii) If Ralph needs the entire amount of $20,000 at the beginning of the year and chooses the terms under (c), how much should he borrow? How much interest would he have to pay at the end of the year?

20-4. If Joyce Heath borrows $14,000 for three months at an annual interest rate of 16 percent paid up-front with a compensating balance of 10 percent, compute the effective annual interest rate of the loan.

Loans with a Life of Less than a Year

20-5. You are planning to borrow $10,000 from a bank for two weeks. The bank's terms are 7 percent annual interest, collected on a discount basis, with a 10 percent compensating balance. Compute the effective annual interest rate of the loan.

Discount Loans with Compensating Balance for Less than a Year

20-6. Bud Baxter is planning a $1 million issue of commercial paper to finance increased sales from easing the credit policy. The commercial paper note has a 60-day maturity and 6 percent discount yield. Calculate:

Commercial Paper

   a. the dollar amount of the discount
   b. the price
   c. the effective annual interest rate for the issue

20-7. Carmen Velasco, an analyst at Smidgen Corporation, is trying to calculate the effective annual interest rate for a $2 million issue of a Smidgen 60-day commercial paper note. The

Commercial Paper

commercial paper dealer is prepared to offer a 4 percent discount yield to the issue. Calculate the effective annual interest rate for Carmen.

Trade Credit

20-8. Bathseba Everdene, the sales manager of Gordon's Bakery, Inc., wants to extend trade credit with terms of 2/15, n45 to your company to boost sales. Calculate the cost of forgoing the discount and paying on the forty-fifth day.

Trade Credit

20-9. Calculate the cost of forgoing the following trade credit discounts and paying on the last day allowed:

   a. 3/10, n60
   b. 2/15, n30

Recalculate the costs assuming payments were made on the fortieth day in each of the preceding cases without any penalty. Compare your results.

Comparing Costs of Alternative Short-Term Financing

20-10. To sustain its growth in sales, Monarch Machine Tools Company needs $100,000 in additional funds next year. The following alternatives for financing the growth are available:

   a. forgoing a discount available on trade credit with terms of 1/10, n45 and, hence, increasing its accounts payable,
   b. obtaining a loan from a bank at 10 percent interest paid up front.

Calculate the cost of financing for each option and select the best source.

Comparing Costs of Alternative Short-Term Financing

20-11. If the bank imposes an additional requirement of a 12 percent compensating balance on Monarch in problem 20-10 and the company could negotiate more liberal credit terms of 1/15, n60 from its supplier, would there be any change in Monarch's choice of short-term financing?

Answers to Self-Test

ST-1. The cost is found using Equation 20-1. The discount percentage is 3 percent, the discount period is 15 days, and payment is to be made on the fortieth day. The calculations follow:

$$k = \left(1 + \frac{3}{100-3}\right)^{\left(\frac{365}{40-15}\right)} - 1$$
$$= (1 + .0309278)^{(14.6)} - 1$$
$$= (1.0309278)^{(14.6)} - 1$$
$$= 1.56 - 1$$
$$= .56$$
$$\times 100 = 56\%$$

ST-2. Use the three-step process described in the text to find the effective annual interest rate as follows:

- *Step 1:* Compute the discount using Equation 20-2.

$$D = \frac{DY \times Par \times DTG}{360}$$
$$= \frac{.04 \times \$1,000,000 \times 60}{360}$$
$$= \frac{\$2,400,000}{360}$$
$$= \$6,667$$

- *Step 2:* Compute the price using Equation 20-3.

$$\text{Price} = \text{Par} - D$$
$$= \$1,000,000 - \$6,667$$
$$= \$993,333$$

- *Step 3:* Compute the effective annual interest rate using Equation 20-4.

$$\text{Effective Annual Interest Rate} = \left(\frac{\text{Par}}{\text{Price}}\right)^{\left(\frac{365}{\text{DTG}}\right)} - 1$$
$$= \left(\frac{\$1,000,000}{\$993,333}\right)^{\left(\frac{365}{60}\right)} - 1$$
$$= (1.00671)^{6.083} - 1$$
$$= 1.0415 - 1$$
$$= .0415$$
$$\times 100 = 4.15\%$$

ST-3. The amount your firm would pay in interest with the loan is:

$$.08/2 = .04 \text{ for six months}$$
$$.04 \times \$20,000 = \$800$$

The amount your firm would be able to use during the life of the loan is the principal less the compensating balance:

$$\$20,000 - (.10 \times \$20,000) = \$18,000$$

The loan is for six months, so we use Equation 20-6 to solve for the effective annual interest rate:

$$k = \left(1 + \frac{\$800}{\$18,000}\right)^{(2)} - 1$$
$$= (1 + .0444)^{(2)} - 1$$
$$= (1.0444)^{(2)} - 1$$
$$= 1.0908 - 1$$
$$= .0908$$
$$\times 100 = 9.08\%$$

ST-4. Let X = the amount to borrow.

$$X - .10X = \$20,000$$
$$.09X = \$20,000$$
$$X = \$22,222$$

PART 5:
FINANCE IN THE GLOBAL ECONOMY

CHAPTER 21

INTERNATIONAL FINANCE

A man's feet must be planted in his country, but his eyes should survey the world.
—George Santayana

BUSINESS WITHOUT BORDERS

Daimler-Benz and Chrysler announced in May 1998 a $43 billion merger. It was the largest cross-border merger in history at that time. Daimler-Benz paid $43 billion in stock and assumed debt to acquire Chrysler. The offer was an amount that was 50 percent above the closing price of Chrysler's stock on May 5, 1998. In the prior year Daimler-Benz had revenues of $68.9 billion and operating profit of $2.4 billion whereas Chrysler had $61.1 billion in revenues and $4.7 billion in operating profit during that year. These were clearly two very large companies joining together.

There are definite legal and cultural differences that will have to be overcome as the large German and American companies combine their operations. Chrysler has had to live up to the strict financial reporting standards for U.S. publicly traded companies. The German disclosure standards are somewhat more lax. Chrysler shareholders who had been used to having copies of financial statements well in advance of stockholder meetings may not find that to be the case when the Daimler schedule is used. American executives are typically compensated more generously than their German counterparts. Robert Eaton, CEO of Chrysler at the time of the merger, had made $16 million in 1997. This compares to $2.5 million made by the top executive at Daimler, Jürgen Schrempp. Mr. Schrempp was the chair of the Daimler management board.

CHAPTER OBJECTIVES

After reading this chapter, you should be able to:

- Define a multinational corporation and explain its importance.

- Demonstrate how the law of comparative advantage leads to international trade.

- Describe exchange rates and their effects on firms.

- Show how firms manage the risks of fluctuating exchange rates.

- Discuss exchange rate theories.

- Describe the political and cultural risks that affect MNCs.

- Explain how international trade agreements affect international business.

It is getting more difficult to classify a company as American, German, or Japanese. It is not just business that is becoming more global but companies are going global as well. Legal, cultural, accounting, and other barriers will have to continue to be bridged.

Source: Joseph Sargent, "What Daimler-Chrysler Means to Shareholders," *Global Finance* (August 1998): 8–9.

Chapter Overview

This chapter addresses the financial issues that companies face when they maintain operations in, or sell goods and services to, other countries. These issues include differences in currency, language, politics, and culture. We also explore the potential benefits and risks of doing business in other countries. We closely examine the risk of transacting business in either a domestic or foreign currency and how to manage that risk. Finally, we look at international trade agreements and their effect on business.

Multinational Corporations

A **multinational corporation (MNC)** is a corporation that has operations in more than one country. Most large corporations conduct at least some of their business in countries other than the one they call home. In fact, it is getting difficult to accurately describe whether companies are U.S., German, or Japanese, in spite of what their names might suggest.[1] Table 21-1 shows the top ten U.S.-based multinational corporations.

Financial Advantages of Foreign Operations

You may have read stories about McDonald's fast-food restaurants expanding into Russia and Japan, or Kentucky Fried Chicken opening a branch in the People's Republic of China. Coca-Cola and Pepsi-Cola executives count on the billion-dollar growth potential of their popular products in overseas markets. Boeing, once a leading supplier for U.S. firms, is now primarily an exporter. Companies are always looking for growth opportunities, and many of these opportunities are in other countries.

In addition to the potential demand for products and services, operating abroad can decrease production costs. Labor may be less costly in other countries. For instance, much of the production of clothing by companies based in the United States is done in other countries where lower wages are paid to workers.

Companies can often obtain political advantages by shifting some operations away from home. For example, Toyota, Honda, and Nissan of Japan, and the German company BMW all produce some of their cars in the United States. This lessens

---

[1] A few years ago, the Volkswagen Rabbit, made by a German-based company, was the only car sold in the United States that was made in the United States, by U.S. workers, with all U.S.-made parts.

TABLE 21-1

Largest U.S.-Based Multinational Corporations

		REVENUE		
Rank	Company	Foreign ($ mil)	Total ($ mil)	Foreign as % of Total
1	Exxon	92,540	120,279	76.9
2	General Motors	51,046	178,174	28.6
3	Ford Motor	46,991	145,348	32.3
4	IBM	45,845	78,508	58.4
5	Mobil	35,606[1]	59,978[1]	59.4
6	Texaco[2]	33,292	59,828	55.6
7	General Electric	26,981	90,840	29.7
8	Hewlett-Packard	23,819	42,895	55.5
9	Chevron[2]	23,055	48,836	47.2
10	Citicorp	21,566	34,697	62.2

[1]Includes other income.
[2]Includes proportionate interest in unconsolidated subsidiaries or affiliates.

Source: "Top 100 Multinationals," Forbes Magazine (July 27, 1998): 163.
Reprinted by permission of Forbes Magazine © Forbes Inc., 1998.

Table 21-1 shows the ten largest U.S.-based MNCs ranked by revenue of foreign subsidiaries (as of July, 1998).

the risk of the U.S. Congress passing laws that would restrict imports of, or increase the tax on, sales of these cars. Firms based in the United States often have the same import and tax reasons to open production plants in other countries.

Ethical Issues Facing Multinational Corporations

Using foreign labor may pose difficult ethical questions for U.S. companies. Are very poor people from other countries being exploited because they can be employed at lower wages than would be paid to U.S. workers? Are overseas plants as safe as U.S.

Source: CALVIN AND HOBBES © 1992, 1993 Watterson. Dist. by Universal Press Syndicate. Reprinted with permission. All rights reserved.

plants, where OSHA inspectors check up on employers?[2] Does a corporation have the moral responsibility to meet the same health and safety standards for its workers in other countries as it applies to its U.S. workers, even if not legally required to do so? Calvin Klein, Liz Claiborne, The Gap, and other garment industry companies that manufacture clothing outside of the United States to control labor costs must balance the moral issues of human rights against their obligations to stockholders.

Another ethical issue arises when operations are moved to countries where environmental laws are less strict than the laws of the company's home country. Firms may lower costs by disposing of waste products or emitting pollutants in a potentially dangerous manner. Even though the laws of the other country may not have pollution control or disposal guidelines, should a business harm the environment or create a danger that could harm future generations? How should managers weigh the morality of business procedures against their fiduciary duty to the owners of the company's common stock to maximize value?

Unfortunately, there are no clear-cut answers to many of these questions. Certainly, worker exploitation and the destruction of the environment would be condoned by few people. But what if working conditions are only slightly less safe, or the environmental practices in that other country only slightly more offensive by the home country's standards? What if these slightly offensive environmental practices are commonplace and accepted in that other country? Where should we draw the line? Facing these questions, even when the answers are difficult to agree on, is an important first step in making ethical decisions.

Comparative Advantage

International trade creates significant financial benefits because not all countries can produce goods and services with the same degree of efficiency. The law of comparative advantage says that each country should concentrate on that which it does well.

For example, coffee does not grow well in the United States. If coffee drinkers in the United States had to depend on locally grown coffee for their entire supply, coffee would be scarce, of relatively poor quality, and extremely expensive. Coffee grows very well in Brazil, however. Brazilians can grow much more than they can consume themselves, and at a relatively low cost. It makes sense, then, that U.S. coffee drinkers should buy Brazilian coffee.

Brazil, in contrast, cannot easily make major motion pictures. If Brazilians could only watch movies made in Brazil, they wouldn't see very many. Filmmakers in the United States make many movies, most with sophisticated technology. Film companies can (and do) export films easily and cheaply to foreign countries like Brazil. It makes sense, then, for Brazilians to stick to growing coffee and to watch U.S. movies. Brazil has a comparative advantage over the United States with respect to coffee growing, whereas the United States has a comparative advantage over Brazil with respect to movie making. International trade makes it possible for U.S. citizens to enjoy high-quality, affordable coffee, and for Brazilians to enjoy numerous, technologically advanced U.S. movies.[3]

---

[2] OSHA stands for Occupational Safety and Health Administration, a federal agency responsible for enforcing national laws intended to protect workers employed by U.S. companies.
[3] The quality of U.S.-produced movies is a subjective judgment, but it is clear from film company revenues that the world seems to enjoy watching them.

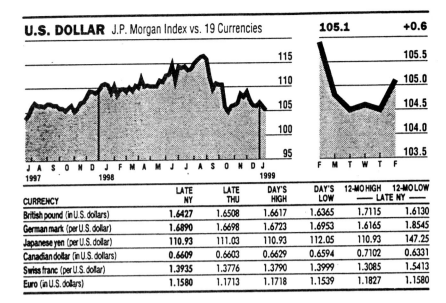

Figure 21-1 Exchange Rates Relative to the U.S. Dollar

Figure 21-1 shows various exchange rates relative to the U.S. dollar on January 8, 1999.
NOTE: *The letters on the horizontal axis of the graph on the left stand for months of the year, beginning with July. The letters on the horizontal axis of graph on the right stand for weekdays, excluding Saturday and Sunday.*

Source: *Reprinted by permission of* The Wall Street Journal, © *1999 Dow Jones & Company, Inc. All rights reserved worldwide.*

EXCHANGE RATES AND THEIR EFFECTS

Given that countries can benefit from trading with each other, how do they overcome the problem of having different units of currency? An **exchange rate** is an expression of the value of one country's currency in terms of another country's currency. It specifies how many units of one country's currency can be exchanged for one unit of the other country's currency. For example, the exchange rate between the U.S. dollar and the Mexican peso might be ten Mexican pesos per one U.S. dollar. The exchange rate between the U.S. dollar and the Japanese yen might be 100 Japanese yen per one U.S. dollar.

Exchange rates can be expressed as the number of units of the foreign currency per one unit of domestic currency, as in the previous paragraph, or as the number of units of the domestic currency per one unit of the foreign currency. For example, if ten Mexican pesos can be exchanged for one U.S. dollar, then one Mexican peso is worth one-tenth of a U.S. dollar, or $0.10. The exchange rate can be expressed as ten pesos per dollar or one-tenth of a dollar per peso. The one exchange rate is the reciprocal of the other.

A sampling of actual exchange rates as of January 8, 1999 is shown in Figure 21-1. We see from Figure 21-1 that the exchange rate is expressed as how many U.S. dollars (the number of units of the domestic currency) per one unit of the foreign currency. That is, one British pound was worth a little more than 1.64 U.S. dollars on January 8, 1999.

Fluctuating Exchange Rates

Exchange rates fluctuate every day due to changing world conditions. Currency traders take advantage of these fluctuations by "buying" and "selling" currencies. This means they exchange a certain amount of one currency for a certain amount of another. The currency market is one of the largest financial markets in the world. Currency traders buy and sell the equivalent of over $1 trillion in an average trading day. The "price" of buying or selling a currency is its exchange rate. So, if a trader had wanted to sell U.S. dollars in exchange for British pounds on January 8, 1999, the trader would have paid 1.64 dollars for each pound.

If the value of one currency decreases relative to the value of another currency, the currency with the falling value is called a **weakening currency.** For example, if the exchange rate between the U.S. dollar and the Mexican peso were to change from ten Mexican pesos per dollar to eight Mexican pesos per dollar, then the U.S. dollar has weakened relative to the Mexican peso, and it now takes more dollars to buy a given number of Mexican pesos. At the same time, of course, the Mexican peso is said to be a **strengthening currency** relative to the U.S. dollar because the exchange rate would have changed from $0.10 dollar per peso to $0.125 dollar per peso. A given number of Mexican pesos now buys more U.S. dollars.

When a country's currency weakens relative to the currencies of other countries, imported goods become more expensive for citizens of the country with the weakened currency. Say that one pound of Mexican limes costs ten pesos. If the exchange rate between pesos and U.S. dollars is ten pesos per dollar, then one dollar (ten pesos) will buy a pound of Mexican limes. However, if the exchange rate changes from ten pesos per dollar to eight per dollar, then the importer could no longer buy the pound of limes with one dollar: The limes cost ten pesos and the dollar is equivalent to eight pesos. Each peso is now worth $0.125, so the importer would need $0.125 × 10 = $1.25 to buy a pound of limes.

When the U.S. dollar weakens against the Mexican peso, U.S. importers of Mexican goods will require more dollars to purchase Mexican goods that are sold in pesos. Similarly, people in the United States will find that Mexican goods and services are now more expensive in general. U.S. vacationers in Mexico learn this lesson quickly.

Conversely, a weakened dollar relative to the Mexican peso means that U.S. goods and services are less expensive for Mexican citizens to buy. This is why Mexican tourists tend to come to the United States in greater numbers when the dollar is weakening against the Mexican peso, and why business improves for U.S. companies exporting goods to Mexico. Generally, these same effects occur when the U.S. dollar weakens compared to any other country's currency.

What happens when the U.S. dollar strengthens against the currency of another country? Using our Mexican peso example, if the dollar strengthens against the peso, more U.S. consumers would buy Mexican limes and other goods, and vacation in greater numbers in Mexico. At the same time, Mexican people would buy fewer U.S. goods and take fewer U.S. vacations.[4]

These changes in spending patterns usually occur whenever the exchange rate fluctuates. The relative attractiveness of one country's goods and services abroad fluctuates with the relative value of the currency changes.

Cross Rates

If we know the exchange rate between the currencies of Country A and Country B, and also between the currencies of Country A and Country C, then we can determine the exchange rate between the currencies of Country B and Country C. For example, if we know that one U.S. dollar is worth ten Mexican pesos and that one U.S.

---

[4] In our running example, we assume that the domestic prices in the two countries stay the same.

dollar is worth 100 Japanese yen, we can determine how many Japanese yen a person would receive in exchange for each Mexican peso. An exchange rate of two currencies found by using a common third currency is known as a currency **cross rate.**

To calculate a currency cross rate, multiply the ratio of Currency A to Currency B exchange rate by the ratio of Currency B to Currency C exchange rate as shown in Equation 21-1.

$$\text{Calculating Cross Rates}$$
$$\frac{\text{Currency A}}{\text{Currency B}} \times \frac{\text{Currency B}}{\text{Currency C}} = \frac{\text{Currency A}}{\text{Currency C}} \qquad (21\text{-}1)$$

The calculation of the Japanese yen to the Mexican peso cross rate using the cross rate formula in Equation 21-1 is shown next.

$$\frac{\text{Japanese Yen}}{\text{U.S. Dollar}} \times \frac{\text{U.S. Dollar}}{\text{Mexican Peso}} = \frac{\text{Japanese Yen}}{\text{Mexican Peso}}$$

Suppose 100 Japanese yen are worth one dollar, as we said earlier. Then suppose one U.S. dollar is worth ten Mexican pesos. Substituting these values into Equation 21-1 produces the following cross rate:

$$\frac{100 \text{ Japanese Yen}}{1 \text{ U.S. Dollar}} \times \frac{1 \text{ U.S. Dollar}}{10 \text{ Mexican Pesos}} = \frac{10 \text{ Japanese Yen}}{1 \text{ Mexican Peso}}$$

We find that the U.S. dollar in the first term of the equation cancels out the U.S. dollar in the second term, leaving the yen-per-peso exchange rate, which in this case is 100 to 10, reduced to 10 to 1.

A sampling of currency cross rates is shown in Table 21-2.

Cross rates are a useful tool. Financial publications of many countries offer exchange rates relative to the domestic currency only. So, if you were on a business trip in Europe but were managing a project for a U.S. firm that had a Japanese supplier, you might want to know the exchange rate between the dollar and yen but would only have information about the euro. The cross rate formula allows you to quickly calculate the dollar-to-euro exchange rate.

TABLE 21-2

Currency Cross Rates on January 8, 1999

Key Currency Cross Rates Late New York Trading Jan 8, 1999										
	Dollar	Euro	Pound	SFranc	Guilder	Peso	Yen	Lira	D-Mark	FFranc	CdnDlr
Canada	1.5132	1.7523	2.4857	1.0859	.79515	.15355	.01364	.00090	.89591	.26713	....
France	5.6646	6.5596	9.3052	4.0650	2.9766	.57479	.05106	.00339	3.3538	....	3.7434
Germany	1.6890	1.9559	2.7745	1.2121	.88753	.17139	.01523	.00101	....	.29817	1.1162
Italy	1672.1	1936.3	2746.7	1199.9	878.64	169.67	15.073	....	989.98	295.18	1105.0
Japan	110.93	128.46	182.22	79.605	58.291	11.256	....	.06634	65.678	19.583	73.308
Mexico	9.8550	11.412	16.189	7.0721	5.1786	....	.08884	.00589	5.8348	1.7398	6.5127
Netherlands	1.9030	2.2037	3.1261	1.3656	....	.19310	.01716	.00114	1.1267	.33595	1.2576
Switzerland	1.3935	1.6137	2.2891	....	.73225	.14140	.01256	.00083	.82504	.24600	.92090
U.K.	.60875	.70494	....	.43685	.31989	.06177	.00549	.00036	.36042	.10747	.40230
Euro	.86356	....	1.4186	.61970	.45378	.08763	.00778	.00052	.51128	.15245	.57068
U.S.	....	1.1580	1.6427	.71762	.52548	.10147	.00901	.00060	.59207	.17654	.66085

Source: Telerate

Source: Reprinted by permission of The Wall Street Journal, © 1999 Dow Jones & Company, Inc. All rights reserved worldwide.

Table 21-2 shows the cross exchange rates for 11 major currencies on January 8, 1999.

Exchange Rate Effects on MNCS

Fluctuating exchange rates present special risks and opportunities for multinational corporations. As the home currency of an MNC strengthens or weakens against currencies of other countries where the MNC has operations, the firm will feel a financial impact.

Suppose, for example, that McDonald's Corporation, a U.S.-based MNC, realized profits of 80 million euros from its operations in Germany during 1999. At some point, McDonald's Corporation is likely to convert its euro profits to U.S. dollars and bring the dollars to the United States. Converting foreign currency profits into domestic currency to send to the home country of the business is known as *repatriating* the profits. Why repatriate profits? McDonald's stockholders expect U.S. dollar dividends, not euro dividends.

To demonstrate how to report repatriated profits to shareholders, let's revisit the McDonald's example. Suppose the exchange rate were .80 euros per U.S. dollar at the start of the year. Also assume the exchange rate held steady throughout the year. When McDonald's repatriates its profits at year's end, it would report on the income statement 80 million divided by .80 = 100 million in U.S. dollar profits from its German operations.

Now suppose the exchange rate did not hold steady during the year. Suppose instead that the exchange rate fell from .80 euros per dollar to .75 euros per dollar by year's end. Now when the profits are repatriated, they are worth 80 million/.75 = 106.67 million U.S. dollars. Clearly, the weakening of the U.S. dollar during the year helped the McDonald's Corporation. Its U.S. dollar profits were nearly 7 million U.S. dollars higher ($106.67 million − $100 million) than they would have been had the dollar not weakened.

A weakening U.S. dollar has a positive effect on an MNC that repatriates its profits from the country whose currency is strengthening. However, if the U.S. dollar strengthens relative to the foreign currency, profits tumble. Going back to our McDonald's example, suppose that the U.S. dollar had strengthened relative to the euro during the year. McDonald's U.S. dollar profits would have been lower after converting the euros to dollars and repatriating the profits back to the United States.

Exchange Rate Effects on Foreign Stock and Bond Investments

A corporation does not have to have actual business operations in another country to be affected by exchange rate fluctuations. If a company holds stocks or bonds denominated in the currencies of other countries, the fluctuations in the value of these currencies will affect the dollar value of the stocks and bonds. This applies to individual investors' holdings as well.

If Betsy Ross, a U.S. investor, held 100 shares of Lufthansa common stock valued at 14 euros per share, a strengthening of the euro (weakening of the U.S. dollar) would help Ross just as it helped McDonald's in our earlier example. If the exchange rate were, say, .80 euro per dollar, then Ross's investment would be worth 14/.80 = 17.50 U.S. dollars per share.

However, if the exchange rate went from .80 to .60 euro per dollar, the U.S. dollar value of Ross's stock would increase from $17.50 per share to $23.33 (14 euros/.60 = $23.33). Ross experienced a gain of $5.83/$17.50 = .33, or 33 percent even though the stock's price on the German stock exchange did not change at all.

Of course, if the dollar had strengthened, there would have been a negative financial impact on Betsy. A strengthening dollar is one of the risks of investing outside your home country.

Managing Risk

International operations provide special challenges and opportunities but also special risks. Exchange rate risk can be hedged. Investors and MNCs often find diversification benefits when foreign investments are added to an all-domestic portfolio. Risk of foreign securities held by U.S. citizens, caused by fluctuating exchange rates, can be managed by purchasing foreign claims on foreign securities denominated in U.S. dollars.

Hedging

The risk that a multinational corporation faces due to fluctuating exchange rates is one that can be managed. *Forward contracts, futures contracts,* and *currency swaps* are all available to help an MNC hedge currency risk. A **hedge** is a financial agreement used to offset or guard against risk. A company may choose to hedge against adverse changes in interest rates, commodity prices, or currency exchange rates.

Forward contracts are contracts where one party agrees to buy, and the other party agrees to sell, a certain amount of an asset (a currency for example) at a specified price (exchange rate) at a specified future time. *Futures contracts* are similar except they are standardized contracts and can be traded on organized exchanges. *Swaps* are directly negotiated contracts, like forward contracts, in which each party agrees to swap payments at specified points in time according to a predetermined formula. For instance, one party could pay U.S. dollars and the other Japanese yen, each according to the amounts called for in the swap contract. By agreeing to a forward, futures, or swap contract, an MNC can protect against a loss that will occur if the feared change in exchange rate occurs. The firm using these hedging instruments insulates itself from this risk.

Diversification Benefits of Foreign Investments

There are often significant diversification benefits to investing in a variety of countries, both for MNCs and for individual investors. Instead of putting all your money in one country, spreading your investment around several countries will often prove beneficial. If the economy or market is weak in one country, it may be stronger in another. If your money is spread around in several countries, the good news in one country will often cancel out the bad news from another.

The returns earned from investments in other countries often have low correlations with the returns earned from investments in the home country. Chapter 7 showed the risk-reducing potential of creating portfolios where the individual assets have returns with low correlations relative to each other. Figure 21-2 shows how diversification benefits are greater with a portfolio that contains both domestic and foreign securities (average correlation of .4 in our example), rather than domestic securities alone (average correlation of .8 in our example).

If you review chapter 7, you will see specifically how the correlation between the returns of assets affects the risk of a portfolio. Because the correlation of returns

Figure 21-2 Portfolio Risk as Diversification Changes

Figure 21-2 shows the diversification effects for a sample domestic-only portfolio and a mixed (domestic and foreign) portfolio. The upper curve represents a domestic security only portfolio. The lower curve represents a portfolio containing both domestic and foreign securities.

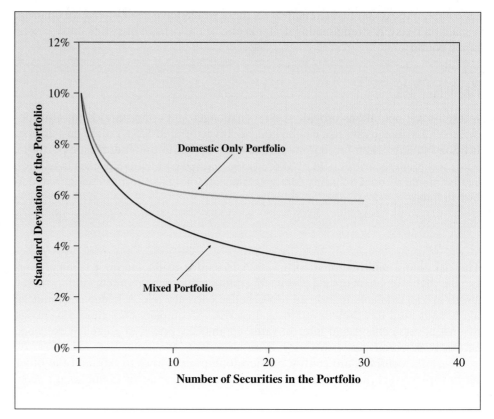

between a U.S. stock and a foreign stock tends to be lower than between two U.S. stocks, international diversification is often an important risk-management tool.

Figure 21-2 shows how the diversification effects differ for a portfolio with U.S. stock only compared to one with U.S. and foreign stock. The sample domestic portfolio has an average correlation coefficient of returns of .8. The sample portfolio containing both domestic and foreign securities has an average correlation coefficient of returns of .4. The mixed portfolio has a lower correlation coefficient, so the risk of that portfolio is lower than that of the domestic only portfolio.

American Depository Receipts

The common stock of many major foreign companies is traded in the United States in a form that is denominated in U.S. dollars. Some examples include Nestlé (Switzerland), Volvo (Sweden), and British Gas (United Kingdom).[5] Special trusts are created, and foreign stock is purchased and placed in these trusts. These trusts then issue their own securities, called **American depository receipts (ADRs).** Many ADRs are traded on the New York Stock Exchange and the NASDAQ over-the-counter market.

In this section, we examined exchange rate risk management tools, the diversification benefits of foreign investments, and American depository receipts. We turn to exchange rate theories next.

---

[5]Sam Jaffe, "World-Class Investments," *SmartMoney* (February 1996): 88.

FINANCE AT WORK

Interview with Don Burton, International Import–Export Institute

Don Burton

Don Burton is on the board of directors for the International Import–Export Institute, which certifies import–export professionals around the globe. Don brought a strong business background to the job. The president and CEO of Midland Cascade Corporation, Don has also developed export education programs for major universities. The International Import–Export Institute provides education and certification testing to import–export professionals in a variety of areas, ranging from finance and documentation to letters of credit and marketing.

Q. What risks do businesses face when they trade internationally?

There are a lot of different risks that people and businesses don't really think about when they start dealing in the international trade arena. In the United States, we take stability for granted, but many countries are economically and politically unstable. For instance, a firm may not get paid because a country might nationalize either the bank or company with which you were dealing. Once a bank or company has been nationalized, the firm usually has worthless contracts that the country won't recognize as a binding obligation.

Q. How safe are letters of credit?

When a business deals in international trade, accepting letters of credit creates a big risk. A letter of credit is an obligation on the part of a foreign bank to pay the firm. Sometimes the letters are forged or fake; the bank doesn't even exist. Nigeria is famous for invalid letters of credit. Companies present a letter of credit with names of banks that don't exist, trying to encourage U.S. firms to send goods. Once the goods are in Nigeria, the U.S. firm's invoice is ignored.

Q. If a company is unsure about the buyer or their bank, how can a firm greatly reduce its trade credit risk?

The firm should get a confirmed letter of credit. This means that the company's local bank has confirmed the letter of credit. Even if the other company doesn't pay, the local bank will. The payment responsibility shifts from the foreign bank or company to the local bank. The bank charges only a small fee for this service.

Q. What are some of the risks of fluctuating exchange rates?

Exchange rates affect companies in many ways, but timing is a crucial factor. Let's say you ship some goods overseas. They go by boat. That means it may be 28 to 45 days before your customer receives the goods. That's often when the obligation to pay begins. If you have given the customer trade credit terms of 30 days net of receipt, that means that you may not get paid for up to 75 days after the day you ship the goods. If during this time the exchange rate drops—meaning the dollar strengthens if you're dealing in a foreign currency—you're at risk.

Q. Is there any way to avoid getting hurt by dropping exchange rates?

Agree to the exchange rate at the inception of the deal in a contract. Or buy insurance to guard against dropping exchange rates. Domestic banks and the federal government offer exchange rate insurance at fairly cheap rates. If the business transaction goes through the Export–Import Bank agency of the federal government, the government automatically insures a business for the current exchange rate.

Source: Interview with Don Burton.

Exchange Rate Theories

To aid business decision makers, financial theorists try to explain exchange rate levels and fluctuations. Why is the current exchange rate between two currencies at the level it is? What causes that exchange rate to change? Among the most popular exchange rate theories are the *purchasing power parity theory,* the *international Fisher effect,* and the *interest rate parity theory.*

Purchasing Power Parity Theory

Economists have studied the question of how the financial market prices one country's currency in terms of another country's currency. One explanation, the *purchasing power parity (PPP) theory,* says that it is the relative prices in two countries that determine the exchange rate of their currencies. There are two versions of the PPP theory: the *absolute PPP* and the *relative PPP.*

The **absolute purchasing power parity theory** posits that exchange rates are determined by the differences in the prices of a given market basket of traded goods and services when there are no trade barriers. For instance, if a given basket of traded goods and services available in Japan and in the United States costs 80,000 yen in Japan and 1,000 dollars in the United States, then the exchange rate should be 80 yen per dollar. The calculation follows:

$$\frac{¥80{,}000}{\$1{,}000} = 80 \text{ Yen per Dollar}$$

Another version of the theory is more realistically applicable in an economic world with transportation costs, tariffs, quotas, and other trade barriers. The **relative purchasing power parity theory** focuses on the *changes over time* in the relative prices of traded baskets of similar goods and services in the two countries. At any given time, the exchange rate between the two currencies is related to the rate of change in the prices of the similar market baskets. According to the relative PPP theory, as prices change in one country relative to those prices in another country for a traded basket of similar goods and services, the exchange rate will tend to change proportionately but in the opposite direction.

The rationale for this theory is that if one country experiences rising prices while its international trading partners do not, its exports will become less competitive. Similarly, imports will become more attractive because of their relatively lower prices. The exchange rate will change as citizens buy the currency of the country with falling prices and sell the currency of the country with rising prices.

International Fisher Effect

Another theory used to explain currency exchange rates is the international Fisher effect, named for economist Irving Fisher. The *domestic Fisher effect* says simply that the nominal rate of interest equals the real rate plus the expected inflation rate. When this concept is used in an international setting, however, the **international Fisher effect** states that changes in the nominal interest rates for two countries will be offset by equal changes, in the opposite direction, in the exchange rate. The difference in nominal interest rates across countries reflects the difference in expected rates of inflation in those countries. According to Irving Fisher, the exchange rate would change by the same amount, but in the opposite direction, as the difference between the nominal interest rates of the two countries.

Changes in nominal interest rates are determined by changes in expected inflation, and exchange rates change for the same reason. The rationale is that investors must be compensated, or will offer compensation, to accommodate the expected change in the exchange rate.

Interest Rate Parity Theory

The **interest rate parity theory** says that the percentage difference between the exchange rate specified for future delivery (the **forward rate**), and for current delivery (the **spot rate**), equals the difference in the interest rates for equal maturity securities in the two countries.

If the difference between the spot and forward rates for two countries' currencies did not equal the difference between the interest rates on equal maturity securities in those countries, an *arbitrage* opportunity would exist. **Arbitrage** is the process whereby equivalent assets are bought in one place and simultaneously sold in another, making a risk-free profit. Arbitragers would buy in the spot market and sell in the forward market, or vice versa, depending on which was undervalued and which overvalued, until interest rate parity is achieved.

Other Factors Affecting Exchange Rates

Currency exchange rates fluctuate daily. Traders often buy and sell the dollar equivalent of $1 trillion per day, continually reevaluating how many units of one country's currency should be exchanged for a given number of units of another country's currency.

The purchasing power parity theory, the international Fisher effect, and interest rate parity theory focus on rational economic explanations for exchange rates. But, like any market, the foreign exchange market is affected by both real economic factors and psychological factors.

If a country is experiencing political trouble, for example, currency traders may fear that this will spill over to that country's economy. If investments in the foreign country are expected to suffer in the future because of the political turmoil, traders will dump that country's currency in the foreign exchange market. To liquidate investments in the troubled country, a firm must not only sell the investment but also the foreign currency it receives through the sale of the security. Currencies from other countries, where the political climate is better, will be preferred. These other currencies can then be invested in countries where the political climate is more hospitable.

When the Soviet Union broke up in 1992, the German mark dropped precipitously in the foreign exchange market against other major currencies, particularly against the dollar. The reason given by most experts is that refugees from many of the states that had been part of the Soviet Union were fleeing their home countries and moving to Germany. (The German constitution established an open door policy for foreign refugees.) This influx of poor people into Germany was expected to strain the German economy as government money was spent to take care of them. The German mark went down accordingly, in anticipation of this drain.

The U.S. dollar often increases in value against other currencies on the foreign exchange market when there is a significant threatening world event. For instance, when Iraq invaded Kuwait in 1990, which subsequently led to the Persian Gulf War, the dollar soared against most other currencies in value. The same was true in 1998 when there were fears that the Asian economic crises would spread to other parts of the world. The U.S. dollar has become something of a "safe haven" for investors during difficult times.

The rationale seems to be that U.S. investments are safer because the United States has considerable military and economic power and secure location compared to countries with limited means to protect themselves from unfriendly neighbors. South Korea, for example, has a powerful economy that could crumble overnight if North Korea invaded. Similarly, Taiwan has a powerful economy that would be threatened if the People's Republic of China invaded.

Government Intervention in Foreign Exchange Markets

Sometimes the central banks of countries will enter the foreign exchange market in an attempt to influence exchange rates. If the Fed, for example, were unhappy with market conditions where the U.S. dollar was weakening against other currencies, it could buy U.S. dollars, using its holdings of other currencies such as German marks or British pounds, in an attempt to bid up (strengthen) the value of the dollar. Central banks of other countries can also buy or sell currencies in the foreign exchange market to further their own policy goals. Although the central banks of major countries have great market power, it is difficult to move a trillion-dollar-plus market when it doesn't want to be moved.

Sometimes central banks will act in concert with each other to pursue a common policy objective. The major economic powers of the world, known as the Group of Seven (G-7), have sometimes pursued a common effort to alter exchange rates. The Group of Seven countries are the United States, the United Kingdom, Germany, Japan, Canada, France, and Italy.

Psychological and political factors surely affect exchange rates. Like any other market, the foreign exchange market is influenced by logical and illogical factors alike, some of which are unidentifiable. The end result, however, is felt by multinational corporations and by individual consumers alike.

In this section we explored exchange rate theories that help decision makers understand exchange rate risks and fluctuations. In the next section, we address the main political and cultural risks that multinational corporations may confront.

Political and Cultural Risks Facing MNCs

MNCs must deal with political and cultural factors when engaging in international business. Awareness and sensitivity to these factors are crucial to successful international business activities.

Political Risk

Political risk is the risk that a country's government may take some action that would harm a foreign-owned company doing business in that country. For example, a foreign government might expropriate (take) the assets of the company for its own reasons. Expropriation of assets is an extreme example of political risk. Sometimes funds are blocked by the host country, making it impossible for an MNC to repatriate profits earned in the foreign country. For instance, the U.S. government froze the bank accounts of Iranian companies when tension developed after U.S. citizens were taken hostage by Iranian nationals in 1979.

Another political risk is the chance that special taxes could be imposed on profits made by foreign companies. An example of such a tax is the unitary tax imposed by many state governments in the United States. A **unitary tax** is a tax on a portion of the profits earned by non-U.S. companies.

Another political risk is the chance that a foreign government may impose a minimum number or minimum percentage of domestic workers that must be employed by foreign companies operating in that country. These workers may or may not have the same level of training and ability as the workers of the home country. In Bermuda, for instance, companies are not allowed to hire foreign workers unless they can demonstrate that they cannot obtain Bermudians for the job positions. If a Bermudian wants a job, then, he or she gets it even though there may be many more qualified foreigners willing to move to Bermuda to work.

Foreign governments may also impose a requirement for the use of a certain amount of raw materials or parts manufactured in the country where the foreign operation is located. Again, these materials and parts may or may not meet the standards of the home country.

These political risks must be weighed, along with other risks unique to international operations, against the special opportunities these international operations present.

Cultural Risk

When a company operates in another country, cultural issues often affect business. These differences include language barriers as well as differences in attitudes, values, and business protocol. Cultural risk is the risk that foreigners doing business in another country will fail to adapt to cultural differences, and this failure will affect the firm's success.

To offset cultural risk, companies that operate abroad often train their employees in cultural differences. A lack of awareness about differences in business practices could cause an unintentional insult or jeopardize a negotiation about a major project. For example, some hand gestures that have innocent connotations in one country are considered offensive or obscene in another country. What is called a bribe in one country may be considered a routine gratuity in another country.

Taking cultural risks can generate high returns. For example, take James Hickman, a U.S. entrepreneur who founded Rustel, a $10 million telephone company operating in Moscow after amassing a pile of rubles in the late 1980s by promoting American rock concerts. Rustel succeeded in large part because Hickman and his partner learned about the Russian culture and searched for employees who could both work with the Russian bureaucracy and act entrepreneurial. Finding such talented employees was tough, however.

Through trial and error, Hickman learned that the totalitarian regime in the former Soviet Union had created a work culture of secrecy and a lack of accountability. Few workers kept records of their business activities, and many dodged responsibility to avoid the risk of blame for failure. Because of the slow pace of the former Communist bureaucracy, few workers had a sense of urgency about deadlines. Project planners had no understanding of business in a market economy, so they did not value investments, estimate costs, or assess market potential. Fortunately, Hickman learned these cultural lessons quickly and was able to effectively hire and train employees. Otherwise, his business would not have succeeded.[6]

The risks of doing business in a foreign country include exchange rate fluctuations, and political and cultural risk. Yet the return potential is high. Recognizing the risk and potential return, governments have forged trade agreements to promote and regulate international business. We look at trade agreements next.

International Trade Agreements

Groups of countries sometimes form alliances and agreements that both regulate and foster international trade. NAFTA, EU, and GATT are the major international agreements we discuss in this section. Read on to learn the details of this alphabet soup.

NAFTA

Canada, Mexico, and the United States signed an agreement in 1994 called the *North American Fair Trade Agreement (NAFTA)*. NAFTA breaks down some of the barriers to trade between these countries, such as tariffs and quotas.

Tariffs are taxes assessed by Country A on the goods of Country B that are sold in Country A. Mexico, for example, might impose a tariff on Canadian goods sold in Mexico. **Quotas** are quantity restrictions imposed by Country A on certain goods imported from Country B. Both are barriers to international trade. When Country A imposes tariffs and quotas on the goods of Country B, Country B is likely to impose tariffs and quotas on goods from Country A. This can lead to a trade war wherein each country increases the barriers imposed on the goods of the other. Trade wars can lead to economic decline of both countries.

NAFTA is, in effect, a pact to end such wars. It should increase business for all three countries as tariffs, quotas, and other trade barriers are eliminated. NAFTA is politically controversial, however, because its opponents believe that U.S. jobs will be lost as U.S.-based companies shift some operations to Mexico where labor costs tend to be lower. Proponents claim that jobs in exporting industries are likely to increase and create ripple effect benefits for the economies of all three countries.

GATT

The *General Agreement on Tariffs and Trade (GATT)* is a treaty that provides for ongoing discussions among participating nations to find ways to minimize international trade barriers. The Uruguay round of GATT talks established the World

---

[6]James Hickman, "Red Stars Rising," *Inc. Magazine* (August 1995): 19–20.

Trade Organization (WTO) on April 15, 1994. When one WTO country has a trade complaint against another, the WTO court can hear the complaint and impose economic sanctions if the accused country is found guilty. In 1995, the United States accused Japan of unfairly placing barriers to the sale of U.S. cellular phones, automobiles, and photographic film in Japan. Japan claimed that no such barriers existed. These are the types of disputes that the WTO adjudicates.

European Union

The *European Union (EU)* created a different type of association among the participating nations. The alliance was formed in late 1993 after the framework was negotiated by nations who signed the Maastricht Treaty, so named because Maastricht is the city (in the Netherlands) where the EU leaders signed the treaty creating the integrated economic plan. The original country members of the EU are Belgium, Denmark, France, Germany, Greece, Britain, Ireland, Italy, Luxembourg, Netherlands, Portugal, and Spain. Most of the barriers to trade among these countries have been minimized. Austria, Finland, and Sweden have since joined the EU.

The EU encourages joint business ventures and coordinated economic policies. All EU members now have a common passport. Furthermore, each country recognizes professional and educational degrees of other member countries so that doctors, lawyers, and other licensed professionals can practice in any country within the EU.

Perhaps the most ambitious goal of the EU was to replace domestic currencies in use now (francs, marks, pounds, and so on) with a common currency, the **euro**, for all business transactions between EU members. On January 1, 1999 eleven European countries fixed their exchange rates against each other and against the euro. The euro began trading against other currencies on January 4, 1999. These European Monetary Union (EMU) countries are Austria, Belgium, Finland, France, Germany, Ireland, Italy, Luxembourg, the Netherlands, Portugal, and Spain. European Union members Denmark, Sweden, and Britain chose not to join the monetary union. Greece did not qualify.

The euro has been used for electronic transactions in the EMU countries since January 1999. This includes stock and government bond markets. The new euro notes and coins are scheduled to appear on January 1, 2002. The European Central Bank (ECB) began operating June 1, 1998. It sets monetary policy for EMU countries.

Free Trade versus Fair Trade

"Free trade" and "fair trade" differ. *Free trade* suggests an unconditional lowering of trade barriers. *Fair trade* suggests lowering trade barriers only if the other country lowers its barriers, and perhaps only if the other country meets additional conditions. Examples of other conditions include meeting minimum wage, worker safety, human rights, or environmental standards.

NAFTA, GATT, EMU, EU, and ECB are signs of an ever-increasing global economy. As the world moves into the twenty-first century, international business will become more significant for all countries. An understanding of the financial risks, potential returns, and basic rules of international business will hopefully lead to greater financial success.

Summary

1. Define a multinational corporation (MNC) and explain its importance.

A multinational corporation is a corporation that operates in more than one country. It is becoming increasingly difficult to accurately classify some corporations as U.S., Swedish, Canadian, and so forth because many large corporations operate worldwide. International concerns are a vital concern for many U.S. businesses because of the potential market for products in countries throughout the world.

2. Demonstrate how the law of comparative advantage leads to international trade that benefits individuals and firms.

The law of comparative advantage says that countries (and individuals) should do that which they do best. Countries can export goods and services they produce well and import those goods and services that others develop better. By trading, both countries should see quality, quantity, or price benefits, or a combination of all three.

3. Describe exchange rates and their effects on businesses.

An exchange rate is an expression of the value of one country's currency relative to another country's currency. Fluctuating currency exchange rates affect the prices U.S. citizens pay for goods and services of other countries, and the prices that citizens of other countries pay for U.S. goods and services. When a country's currency weakens relative to the currencies of other countries, goods and services imported into that country become more expensive. That country's exported goods and services become cheaper to those countries that import them.

MNCs are also affected when their profits are earned in one country, then converted to the currency of the home country. Converting foreign currency profits into domestic currency to send to the home country of the business is known as *repatriating* the profits. If the U.S. dollar weakens relative to the foreign currency, the U.S. company's profits increase due to the change in exchange rates. However, if the dollar strengthens, profits decrease.

4. Show how firms manage the risks of fluctuating exchange rates.

Diversifying investments across several countries often reduces risk. Foreign securities can be bought in the United States by purchasing American depository receipts, which are U.S. dollar securities issued by a trust holding foreign currency denominated securities.

Firms can also manage risk through hedging—entering into a financial agreement that offsets or guards against risk. Three common types of hedging instruments are forward contracts, futures contracts, and swaps. Forward contracts are contracts where one party agrees to buy, and the other party agrees to sell, a certain amount of a currency at a specified exchange rate at a specified future time. Futures contracts are similar except they can be traded on organized exchanges. Swaps are negotiated contracts, like forward contracts, in which each party agrees to swap payments at specified points in time according to a predetermined formula.

5. Discuss exchange rate theories.

Currency exchange rates fluctuate for a variety of reasons. Among the most popular theories seeking to explain these changes are those that focus on relative prices in the two countries (purchasing power parity), relative interest rates (international Fisher effect), and the difference between spot and forward rates relative to the exchange rate (interest rate parity).

6. Describe the political and cultural risks that affect MNCs.

MNCs run the political risk of government trade restrictions, confiscation of assets by foreign governments, and even wars. Cultural risks include the risk of jeopardizing a deal or business due to insensitivity to differences in language, values, and attitudes. Training employees to recognize and respect cultural differences can reduce cultural risk.

7. Explain how major international trade agreements affect international business.

International agreements such as NAFTA, GATT, and the Maastricht Treaty that created the EU can bring down trade barriers and potentially create wide-ranging benefits. Opponents of these agreements claim, however, that they can result in job loss or harm a country's identity and economy.

You can access your FREE PHLIP/CW On–Line Study Guide, Current Events, and Spreadsheet Problems templates, which may correspond to this chapter either through the Prentice Hall Finance Center CD–ROM or by directly going to: **http://www.prenhall.com/gallagher.**

PHLIP/CW — *The Prentice Hall Learning on the Internet Partnership–Companion Web Site*

Equations Introduced in This Chapter

Equation 21-1. Calculation of a Currency Cross Rate:

$$\frac{\text{Currency A}}{\text{Currency B}} \times \frac{\text{Currency B}}{\text{Currency C}} = \frac{\text{Currency A}}{\text{Currency C}}$$

Self-Test

ST-1. What is the law of comparative advantage?

ST-2. What is the rationale given for trade agreements such as the North American Free Trade Agreement (NAFTA)?

ST-3. If you can get 1.25 Canadian dollars for one U.S. dollar and one U.S. dollar gets you 113 Japanese yen, how many Japanese yen do you get for one Canadian dollar?

ST-4. What are American depository receipts (ADRs)?

Review Questions

1. What does it mean when the U.S. dollar weakens in the foreign exchange market?
2. What kinds of U.S. companies would benefit most from a stronger dollar in the foreign exchange market? Explain.
3. Under what circumstance would the U.S. dollar and the Canadian dollar be said to have achieved purchasing power parity?
4. What are some of the primary advantages when a corporation has operations in countries other than its home country? What are some of the risks?
5. What is GATT, and what is its goal?

Build Your Communication Skills

CS-1. Form two groups to debate the issue of free trade versus fair trade. Some questions to address in the discussion include: (1) Should a country allow other countries to sell goods to its citizens with few or no restrictions (free trade)? (2) Should opening markets to companies from other countries be conditional on those other countries showing reciprocating openness (fair trade)? (3) Who decides which countries are engaging in fair trading practices?

CS-2. Write a short report on the current state of exchange rates. Which currencies are strengthening in foreign exchange markets and which are weakening? Why? Incorporate what you learned in this chapter in your analysis. What are the implications for international trade that can be drawn from the direction of the change in exchange rate values you observe?

Problems

Calculating Exchange Rates

21-1. The New York foreign exchange selling rates on January 19, 1999, as reported in *The Wall Street Journal*, are shown next for a few selected currencies:

Country/(Currency)	U.S. $ Equivalent
Australia (dollar)	.6340
Britain (pound)	1.6560
Canada (dollar)	.6498
Jordan (dinar)	1.4114
Hong Kong (dollar)	.1291
India (rupee)	.02353
Chile (peso)	.002086
Japan (yen)	.008818
Mexico (peso)	.09785
Israel (shekel)	.2464
Singapore (dollar)	.5947
Thailand (baht)	.02714
Turkey (lira)	.00000311
(Euro)	1.1605

Calculate the number of the following foreign currencies that can be bought with 1 million U.S. dollars.

 a. British pounds
 b. Indian rupees
 c. Japanese yen
 d. Australian dollars
 e. Mexican pesos
 f. Turkish lira
 g. Israeli shekels

Calculating Exchange Rates

21-2. Using the information given in problem 21-1, calculate the number of the following foreign currencies that can be bought with 1 million U.S. dollars.

 a. Chilean pesos
 b. Hong Kong dollars
 c. Singaporean dollars
 d. euros
 e. Indian rupees
 f. Mexican pesos
 g. Thai bahts

21-3. Using the information given in problem 21-1, calculate the number of U.S. dollars required to buy *Calculating Exchange Rates*

   a. 2 million Australian dollars
   b. 1.6 million Singaporean dollars
   c. 5 million euros
   d. 2.6 million Mexican pesos
   e. 2 million Japanese yen
   f. 25 million Thai bahts

21-4. If the Canadian dollar is selling at U.S. $ 0.6471 and the Israeli shekel at U.S. $0.2003, how many shekels are equal to one Canadian dollar? *Cross Rates*

21-5. If 58 rupees or 9.67 Hong Kong dollars could be purchased with one euro, how many rupees are equal to one Hong Kong dollar? *Cross Rates*

21-6. If one British pound is equivalent to 16.9 Mexican pesos or 2.8 Singapore dollars, how many Singapore dollars can one purchase with 10 million Mexican pesos? *Cross Rates*

21-7. If one British pound is equivalent to 1.5 euros, and one euro can purchase 60 baht, how many baht can one purchase with 1 million British pounds? *Cross Rates*

21-8. If one British pound is equivalent to 1.5 euros, and one euro can purchase .8 dinars, and one dinar is worth 160 yen, how many yen can one purchase with 1 million British pounds? *Cross Rates*

21-9. Assume that you invested $100,000 in a Japanese security a year back when the exchange rate was 119 yen per one U.S. dollar. However, the U.S. dollar depreciated against the yen throughout the year, and the current exchange rate is 100 yen per U.S. dollar. Calculate your percentage return on the investment due to this depreciation of the dollar. *Exchange Rate Effects*

21-10. An Indian investor bought 1,000 shares of General Motors at $37 per share on January 1, 1998, when the exchange rate was 42 rupees per one U.S. dollar. At the end of the year, the U.S. dollar appreciated against the rupee and the present exchange rate is 44 rupees per one U.S. dollar. Calculate the yearly return on investment for the Indian investor assuming the stock price remained the same. *Challenge Problem*

Answers to Self-Test

ST-1. The law of comparative advantage says that all will prosper when each party (or nation) does that at which it excels and then trades with others who do that at which they excel.

ST-2. NAFTA is a trade agreement between the United States, Canada, and Mexico that eliminated major barriers so the countries could freely trade goods and services to the benefit of the citizens of all three countries. Whether this has occurred is a matter of interpretation.

ST-3.
$$\frac{113\ ¥}{\$1.00\ U.S.} \times \frac{\$1.00\ U.S.}{1.25\ CD} = 90.4\ ¥\ per\ 1.00\ CD$$

ST-4. American depository receipts (ADRs) are dollar-denominated securities traded in the United States that represent a claim on a special trust that is created to hold foreign stock.

APPENDIX

TABLE I
Future Value Interest Factors, FVIF, Compounded at k Percent for n Periods: $FVIF_{k,n} = (1 + k)^n$

INTEREST RATE, K

Number of Periods, n	0%	1%	2%	3%	4%	5%	6%	7%	8%	9%	10%	12%	14%	16%	18%	20%	25%	30%	35%	40%	45%	50%
0	1.0000	1.0000	1.0000	1.0000	1.0000	1.0000	1.0000	1.0000	1.0000	1.0000	1.0000	1.0000	1.0000	1.0000	1.0000	1.0000	1.0000	1.0000	1.0000	1.0000	1.0000	1.0000
1	1.0000	1.0100	1.0200	1.0300	1.0400	1.0500	1.0600	1.0700	1.0800	1.0900	1.1000	1.1200	1.1400	1.1600	1.1800	1.2000	1.2500	1.3000	1.3500	1.4000	1.4500	1.5000
2	1.0000	1.0201	1.0404	1.0609	1.0816	1.1025	1.1236	1.1449	1.1664	1.1881	1.2100	1.2544	1.2996	1.3456	1.3924	1.4400	1.5625	1.6900	1.8225	1.9600	2.1025	2.2500
3	1.0000	1.0303	1.0612	1.0927	1.1249	1.1576	1.1910	1.2250	1.2597	1.2950	1.3310	1.4049	1.4815	1.5609	1.6430	1.7280	1.9531	2.1970	2.4604	2.7440	3.0486	3.3750
4	1.0000	1.0406	1.0824	1.1255	1.1699	1.2155	1.2625	1.3108	1.3605	1.4116	1.4641	1.5735	1.6890	1.8106	1.9388	2.0736	2.4414	2.8561	3.3215	3.8416	4.4205	5.0625
5	1.0000	1.0510	1.1041	1.1593	1.2167	1.2763	1.3382	1.4026	1.4693	1.5386	1.6105	1.7623	1.9254	2.1003	2.2878	2.4883	3.0518	3.7129	4.4840	5.3782	6.4097	7.5938
6	1.0000	1.0615	1.1262	1.1941	1.2653	1.3401	1.4185	1.5007	1.5869	1.6771	1.7716	1.9738	2.1950	2.4364	2.6996	2.9860	3.8147	4.8268	6.0534	7.5295	9.2941	11.3906
7	1.0000	1.0721	1.1487	1.2299	1.3159	1.4071	1.5036	1.6058	1.7138	1.8280	1.9487	2.2107	2.5023	2.8262	3.1855	3.5832	4.7684	6.2749	8.1722	10.5414	13.4765	17.0859
8	1.0000	1.0829	1.1717	1.2668	1.3686	1.4775	1.5938	1.7182	1.8509	1.9926	2.1436	2.4760	2.8526	3.2784	3.7589	4.2998	5.9605	8.1573	11.0324	14.7579	19.5409	25.6289
9	1.0000	1.0937	1.1951	1.3048	1.4233	1.5513	1.6895	1.8385	1.9990	2.1719	2.3579	2.7731	3.2519	3.8030	4.4355	5.1598	7.4506	10.6045	14.8937	20.6610	28.3343	38.4434
10	1.0000	1.1046	1.2190	1.3439	1.4802	1.6289	1.7908	1.9672	2.1589	2.3674	2.5937	3.1058	3.7072	4.4114	5.2338	6.1917	9.3132	13.7858	20.1066	28.9255	41.0847	57.6650
11	1.0000	1.1157	1.1234	1.3842	1.5395	1.7103	1.8983	2.1049	2.3316	2.5804	2.8531	3.4785	4.2262	5.1173	6.1759	7.4301	11.6415	17.9216	27.1439	40.4957	59.5728	86.4976
12	1.0000	1.1268	1.2682	1.4258	1.6010	1.7959	2.0122	2.2522	2.5182	2.8127	3.1384	3.8960	4.8179	5.9360	7.2876	8.9161	14.5519	23.2981	36.6442	56.6939	86.3806	129.7463
13	1.0000	1.1381	1.2936	1.4685	1.6651	1.8856	2.1329	2.4098	2.7196	3.0658	3.4523	4.3635	5.4924	6.8858	8.5994	10.6993	18.1899	30.2875	49.4697	79.3715	125.2518	194.6195
14	1.0000	1.1495	1.3195	1.5126	1.7317	1.9799	2.2609	2.5785	2.9372	3.3417	3.7975	4.8871	6.2613	7.9875	10.1472	12.8392	22.7374	39.3738	66.7841	111.1201	181.6151	291.9293
15	1.0000	1.1610	1.3459	1.5580	1.8009	2.0789	2.3966	2.7590	3.1722	3.6425	4.1772	5.4736	7.1379	9.2655	11.9737	15.4070	28.4217	51.1859	90.1585	155.5681	263.3419	437.8939
16	1.0000	1.1726	1.3728	1.6047	1.8730	2.1829	2.5404	2.9522	3.4259	3.9703	4.5950	6.1304	8.1372	10.7480	14.1290	18.4884	35.5271	66.54171	121.7139	217.7953	381.8458	656.8408
17	1.0000	1.1843	1.4002	1.6528	1.9479	2.2920	2.6928	3.1588	3.7000	4.3276	5.0545	6.8660	9.2765	12.4677	16.6722	22.1861	44.4089	86.5042	164.3138	304.9135	553.6764	985.2613
18	1.0000	1.1961	1.4282	1.7024	2.0258	2.4066	2.8543	3.3799	3.9960	4.7171	5.5599	7.6900	10.5752	14.4625	19.6733	26.6233	55.5112	112.4554	221.8236	426.8789	802.8308	1477.8919
19	1.0000	1.2081	1.4568	1.7535	2.1068	2.5270	3.0256	3.6165	4.3157	5.1417	6.1159	8.6128	12.0557	16.7765	23.2144	31.9480	69.3889	146.1920	299.4619	597.63041	164.1047	2216.8378
20	1.0000	1.2202	1.4859	1.8061	2.1911	2.6533	3.2071	3.8697	4.6610	5.6044	6.7275	9.6463	13.7435	19.4608	27.3930	38.3376	86.7362	190.0496	404.2736	836.6826	1687.9518	3325.2567
25	1.0000	1.2824	1.6406	2.0938	2.6658	3.3864	4.2919	5.4274	6.8485	8.6231	10.8347	17.0001	26.4619	40.8742	62.6686	95.3962	264.698	705.641	1812.78	4499.88	10819.3	2525.2
30	1.0000	1.3478	1.8114	2.4273	3.2434	4.3219	5.7435	7.6123	10.0627	13.2677	17.4494	29.9599	50.9502	85.8499	143.371	237.376	807.794	2620.00	8128.55	24201.4	69349.0	191751
35	1.0000	1.4166	1.9999	2.8139	3.9461	5.5160	7.6861	10.6766	14.7853	20.4140	28.1024	52.7996	98.1002	180.314	327.997	590.668	2465.19	9727.86	36448.7	130161	444509	1456110
40	1.0000	1.4889	2.2080	3.2620	4.8010	7.0400	10.2857	14.9745	21.7245	31.4094	45.2593	93.0510	188.884	378.721	750.378	1469.77	7523.16	36118.9	163437	700038	2849181	11057332
45	1.0000	1.5648	2.4379	3.7816	5.8412	8.9850	13.7646	21.0025	31.9204	48.3273	72.8905	163.988	363.679	795.444	1716.68	3657.26	22958.9	134107	732858	3764971	18262495	83966617
50	1.0000	1.6446	2.6916	4.3839	7.1067	11.4674	18.4202	29.4570	46.9016	74.3575	117.391	289.002	700.233	1670.70	3927.36	9100.44	70064.9	497929	3286158	0248916	117057734	637621500

TABLE II
Present Value Interest Factors, PVIF, Discounted at k Percent for n Periods: $PVIF_{k,n} = \dfrac{1}{(1+k)^n}$

DISCOUNT RATE, k

Number of Periods, n	0%	1%	2%	3%	4%	5%	6%	7%	8%	9%	10%	12%	14%	16%	18%	20%	25%	30%	35%	40%	45%	50%
0	1.0000	1.0000	1.0000	1.0000	1.0000	1.0000	1.0000	1.0000	1.0000	1.0000	1.0000	1.0000	1.0000	1.0000	1.0000	1.0000	1.0000	1.0000	1.0000	1.0000	1.0000	1.0000
1	1.0000	0.9901	0.9804	0.9709	0.9615	0.9524	0.9434	0.9346	0.9259	0.9174	0.9091	0.8929	0.8772	0.8621	0.8475	0.8333	0.8000	0.7692	0.7407	0.7143	0.6897	0.6667
2	1.0000	0.9803	0.9612	0.9426	0.9246	0.9070	0.8900	0.8734	0.8573	0.8417	0.8264	0.7972	0.7695	0.7432	0.7182	0.6944	0.6400	0.5917	0.5487	0.5102	0.4756	0.4444
3	1.0000	0.9706	0.9423	0.9151	0.8890	0.8638	0.8396	0.8163	0.7938	0.7722	0.7513	0.7118	0.6750	0.6407	0.6086	0.5787	0.5120	0.4552	0.4064	0.3644	0.3280	0.2963
4	1.0000	0.9610	0.9238	0.8885	0.8548	0.8227	0.7921	0.7629	0.7350	0.7084	0.6830	0.6355	0.5921	0.5523	0.5158	0.4823	0.4096	0.3501	0.3011	0.2603	0.2262	0.1975
5	1.0000	0.9515	0.9057	0.8626	0.8219	0.7835	0.7473	0.7130	0.6806	0.6499	0.6209	0.5674	0.5194	0.4761	0.4371	0.4019	0.3277	0.2693	0.2230	0.1859	0.1560	0.1317
6	1.0000	0.9420	0.8880	0.8375	0.7903	0.7462	0.7050	0.6663	0.6302	0.5963	0.5645	0.5066	0.4556	0.4104	0.3704	0.3349	0.2621	0.2072	0.1652	0.1328	0.1076	0.0878
7	1.0000	0.9327	0.8706	0.8131	0.7599	0.7107	0.6651	0.6227	0.5835	0.5470	0.5132	0.4523	0.3996	0.3538	0.3139	0.2791	0.2097	0.1594	0.1224	0.0949	0.0742	0.0585
8	1.0000	0.9235	0.8535	0.7894	0.7307	0.6768	0.6274	0.5820	0.5403	0.5019	0.4665	0.4039	0.3506	0.3050	0.2660	0.2326	0.1678	0.1226	0.0906	0.0678	0.0512	0.0390
9	1.0000	0.9143	0.8368	0.7664	0.7026	0.6446	0.5919	0.5439	0.5002	0.4604	0.4241	0.3606	0.3075	0.2630	0.2255	0.1938	0.1342	0.0943	0.0671	0.0484	0.0353	0.0260
10	1.0000	0.9053	0.8203	0.7441	0.6756	0.6139	0.5584	0.5083	0.4632	0.4224	0.3855	0.3220	0.2697	0.2267	0.1911	0.1615	0.1074	0.0725	0.0497	0.0346	0.0243	0.0173
11	1.0000	0.8963	0.8043	0.7224	0.6496	0.5847	0.5268	0.4751	0.4289	0.3875	0.3505	0.2875	0.2366	0.1954	0.1619	0.1346	0.0859	0.0558	0.0368	0.0247	0.0168	0.0116
12	1.0000	0.8874	0.7885	0.7014	0.6246	0.5568	0.4970	0.4440	0.3971	0.3555	0.3186	0.2567	0.2076	0.1685	0.1372	0.1122	0.0687	0.0429	0.0273	0.0176	0.0116	0.0077
13	1.0000	0.8787	0.7730	0.6810	0.6006	0.5303	0.4688	0.4150	0.3677	0.3262	0.2897	0.2292	0.1821	0.1452	0.1163	0.0935	0.0550	0.0330	0.0202	0.0126	0.0080	0.0051
14	1.0000	0.8700	0.7579	0.6611	0.5775	0.5051	0.4423	0.3878	0.3405	0.2992	0.2633	0.2046	0.1597	0.1252	0.0985	0.0779	0.0440	0.0254	0.0150	0.0090	0.0055	0.0034
15	1.0000	0.8613	0.7430	0.6419	0.5553	0.4810	0.4173	0.3624	0.3152	0.2745	0.2394	0.1827	0.1401	0.1079	0.0835	0.0649	0.0352	0.0195	0.0111	0.0064	0.0038	0.0023
16	1.0000	0.8528	0.7284	0.6232	0.5339	0.4581	0.3936	0.3387	0.2919	0.2519	0.2176	0.1631	0.1229	0.0930	0.0708	0.0541	0.0281	0.0150	0.0082	0.0046	0.0026	0.0015
17	1.0000	0.8444	0.7142	0.6050	0.5134	0.4363	0.3714	0.3166	0.2703	0.2311	0.1978	0.1456	0.1078	0.0802	0.0600	0.0451	0.0225	0.0116	0.0061	0.0033	0.0018	0.0010
18	1.0000	0.8360	0.7002	0.5874	0.4936	0.4155	0.3503	0.2959	0.2502	0.2120	0.1799	0.1300	0.0946	0.0691	0.0508	0.0376	0.0180	0.0089	0.0045	0.0023	0.0012	0.0007
19	1.0000	0.8277	0.6864	0.5703	0.4746	0.3957	0.3305	0.2765	0.2317	0.1945	0.1635	0.1161	0.0829	0.0596	0.0431	0.0313	0.0144	0.0068	0.0033	0.0017	0.0019	0.0005
20	1.0000	0.8195	0.6730	0.5537	0.4564	0.3769	0.3118	0.2584	0.2145	0.1784	0.1486	0.1037	0.0728	0.0514	0.0365	0.0261	0.0115	0.0053	0.0025	0.0012	0.0006	0.0003
25	1.0000	0.7798	0.6095	0.4776	0.3751	0.2953	0.2330	0.1842	0.1460	0.1160	0.0923	0.0588	0.0378	0.0245	0.0160	0.0105	0.0038	0.0014	0.0006	0.0002	0.0001	0.0000
30	1.0000	0.7419	0.5521	0.4120	0.3083	0.2314	0.1741	0.1314	0.0994	0.0754	0.0573	0.0334	0.0196	0.0116	0.0070	0.0042	0.0012	0.0004	0.0001	0.0000	0.0000	0.0000
35	1.0000	0.7059	0.5000	0.3554	0.2534	0.1813	0.1301	0.0937	0.0676	0.0490	0.0356	0.0189	0.0102	0.0055	0.0030	0.0017	0.0004	0.0001	0.0000	0.0000	0.0000	0.0000
40	1.0000	0.6717	0.4529	0.3066	0.2083	0.1420	0.0972	0.0668	0.0460	0.0318	0.0221	0.0107	0.0053	0.0026	0.0013	0.0007	0.0001	0.0000	0.0000	0.0000	0.0000	0.0000
45	1.0000	0.6391	0.4102	0.2644	0.1712	0.1113	0.0727	0.0476	0.0313	0.0207	0.0137	0.0061	0.0027	0.0013	0.0006	0.0003	0.0000	0.0000	0.0000	0.0000	0.0000	0.0000
50	1.0000	0.6080	0.3715	0.2281	0.1407	0.0872	0.0543	0.0339	0.0213	0.0134	0.0085	0.0035	0.0014	0.0006	0.0003	0.0001	0.0000	0.0000	0.0000	0.0000	0.0000	0.0000

TABLE III
Future Value Interest Factors for an Annuity, FVIFA, Compounded at k Percent for n Periods: $FVIFA_{k,n} = \sum_{t=1}^{n}(1+k)^{n-t} = \frac{(1+k)^n - 1}{k}$ (for non-zero k)

INTEREST RATE, K

Number of Annuity Pmts., n	0%	1%	2%	3%	4%	5%	6%	7%	8%	9%	10%	12%	14%	16%	18%	20%	25%	30%	35%	40%	45%	50%
1	1.0000	1.0000	1.0000	1.0000	1.0000	1.0000	1.0000	1.0000	1.0000	1.0000	1.0000	1.0000	1.0000	1.0000	1.0000	1.0000	1.0000	1.0000	1.0000	1.0000	1.0000	1.0000
2	2.0000	2.0100	2.0200	2.0300	2.0400	2.0500	2.0600	2.0700	2.0800	2.0900	2.1000	2.1200	2.1400	2.1600	2.1800	2.2000	2.2500	2.3000	2.3500	2.400	2.4500	2.5000
3	3.0000	3.0301	3.0604	3.0909	3.1216	3.1525	3.1836	3.2149	3.2464	3.2781	3.3100	3.3744	3.4396	3.5056	3.5724	3.6400	3.8125	3.9900	4.1725	4.3600	4.5525	4.7500
4	4.0000	4.0604	4.1216	4.1836	4.2465	4.3101	4.3746	4.4399	4.5061	4.5731	4.6410	4.7793	4.9211	5.0665	5.2154	5.3680	5.7656	6.1870	6.6329	7.1040	7.6011	8.1250
5	5.0000	5.1010	5.2040	5.3091	5.4163	5.5256	5.6371	5.7507	5.8666	5.9847	6.1051	6.3528	6.6101	6.8771	7.1542	7.4416	8.2070	9.0431	9.9544	10.9456	12.0216	13.1875
6	6.0000	6.1520	6.3081	6.4684	6.6330	6.8019	6.9753	7.1533	7.3359	7.5233	7.7156	8.1152	8.5355	8.9775	9.4420	9.9299	11.2588	12.7560	14.4384	16.3238	18.4314	20.7813
7	7.0000	7.2135	7.4343	7.6625	7.8983	8.1420	8.3938	8.6540	8.9228	9.2004	9.4872	10.0890	10.7305	11.4139	12.1415	12.9159	15.0735	17.5828	20.4919	23.8534	27.7255	32.1719
8	8.0000	8.2857	8.5830	8.8923	9.2142	9.5491	9.8975	10.2598	10.6366	11.0285	11.4359	12.2997	13.2328	14.2401	15.3270	16.4991	19.8419	23.8577	28.6640	34.3947	41.2019	49.2578
9	9.0000	9.3685	9.7546	10.1591	10.5828	11.0266	11.4913	11.9780	12.4876	13.0210	13.5795	14.7757	16.0853	17.5185	19.0859	20.7989	25.8023	32.0150	39.6964	49.1526	60.7428	74.8867
10	10.0000	10.4622	10.9497	11.4639	12.0061	12.5779	13.1808	13.8164	14.4866	15.1929	15.9374	17.5487	19.3373	21.3215	23.5213	25.9587	33.2529	42.6195	54.5902	69.8137	89.0771	113.330
11	11.0000	11.5668	12.1687	12.8078	13.4864	14.2068	14.9716	15.7836	16.6455	17.5603	18.5312	20.6546	23.0445	25.7329	28.7551	32.1504	42.5661	56.4053	74.6967	98.7391	130.162	170.995
12	12.0000	12.6825	13.4121	14.1920	15.0258	15.9171	16.8699	17.8885	18.9771	20.1407	21.3843	24.1331	27.2707	30.8502	34.9311	39.5805	54.2077	74.3270	101.841	139.235	189.735	257.493
13	13.0000	13.8093	14.6803	15.6178	16.6268	17.7130	18.8821	20.1406	21.4953	22.9534	24.5227	28.0291	32.0887	36.7862	42.2187	48.4966	68.7596	97.6250	138.485	195.929	276.115	387.239
14	14.0000	14.9474	15.9739	17.0863	18.2919	19.5986	21.0151	22.5505	24.2149	26.0192	27.9750	32.3926	37.5811	43.6720	50.8180	59.1959	86.9495	127.913	187.954	275.300	401.367	581.859
15	15.0000	16.0969	17.2934	18.5989	20.0236	21.5786	23.2760	25.1290	27.1521	29.3609	31.7725	37.2797	43.8424	51.6595	60.9653	72.0351	109.687	167.286	254.738	386.420	582.982	873.788
16	16.0000	17.2579	18.6393	20.1569	21.8245	23.6575	25.6725	27.8881	30.3243	33.0034	35.9497	42.7533	50.9804	60.9250	72.9390	87.4421	138.109	218.472	344.897	541.988	846.324	1311.68
17	17.0000	18.4304	20.0121	21.7616	23.6975	25.8404	28.2129	30.8402	33.7502	36.9737	40.5447	48.8837	59.1176	71.6730	87.0680	105.931	173.636	285.014	466.611	759.784	1228.17	1968.52
18	18.0000	19.6147	21.4123	23.4144	25.6454	28.1324	30.9057	33.9990	37.4502	41.3013	45.5992	55.7497	68.3941	84.1407	103.740	128.117	218.045	371.518	630.925	1064.70	1781.85	2953.78
19	19.0000	20.8109	22.8406	25.1169	27.6712	30.5390	33.7600	37.3790	41.4463	46.0185	51.1591	63.4397	78.9692	98.6032	123.414	154.740	273.556	483.973	852.748	1491.58	2584.68	4431.68
20	20.0000	22.0190	24.2974	26.8704	29.7781	33.0660	36.7856	40.9955	45.7620	51.1601	57.2750	72.0524	91.0249	115.380	146.628	186.688	342.945	630.165	1152.21	2089.21	3748.78	6648.51
25	25.0000	28.2432	32.0303	36.4593	41.6459	47.7271	54.8645	63.2490	73.1059	84.7009	98.3471	133.334	181.871	249.214	342.603	471.981	1054.79	2348.80	5176.50	11247.1990	24040.7	50500.3
30	30.0000	34.7849	40.5681	47.5754	56.0849	66.4388	79.0582	94.4608	113.283	136.308	164.494	241.333	356.787	530.312	790.948	1181.88	3227.17	8729.99	23221.6	60501.1	154107	383500
35	35.0000	41.6603	49.9945	60.4621	73.6522	90.3203	111.435	138.237	172.317	215.711	271.024	431.663	693.573	1120.71	1816.65	2948.34	9856.76	32422.9	104136	325400	987794	2912217
40	40.0000	48.8864	60.4020	75.4013	95.0255	120.800	154.762	199.635	259.057	337.882	442.593	767.091	1342.03	2360.76	4163.21	7343.86	30088.7	120393	466960	1750092	6331512	22114663
45	45.0000	56.4811	71.8927	92.7199	121.029	159.700	212.744	285.749	386.506	525.859	718.905	1358.23	2590.56	4965.27	9531.58	18281.3	91831.5	447019	2093876	9412424	40583319	167933233
50	50.0000	64.4632	84.5794	112.797	152.667	209.348	290.336	406.529	573.770	815.084	1163.91	2400.02	4994.52	10435.6	21813.1	45497.2	280256	1659761	9389020	50622288	260128295	1275242998

TABLE IV
Present Value Interest Factors for an Annuity, PVIFA, Discounted at k Percent for n Periods:

$$\text{PVIFA}_{k,n} = \sum_{t=1}^{n} \frac{1}{(1+k)^t} = \frac{1 - \frac{1}{(1+k)^n}}{k} = \frac{1}{k} - \frac{1}{k(1+k)^n} \quad \text{(for non-zero k)}$$

DISCOUNT RATE, k

Number of Annuity Pmts., n	0%	1%	2%	3%	4%	5%	6%	7%	8%	9%	10%	12%	14%	16%	18%	20%	25%	30%	35%	40%	45%	50%
1	1.0000	0.9901	0.9804	0.9709	0.9615	0.9524	0.9434	0.9346	0.9259	0.9174	0.9091	0.8929	0.8772	0.8621	0.8475	0.8333	0.8000	0.7692	0.7407	0.7413	0.6897	0.6667
2	2.0000	1.9704	1.9416	1.9135	1.8861	1.8594	1.8334	1.8080	1.7833	1.7591	1.7355	1.6901	1.6467	1.6052	1.5656	1.5278	1.4400	1.3609	1.2894	1.2245	1.1653	1.1111
3	3.0000	2.9410	2.8839	2.8286	2.7751	2.7232	2.6730	2.6243	2.5771	2.5313	2.4869	2.4018	2.3216	2.2459	2.1743	2.1065	1.9520	1.8161	1.6959	1.5889	1.4933	1.4074
4	4.0000	3.9020	3.8077	3.7171	3.6299	3.5460	3.4651	3.3872	3.3121	3.2397	3.1699	3.0373	2.9137	2.7982	2.6901	2.5887	2.5616	2.1662	1.9969	1.8492	1.7195	1.6049
5	5.0000	4.8534	4.7135	4.5797	4.4518	4.3295	4.2124	4.1002	3.9927	3.8897	3.7908	3.6048	3.4331	3.2743	3.1272	2.9906	2.6893	2.4356	2.2200	2.0352	1.8755	1.7366
6	6.0000	5.7955	5.6014	5.4172	5.2421	5.0757	4.9173	4.7665	4.6229	4.4859	4.3553	4.1114	3.8887	3.6847	3.4976	3.3255	2.9514	2.6427	2.3852	2.1680	1.9831	1.8244
7	7.0000	6.7282	6.4720	6.2303	6.0021	5.7864	5.5824	5.3893	5.2064	5.0330	4.8684	4.5638	4.2883	4.0386	3.8115	3.6046	3.1611	2.8021	2.5075	2.2628	2.0573	1.8829
8	8.0000	7.6517	7.3255	7.0197	6.7327	6.4632	6.2098	5.9713	5.7466	5.5348	5.3349	4.9676	4.6389	4.3436	4.0776	3.8372	3.3289	2.9247	2.5982	2.3306	2.1085	1.9220
9	9.0000	8.5660	8.1622	7.7861	7.4353	7.1078	6.8017	6.5152	6.2469	5.9952	5.7590	5.3282	4.9464	4.6065	4.3030	4.0310	3.4631	3.0190	2.6653	2.3790	2.1438	1.9480
10	10.0000	9.4713	8.9826	8.5302	8.1109	7.7217	7.3601	7.0236	6.7101	6.4177	6.1446	5.6502	5.2161	4.8332	4.4941	4.1925	3.5704	3.0915	2.7150	2.4136	2.1681	1.9653
11	11.0000	10.3676	9.7868	9.2526	8.7605	8.3064	7.8869	7.4987	7.1390	6.8052	6.4951	5.9377	5.4527	5.0286	4.6560	4.3271	3.6564	3.1473	2.7519	2.4383	2.1849	1.9769
12	12.0000	11.2551	10.5753	9.9540	9.3851	8.8633	8.3838	7.9427	7.5361	7.1607	6.8137	6.1944	5.6603	5.1971	4.7932	4.4392	3.7251	3.1903	2.7792	2.4559	2.1965	1.9846
13	13.0000	12.1337	11.3484	10.6350	9.9856	9.3936	8.8527	8.3577	7.9038	7.4869	7.1034	6.4235	5.8424	5.3423	4.9095	4.5327	3.7801	3.2233	2.7994	2.4685	2.2045	1.9897
14	14.0000	13.0037	12.1062	11.2961	10.5631	9.8986	9.2950	8.7455	8.2442	7.7862	7.3667	6.6282	6.0021	5.4675	5.0081	4.6106	3.8241	3.2487	2.8144	2.4775	2.2100	1.9931
15	15.0000	13.8651	12.8493	11.9379	11.1184	10.3797	9.7122	9.1079	8.5595	8.0607	7.6061	6.8109	6.1422	5.5755	5.0916	4.6755	3.8593	3.2682	2.8255	2.4839	2.2138	1.9954
16	16.0000	14.7179	13.5777	12.5611	11.6523	10.8378	10.1059	9.4466	8.8514	8.3126	7.8237	6.9740	6.2651	5.6685	5.1624	4.7296	3.8874	3.2832	2.8337	2.4885	2.2164	1.9970
17	17.0000	15.5623	14.2919	13.1661	12.1657	11.2741	10.4773	9.7632	9.1216	8.5436	8.0216	7.1196	6.3729	5.7487	5.2223	4.7746	3.9099	3.2948	2.8398	2.4918	2.2182	1.9980
18	18.0000	16.3983	14.9920	13.7535	12.6583	11.6896	10.8276	10.0591	9.3719	8.7556	8.2014	7.2497	6.4674	5.8178	5.2732	4.8122	3.9279	3.3037	2.8443	2.4941	2.2195	1.9986
19	19.0000	17.2260	15.6785	14.3238	13.1339	12.0853	11.1581	10.3356	9.6036	8.9501	8.3649	7.3658	6.5504	5.8775	5.3162	4.8435	3.9424	3.3105	2.8476	2.4958	2.2203	1.9991
20	20.0000	18.0456	16.3514	14.8775	13.5903	12.4622	11.4699	10.5940	9.8181	9.1285	8.5136	7.4694	6.6231	5.9288	5.3527	4.8696	3.9539	3.3158	2.8501	2.4970	2.2209	1.9994
25	25.0000	22.0232	19.5235	17.4131	15.6221	14.0939	12.7834	11.6536	10.6748	9.8226	9.0770	7.8431	6.8729	6.0971	5.4669	4.9476	3.9849	3.3286	2.8556	2.4994	2.2220	1.9999
30	30.0000	25.8077	22.3965	19.6004	17.2920	15.3725	13.7648	12.4090	11.2578	10.2737	9.4269	8.0552	7.0027	6.1772	5.5168	4.9789	3.9950	3.3321	2.8568	2.4999	2.2222	2.0000
35	35.0000	29.4086	24.9986	21.4872	18.6646	16.3742	14.4982	12.9477	11.6546	10.5668	9.5442	8.1755	7.0700	6.2153	5.5386	4.9915	3.9984	3.3330	2.8571	2.5000	2.2222	2.0000
40	40.0000	32.8347	27.3555	23.1148	19.7928	17.1591	15.0463	13.3317	11.9246	10.7574	9.7791	8.2438	7.1050	6.2335	5.5482	4.9966	3.9995	3.3332	2.8571	2.5000	2.2222	2.0000
45	45.0000	36.0945	29.4902	24.5187	20.7200	17.7741	15.4558	13.6055	12.1084	10.8812	9.8628	8.2825	7.1232	6.2421	5.5523	4.9986	3.9998	3.3333	2.8571	2.5000	2.2222	2.0000
50	50.0000	39.1961	31.4236	25.7298	21.4822	18.2559	15.7619	13.8007	12.2335	10.9617	9.9148	8.3045	7.1327	6.2463	5.5541	4.9995	3.9999	3.3333	2.8571	2.5000	2.2222	2.0000

GLOSSARY

ABC system An inventory system in which items are classified according to their value for inventory control purposes.

absolute purchasing power parity theory Theory that claims the current exchange rate is determined by the relative prices in two countries of a similar basket of traded goods and services.

actuaries People who use applied mathematics and statistics to predict claims on insurance companies and pension funds.

additional funds needed The additional external financing required to support asset growth when forecasted assets exceed forecasted liabilities and equity.

after-tax cost of debt (ATk$_d$) The after-tax cost to a company of obtaining debt funds.

agency costs Costs incurred to monitor agents to reduce the conflict of interest between agents and principals.

agency problem The possibility of conflict between the interests of a firm's managers and those of the firm's owners.

agent A person who has the implied or actual authority to act on behalf of another.

aggressive working capital financing approach The use of short-term funds to finance all temporary current assets, possibly all or some permanent current assets, and perhaps some fixed assets.

American depository receipts (ADRs) Securities denominated in U.S. dollars that represent a claim on foreign currency—denominated stocks held in a trust.

amortized loan A loan that is repaid in regularly spaced, equal installments, which cover all interest and principal owed.

annuitant A person who is entitled to receive annuity payments.

annuity A series of equal cash payments made at regular time intervals.

arbitrage The process whereby equivalent assets are bought in one place and simultaneously sold in another, making a risk-free profit.

average tax rate The amount of tax owed, divided by the amount of taxable income.

balance sheet The financial statement that shows an economic unit's assets, liabilities, and equity at a given point in time.

banker's acceptance A security that represents a promise by a bank to pay a certain amount of money, if the original note maker doesn't pay.

bearer The owner of a security.

best efforts basis An arrangement in which the investment banking firm tries its best to sell a firm's securities for the desired price, without guarantees. If the securities must be sold for a lower price, the issuer collects less money.

beta (β) The measure of nondiversifiable risk. The stock market has a beta of 1.0. Betas higher than 1.0 indicate more nondiversifiable risk than the market, and betas lower than 1.0 indicate less. Risk-free portfolios have betas of 0.0.

bird-in-the-hand theory A theory that says that investors value a dollar of dividends more highly than a dollar of reinvested earnings because uncertainty is resolved.

board of directors A group of individuals elected by the common stockholders to oversee the management of the firm.

bond maturity date The date when a bond issuer makes the final payment promised.

bonds Securities that promise to pay the bearer a certain amount at a time in the future and may pay interest at regular intervals over the life of the security.

book value, net worth The total amount of common stockholder's equity on a company's balance sheet.

broker A person who brings buyers and sellers together.

business risk The uncertainty a company has due to fluctuations in its operating income.

call premium The premium the issuer pays to call in a bond before maturity. The excess of the call price over the face value.

call provision A bond indenture provision that allows the issuer to pay off a bond before the stated maturity date at a specified price.

capital Funds supplied to a firm.

capital asset pricing model (CAPM) A financial model that can be used to calculate the appropriate required rate of return for an investment project, given its degree of risk as measured by beta (β).

capital budget A document that shows planned expenditures for major asset acquisitions items, such as equipment or plant construction.

capital budgeting The process of evaluating proposed projects.

capital gains The profit made when an asset price is higher than the price paid.

capital market The market where long-term securities are traded.

capital rationing The process whereby management sets a limit on the amount of cash available for new capital budgeting projects.

capital structure The mixture of sources of capital that a firm uses (for example, debt, preferred stock, and common stock).

cash budget A detailed budget plan that shows the cash flows expected to occur during specific time periods.

chief financial officer The manager who directs and coordinates the financial activities of the firm.

clientele dividend theory The theory that says that a company should attempt to determine the dividend wants of its stockholders and maintain a consistent policy of paying stockholders what they want.

coefficient of variation The standard deviation divided by the mean. A measure of the degree of risk used to compare alternatives with possible returns of different magnitudes.

collateral Assets a borrower agrees to give to a lender if the borrower defaults on the terms of a loan.

collection policy The firms' plans for getting delinquent credit customers to pay their bills.

combined leverage The phenomenon whereby a change in sales causes net income to change by a larger percentage because of fixed operating and financial costs.

commercial paper A short-term, unsecured debt instrument issued by a large corporation or financial institution.

common stock A security that indicates ownership of a corporation.

compensating balance A specified amount that a lender requires a borrower to maintain in a non-interest-paying account during the life of a loan.

component cost of capital The cost of raising funds from a particular source, such as bondholders or common stockholders.

compound interest Interest earned on interest in addition to interest earned on the original principal.

conservative working capital financing approach The use of long-term debt and equity to finance all long-term fixed assets and permanent current assets, in addition to some part of temporary current assets.

continuous compounding A process whereby interest is earned on interest every instant of time.

contribution margin Sales price per unit minus variable cost per unit.

controller The manager responsible for the financial and cost accounting activities of a firm.

conversion ratio The number of shares (usually of common stock) that the holder of a convertible bond would receive if he or she exercised the conversion option.

conversion value The value of the stock that would be received if the conversion option on a convertible bond were exercised.

convertible bond A bond that may be converted, at the option of the bond's owner, into a certain amount of a different type of security issued by the company.

corporate bond A security that represents a promise by the issuing corporation to make certain payments, according to a certain schedule, to the owner of the bond.

corporation A business chartered by the state, that is a separate legal entity having some of, but not all of, the rights and responsibilities of a natural person.

correlation The degree to which one variable is linearly related to another.

correlation coefficient The measurement of degree of correlation, represented by the letter r. Its values range from $+1.0$ (perfect positive correlation) to -1.0 (perfect negative correlation).

cost of debt (k_d) The lender's required rate of return on a company's new bonds or other instrument of indebtedness.

cost of equity from new common stock (k_n) The cost of external equity, including the costs incurred to issue new common stock.

cost of internal equity (k_s) The required rate of return on funds supplied by existing common stockholders through earnings retained.

cost of preferred stock (k_p) Investors' required rate of return on a company's new preferred stock.

coupon interest payments Periodic interest payments promised at the time of a bond's issuance.

credit policy Credit standards a firm has established and the credit terms it offers.

credit scoring A process by which candidates for credit are compared against indicators of creditworthiness and scored accordingly.

credit standards Requirements customers must meet in order to be granted credit.

credit unions Financial institutions owned by members who receive interest on shares purchased and who obtain loans.

cross rate An exchange rate determined by examining how each of two currencies is valued in terms of a common third currency.

cross-sectional analysis Comparing variables for different entities (such as ratio values for different companies) for the same point in time or time period.

current assets Liquid assets of an economic entity (e.g., cash, accounts receivable, inventory, etc.) usually converted into cash within one year.

current liabilities Liabilities that are coming due soon, usually within one year.

date of record The date on which stockholder records are checked for the purpose of determining who will receive the dividend that has been declared.

dealer A person who makes his or her living buying and selling assets.

debenture A bond that is unsecured.

declaration date The date on which the board of directors announces a divided is to be paid.

default risk premium The extra interest lenders demand to compensate for assuming the risk that promised interest and principal payments may be made late or not at all.

deficit economic unit A government, business, or household unit with expenditures greater than its income.

degree of combined leverage (DCL) The percentage change in net income divided by the percentage change in sales.

degree of financial leverage (DFL) The percentage change in net income divided by the percentage change in operating income.

degree of operating leverage (DOL) The percentage change in operating income divided by the percentage change in sales.

depreciation basis The total value of an asset upon which depreciation expense will be calculated, a part at a time, over the life of the asset.

discount loan A loan with terms that call for the loan interest to be deducted from the loan proceeds at the time the loan is granted.

discount rate The interest rate used when calculating a present value representing the required rate of return.

discount yield The return realized by an investor who purchases a security for less than face value and redeems it at maturity for face value.

discounted cash flow (DCF) model A model that estimates the value of an asset by calculating the sum of the present values of all future cash flows.

diversification effect The effect of combining assets in a portfolio such that the fluctuations of the assets' returns tend to offset each other, reducing the overall volatility (risk) of the portfolio.

dividend payout ratio Dividends paid divided by net income.

dividend reinvestment plan An arrangement offered by some corporation where cash dividends are held by the company and used to purchase additional shares of stock for the investor.

dividend yield A stock's annual dividend divided by its current market price.

dividends Payments made to stockholders at the discretion of the board of directors of the corporation.

dividends payable The liability item on a firm's balance sheet that reflects the amount of dividends declared but not yet paid.

economic value added (EVA) The amount of profit remaining after accounting for the return expected by the firm's investors.

effective interest rate The annual interest rate that reflects the dollar interest paid divided by the dollar obtained for use.

electronic funds transfers The act of crediting one account and simultaneously debiting another by a computer.

equity multiplier The total assets to total common stockholders' equity ratio.

euro The common currency used by the European Monetary Union countries.

Eurodollars Dollar-denominated deposits in banks located outside the United States.

European Currency Unit (ECU) An index reflecting the value of a market basket of currencies from European Union countries.

excess financing The amount of excess funding available for expected asset growth when forecasted liabilities and equity exceed forecasted assets.

exchange rate The number of units of one country's currency that is needed to purchase one unit of another country's currency.

ex-dividend A characteristic of common stock such that it is trading *without* entitlement to an upcoming dividend.

ex-rights A characteristic of common stock such that it is trading *without* the entitlement to an upcoming rights offering.

externalities Positive or negative effects that will occur to existing projects if a new capital budgeting project is accepted.

face value, or par value, or principal The amount the bond issuer promises to pay to the investor when the bond matures. The terms *face value, par value,* and *principal* are often used interchangeably.

factoring The practice of selling accounts receivable to another firm.

factors Firms that buy and administer another firm's accounts receivable.

Federal Reserve System The government-sponsored entity that acts as the central bank of the United States, and that examines and regulates banks and other financial institutions.

fiduciary responsibility The legal requirement that those who are managing assets owned by someone else do so in a prudent manner and in accordance with the interests of the person(s) they represent.

financial (capital) lease A lease that is generally long term and noncancelable with the lessee using up most of the economic value of the asset by the end of the lease's term.

financial leverage The phenomenon whereby a change in operating income causes net income to change by a larger percentage because of the presence of fixed financial costs.

financial ratio A number that expresses the value of one financial variable relative to the value of another.

financial risk The additional volatility of a firm's net income caused by the presence of fixed financial costs.

financing cash flows Cash flows that occur as creditors are paid interest and principal and stockholders are paid dividends.

first mortgage A mortgage bond (a bond secured by real property) that gives the holder first claim on the real property pledged as security if there is a foreclosure.

fixed assets Assets that would not normally be sold or otherwise disposed of for a long period of time.

fixed costs Costs that do not vary with the level of production.

flotation costs Fees that companies pay (to investment bankers and to others) when new securities are issued.

forward rate The exchange rate for future delivery.

future value The value money or another asset will have in the future.

future value interest factor for a single amount (FVIF) The $(1 + k)^n$ factor that is multiplied by the original value to solve for future value.

future value interest factor for an annuity (FVIFA) The factor, which when multiplied by an expected annuity, gives you the sum of the future values of the annuity stream:

$$\frac{(1 + k)^n - 1}{k}$$

going concern value That value that comes from the future earnings and cash flows that can be generated by a company if it continues to operate.

going out of business value The value that is the amount that would be left if all the assets were sold, the liabilities and preferred stock paid off, and the remainder distributed among the owners.

hedge A financial arrangement used to offset or protect against negative effects of something else, such as fluctuating exchange rates.

hurdle rate The minimum rate of return that management demands from a proposed project before that project will be accepted.

income statement A financial statement that presents the revenues, expenses, and income of a business over a specific time period.

incremental cash flows Cash flows that will occur only if an investment is undertaken.

incremental depreciation expense The change in depreciation expense that a company will incur if a proposed capital budgeting project is accepted.

indenture The contrast between the issuing corporation and the bond's purchaser.

independent projects A group of projects such that any or all could be accepted.

industry comparison The process whereby financial ratios of a firm are compared to those of similar firms to see if the firm under scrutiny compares favorably or unfavorably with the norm.

inflation premium The extra interest that compensates lenders for the expected erosion of purchasing power of funds due to inflation over the life of the loan.

initial public offering (IPO) The process whereby a private corporation issues new common stock to the general public, thus becoming a publicly traded corporation.

institutional investors Financial institutions that invest in the securities of other companies.

interest The compensation lenders demand and borrowers pay when money is lent.

interest rate parity A theory that states that the difference between the exchange rate specified for future delivery, and that for current delivery, equals the difference in the interest rates for securities of the same maturity.

interest rate spread The rate a bank charges for loans minus the rate it pays for deposits.

intermediation The process by which funds are channeled from surplus to deficit economic units through a financial institution.

internal rate of return (IRR) The estimated rate of return for a proposed project, given the size of the project's incremental cash flows and their timing.

international bonds Bonds that are sold in countries other than where the issuer is domiciled.

international Fisher effect A theory developed by economist Irving Fisher that claims that changes in interest rates for two countries will be offset by equal changes, in the opposite direction, in the exchange rate.

inventory financing A type of financing that uses inventory as loan collateral.

investment banking firm A firm that helps issuers sell their securities and that provides other financial services.

investment-grade bond Bonds rated Baa3 or above by Moody's bond rating agency and BBB− or above by Standard & Poor's.

JIT An inventory system in which inventory items are scheduled to be delivered "just in time" to be used as needed.

junk bond bonds with lower than investment-grade ratings.

lease A contract between an asset owner (lessor) and another party who uses that asset (lessee) that allows the use of the asset for a specified period of time, specifies payment terms, and does not convey legal ownership.

lessee The party in a lease contract who uses the asset.

lessor The party in a lease contract who owns the asset.

leverage effect The result of one factor causing another factor to be magnified, such as where debt magnifies the return stockholders earn on their invested funds over the return on assets.

liability insurance Insurance that pays obligations that may be incurred by the insured as a result of negligence, slander, malpractice, and similar actions.

line of credit An informal arrangement between a lender and borrower in which the lender sets a limit on the maximum amount of funds the borrower may use at any one time.

liquidation value The amount that would be received by the owners of a company that sold all its assets as market value, paid all its liabilities and preferred stock, and distributed what was left to the owners of the firm.

liquidity risk premium The extra interest lenders demand to compensate for holding a security that is not easy to sell at its fair value.

lockbox A way station (typically a post office box) at which customers may send payments to a firm.

MACRs, modified accelerated cost recovery system The depreciation rules established by the Tax Reform Act of 1986 that allow owners to take accelerated depreciation with greater deductions in earlier years than in later years.

managing investment banker The head of an investment banking underwriting syndicate.

marginal cost of capital (MCC) The weighted average cost of capital for the next dollar of funds raised.

marginal tax rate The tax rate applied to the next dollar of taxable income.

market efficiency The relative ease, speed, and cost of trading securities. In an efficient market, securities can be traded easily, quickly, and at low cost. In an inefficient market, one or more of these qualities is missing.

market risk premium The additional return above the risk-free rate demanded by investors for assuming the risk of investing in the market.

market value added (MVA) The market value of the firm, debt plus equity, minus the total amount of capital invested in the firm.

maturity date The date the bearer of a security is to be paid the principal, or face value, of a security.

maturity risk premium The extra (or sometimes lesser) interest that lenders demand on longer-term securities.

mixed ratio A financial ratio that includes variables from both the income statement and from the balance sheet.

moderate working capital financing approach An approach in which a firm finances temporary current assets with short-term funds and permanent current assets and fixed assets with long-term funds.

Modigliani and Miller dividend theory A theory developed by financial theorists Franco Modigliani and Merton Miller that says the amount of dividends paid by a firm does not affect the firm's value.

money market The market where short-term securities are traded.

mortgage bond A bond secured by real property.

multinational corporation (MNC) A corporation that has operations in more than one country.

municipal bonds Bonds issued by a state, city, county, or other nonfederal government authority, including specially created municipal authorities such as a toll road or industrial development authority.

mutually exclusive projects A group of projects that compete against each other; only one of the mutually exclusive projects may be chosen.

mutuals Institutions (savings and loans or insurance companies e.g.,) that are owned by their depositors or policy holders.

negotiable CDs A deposit security issued by financial institutions that comes in minimum denominations of $100,000 and can be traded in the money market.

net present value (NPV) The estimated change in the value of the firm that would occur if a project is accepted.

net working capital The amount of current assets minus the amount of current liabilities of an economic unit.

nominal interest rate The rate observed in the financial marketplace that includes the real rate of interest and various premiums.

nominal risk-free rate of interest The interest rate without any premiums for the uncertainties associated with lending.

nondiversifiable risk The portion of a portfolio's total risk that cannot be eliminated by diversifying. Factors shared to a greater or lesser degree by most assets in the market, such as inflation and interest rate risk, cause nondiversifiable risk.

nonsimple project A project that has a negative initial cash flow, and also has one or more negative future cash flows.

NPV profile A graph that displays how a project's net present value changes as the discount rate, or required rate of return changes.

open-market operations The buying and selling of U.S. Treasury securities or foreign currencies to achieve some economic objective.

operating lease A lease that has a term substantially shorter than the useful life of the asset and is usually cancelable by the lessee.

operating leverage The effect of fixed operating costs on operating income; because of fixed costs, any change in sales results in a larger percentage change in operating income.

opportunity cost The cost of forgoing the best alternative to make a competing choice.

optimal capital budget The list of all accepted projects and the total amount of their initial cash outlays.

over-the-counter market A network of dealers around the world who purchase securities and maintain inventories of securities for sale.

par value The stated value of a share of common stock.

partnership An unincorporated business owned by two or more people.

payback period The expected amount of time a capital budgeting project will take to generate cash flows that cover the project's initial cash outlay.

payment date The date the transfer agent sends out a company's dividend checks.

pension fund A financial institution that takes in funds for workers, invests those funds, and then provides for a retirement benefit.

permanent current assets The minimum level of current assets maintained.

perpetuity An annuity that has an infinite maturity.

portfolio A collection of assets that are managed as a group.

preemptive right A security given by some corporations to existing stockholders that gives them the right to buy new shares of common stock at a below-market price until a specified expiration date.

preferred stock A security issued by a corporation with a given dividend to be paid before any dividends are paid to holders of the firm's common stock.

present value Today's value of promised or expected future value.

present value interest factor for a single amount (PVIF) The $1/(1 + k)^n$ factor that is multiplied by a given future value, to solve for the present value.

present value interest factor for an annuity (PVIFA) The factor, which when multiplied by an expected annuity payment, gives you the present value of the annuity stream:

$$\frac{1 - \frac{1}{(1+k)^n}}{k}$$

primary market The market in which newly issued securities are sold to the public.

primary reserves Vault cash and deposits at the Fed that go toward meeting a bank's reserve requirements.

principal A person who authorizes an agent to act for him or her.

private corporation A corporation that does not offer its shares to the general public and that can keep its financial statements confidential.

***pro forma* financial statements** Projected financial statements.

progressive tax rate structure A tax structure under which the tax rate increases as taxable income increases usually in a pattern of several steps.

promissory note A legal document the borrower signs indicating agreement to the terms of a loan.

proprietorship A business that is not incorporated and is owned by one person.

prospectus A disclosure document given to a potential investor when a new security is issued.

publicly traded corporations Corporations that have common stock that can be bought in the marketplace by any interested party and that must release audited financial statements to the public.

pure time value of money The value demanded by an investor to compensate for the postponement of consumption.

putable bonds Bonds that can be redeemed before the scheduled maturity date, at the option of the bondholder.

quota A quantity limit imposed by one country on the amount of a given good that can be imported from another country.

real option A valuable characteristic of some projects where revisions to that project at a later date are possible.

real rate of interest The rate that the market offers to lenders to compensate for postponing consumption.

refunding Issuing new bonds to replace old bonds.

relative purchasing power parity theory A theory that states that as prices change in one country relative to those prices in another country, for a given traded basket of similar goods and services, the exchange rate will tend to change proportionately but in the opposite direction.

required reserve ratio The percentage of deposits that determines the amount of reserves a financial institution is required to hold.

residual income Income left over, and available to common stockholders, after all other claimants have been paid.

restrictive covenants Promises made by the issuer of a bond to the investor, to the benefit of the investor.

rights-on A characteristic of common stock such that it is trading *with* the entitlement to an upcoming right.

risk The potential for unexpected events to occur.

risk-adjusted discount rate (RADR) A required rate of return adjusted to compensate for the effect a project has on a firm's risk.

risk aversion A tendency to avoid risk that explains why most investors require a higher expected rate of return the more risk they assume.

risk-free rate of return The rate of return that investors demand in order to take on a

FOR THE STUDENT

- **Student Workbook and Study Guide**—The workbook contains an overview of key points for each chapter, additional problems with detailed solutions, and self-tests to aid students in preparing for exams and other assignments.

- **The Collegiate Investment Challenge**—A real-time on-line portfolio management simulation allows students to experience the stock market. (A 33% discount coupon is included in the text!)

- **FREE On-Line Study Guide**—Free on the Internet site **www.prenhall.com/gallagher**.

- **FREE Prentice Hall Finance Center CD-ROM**—The CD-ROM is provided in the back of the text and helps bring financial management to life. The following features are contained on the CD-ROM:

  - ✔ **The Careers Center**—Provides a wealth of details pertaining to the broad professional opportunities in the world of finance. Also includes a self-test, 32 interviews with actual people in finance positions, tips for résumé writing, and much more!

  - ✔ **FinCoach**—This financial math practice software is designed to help students learn the math of finance through interactive practice.

  - ✔ **PHLIP/CW** Internet Access—Accessed directly from the CD-ROM or remotely at **www.prenhall.com/gallagher**, PHLIP/CW is the most advanced, text-specific Web site for finance available.

THE SUPPORT PACKAGE

FOR THE INSTRUCTOR

◆ **Instructor's Manual with Solutions**—Prepared by the authors, this manual contains concise chapter orientations, detailed chapter outlines, complete answers to the end-of-chapter questions, and worked-out solutions to all problems in the text.

◆ **Prentice Hall Custom Test Computerized Test Bank** (Mac and Windows)—Written by Professor Philip Thames of California State University-Fullerton, the test item file contains over 5,000 true/false, multiple choice, and short-answer questions. Instructors can edit, add, or delete questions from the file, as well as generate their own custom exams.

◆ **Text Bank**—Written by the authors, the test item file contains over 5,000 true/false, multiple choice, and short-answer questions. Instructors can edit, add, or delete questions from the file, as well as generate their own custom exams.

◆ **PHLIP/CW**—Use the Internet to bring real business practices into your finance course. Prentice Hall's "PHLIP/CW" Web site contains an assortment of rich resources for both students and instructors, making the course fun and relevant. There are downloadable PowerPoint presentations (prepared by the authors), an instructor's manual, and Excel spreadsheet solutions (prepared by the authors).

◆ **Color Transparencies**—These full-color images contain all figures and tables from the text.

◆ **Take the Collegiate Investment Challenge!**—Imagine giving your students half a million dollars to experience the world of investing while assuming absolutely no risk! Consider the valuable lessons they'll learn from buying and selling equities, trading options, short selling, and purchasing on the margin—all at current market prices. Mimicking the daily operations of real brokerage firms, The Collegiate Investment Challenge, a real-time, on-line portfolio management simulation, allows instructors to monitor trades and progress of all participating students. An excellent teaching companion for the investments sections of this text, The Collegiate Investment Challenge provides all the resources necessary to easily and effectively implement this program into any classroom, including:
  ✔ Free Professor Portfolio
  ✔ Classroom Reports
  ✔ Toll-free Professor Support Line
  ✔ Professor Chat Room

See other side for the Student Support Package.

project that contains no risk other than an inflation premium.

risk–return relationship The positive relationship between the risk of an investment and an investor's required rate of return.

sales breakeven point The level of sales that must be achieved such that operating income equals zero.

savings and loan associations (S&Ls) Financial institutions that take in deposits and make loans (primarily mortgage loans).

second mortgage A mortgage bond (a bond secured by real property) that gives the holder second claim (after the first-mortgage bondholder) on the real property pledged as security.

secondary market The market in which previously issued securities are traded from one investor to another.

secondary reserves Marketable securities that can be readily sold to obtain cash.

secured bond A bond backed by specific assets that the investor may claim if there is a default.

security A document that establishes the bearer's claim to receive funds in the future.

self-liquidating loan A loan that is used to acquire assets that generate enough cash to pay off the loan.

senior debenture An unsecured bond having a superior claim on the earnings and assets of the issuing firm relative to other debentures.

serial payments A mode of payment in which the issuer pays off bonds according to a staggered maturity schedule.

signaling The message sent by managers, or inferred by investors, when a financial decision is made.

signaling dividend theory A theory that says that dividend payments often send a signal from the management of a firm to market participants.

simple project A project that has a negative initial cash flow, followed by positive cash flows only.

sinking fund A method for retiring bonds. The bond issuer makes regular contributions to a fund that the trustee uses to buy back outstanding bonds and retire them.

spot rate The exchange rate for current delivery.

stakeholder A party having an interest in a firm (for example, owners, workers, management, creditors, suppliers, customers, and the community as a whole).

standard deviation A statistic that indicates how widely dispersed actual or possible values are distributed around a mean.

stated interest rate The interest rate advertised by the lender. Depending on the terms of the loan, the stated rate may or may not be the same as the effective interest rate.

statement of retained earnings A financial statement that shows how the value of retained earnings changes from one point in time to another.

stock Certificates of ownership interest in a corporation.

stock dividend A firm sends out new shares of stock to existing stockholders and makes an accounting transfer from retained earnings to the common stock and capital in excess of par accounts of the balance sheet.

stock split A firm gives new shares of stock to existing shareholders; on the balance sheet, they decrease the par value of the common stock proportionately to the increase in the number of shares outstanding.

straight bond value The value a convertible bond would have if it did not offer the conversion option to the investor.

straight-line depreciation A depreciation rule that allows equal amounts of the cost of an asset to be allocated over the asset's life.

strengthening currency A currency that is now convertible into a larger number of units of another currency than previously.

Subchapter S corporation A small corporation whose income is taxed like a partnership.

subordinated debenture An unsecured bond having an inferior claim on the earnings and assets of the issuing firm relative to other debentures.

sunk cost A cost that must be borne whether a proposed capital budgeting project is accepted or rejected.

surplus economic unit A business, household, or government unit with income greater than its expenditures.

syndicate A temporary alliance of investment banking firms that is formed for the purpose of underwriting a new security issue.

target stock A class of common stock that represents a claim on a part of company.

tariff A tax imposed by one country on imports from another country.

temporary current assets The portion of current assets that fluctuates during the company's business cycle.

10-K reports An audited set of financial statements submitted annually by all public corporations to the Securities and Exchange Commission (SEC).

10-Q reports An unaudited set of financial statements submitted quarterly by all public corporations to the Securities and Exchange Commission (SEC).

time value of money The phenomenon whereby money is valued more highly the sooner it is received.

trade credit Funds obtained by delaying payment to suppliers.

transaction cost The cost of making a transaction, usually the cost associated with purchasing or selling a security.

transfer agent A party, usually a commercial bank, that keeps track of changes in stock ownership, collects cash from a company, and pays dividends to its stockholders.

treasurer The manager responsible for financial planning, fund raising, and allocation of money in a business.

Treasury bills Securities issued by the federal government in minimum denominations of $10,000 in maturities of three, six, or twelve months.

Treasury bonds and notes Securities issued by the federal government that make semi-annual coupon interest payments and pay the face value at maturity. Treasury notes come in maturities of one to ten years. Treasury bonds come in maturities of more than ten years.

trend analysis An analysis in which something (such as a financial ratio) is examined over time so as to discern any changes.

trustee The party that oversees a bond issue and makes sure all the provisions set forth in the indenture are carried out.

uncertainty The chance, or probability, that outcomes other than what is expected will occur.

underwriting The process by which investment banking firms purchase a new security issue in its entirety and resell it to investors. The risk of the new issue is transferred from the issuing company to the investment bankers.

unitary tax A tax assessed by a state on the earnings or a foreign corporation.

variable costs Costs that vary with the level of production.

variable-rate bonds Bonds that have periodic changes in their coupon rates, usually tied to changes in market interest rates.

warrant A security that gives the holder the option to buy a certain number of shares of common stock of the issuing company, at a certain price, for a specified period of time.

weakening currency A currency that is now convertible into a smaller number of units of another currency than previously.

wealth Assets minus liabilities.

weighted average cost of capital (k_a) or (WACC) The average of all the component costs of capital, weighted according to the percentage of each component in the firm's optimal capital structure.

working capital Another name for the current assets on a firm's balance sheet.

yield to maturity (YTM) The investor's return on a bond, assuming that all promised interest and principal payments are made on time and the interest payments are reinvested at the YTM rate.

zero-coupon bonds Bonds that pay face value at maturity and that pay no coupon interest.

INDEX

A

ABC inventory classification system, 478
Absolute purchasing power parity theory, 528
Accept/reject decision, 236–37
    IRR, 269
    NPV, 242–43
Accounting, 59–82
    depreciation, 71–73
    financial statements, 61–71
        balance sheet, 66–68
        income statement, 61–65
        statement of cash flows, 68–71
    fundamentals of, 61
    Hollywood practices, 64
    income taxes, 73–76
    treatment of leases in, 367
Accounting depreciation, 71–72
Accounts payable, 67, 422
    on *pro forma* balance sheet, 127–28
Accounts receivable, 67, 422, 463–93
    in average collection period, 92
    as collateral, 508
    collection policies and, 480–83
    conflicting forces in, 467
    credit decisions, making, 479–80
    factoring, 465n, 483
    fluctuating, 423–24
    liquidity versus profitability of, 425
    optimal levels of, 425–26, 467–72, 482
        analyzing levels of, 468–72
        credit policy and, 467–68
    profitability and liquidity and, 465–67
    on *pro forma* balance sheet, 127
    reasons for accumulating, 465
Accrued expenses, 67
Accumulated depreciation, 67
Actuaries, 54
Additional funds needed (AFN), 129–31
Additions to retained earnings in *pro forma* income statement, 126
Adjustment for depreciation expense, 68–69
Adolph Coors Brewing Company, 378
ADRs (American depository receipts), 526
Aetna Life and Casualty, 381
After-tax cost of debt (AT k$_d$), 297–98
    as discount rate, 373, 375
Agency, 8–10

Note: Each term that is defined in the text appears in boldface in this index, as does the page number on which the definition appears.

Agency costs, 10
Agency problem, 9–10, 380
Agent, 8–10
    transfer, 409
Aggressive working capital financing approach, 427, 430
Alkalay, Yael, 351
Allen, Woody, 83
Alternative minimum tax (AMT), 75
Amcor Ltd., 295–96
American depository receipts (ADRs), 526
American Society of Appraisers (ASA), 208
American Stock Exchange (AMEX), 24
Amortization, 72
Amortization table, 194, 195
Amortized loan, 191, 193
Annual coupon interest payments, 210–11
Annualization of interest rates, 505–7
Annuitant, 55
Annuity(ies), 55, 182c
    coupon interest payments as, 210–11
    equivalent annual (EAA), 271–72
    perpetuity, 187, 470–72, 477
    present value of ordinary (PVA), 183–85
Annuity compounding periods, 196–97
Annuity due, 185–87
Arbitrage, 529
Articles of incorporation, 13
Articles of partnership, 12–13
Ascendix Publishing, 463–64
Ask (offer) price, 22
Asset(s), 61
    on balance sheet, 66, 67
        changes in balance sheet asset accounts, 69, 70
    current. *See* Current assets
    debt to, 103
    debt to total, 91, 103
    expropriation of, 531
    fixed, 67, 424
    incremental cash flows associated with disposal of old, 286
    market values of, 221
    noncurrent, 425
    return on (ROA), 89, 90, 97–101
    short-term, as collateral, 508–10
Asset activity ratios, 86, 92–93, 99, 102
    average collection period, 92, 103, 469
    inventory turnover, 92–93, 103
    summary analysis of, 102, 103
    total asset turnover, 93, 97–100, 103
Asset replacement decision, 285–86
AT&T Corp., 19
Audit, 10

Australian Society of Certified Practicing Accountants, 439
Average collection period (ACP), 92, 103, 469
Average daily credit sales in average collection period, 92
Average tax rate, 74, 76

B

"Baby-boomer" generation, 56
Bad debts, collection policies to handle, 480–83
Balance sheet, 66–68
    changes in asset accounts on, 69, 70
    changes in liability and equity section of, 69, 70
    mixed ratio using, 89–90
    pro forma, 127–29, 131
    ratios from. *See* Financial ratios
Balance sheet valuation approaches, 218, 221–22
Balancing problem, 131
Bank(s). *See also* Financial institutions
    commercial, 44–46
    credit unions compared to, 52
    Federal Home Loan, 49
    Federal Reserve, 47, 48
    financial risk of, 155
    Internet, 41
    short-term loans from, 497–99
Bank charter, 44
Banker's acceptance, 26–27
Bantel, Karen, 352
Barings Bank, 46
Batra, Ravi, 120
Bearer, 21
Bearer bonds, 27n
Beatrice Company, 337, 338
Beebe Institute, 118
Berkshire Hathaway stock split of 1996, 414–15
Best efforts basis, 22
Best efforts offering, underwriting versus, **387**
Beta (β), 160–61
    in CAPM, 163–65
Beta risk, 253
BIA Consulting, 207
Bid price, 22
Bird-in-the-hand theory, 408
Black, Dan, 176
Blanket lien, 509
BLC Financial Services of New York, 496
BMW, Inc., 518

I-1

Board of directors, 337, **380**
    dividends and, 379, 403
    elections for, 382–84
    fiduciary responsibility of, 380
Board of Governors, 47
Boeing Corporation, 518
Boli, John, 177
Bond(s), 27–29, 352–64
    bond indentures, 352, 353–57
    call provisions, 354–55
    plans for paying off bond issue, 354
    refunding, 355, 372–75
    source of information on, 104
    straight bond value, 360–61
    terminology, 27
    types of, 27–29
        bearer, $27n$
        convertible, 358–61
        corporate, 28–29, 352, 354–55
        international, 363–64
        investment-grade, 29, 352–53
        junk, 29, 145, 353, 362–63
        mortgage, 358
        municipal, 28
        putable, 361
        secured, 354, 357–58
        super long-term, 364
        unsecured (debentures), 354, 358
        variable-rate, 361
        zero-coupon, 27
Bondholders, leveraged buyouts and, 337, 338
Bond valuation, 210–15
    annual coupon interest payments, 210–11
    formula, 210
    importance of, 208–9
    semiannual coupon interest payments, 212
    yield to maturity (YTM), 212–15
Book value, ratio of market to, 94–95
Book value per share, 94, **221**
Borrowers, use of yield curve by, 34–35
Borrowing, cost of. *See* Cost of debt
"Bottom-line" measures, 89
Breakeven analysis, 324–30
    applying, 327–30
    constructing sales breakeven chart, 324–27
Breakeven chart, sales, 324–27
Break points, 308–11
    debt, 308–9
    equity, 309–10
    marginal cost of capital up to first, 310–11
Brokers, 22
    buying Treasury securities without using, 30
Bruner, Jim, 252–53
Budget(s)
    capital, 123, 313
    cash, 122–23, 127, 446–51, 452, 453
    to produce *pro forma* financial statements, 122–23
Budgeting, capital. *See* Capital budgeting
Burton, Don, 527
Busby, Jheryl, 11
Buschman, Tom, 421
Business Loan Center, 495–96
Business news, sources of, 104
Business organization, forms of, 11–14
Business risk (risk of operating leverage), 151–53, 333
    fixed operating costs and, 152–53
    measuring, 151
    sales volatility and, 151–52

Buy versus lease decision, 368
Byron, Lord, 421

C

Calculator. *See* Financial calculator
Call date, $354n$
Call premium, 355, 373
Call price, 354
Call provisions, bond, 354–55
Cameron, James, 85
Canada, NAFTA and, 532
Capacity, credit standards based on, 479
Capital
    cost of. *See* Cost of capital
    credit standards based on, 479–80
    debt, $96n$, 307
    defined, 296
    invested (IC), 95, 96
    mezzanine, 366
    raising, 307
    sources of, **61, 296,** 297–306
    working. *See* Working capital
Capital asset pricing model (CAPM), 163–65
    cost of internal common equity estimated using, 301–2
Capital budget, 123
    optimal, 313
Capital budgeting, 235, 236–72. *See also* Incremental cash flows
    bond refunding, 372–75
    capital rationing, 249–51
    cash flows in, 237
        estimating, 287
    decision methods in, 238–49. *See also* Internal rate of return (IRR); Net payback method, 238–39 present value (NPV)
    defined, 236
    incremental depreciation expense of, 277–78
    MCC schedule and, 311–14
    multiple internal rates of return and, 268–69
    mutually exclusive projects, 237
        conflict among, 248–49
        with unequal project lives, 270–72
    nonsimple projects, 267–68
    for private versus public project, 252–53
    process of, 236–38
        decision practices, 236–37
        stages in, 237–38
    residual theory of dividends and, 407
    risk in, 251–55
        adjusting for, 255
        measuring, 251–54
    types of projects, 237
    working capital and, 275–76
Capital Cities/ABC, 2
Capital Electric Supply, 482
Capital (financial) lease, 366–67
Capital gains, 30, 75, 279, 404
Capital in excess of par, 68, 128
Capital market, 22, 23
    securities in, 27–31
Capital rationing, 249–51
Capital structure, 304, 323–49
    assessing, 304
    breakeven analysis, 324–30
        applying, 327–30
        constructing sales breakeven chart, 324–27

    defined, 338
    leverage and, 324, 330–36
        combined, 335–36
        financial, 324, 333–35
        operating, 324, 330–33
    leveraged buyouts (LBOs), 336–38
    theory, 338–41
        Modigliani and Miller on, 339
        toward optimal capital structure, 339–41
        tax deductibility of interest and, 338–39
CAPM. *See* Capital asset pricing model (CAPM)
Career paths, 3
Carrying costs, 472, 473, 478
Cash, 67, 422
    earnings versus, 406
    fluctuating, 424
    forecasting cash needs, 442, 446–51
    liquidity versus profitability of, 425
    optimal level of, 425–26, 441–46
    on *pro forma* balance sheet, 127
Cash budget, 122–23, 127, **446**–51
    defined, 446
    developing, 447–51, 452, 453
Cash flow(s). *See also* Cash management
    bond refunding and, 355
    capital budgeting, 237
    discounted cash flow model, 209, 218–19
    financing, 286–87
    importance of, 6
    incremental. *See* Incremental cash flows
    initial investment, 275–76, 280–81, 282
    managing, 451–55
    of mutually exclusive projects with unequal lives, comparing, 270–72
    net, 71, 449–51
    for nonsimple versus simple project, 267–68
    operating, 276–79, 281–82, 283
    payback method and, 238, 239
    present value of investment with uneven, 187–89
    profits versus, 7–8, 439–40
    in refunding capital budgeting problem, 373–75
    shutdown, 279–80, 282
    statement of, 68–71
    supernormal growth of, 232–33
    timing of, effect of, 6–7, 8
Cash flow approach to valuing companies, 207
Cash inflows and outflows. *See under* Cash management
Cash management, 439–61
    cash inflows (cash collections), 6
        computing, for cash budget, 447, 448
        increasing, 451–54
        speeding up, 454–55
    cash needs
        dividend policy and, 403
        forecasting, 442, 446–51
    cash outflows, 6
        decreasing, 454
        developing cash budget and, 447–49
        slowing down, 455
    concepts in, 440–41
    optimal cash balance, 425–26, 441–46
        determining, 443–46
        maximum cash balance, 442–43
        minimum cash balance, 441–42
Central banks, influence on exchange markets, 530. *See also* Federal Reserve System

Central Liquidity Facility (CLF), 53
Certificates of deposit, negotiable (CDs), 25
Character, credit standards based on, 479
Charter
  bank, 44
  credit union, 53
Checking account, compensating balance in, 503–4
Chief financial officer, 4, 5
Chrysler Corporation, 516
Citron, Robert L., 26
Clientele dividend theory, 407–8
Clinton, Bill, 28, 75
Coady, Roxanne, 421
Coca-Cola Company, 119, 518
Coefficient of variation, 150
  business risk and, 151, 152
  computing changes in, 253–55
  financial risk and, 154
  formula for, 150
  measuring risk with, 149–51
  of net income with versus without interest expense, 153–54
Coleman, Michael, 431
Collateral, 425
  credit standards based on, 480
  for short-term financing, 507–10
    accounts receivable, 508
    inventory, 508–10
Collection agencies, 481
Collection period, average, 92, 103, 469
Collection policies, 480–83
College
  incremental costs of studying abroad, 281
  investment in, 175–77
Columbia University, 351
Combined leverage, 335–36
  degree of (DCL), 335–36
  fixed costs and, 336
Commercial banks, 44–46
  operations, 44–46
  regulation of, 44, 45
  reserves of, 45–46
Commercial finance companies, 53
Commercial paper, 25–26, **500**–502
Common bond requirement of credit union, 52
Common stock, 30, **378**–400
  characteristics of, 378–82
  classes of, 378
  cost of equity from new (k_n), 302–3
  cost of internal common equity (k_s), 300–302
  defined, 378
  dividends, 65, 357, 379
  equity break point analysis and, 309–10
  institutional ownership of, 380–82
  issued by private corporations, 379
  issued by publicly traded corporations, 380
  issuing, 386–89
    initial public offering (IPO), 377, 386
    investment bankers, function of, 387
    pricing new issues, 387–89
  on pro forma balance sheet, 128
  pros and cons of equity financing, 384–86
  return on equity and, 89
  rights and warrants, 389–94
  preemptive rights, 390–92
  warrants, 392–94
  voting rights of common stockholders, 378, 382–84

Common stock equity section of balance sheet, 68
Common stockholders, voting rights of, 378, 382–84
Common stock valuation, 218–22
  balance sheet valuation approaches, 218, 221–22
  deciding on method of, 222
  going concern valuation models, 218–21
  supernormal growth, 232–33
  yield on, 222
Companies, source of information on, 104
Comparative advantage, 520
Compensating balance, 503–4
Compensation, shares of stock as, 10
Competition, forecasting and, 132
Component cost of capital, 297
Compounding periods, 180n
  compounding more than once per year, 195–98
    annuity compounding periods, 196–97
    continuous compounding, 198
    semiannual compounding, 195–96
  sensitivity to changes in number of, 181
Compound interest, 178–80, 182
Conditions, credit standards based on, 480
Conoco, Inc., 377
Conservative working capital financing approach, 428, 430
Consolidated Edison Company of New York (Con Ed), 406
Constant growth dividend model, 219, 222
  adjusted for supernormal growth, 232–33
  to estimate cost of internal common equity, 300
Consumer finance companies, 53
Continuous compounding, 198
Continuous distribution, 146n
Contribution margin, 325
Contributions to sinking fund, 354
Controller, 4, 5
Controlling interest, stockholder with, 382
Conversion ratio, 359
Conversion value, 360
Convertible bonds, 358–61
Corporate bonds, 28–29, **352**
  call provisions, 354–55
Corporate public relations, 381
Corporate stock, 29–31
Corporation(s), 13–14
  benefits of, 13–14
  board of directors of, 337, 379, 380, 382–84, 403
  as major investor in preferred stock, 365
  marginal tax rates for, 74–76
  multinational. See Multinational corporations (MNCs)
  organization of typical, 4
  private, 379
  publicly traded, 380
  Subchapter C, 14
  Subchapter S, 14
Correlation
  forecasting based on, 120
  portfolio risk and, **156**–57
Correlation coefficient, 157
Cost(s)
  agency, 10
  carrying, 472, 473, 478
  cutting, to decrease cash outflows, 454
  delivery, 275

fixed. See Fixed costs
flotation, 299, 302–3, 385
installation, 275
labor, 324
of maintaining inventory, 472–73
opportunity, 178, 278, 465
ordering, 473
sunk, 274
transaction, 443
variable, 324–27
Cost approach to valuing companies, 207
Cost of capital, 295–322
  capital structure and, 339–40
  component, 297
  cost of debt, 213n, 297–98
    after-tax, 297–98, 373, 375
    lower, 340
  cost of equity from new common stock (k_n), 302–3
  cost of internal common equity (k_s), 300–302
  cost of preferred stock (k_p), 298, 299–300
  outside factors affecting, 307
  marginal (MCC), 306–14
    break points, 308–11
    calculating amount of changes in, 310–11
    capital budgeting and, 311–14
    defined, 306
    firm's MCC schedule, 307–14
  weighted average (WACC), 95, 303–6, 310–11, 314, 339, 340
Cost of debt, 213n, **297**–98. See also Required rate of return
  after-tax (AT k_d), 297–98
  as discount rate, 373, 375
  lower, 340
Cost of equity from new common stock (k_n), 302–3
Cost of goods sold (COGS), 62, 132
  in inventory turnover ratio, 93n
  in pro forma income statement, 125
Cost of internal common equity (k_s), 300–302
Cost of preferred stock (k_p), 298, 299–300
Cost of retained earnings, 300
Coupon interest, 27
Coupon interest payments, 210–12
  annual, 210–11
  semiannual, 212
Coupon interest rate, 27
Covenants, restrictive, 355–57
Cray Computer Corporation, 119
Credit
  extension of, 455, 465
    average collection period and, 92
  Five Cs of, 479–80
  letter of, 527
  line of, 498–99
  trade, 499–500
Credit cards, 509
Credit decisions, making, 479–80
Credit policy, 467–68, 470
Credit risk, 43
Credit scoring, 480, 481
Credit standards, 468, 479–80
Credit terms, 455, 499–500
Credit unions, 52–53
Critchlow, Paul, 26
Cross-border merger, 516
Cross rates, 522–23
Cross-sectional analysis, 101

Index I-3

Cultural risk, 531–32
Cumulative preferred stock, 364
Cumulative voting rules, 382, 383–84
Currency, euro, 533
Currency cross rate, 522–23
Currency market, 521–22. *See also* Exchange rates
Current assets, 66, 67, 422
   capital budgeting project and, 275–76
   in current ratio, 90
   fluctuating, 423–24
   optimal level of, establishing, 425–26
   permanent, 424
   in quick ratio, 91
   temporary, 424
Current liabilities, 66, 422
   capital budgeting project and, 275–76
   in current ratio, 90
   managing, 426
   in quick ratio, 91
Current ratio, 90, 103, 466

D

Daimler-Benz, 516
Date of record, 409, 410
Dealers, 22
Debentures, 354, 358
Debt. *See also* Capital structure
   cost of, 213n, 297–98
      after-tax, 297–98, 373, 375
      lower, 340
   financial risk and, 339
   interest payments on, 385
   long-term, 128, 426, 428, 497
   mezzanine, 366
   restrictive covenants limiting future borrowings, 355–57
   short-term. *See* Short-term financing
   tax deductibility of interest on, 338–39
Debt break point, 308–9
Debt capital, 96n, 307
Debt ratios, 86, 91–92, 99, 102
   debt to asset, 103
   debt to equity, 91–92, 99, 103
   debt to total assets, 91, 103
   summary analysis of, 102, 103
   times interest earned, 92, 103
Decision methods, capital budgeting, 238–49. *See also* Internal rate of return (IRR); Net present value (NPV)
   payback method, 238–39
Decision practices, capital budgeting, 236–37
Declaration date, 409, 410
Default, 32
Default risk premium, 32
Deferred call provision, 354n
Deficit economic units, 21
   intermediation between surplus economic units and, 42–46
Defined benefit plan, 55
Defined contribution plan, 55
Degree of combined leverage (DCL), 335–36
Degree of financial leverage (DFL), 333–34, 336
Degree of operating leverage (DOL), 330–32
Delivery costs, 275
Delta Air Lines, 117–18
Denomination matching, 42–43
Depository Institutions Deregulation and Monetary Control Act of 1980 (MCA), 49–50

Depreciation, 71–73
   accounting, 71–72
   accumulated, 67
   disposal of old asset and, 286
   economic, 72
   methods of, 72–73
Depreciation basis, 72
Depreciation book value, salvage value in relation to, 279–80
Depreciation expense, 63, 282
   adjustment for, 68–69
   calculating amount of, 72–73
   incremental, 277–78
   as part of operating cash flow, 276–77
   in *pro forma* income statement, 125
Derivatives, 26
Directors. *See* Board of directors
Disability benefits, 56
Discount
   bond price at, 215
   on commercial paper note, dollar amount of, 501
   credit terms offering, 455, 499–500
Discounted cash flow model (DCF), 209
   applied to common stock, 218–19
Discount loans, 503, 506
   effective interest rate and, 503, 504
Discount rate(s), 182
   after-tax cost of debt as, 373, 375
   as cost of capital, 296
   in **discounted cash flow valuation model, 209**
   financing costs factored into, 287
   multiple IRRs and, 268–69
   NPV profile and, 243–45
   risk-adjusted (RADRs), 255
   trial-and-error IRR method and, 246–47
Discount window, 49
Discount yield, 500–501
Discrete distribution, 146n
Distribution(s)
   sales forecast, 145–46
   standard deviation of, 145–49
Diversifiable risk, 165
Diversification effect, 156
   calculating standard deviation of two-asset portfolio and, 158
   correlation and, 156–57
   foreign investments and, 525–26
   risk-reduction and, 162
Dividend(s), 30, 65, 402–3
   advantages over interest payment on debt, 385
   board of directors and, 379, 403
   common stock, 65, 357, 379
   constant growth dividend valuation model, 219, 222, 232–33, 300
   defined, 402
   double taxation of, 13
   mechanics of paying, 409–10
   no-growth dividend model, 220
   preferred stock, 65, 364
   present value of, 216–17
   restrictions on, 357, 404
   stock, 411–13
Dividend growth models
   constant growth version, 219, 222, 232–33, 300
   to estimate cost of internal common equity, 300–301
   including flotation costs, 302–3
   no-growth model, 220

Dividend payout ratio, 404–6
Dividend policy, 401–20
   alternatives to cash dividends, 410–15
      stock dividends, 411–13
      stock splits, 413–15
   dividend payout ratio, 404–6
   factors affecting, 403–4, 405
   mechanics of paying dividends, 409–10
   reasons for, 403
   restructuring plan, example of, 401–2
   theories of effect of, 406–9
      bird-in-the-hand theory, 408
      clientele dividend theory, 407–8
      M & M dividend theory, 409
      residual theory of dividends, 406–7
      signaling dividend theory, 408
Dividend reinvestment plan (DRIP), 410, 411
Dividends paid, in *pro forma* income statement, 126
Dividends payable, 409, 410
Dividend yield, 222, 301
Domestic Fisher effect, 528
Domestic workers, political risk from required, 531
Double taxation, 13, 14
DRIP (dividend reinvestment plan), 410, 411
Dunlap, Albert, 59–60
Du Pont Corporation, 377
Du Pont system of ratio analysis, 97–100

E

e, defined, 198
Earnings
   cash versus, 406
   price to earnings ratio, 93–94, 103, 220–21
   retained. *See* Retained earnings
Earnings before interest, taxes, depreciation, and amortization (EBITDA), 63
Earnings before interest and taxes (EBIT), 63
   business risk and, 151–53
   operating leverage and, change in, 330–33
   in operating profit margin, 88–89
Earnings before taxes (EBT), 63
Earnings per share (EPS), 65
   in price to earnings ratio, 93–94
Eaton, Robert, 516
EBIT. *See* Earnings before interest and taxes (EBIT)
Economic depreciation, 72
Economic Forecasting Center (EFC), at Georgia State University, 117–18
Economic order quantity (EOQ) model, 474–75
Economic Recovery Tax Act (1981), 75
Economic units, 21
   intermediation between, 42–46
Economic value added (EVA), 95–96
Economy, source of information on, 104
Effective interest rate, 497–98
   annualizing interest rates and, 505–7
   of commercial paper note, 501–2
   formula for, 502
   loan terms and, 502–7
      compensating balance and, 503–4
      comprehensive example of, 506–7
      computing amount to borrow, 507
      discount loans and, 503, 504
      loan maturities shorter than one year and, 505–6

trade credit effective annual interest rate formula, 500
Efficiency, market, 24
Elections, board of directors, 382–84
Electronic funds transfer, 455
Emergencies, cash balances to cope with, 442
Employee education expenses, 75
Employees, delaying payment to, 455
Ender, Keith, 132–33
Entrepreneurs, 351
Environmental issues, MNCs and, 520
Environmental Protection Agency, 307
Equity. *See also* Capital structure
    changes in balance sheet equity section, 69, 70
    debt to equity, 91–92
    internal common, cost of (k_s), 300–302
    from new common stock, cost of (k_n), 302–3
    return on (ROE), 89, 98
Equity break point, 309–10
Equity financing, 426. *See also* Common stock; Preferred stock
    conservative working capital financing approach using, 428
    pros and cons of, 384–86
Equity multiplier, 98–99
Equity section of balance sheet, 66, 68
    changes in, 69, 70
Equivalent annual annuity (EAA), 271–72
Ethics, 8–11
    of boards of directors, 337
    excuse making, trend in, 26
    of financial institutions, 46
    fraud in muni-bond market, 363
    of Hollywood accounting practices, 64
    issues facing multinational corporations, 519–20
Eugene Lang Entrepreneurial Fund, 351
Euro, 533
Eurobonds, 363–64
Euro Disney, 86
European Central Bank (ECB), 533
European Monetary Union (EMU), 533
European Union (EU), 533
Excess financing, 130
Exchange rates, 521–25
    cross rates, 522–23
    defined, 521
    effects on foreign stock and bond investments, 524–25
    effects on MNCs, 524
    fluctuating, 521–22, 524, 527
    theories of, 528–30
Exchanges, security, 23–24
Ex-dividend, 409, 410
Exercise price, 392–93
Exercise value of warrant, 392–94
Expansion project, incremental cash flows of, 280–85
Expectations, dividend policy and management, 403–4
Expected rate of return of portfolio, 155–56
Expected return on investments, 442–43
Expected value (mean)
    business risk and, 151, 152
    coefficient of variation and, 150–51
    financial risk and, 154
    formula for, 146–47
Expense(s)
    accrued, 67
    depreciation, 63, 282

adjustment for, 68–69
    calculating amount of, 72–73
    incremental, 277–78
    as part of operating cash flow, 276–77
    in *pro forma* income statement, 125
    on income statement, 62–64
    prepaid, 67
    tax deductible, 64
Experience, forecasting based on, 119–20
Export-Import Bank, 527
Expropriation of assets, 531
Ex-rights, 390, 391–92
External financing, factors influencing duration of, 497, 498. *See also* Long-term financing; Short-term financing
Externalities, 278

F

Face value, bond, 27
Factoring, 53, 465*n*, **483**
Factors, 53, 483
Fair trade, free trade versus, 533
Fallen angels, 362–63
FASB (Financial Accounting Standards Board), 61, 68, 367
Federal Deposit Insurance Corporation (FDIC), 41, 44
Federal funds market, 45
Federal Home Loan banks, 49
Federal Open Market Committee (FOMC), 47
Federal Reserve Bank of New York, 47, 48
Federal Reserve System, 44, 47–49
    discount window, 49
    money supply, control of, 47–49
    organization of, 47
    reserves requirements, 45–46
Fidelity Investments, 381–82
Fiduciary responsibility, 380
Finance, field of, 2–3
    careers in, 3
Finance companies, 53–54
Finance team, organization of, 4–5
Financial Accounting Standards Board (FASB), 61, 68, 367
Financial analyst, 85*n*
Financial calculator
    annuities on
        annuity compounding periods, 197
        annuity due problems, 186
        present value, 184–85
    bond valuation with, 210, 211
        yield to maturity, 214
    finding interest rate using, 190, 191
    finding number of periods using, 192
    future value on, 180–81
    IRR calculated using, 247, 284–85
    NPV calculated on, 240, 241, 242
        of expansion project, 284–85
    present value on, 183
        of annuity, 184–85
        of uneven series of cash flows, 188–89
    semiannual compounding on, 196
    solving for payment on, 194
Financial (capital) leases, 366–**367**
Financial goal of firm, 5–8
Financial institutions, 41–58
    annuities, 55
    commercial banks, 44–46
        operations, 44–46
        regulation of, 44, 45

reserves of, 45–46
    credit unions, 52–53
    defined, 22
    Federal Reserve System, 44, 47–49
        discount window, 49
        money supply, control of, 47–49
        organization of, 47
    finance companies, 53–54
    financial risk of, 155
    insurance companies, 54–55, 307
    intermediation by, 42–43
        credit risk absorption, 43
        denomination matching, 42–43
    pension funds, 55, 56
    saving and loan associations (S&Ls), 49–52
        legislation affecting, 49–50
        matching loan and deposit maturities, problem of, 51
        mutual versus stockholder-owned, 50
        real assets of, 52
        regulation of, 50
Financial Institutions Reform, Recovery, and Enforcement Act (FIRRE Act), 50
Financial intermediaries, 21–22
Financial leverage, 153, 324, **333**–35
    degree of, 333–34, 336
    interest expense and, 334–35
    leveraged buyout (LBO) and, 336–38
    risk of (financial risk), 153–55, 335, 339
Financial management, 3
    legal and ethical challenges in, 8–11
Financial manager
    influences on, 12
    role of, 3–4
Financial markets, 22–24
    careers in, 3
    securities in, 25–31
Financial ratios, 85–100
    asset activity ratios, 86, 92–93, 99, 102
    debt ratios, 86, 91–92, 99, 102
    defined, 85
    industry comparisons using, 100, 101
    liquidity ratios, 86, 90–91, 99, 102
    locating information about, 103–5
    market value ratios, 87, 93–97, 102
    profitability ratios, 86, 87–90, 99, 102
    relationship among, 97–100
    summary analysis using, 101–3
    trend analysis using, 100–101
    uses of, 85
    working capital policy and, 429–30
Financial risk, 153, 335
    debt and, 339
    measuring, 153–55
Financial statement analysis, 83–116
    assessing financial health, 84–86
    financial ratios. *See* Financial ratios
    industry comparisons, 100, 101
    misleading numbers in, 84–85
    summary analysis, 101–3
    trend analysis, 100–101
Financial statements, 61–71. *See also Pro forma* financial statements
    balance sheet, 66–68
    income statement, 61–65
    statement of cash flows, 68–71
Financial system, 21–22. *See also* Financial intermediaries; Securities
Financial tables
    annuity compounding periods on, 197
    bond valuation using, 210

Index **I-5**

Financial tables (cont.)
 finding interest rate using, 189–90, 191
 future value on, 180
 net present value solved from, 240
 present value interest factor for an annuity (PVIFA), 184
 present value on, 182
 solving for payment using, 193–94
Financing
 equity. See Common stock; Preferred stock
 excess, 130
 inventory, 508–10
 long-term, 128, 426, 428, 497
 short-term. See Short-term financing
Financing activities on statement of cash flows, 71
Financing cash flows, 286–287
Firm
 finance in organization of, 4
 financial goal of, 5–8
 forms of business organization, 11–14
Firm risk, 253
First-mortgage bond, 358
Fisher, Irving, 528–29
Five Cs of credit, 479–80
Fixed assets, 67, 423
Fixed costs, 324
 in breakeven chart, 324–27
 combined leverage and, 336
 degree of operating leverage and, effect on, 331
 operating leverage effect with, 330
 reducing, 161
Fixed operating costs, 152–53
Flotation costs, 299, 302–3, 385
Fluctuating exchange rates, 521–22, 524, 527
Ford Motor Credit Company, 54
Forecasting, 117–42
 approaches to, 119–20
 cash needs, 442, 446–51
 example of, 132–33
 importance of, 119, 132
 pro forma financial statements, 122–33
  additional funds needed (AFN), 129–31
  analyzing, 131–33
  balance sheet, 127–29, 131
  budgets to produce, 122–23
  choosing forecasting basis, 123–24
  for credit policies, 468, 470, 471
  defined, 122
  income statement, 124–26, 130
  inventory changes reflected on, 475, 476
  reasons for mistakes in, 120, 132
  sales, 120–22
   distributions of forecast sales, 145–46
Foreign operations, financial advantages of, 518–19. See also International finance
Forward contracts, 525
Forward rate, 529
Founders National Bank of Los Angeles, 11
4-1 (four-for-one) stock split, 413
Franklin, Benjamin, 175, 495
Fraud in muni-bond market, 363
Free trade versus fair trade, 533
Friedman, Milton, 11n, 295
Future, forecasting. See Forecasting
Futures contracts, 525
Future value interest factor (FVIF), 180
 table
  finding interest rate using, 189–90
  finding number of periods using, 192

Future value interest factor for an annuity (FVIFA), 182f
Future value of single amount, **178**–81
 of annuity due, 185–87
 defined, 178
 formula for, 180
 sensitivity to interest rate changes or number of compounding periods, 181

G

Galbreath, John, 509
Garn-St. Germain Act (1982), 50
Garrison, Jim, 64
Gates, Bill, 161
General Agreement on Tariffs and Trade (GATT), 532–33
General and administrative expenses, in *pro forma* income statement, 125
Generally Accepted Accounting Principles (GAAP), 61
General Motors, 378
 strike (1996), 479
General Motors Acceptance Corporation (GMAC), 54
General obligation bonds (GOs), 28, 29
General partners, 13
Geocities Corporation, 387, 388
Georgia State University
 Beebe Institute, 118
 Economic Forecasting Center (EFC), 117–18
Gibbs, Philip, 19
Gibson Greetings Inc., 26
Going concern valuation models, 218–21
Going concern value, 94
Goldman Sachs, 377
Gordon, Myron, 219
Gordon growth model, 219
Government intervention in international finance, 530
 political risk from, 527, 530–31
Great Depression of 1990, The (Batra), 120
Groom, Winston, 64
Gross profit, 62
Gross profit margin, 87–88
Group of Seven (G-7), 530

H

Half-year convention, 72n, 282n
Hanks, Tom, 64
Hayes, Greg, 439
Hedge, 525
Heller, Chris, 381
Hickman, James, 531–32
Hickman, Robert, 495–96
Higgins, Fred, 307
High-yield (junk) bonds, 29, 145, 353, 362–63
Hollywood accounting practices, 64
Home office deduction rules, 75
Honda Motor Corporation, 518
Household sector, 42
Hughes Aircraft, 378
Human rights issues, MNCs and, 520
Hurdle rate, 245, 247, 287, 296
Hybrid financing, mezzanine capital as, 366
Hybrid security, preferred stock as, 364

I

IBM, 2
Income
 net. See Net income
 operating. See Operating income
 residual, 378
Income statement, 61–65
 mixed ratio using, 89–90
 pro forma, 124–26, 130, 446
 ratios from. See Financial ratios
Income taxes, 73–76
 in *pro forma* income statement, 126
Incorporation, articles of, 13
Incremental cash flows, 237, 273–93
 asset replacement decision and, 285–86
 cash flow estimation process, 287
 of changing average inventory level, 475, 477
 credit policy change and, 470
 defined, 274
 of expansion project, 280–85
 financing cash flows, 286–87
 initial investment cash flows, 275–76, 280–81, 282
 operating cash flows, 276–79, 281–82, 283
 shutdown cash flows, 279–80, 282
Incremental depreciation expense, 277–78
Indenture, bond, 352, 353–57
Independent projects, 237
Industries, sources of information on, 104
Industry comparisons, 100, 101
 summary analysis using, 101–3
Inflation premium, 32
Initial investment cash flows, 275–76, 280–81, 282
Initial public offering (IPO), 377, **386**
Installation costs, 275
Institutional investors, 380–82
 as major buyers of new equity issues, 386
Institutions. See Financial institutions
Insurance, as risk-reduction method, 161
Insurance companies, 54–55, 307
Interest, 31–35. See also Interest rate
 compound, 178–80, 182
 compounding more than once per year, 195–98
 coupon, 27
 defined, 31
 nominal risk-free rate of, 32
 real rate of, 31–32
 tax deductibility of, 338–39
 times interest earned ratio, 92, 103
 yield curves, 33, 34–35
Interest expense, 63, 130–31
 balancing problem and, 131
 financial leverage and, 334–35
 financial risk caused by fixed, 153–55
 in *pro forma* income statement, 125–26
Interest payments on debt, 385
 coupon interest payments, 210–12
Interest rate, 31
 annualizing, 505–7
 as cost of debt, 297
 coupon, 27
 credit card, 509
 determinants of, 31–33
 effective. See Effective interest rate
 nominal, 31–33
 difference across countries, 528–29
 nominal risk-free, 32
 real, 31–32

of single-amount investment, finding, 189–90
stated, 498
time value of money and, 181, 189–92
on variable-rate bonds, 361
Interest rate parity theory, 529
Interest rate spread, 44
Interests
of other groups, 10–11
of society, 11
Intermediaries, financial, 21–22
Intermediation, financial, 42–43
credit risk absorption, 43
denomination matching, 42–43
Internal common equity, cost of (k_s), 300–302
Internal rate of return (IRR), 245–49, 268–69
benefits of, 248
calculating, 246–47
with financial calculator, 247, 284–85
trial-and-error method, 246–47
of credit terms proposal, 472n
decision rule, 247, 269
defined, 245
of expansion project, 284
of inventory proposal, 477n
multiple, 268–69
NPV profile and, 247
problems with, 248
ranking capital budgeting project with, 312–13
conflicting rankings between NPV and, 248–49
Internal Revenue Service (IRS), 73
International bonds, 363–64
International finance, 517–37
exchange rates, 521–25
cross rates, 522–23
defined, 521
effects on foreign stock and bond investments, 524–25
effects on MNCs, 524
fluctuating, 521–22, 524, 527
theories of, 528–30
government intervention in, 530
international trade agreements, 532–33
multinational corporations (MNCs), 518–20, 524
comparative advantage of, 520
cultural risk facing, 531–32
ethical issues facing, 519–20
financial advantages of foreign operations, 518–19
political risk facing, 527, 530–31
top-ten U.S.-based, 519
risk, managing, 525–27
International Fisher effect, 528–29
International Import-Export Institute, 527
International trade agreements, 532–33
Internet bank, 41
Intralase, 352
Inventory, 67, 422, 463–93
as collateral, 508–10
conflicting forces in, 467
fluctuating, 423–24
inventory management approaches, 478–79
ABC inventory classification system, 478
just-in-time inventory control (JIT), 464, 478–79
liquidity versus profitability of, 425
optimal levels of, 425–26, 467, 472–77
analyzing levels of, 473–77
costs of maintaining inventory, 472–73

profitability and liquidity and, 465–67
on *pro forma* balance sheet, 127
in quick ratio, 91
reasons for accumulating, 465
Inventory financing, 508–10
Inventory management approaches, 464, 478–79
Inventory turnover ratio, 92–93, 103
Invested capital (IC), 95, 96
Investment(s)
accounts receivable and inventory as, 465
available investment opportunities, 442
careers in, 3
diversification benefits of foreign, 525–26
exchange rate effects on foreign stock and bond, 524–25
expected return on, 442–43
finding interest rate of single-amount, 189–90
initial investment cash flows, 275–76, 280–81, 282
transaction cost of making, 443
with uneven cash flows, present value of, 187–89
Investment activities on statement of cash flows, 71
Investment banker, 387
Investment banking firms, 22, 337
Investment-grade bonds, 29, 352–53
Investment opportunity schedule (IOS), 312–13
Investors
clientele dividend theory based on type of, 407–8
institutional, 380–82, 386
preferred stock, 365
IRR. *See* Internal rate of return (IRR)
IRS, on genuine versus fake lease, 365–66

J

Jackson, Janet, 11
James River Paper Company, 132
Johnson, Magic, 10–11
Jones, Joann K., 482
Jones, Russell, 295
Julian's Restaurant, 495–96
Junk bonds, 29, 145, **353,** 362–63
Just-in-time inventory control (JIT), 464, 478–79

K

Keating, Charles, 50
Kentucky Fried Chicken, 518
Kitchell Contractors, 99
K-Mart, 362
Kohlberg, Kravis, & Roberts (KKR), 337

L

Labor costs, fixed versus variable, 324
Landman, Frederick A., 235
Lawsuits, collection through, 481
LBOs (leverage buyouts), 336–38
Leases, 365–68
accounting treatment of, 367
defined, 365
financial (capital), 366–67
genuine versus fake, 365–66

lease versus buy decision, 368
operating, 366–67
Lee, Ang, 85
Leeson, Nicholas, 46
Legal considerations, 8–11
Legislation affecting savings and loan associations, 49–50. *See also* Regulation
Lenders, use of yield curve by, 34, 35
Lessee, 365
Lessor, 365
Letter of credit, 527
Leverage, 324, 330–36
combined, 335–36
financial, 153–55, 324, 333–35
equity multiplier and, 98–99
operating, 324, 330–33
risk of (business risk), 151–53, 333
Leveraged buyouts (LBOs), 336–38
Leverage effect, 99
Levy, Steven, 19
Liability(ies), 61
on balance sheet, 66, 67–68
changes in, 69, 70
in corporation, 13–14
current. *See* Current liabilities
dividends payable, 409, 410
in partnership, 13
in proprietorship, 12
total debt capital versus, 96n
unlimited, 12, 13
Liability insurance, 54–55
Lien, blanket, 509
Life insurance companies, 54
Limited liability company (LLC), 14
Limited partners, 13
Lincoln Savings and Loan of California, 50
Line of credit, 498–99
Liquidation value, 221
Liquidity, 66
accounts receivable and, 465–67
inventory and, 465–67
of money market securities, 25
profitability versus, 425, 497
Liquidity function, 21
Liquidity ratios, 86, 90–91, 99, 102
current ratio, 90, 103, 466
quick ratio, 19, 103, 466
summary analysis of, 102, 103
Liquidity risk premium, 33
Live from the Past, 9
Loan(s). *See also* Long-term financing; Short-term financing
amortized, 191
payment on, determining, 193
collateral, 425, 480, 507–10
discount, 503, 504, 506
effective interest rate on, 502–7
mortgage, 49, 51
secured, 507, 508
self-liquidating, 498
unsecured, 507
Lockbox, 455
Long, John Baldwin, 143
Long-term financing, 128, 426, 428
short-term financing versus, 497
Long-term liabilities, 67–68
Long-term projects. *See* Capital budgeting
Long-term securities, 23
Loss, writing off bill as, 483
Lottery, 185
Lotus, 2
Lucent Technologies, Inc., 19–20, 389

Index **I-7**

M

Maastricht Treaty (1993), 533
MCC. See Marginal cost of capital (MCC)
McDonald's Corporation, 518, 524
McGee, Matthew, 496
McGinn, Richard A., 20
MACRS (modified accelerated cost recovery system), 73
Majority interest, stockholder with, 382
Majority voting rules, 382, 383
Management expectations, dividend policy and, 403–4
Managers, financial, 3–4, 12
Manufacturing company, managing working capital for, 431
Marginal cost of capital (MCC), 306–14
  break points, 308–11
  calculating amount of changes in, 310–11
  capital budgeting and, 311–14
  defined, 306
  firm's MCC schedule, 307–14
Marginal tax rate, 74–76
Marketable securities, 67, 127
Market approach to valuing companies, 207
Market (beta) risk, 253
Market data, 104
Market efficiency, 24
Market price of stock
  signaling dividend theory and, 408
  after stock dividend, 413
  after stock split, 414
Market price per share
  in market to book value ratio, 94
  in price to earnings ratio, 93–94
Market risk premium, 163
Markets, financial, 22–24
  careers in, 3
  securities in, 25–31
Market share, 132
Market to book value, 94–95, 103
Market value
  of assets, 221
  of stock of company not publicly traded, estimating, 389
  of warrant, 394
Market value added (MVA), 95, 96–97
Market value ratios, 87, 93–97, 102
  economic value added (EVA), 95–96
  market to book value, 94–95, 103
  market value added (MVA), 95, 96–97
  price to earnings ratio, 93–94, 103, 220–21
  summary analysis of, 102, 103
Matching principle, 72, 429
Matsen, Paul, 117–18
Matthews, Tom, 401–2
Maturity(ies)
  loan maturities shorter than one year, 505–6
  staggered, 354
  yield to (YTM), 212–15
Maturity date, 21, 27
Maturity matching, 43
  between loan and deposit, problem of, 51
Maturity risk premium, 33
Maximum cash balance, 442–43
Mean. See Expected value (mean)
Merger, cross-border, 516
Metropolitan Life Insurance Company, 337, 338
Mexico, NAFTA and, 532

Mezzanine capital, 366
Michigan, University of, 351, 352
Miller, Arthur, 463
Miller, Merton, 339, 409, 443–46
Miller-Orr cash management model, 443–46
Minimum cash balance, 441–42
Minit Mart Foods, Inc., 307
Mixed ratio, 89–90
M & M dividend theory, 409
MNCs. See Multinational corporations (MNCs)
Moderate working capital financing approach, 429, 430
Modified accelerated cost recovery system (MACRS), 73
Modified Du Pont equation, 98
Modigliani, Franco, 339, 409
Monetary Control Act (MCA), 49–50
Money, time value of. See Time value of money
Money market, 22, 23
  securities in, 25–27
Money supply, control of, 47–49
Moody's bond ratings, 352–53, 362
Morgan Stanley, 377
Morley, J. Kenfield, 497n
Mortgage bonds, 358
Mortgage loans, 49, 51
Multinational corporations (MNCs), 518–20
  comparative advantage of, 520
  cultural risk facing, 531–32
  ethical issues facing, 519–20
  exchange rate effects on, 524
  financial advantages of foreign operations, 518–19
  political risk facing, 527, 530–31
  top-ten U.S.-based, 519
Multiple internal rates of return, 268–69
Multiplier, equity, 98–99
Municipal bond market, fraud and abuse in, 363
Municipal bonds, 28
Mutual fund rating services, 162
Mutual funds, 104, 162
Mutually exclusive projects, 237
  conflict among, 248–49
  with unequal project lives, 270–72
Mutual savings and loan associations, 50

N

NAFTA, 532
NASDAQ (National Association of Securities Dealers Automated Quote system), 24
National Credit Union Administration (NCUA), 53
National Credit Union Share Insurance Fund (NCUSIF), 53
Need for funds, dividend policy and, 403
Negative net working capital, 427
Negotiable certificates of deposit (CDs), 25
Net cash flows, 71, 449–51
Net income
  financial leverage and volatility of, 333
  on income statement, 62, 64–65
  in net profit margin, 89
  in return on assets, 89
  in return on equity, 89
  volatility, caused by interest expense, 153–55

Net operating losses (NOLs), 75
Net present value (NPV), 188, 239–45
  calculating, 240–42
  conflicting rankings between IRR and, 248–49
  of credit policy change, 470–72
  decision rules, 242–43
  defined, 239–40
  of expansion project, 283–84
  of inventory policy change, 475–77
  modified to reflect value of real options, 279
  of mutually exclusive projects with unequal project lives, comparing, 270–72
NPV profile, 243–45
  IRR and, 247
  multiple IRRs shown by, 269
  problems with, 245
Net profit margin, 89, 99
  in Du Pont equation, 97–100
  in Modified Du Pont equation, 98
Net profits, 64
Netscape, 386, 389
Net working capital, 67–68, 275–76, **422,** 427
Net worth, 221
New York Stock Exchange (NYSE), 23, 24
New York Times Informational Services Group, 9
New York University, 351
Nice/Goldman Fund, 351–52
Nissan of Japan, 518
Noble, Karen, 445
No-growth common stock dividend valuation formula, 220
Nominal interest rate, 31–33
  difference across countries, 528–29
Nominal risk-free rate of interest, 32
Noncumulative preferred stock, 364
Noncurrent assets, liquidity versus profitability of, 425
Nondiversifiable risk, 159–61
Nonsimple projects, 267–68
Normal probability distribution, 148–49
North American Fair Trade Agreement (NAFTA), 532
Northwestern University, 351
Notes payable, 67, 128, 422
NPV. See Net present value (NPV)
Nunn, Gregory, 273

O

Occupational Safety and Health Administration, 520n
Octel Communications Corp., 20
Offer price, 22
Office of the Comptroller of the Currency (OCC), 44
Office of Thrift Supervision (OTS), 41, 49, 50
Ondrish, Michael, 185
Open-market operations, 47–49
Operating activities on statement of cash flows, 68, 70
Operating cash flows, 276–79, 281–82, 283
Operating costs, fixed, 152–53
Operating income, 63
  business risk and, 151–53
  operating leverage and, change in, 330–33
  in operating profit margin, 88–89

Operating lease, 366–67
Operating leverage, 152–53, 324, 330–33. *See also* Breakeven analysis
   defined, 330
   degree of, 330–32
   risk of (business risk), 151–53, 333
Operating losses, net, 75
Operating profit margin, 88–89
Opportunity costs, 178, 278, 465
Optimal capital budget, 313
Optimal capital structure, 339–41
Optimal cash balance, 425–26, 441–46
   determining, 443–46
   maximum cash balance, 442–43
   minimum cash balance, 441–42
Optimal level of current assets, 425–26
Optimal levels of accounts receivable, 467–72, 482
   analyzing levels of, 468–72
   credit policy and, 467–68
Optimal levels of inventory, 425–26, 467, 472–77
   analyzing levels of, 473–77
   costs of maintaining inventory, 472–73
Option pricing, 390n, 394
Options, real, 278–79
Ordering costs, 473
Ordinary annuity, present value of (PVA), 183–85
Organization
   business, forms of, 11–14
   of finance team, 4–5
   of firm, finance in, 4
Orr, Daniel, 443–46
Outlier, 352
Overdraft accounts, 440
Over-the-counter (OTC) market, 24

P

PanAmSat Corporation, 235–36
Paramount Pictures, 64
Participating preferred stock, 364
Partners, 13
Partnership, 12–13, 14
   articles of, 12–13
Par value, bond, 27, 68
Payback method, 238–39
Payback period, calculating, 238
Payment date, 410
Pension funds, 55, 56
Pepsi-Cola, 518
Permanent current assets, 424
Perpetuities, 187
   present value of, 470–72, 477
   present value of preferred stock dividends as, 216–17
Persian Gulf War, 530
Pi, 198
Pixar, 389
Political risk, 527, 530–31
Political trouble, exchange rates and, 529
Portfolio, 155
   diversification benefits of foreign investments, 525–26
   expected rate of return of, 155–56
Portfolio risk, 155–61
   correlation and, 156–57
   defined, 155
   diversification effect and, 156
   nondiversifiable risk, 159–61

   standard deviation of two-asset portfolio, calculating, 158
Post, the, 24
Preemptive rights, 389–92
Preferred stock, 30–31, 364–65
   cost of (k_p), 298, 299–300
   dividends, 65, 216–17, 364
      present value of, 216–17
   investors, 365
Preferred stock equity section of balance sheet, 68
Preferred stock valuation, 215–18
   present value of dividends, 216–17
   yield on, 217–18
Premium(s), 32–33
   bond price at, 215
   call, 355, 373
   inflation, 32
   insurance, 54
   risk, 32–33
   market, 163
Prepaid expenses, 67
Present value, 178, 181
   of annuity, formula for, 197
   formula for, 182
   net. *See* Net present value (NPV)
   of ordinary annuity
      finding interest rate for, 190–91
      solving for payment on, 193
   of perpetuity, 470–72, 477
   of preferred stock dividends, 216–17
Present value interest factor (PVIF), 182
   for an annuity (PVIFA), 184
   finding interest rate using, 191
Present value interest factor for an annuity (PVIFA), 184
Present value of single amount, 181–89
   of annuity due, 185–87
   defined, 181
   investment with uneven cash flows, 187–89
   of ordinary annuity (PVA), 183–85
   perpetuities, 187, 470–72, 477
Price
   ask (offer), 22
   bid, 22
   call, 354
   exercise, 392–93
   purchase, 275
   stock price as measure of firm's value, 6, 7, 8
   subscription, 390
   yield to maturity and, 215
Price, Bill, 19
Price to earnings ratio, 93–94, 103, 220–21
Pricing
   of new issues of common stock, 387–89
   option, 390n, 394
Primary market, 22, 23
Primary reserves, 45
Principals, 8–10, **27**
Private corporations, 379
Private project, capital budgeting for, 252–53
Probability, forecasting based on, 120
Probability distribution, 145
   normal, 148–49
Production schedule, forecast cash outflows developed using, 447–49
Professional golfer, cash flow volatility of, 445
Profit(s)
   cash flow versus, 7–8, 439–40
   defined in accounting, 7
   gross, 62

Profitability
   accounts receivable and, 465–67
   inventory and, 465–67
   liquidity versus, 425, 497
Profitability ratios, 86, 87–90, 99, 102
   gross profit margin, 87–88
   mixed ratio, 89–90
   net, 64
   repatriated, 524, 531
   return on assets (ROA), 89, 90, 97–101
   return on equity (ROE), 89
   summary analysis of, 101–2
Profit and loss potential, breakeven analysis for, 328–30
Profit margin
   gross, 87–88
   net, 89, 97–100
   operating, 88–89
Pro forma **financial statements, 122**–33
   additional funds needed (AFN), 129–31
   analyzing, 131–33
   balance sheet, 127–29, 131
   budgets to produce, 122–23
   choosing forecasting basis, 123–24
   for credit policies, 468
      reflecting policy changes, 470, 471
   defined, 122
   income statement, 124–26, 130, 446
   inventory changes reflected on, 475, 476
Progressive tax rate structure, 74
Projections. *See* Forecasting
Projects, capital budgeting, 237
   expansion project, incremental cash flows of, 280–85
   independent, 237
   mutually exclusive, 237, 248–49, 270–72
   nonsimple, 267–68
   private versus public, 252–53
   ranking, 312–13
   simple, 267
Project salvage value, 279–80
Project-specific risk, 251–53
Prominet Corp., 20
Promissory note, 498
   commercial paper, 25–26
Property, plant, and equipment, on *pro forma* balance sheet, 127
Property and casualty insurance companies, 54–55
Proprietorship, 11–12, 14
Prospectus, 387
Proxies, 382
Publicly traded corporations, 380
Public project, capital budgeting for, 252–53
Public relations, corporate, 381
Public warehouse, controlling pledged inventory in, 509–10
Purchase price, 275
Purchasing power parity (PPP) theory, 528
Pure time value of money, 178
Putable bonds, 361

Q

Quick ratio, 19, 103, 466
Quotas, 532

R

Rabelais, Francois, 323
Ranking, 237
   NPV ranking decision, 243

Ratajczak, Donald, 117–18
Ratio analysis. *See* Financial ratios
Rationing, capital, 249–51
Reagan, Ronald, 75
Real options, 278–79
Real rate of interest, 31–32
Reduced account receivable, settling for, 483
Refunding, bond, 355, 372–75
Regulation
  of commercial banks, 44, 45
  of credit union, 53
  of savings and loan associations, 50
Relative purchasing power parity theory, 528
Relaxing credit policy, 468
Reminder letters, 480
Repatriated profits, 524, 531
Replacement chain approach, 271
Replacement decision, asset, 285–86
Required rate of return, 210
  adjusting, for risk, 162–63
  CAPM to calculate, 163–65
  in discounted cash flow valuation model, 209
  hurdle rate, 245, 247, 287, 296
  risk hierarchy and, 358
  weighted cost of capital and, 305–6
Required reserve ratio, 45
Reserves, 45–46
Residual income, 378
Residual theory of dividends, 406–7
Resolution Trust Corporation (RTC), 50
Restrictive covenants, 355–57
Retained earnings, 65, 68, 406–7
  additions to, in *pro forma* income statement, 126
  cost of, 300
  on *pro forma* balance sheet, 128–29
  stock dividend and, 412
Retirement
  pension funds and, 55
  Social Security System and, 56
Return. *See also* Internal rate of return (IRR); Required rate of return; Risk
  on investments, expected, 442–43
  risk-free rate of, 163
  risk-return relationship, 144–45, 163–65, 255, 497
Return on assets (ROA), 89, 90, 97–101
  cross-sectional analysis of, 100
  trend analysis using, 100–101
Return on equity (ROE), 89
  in modified Du Pont equation, 98
Revco Drug Stores, 337, 338
Revenue bonds, 28, 29
Revenues on income statement, 61–62
Revolving credit agreement, 499
Rights, preemptive, 390–92
Rights-on period, 390–91
Risk, 144–73
  agency problem and, 10
  business risk, 151–53, 333
  in capital budgeting, 251–55
    adjusting for, 255
    measuring, 251–54
  compensating for presence of, 162–65
    adjusting required rate of return, 162–63
    capital asset pricing model (CAPM), 163–65
  credit, 43
  cultural, 531–32
  decreasing, to decrease cash outflows, 454

defined, 144
financial risk, 153–55, 335, 339
insurance and, 54–55
in international finance, managing, 525–27
managing current liabilities and, 426
measuring, 145–51
  with coefficient of variation, 149–51
  with standard deviation, 145–49
of money market securities, 25
mutual funds and, 162
optimal capital structure and, 340–41
political, 527, 530–31
portfolio risk, 155–61
  correlation and, 156–57
  defined, 155
  diversification effect and, 156
  nondiversifiable risk, 159–61
  standard deviation of two-asset portfolio, calculating, 158
profit-loss potential and, 329–30
risk aversion, 144
risk-reduction methods, 161–62
value and, 7, 8
working capital financing approach and, 427, 428, 429
financial ratios and, 429–30
Risk-adjusted discount rates (RADRs), 255
Risk aversion, 144
Risk-free rate of return (k_{RF}), 163
Risk hierarchy, 359
Risk premiums, 32–33
Risk–return relationship, 144–45, 163–65
  capital asset pricing model (CAPM) and, 163–65
  in risk-adjusted discount rates (RADRs), 255
  short-term versus long-term financing and, 497
Risk tolerance, 330
RJR/Nabisco, 337, 363
Rogers, Steven, 351–52
Rogers, Will, 117
Rosen, Melanie, 9
Rustel, 531

S

Sales
  fluctuating, 423
  in gross profit margin, 87–88
  increasing, to increase cash inflows, 451–54
  in inventory turnover, 93
  in net profit margin, 89
  in operating profit margin, 88–89
  in total asset turnover, 93
Sales breakeven chart, 324–27
Sales breakeven point, 324
Sales finance companies, 53–54
Sales forecasting, 120–22
  distributions of forecast sales, 145–46
Sales projection of *pro forma* income statement, 124–25
Sales revenue
  in breakeven chart, 324–27
  forecasting, 447
Sales volatility
  business risk and, 151–52
  reducing, 161
Sallie Mae bond, variable interest rate on, 361, 362
Salvage value, 72*n*
  project, 279–80

Santayana, George, 516
Savings and loan associations (S&Ls), 49–52
  crisis of late 1980s, 50
  legislation affecting, 49–50
  matching loan and deposit maturities, problem of, 51
  mutual versus stockholder-owned, 50
  real assets of, 52
  regulation of, 50
Savings Association Insurance Fund (SAIF), 50
Schedule C, 12
Schrempp, Jürgen, 516
Schuster, Lee Anne, 99
Secondary market, 22, 23
Secondary reserves, 45–46
Second-mortgage bond, 358
Secured bonds, 354, 357–58
Secured loans, 507, 508
Securities, 21. *See also* Bond(s); Common stock; Preferred stock
  in financial marketplace, 25–31
    capital market, 27–31
    money market, 25–27
  marketable, 67, 127
  preemptive rights, 390–92
  warrants, 392–94
Securities and Exchange Commission (SEC), 61, 363
  EDGAR, Web site of, 83
Securities valuation. *See* Bond valuation; Common stock valuation; Preferred stock valuation; Valuation
Security, bond indenture and, 353–54
Security exchanges, 23–24
Security First Network Bank (SFNB), 41
Self-liquidating loans, 498
Selling and administrative expenses, 62
Selling and marketing expenses in *pro forma* income statement, 125
Selling ex-rights, 391–92
Semiannual compounding, 195–96
Semiannual coupon interest payments, 212
Senior debenture, 358
Serial payments, 354
Share buy-back, 295–96
Shareholders, credit union members as, 52
Short-term financing, 426, 496–515
  aggressive working capital financing approach using, 427
  alternatives, 497–502
    commercial paper, 25–26, 500–502
    short-term loans from banks and other institutions, 497–99
    trade credit, 499–500
  collateral for, 507–10
    accounts receivable, 508
    inventory, 508–10
  effective interest rate, loan terms and, 502–7
    compensating balance and, 503–4
    comprehensive example of, 506–7
    computing amount to borrow, 507
    discount loans and, 503, 504
    loan maturities shorter than one year and, 505–6
  long-term financing versus, 497
  need for, 496
Shutdown cash flows, 279–80, 282
Signaling, 385
Signaling dividend theory, 408
Simple projects, 267

I-10 Index

Sinking funds, 354
Small-loan companies, 53
Smith, Clifford W., Jr., 26
Social Security System, 56
Society, interests of, 11
Software development company, managing working capital for, 431
Sole proprietor, 12
Soliman decision, 75
Sony Corporation, 10–11, 119
Soviet Union
  collapse of, 529
  cultural risk of doing business in, 531–32
S&P 500 stock market index, 163
Spot rate, 529
Spread, 22
  interest rate, 44
Staggered maturities, 354
Stakeholders, 10–11
Standard deviation
  business risk and, 151, 152
  calculating, 146–48
  coefficient of variation and, 150–51
  defined, 146
  financial risk and, 154
  interpreting, 148–49
  measuring risk with, 145–49
  of portfolio, 156
    new portfolio, 254
    two-asset portfolio, calculating, 158
Standard & Poor's
  bond ratings, 352–53, 362
  S&P 500 stock market index, 163
Standards, credit, 468, 479–80
Stated interest rate, 498
Statement of cash flows, 68–71
Statement of retained earnings, 65
Statistical independence, 157
Stern Stewart & Company, 95
Stock, 29–31. *See also* Common stock; Preferred stock
  source of information on, 104
  target, 378
Stock dividends, 411–13
Stockholder-owned savings and loan associations, 50
Stockholders, 6, 29
  changes in types of, 381
  classes of, 378
  dividend policy and preferences of, 404
  voting rights of common, 378, 382–84
Stock price, as measure of value of firm, 6, 7, 8
Stock splits, 413–15
Stock valuation. *See* Common stock valuation; Preferred stock valuation
Straight bond value, 360–61
Straight-line depreciation (SL) method, 72
Strengthening currency, 522
Subchapter C corporation, 14
Subchapter S corporation, 14
Subordinated debenture, 358
Subscription price, 390
Summary analysis, 101–3
Summary of cash flow and financing requirements, 452
Sunbeam Corp., 59–60
Sunk costs, 274
Super long-term bonds, 364
Supernormal growth common stock, 232–33
Supernormal growth of cash flows, 232–33
Suppliers, delaying payment to, 455

Supreme Court, 75
Surplus economic units, 21
  intermediation between deficit economic units and, 42–46
Swaps, 525
Syndicate, 387

T

Tables, financial. *See* Financial tables
Target cash balance, Miller-Orr model formula for, 444
Target stock, 378
Tariffs, 532
Tax(es)
  after-tax cost of debt (AT k_d), 297–98
  average tax rate, 74, 76
  corporate, 13–14
  cost of preferred stock and, 300
  deductibility of interest, 338–39
  deductible expenses, 64
  double taxation, 13, 14
  genuine versus fake lease and, 365–66
  income, 73–76
  incremental depreciation expense and changes in, 277
  lease versus buy decision and, 368
  marginal tax rate, 74–76
  as operating cash flow, 276
  on preferred stock dividends, 365
  progressive tax rate structure, 74
  project salvage value and, 279–80
  proprietorship and, 12
  stockholders' preferences for dividend policy and, 404
  unitary, 531
Taxpayer Relief Act (1997), 75, 404*n*
Tax Reform Act (1986), 73
T-bills (Treasury bills), 25, 32
T-bonds (Treasury bonds), 28
Temporary current assets, 424
10-K reports, 61
10-Q reports, 61
Theibert, Philip, 463–64
Theodosopoulos, Nikos, 20
TI BAII PLUS calculator. *See* Financial calculator
Tightening credit policy, effects of, 468
Times interest earned ratio, 92, 103
Time value of money, 175–206
  compounding more than once per year, 195–98
    annuity compounding periods, 196–97
    continuous compounding, 198
    semiannual compounding, 195–96
  defined, 177
  finding number of periods and, 192–93
  future value of single amount, 178–81
    of annuity due, 185–87
    sensitivity to interest rate changes or number of compounding periods, 181
  interest rate and, 181, 189–92
  measuring, 178
  payback method and, 239
  present value of single amount, 181–89
    of annuity due, 185–87
    defined, 181
    investment with uneven cash flows, 187–89
    of ordinary annuity (PVA), 183–85
    perpetuities, 187, 470–72, 477
    pure, 178

  reasons for, 177–78
  solving for payment, 193–95
Time value of warrant, 393–94
Timing of cash flows, 6–7, 8
Titanic (movie), 85
T-notes (Treasury notes), 28
Todd, Mike, 439
Tombstone ad, 387, 388
Total assets, debt to, 91
Total asset turnover, 93, 103
  in Du Pont equation, 97–100
  in modified Du Pont equation, 98
Total rate of return from common stock, formula for investor's, 222
Toyota Corporation, 518
Trade. *See also* International finance
  fair versus free, 533
  international trade agreements, 532–33
Trade credit, 499–500
Trading rights-on, 390–91
Transaction costs, 443
Transfer agent, 409
Treasurer, 4, 5
Treasury securities
  buying, without using broker, 30
  Treasury bills (T-bills), 25, 32
  Treasury bonds (T-bonds), 28
  Treasury notes (T-notes), 28
  yield curves of, 34
Trend analysis, 100–101
  summary analysis using, 101–3
Trends, forecasting from sales, 121, 122
Trial-and-error method
  calculating IRR using, 246–47
  for finding yield to maturity, 213–14
Trustee of bond issue, independent, 357
Trust receipt, 509
Twain, Mark, 41

U

Uncertainty. *See also* Risk
  risk measurement and, 145
  standard deviation and, 149
Underwriting, 22, 387
Unequal lives, comparing projects with, 271–72
Unitary tax, 531
U.S. Treasury securities. *See* Treasury securities
U.S. West Communications Corporation, 378
Unlimited liability, 12, 13
Unsecured bonds (debentures), 354, 358
Unsecured loans, 507
Upper limit for cash account, Miller-Orr model formula for, 444

V

Valuation, 207–34
  bond, 210–15
    annual coupon interest payments, 210–11
    formula, 210
    importance of, 208–9
    semiannual coupon interest payments, 212
    yield to maturity (YTM), 212–15
  common stock, 218–22
    balance sheet valuation approaches, 218, 221–22
    deciding on method of, 222

Index **I-11**

going concern valuation models, 218–219
   supernormal growth, 232–33
   yield on, 222
general valuation model, 209–10
preferred stock, 215–18
   present value of dividends, 216–17
   yield on, 217–18
warrant, 392–94
Value
   measuring, 5–8
   risk and, 7, 8
Variable costs, 324
   in breakeven chart, 324–27
Variable-rate bonds, 361
Venture capital funds, 351–52
Voting rights of common stockholders, 378, 382–84

W

Walt Disney Company, 2
Warrants, 392–94
Washington Suburban Sanitary District of Maryland, notice of bond redemption, 355, 356
Washington Water Power Company, 401–2
Weakening currency, 522

Wealth, defined, **5**
Web sites, financial information on, 83–84
Wedding Banquet, The (movie), 85–86
Weighted average cost of capital (WACC), 95, **304**–6, 314, 339
   calculating MCC with, 310–11
   change in, as debt is added, 340
   formula for, 305
   required rate of return and, 305–6
Wilde, Oscar, 207
Wolfram, Edward P., Jr., 26
Wolverine Fund, 352
Working capital, 67, 422
   accumulation of, reasons for, 423–25
   defined, 422
   managing, 422–23
   minimum levels of, 357
   need for, 421–22
   net, 67–68, 422, 427
      changes in, 275–76
Working capital policy, 421–38. *See also* Accounts receivable; Inventory
   approaches to, 426–29
      aggressive approach, 426, 427, 430
      conservative approach, 426–27, 428, 430
      moderate approach, 427, 429, 430
   defined, 422–23

   factors affecting, 423–25
   financial ratios and, 429–30
   liquidity versus profitability and, 425
   optimal level of current assets, establishing, 425–26
World events, exchange rates and threatening, 529–30
World Trade Organization (WTO), 532–33

Y

Yahoo Corporation, 6, 386, 389
Yankee bond, 364
Yield
   on common stock, 222
   discount, 500–501
   dividend, 301
   on preferred stock, 217–18
Yield curve, 33, 34–35
Yield to maturity (YTM), 212–15
   calculating, 212–14
   price and, 215

Z

Zemeckis, Robert, 64
Zero-coupon bonds, 27

READ THIS LICENSE CAREFULLY BEFORE OPENING THIS PACKAGE. BY OPENING THIS PACKAGE, YOU ARE AGREEING TO THE TERMS AND CONDITIONS OF THIS LICENSE. IF YOU DO NOT AGREE, DO NOT OPEN THE PACKAGE. PROMPTLY RETURN THE UNOPENED PACKAGE AND ALL ACCOMPANYING ITEMS TO THE PLACE YOU OBTAINED THEM FOR A FULL REFUND OF ANY SUMS YOU HAVE PAID FOR THE SOFTWARE. THESE TERMS APPLY TO ALL LICENSED SOFTWARE ON THE DISK EXCEPT THAT THE TERMS FOR USE OF ANY SHAREWARE OR FREEWARE ON THE DISKETTES ARE AS SET FORTH IN THE ELECTRONIC LICENSE LOCATED ON THE DISK:

1. GRANT OF LICENSE AND OWNERSHIP: The enclosed computer program The Prentice Hall Finance Center CD ("Software") is licensed, not sold, to you by Prentice-Hall, Inc. "We" or the "Company" and in consideration of your payment of the license fee, which is part of the price you paid for your purchase or adoption of the accompanying Company textbooks and/or other materials, and your agreement to these terms. We reserve any rights not granted to you. You own only the disk(s) but we and/or our licensors own the Software itself. This license allows you to use and display your copy of the Software on a single computer (i.e., with a single CPU) at a single location for academic use only, so long as you comply with the terms of this Agreement. You may make one copy for back up, or transfer your copy to another CPU, provided that the Software is usable on only one computer.

2. RESTRICTIONS: You may not transfer or distribute the Software or documentation to anyone else. Except for backup, you may not copy the documentation or the Software. You may not network the Software or otherwise use it on more than one computer or computer terminal at the same time. You may not reverse engineer, disassemble, decompile, modify, adapt, translate, or create derivative works based on the Software or the Documentation. You may be held legally responsible for any copying or copyright infringement which is caused by your failure to abide by the terms of these restrictions.

3. TERMINATION: This license is effective until terminated. This license will terminate automatically without notice from the Company if you fail to comply with any provisions or limitations of this license. Upon termination, you shall destroy the Documentation and all copies of the Software. All provisions of this Agreement as to limitation and disclaimer of warranties, limitation of liability, remedies or damages, and our ownership rights shall survive termination.

4. LIMITED WARRANTY AND DISCLAIMER OF WARRANTY: The Company warrants that for a period of 60 days from the date you purchase this SOFTWARE (or purchase or adopt the accompanying textbook), the Software, when properly installed and used in accordance with the Documentation, will operate in substantial conformity with the description of the Software set forth in the Documentation, and that for a period of 30 days the disk(s) on which the Software is delivered shall be free from defects in materials and workmanship under normal use. The Company does not warrant that the Software will meet your requirements or that the operation of the Software will be uninterrupted or error-free. Your only remedy and the Company's only obligation under these limited warranties is, at the Company's option, return of the disk for a refund of any amounts paid for it by you or replacement of the disk. THIS LIMITED WARRANTY IS THE ONLY WARRANTY PROVIDED BY THE COMPANY AND ITS LICENSORS, AND THE COMPANY AND ITS LICENSORS DISCLAIM ALL OTHER WARRANTIES, EXPRESS OR IMPLIED, INCLUDING WITHOUT LIMITATION, THE IMPLIED WARRANTIES OF MERCHANTABILITY AND FITNESS FOR A PARTICULAR PURPOSE. THE COMPANY DOES NOT WARRANT, GUARANTEE OR MAKE ANY REPRESENTATION REGARDING THE ACCURACY, RELIABILITY, CURRENTNESS, USE, OR RESULTS OF USE, OF THE SOFTWARE.

5. LIMITATION OF REMEDIES AND DAMAGES: IN NO EVENT, SHALL THE COMPANY OR ITS EMPLOYEES, AGENTS, LICENSORS, OR CONTRACTORS BE LIABLE FOR ANY INCIDENTAL, INDIRECT, SPECIAL, OR CONSEQUENTIAL DAMAGES ARISING OUT OF OR IN CONNECTION WITH THIS LICENSE OR THE SOFTWARE, INCLUDING FOR LOSS OF USE, LOSS OF DATA, LOSS OF INCOME OR PROFIT, OR OTHER LOSSES, SUSTAINED AS A RESULT OF INJURY TO ANY PERSON, OR LOSS OF OR DAMAGE TO PROPERTY, OR CLAIMS OF THIRD PARTIES, EVEN IF THE COMPANY OR AN AUTHORIZED REPRESENTATIVE OF THE COMPANY HAS BEEN ADVISED OF THE POSSIBILITY OF SUCH DAMAGES. IN NO EVENT SHALL THE LIABILITY OF THE COMPANY FOR DAMAGES WITH RESPECT TO THE SOFTWARE EXCEED THE AMOUNTS ACTUALLY PAID BY YOU, IF ANY, FOR THE SOFTWARE OR THE ACCOMPANYING TEXTBOOK. BECAUSE SOME JURISDICTIONS DO NOT ALLOW THE LIMITATION OF LIABILITY IN CERTAIN CIRCUMSTANCES, THE ABOVE LIMITATIONS MAY NOT ALWAYS APPLY TO YOU.

6. GENERAL: THIS AGREEMENT SHALL BE CONSTRUED IN ACCORDANCE WITH THE LAWS OF THE UNITED STATES OF AMERICA AND THE STATE OF NEW YORK, APPLICABLE TO CONTRACTS MADE IN NEW YORK, AND SHALL BENEFIT THE COMPANY, ITS AFFILIATES AND ASSIGNEES. THIS AGREEMENT IS THE COMPLETE AND EXCLUSIVE STATEMENT OF THE AGREEMENT BETWEEN YOU AND THE COMPANY AND SUPERSEDES ALL PROPOSALS OR PRIOR AGREEMENTS, ORAL, OR WRITTEN, AND ANY OTHER COMMUNICATIONS BETWEEN YOU AND THE COMPANY OR ANY REPRESENTATIVE OF THE COMPANY RELATING TO THE SUBJECT MATTER OF THIS AGREEMENT. If you are a U.S. Government user, this Software is licensed with "restricted rights" as set forth in subparagraphs (a)-(d) of the Commercial Computer-Restricted Rights clause at FAR 52.227-19 or in subparagraphs (c)(1)(ii) of the Rights in Technical Data and Computer Software clause at DFARS 252.227-7013, and similar clauses, as applicable.

Should you have any questions concerning this agreement or if you wish to contact the Company for any reason, please contact in writing:

Director New Media
Higher Education Division
Business Publishing Group
Prentice Hall, Inc.
One Lake Street
Upper Saddle River, NJ 07458

Should you have any questions concerning technical support of this product, please contact our Technical Support staff in writing at:

New Media Production and Technical Support
Higher Education Division
Prentice Hall, Inc.
One Lake Street
Upper Saddle River, NJ 07458

or call:

201-236-3477

or email:

tech_support@prenhall.com